17th Annual Conference of the Asian Academic Accounting Association (2016 FourA Conference)

SHS Web of Conferences Volume 34 (2017)

Kuching Sarawak, Malaysia
20-22 November 2016

Editors:

Y. Y. Abdul Talib
A. Ahmi
R. H. Raja Mohd Ali
D. Abdul Kadir
S. Ishak
F. H. Rusly
M. H. Mohd Sharif

ISBN: 978-1-5108-3661-7

Printed from e-media with permission by:

Curran Associates, Inc.
57 Morehouse Lane
Red Hook, NY 12571

Some format issues inherent in the e-media version may also appear in this print version.

Printed by Curran Associates, Inc. (2017)

For additional information, please contact EDP Sciences – Web of Conferences
at the address below.

EDP Sciences – Web of Conferences
17, Avenue du Hoggar
Parc d'Activité de Courtabœuf
BP 112
F-91944 Les Ulis Cedex A
France

Phone: +33 (0) 1 69 18 75 75
Fax: +33 (0) 1 69 28 84 91

contact-edps@webofconferences.org

Additional copies of this publication are available from:

Curran Associates, Inc.
57 Morehouse Lane
Red Hook, NY 12571 USA
Phone: 845-758-0400
Fax: 845-758-2633
Email: curran@proceedings.com
Web: www.proceedings.com

TABLE OF CONTENTS

DOI: 10.1051/shsconf/20173400001

Preface

The 2016 FourA Conference is an excellent platform for bringing together a diverse group of accounting researchers, academics, and professionals from, not only locally, but also internationally. This year's conference, with its theme 'Accounting in Developing Economics: Ethical, Regional and Contextual Issues', promises to continue its tradition of being the premier forum for the presentation of research results and reports on leading edge issue in accounting, including corporate governance, financial reporting, taxation, auditing, accounting information systems, environmental accounting, accounting education, management accounting, forensic accounting and Islamic accounting. This effort is greatly benefit all stakeholders in the area of accounting and other related areas.

Special thanks are given to all the reviewers, the scientific committees and the organizing committees for their invaluable efforts in making this event a great success.

Yours faithfully,

Editor-in-chief
Dr. Yurita Yakimin Abdul Talib

Editors
Dr. Suhaimi Ishak
Dr. Aidi Ahmi
Dr. Fariza Hanim Rusly
Dr. Raja Haslinda Raja Mohd Ali
Dr. Mohd Hisham Mohd Sharif
Mr. Dzarfan Abdul Kadir

Corporate Social Responsibility and Corporate Social Innovation: A Conceptual Understanding

Muhamad Nizam Jali[1,*], *Zakaria* Abas[2] , and *Ahmad Shabuddin* Ariffin[3]

[1]UUM, College Of Business, 06010 Sintok Kedah, Malaysia
[2]UUM, College Of Business, 06010 Sintok Kedah, Malaysia
[3]UUM, College Of Business, 06010 Sintok Kedah, Malaysia

Abstract. In decades, various organizations worldwide engaged with Corporate Social Responsibility (CSR) in order to show their corporate commitments and responsibilities towards societies at large. These commitments and responsibilities are coming from monetary and non-monetary resources for example cash, equipment's and human resources whom are used for social purposes and activities that leads to a betterment of society and also to improved organization reputation. However, in today's knowledge and innovation led economy, organizations can no longer affords to get involve in charity and community services merely to fulfil social return without having any sort of economic payoffs. This situation warrants organizations moving beyond CSR to Corporate Social Innovation. This paper explores conceptual understanding between CSR and Corporate Social Innovation. CSR is a traditional philanthropy and old paradigm which is somewhat no longer sufficient in coping with current economic situation. Hence, this paper provides an insight and suggests that corporate social innovation as an emergence new paradigm that perhaps could provide a comprehensive representation in the era of knowledge and innovation led economy that will leads to real change in improving the well-being of people's life, enhance economic and technological growth. Furthermore, this paper also highlighted knowledge resource is the most significant resource of Corporate Social Innovation.

Keywords: CSR, Corporate Social Innovation, Knowledge resource

1 Introduction

For more than five decades, organizations and business entities include CSR in their strategic planning and operational decision making process. By incorporating CSR activities for example philanthropy and volunteering in their main business plan structure, organizations are said fulfilling its obligation towards the welfare of its society. According to [1], sustainability of business growth does not entirely dependent upon its economic performance but most importantly how business deals and interacts with society around

* Corresponding author: mnj_jali@yahoo.com.my

them, which in turn contributes massively towards society and environment betterment. However, present turbulence of business environment, intense competition pressure, globalization and technological change offers organization no other choice but an absolute ultimatum in order to stay survival, to keep abreast with competitive demands and technological change and to ensure long-term business prosperity [2, 3]. From the above challenging and competitive economic situation, [4] stressed that organization and business entities can no longer affords to engage in CSR activities solely to fulfil social responsibilities without reap a commercial return on their contributions on CSR [4]. Organization and business entities must move beyond the traditional nature of CSR to a new paradigm that can cope with the demands and needs of present dynamic economic environment. According to [5], this new paradigm offers organization and business entities a massive return towards both social as well as commercial benefits.

From the above paragraph, corporate social innovation has emerged as the new emerging paradigm of innovation and received an overwhelming interest from organization and business entities worldwide in order to gained commercial driven benefits and at the same time contributes towards the development of human well-being and societal quality and quantity of life [6, 7, 8]. As a result, many multinational organization worldwide among others IBM, Napp Pharmaceuticals, GlaxoSmithKline, Novartis, Siemens, Microsoft, McDonald restaurant and Nike emphasis and focus more on corporate social innovation as compared to CSR in their main business outcome strategy initiatives [9, 10, 11, 12]. For example, GlaxoSmithKline, Novartis and Napp pharmaceutical creating inimitable pharmaceutical products, processes and services in specific diseases area by embedding new knowledge resource through partnership with university and consequently improves wider society quality of life and gained a substantial amount of commercial benefits in return [11]. In addition, IBM and Microsoft through knowledge resource collaboration also with university partners in the various engineering fields for example in computer and mechanical engineering creating new innovation on their products, processes and services which in turn offers society to enjoy a high-end computer and electronic products and at the same times gaining a huge amount of market share as compared to their competitors [13]. Moreover, [5] highlighted organization and business entities must viewed knowledge partnership and social sector activities as the source of opportunities to developed new ideas and innovation, enhance market needs and solving long-standing business problems and leaving behind old perception that viewed social sector activities just as a dumping ground for spare resources i.e. cash, equipment and human resource [14].

On the other hands, present economic environment is moving into knowledge based society where new knowledge resource is regard as the most valuable intangible resource that can be embedded into products, processes and services in order to fulfil private and social needs [15, 16]. With that, the association of corporate social innovation and knowledge resource is immensely important considering its contribution towards social and economic aspects. Thus, a complete and extensive understanding on the insight between CSR and corporate social innovation must be well understood. This is the primary purpose of this paper. This paper is set out in three sections. First, the paper examines the existing CSR literature. Secondly, this paper reviews the literature of social innovation within the context of knowledge and innovation led economy. This paper will also discuss on the knowledge resource as an essential resource for corporate social innovation. Thirdly, the paper ends by enveloping the conclusion of the thought.

2 Review of CSR

The general concept and definition of CSR is describes as the social responsibilities and obligation that organization and business entities have to the societies and environment within which their businesses operate. Specifically, literature highlighted that the classic notion of CSR is associated with the terms among others, social responsible behaviour, social legal responsibility to the citizens and also a charitable contribution to the society and environment [17]. According to [18], CSR encompasses a wide range of voluntarily works that contributes to the better society and cleaner environments. From the above statements, [19] established a formal and organised definition and deepens the understanding on CSR. The authors defined CSR as how organization and business entities socially interacts with employees, suppliers, customers and communities as a whole by integrating monetary and non-monetary resources in a continuous way that can leads to a better society and cleaner environments [20]. Within the literature, CSR have various definition and interpretation, but the common ground of understanding is that CSR is refers to an organization and business entities that made a continuous commitment and obligation through financial and non-financial assets towards social activities and program without having any inclination with their day-to-day business operations i.e. commercial driven benefits, but merely to show legal social responsibility and social connection to the community.

Within the literature of CSR, huge amount of empirical studies shows evidence to support and sums-up the above paragraph, whereby most organizations engaging with CSR only to fulfils non-commercial driven outcomes. Studies by [21, 22] examines the impacts of CSR activities towards organization reputation. Based on their findings, it shows that various multinational organization for example BP, Shell and tobacco related organization engaged with CSR activities i.e. charity, volunteer, philanthropy, only for the purpose of enhancing their organization reputation. For these organizations, CSR successfully creates a favourable corporate image and positive relationship on improving organization image in the eyes of the society at all level [23]. A study by [24] acknowledge that by getting a good corporate reputation through engagement in CSR activities does not necessarily shows organization is having a good financial performance. Furthermore, studies by [25, 26, 27] on CSR activities and environmental sustainability also explained the above notion. In their studies, they highlighted the main reason of multinational organization engaged and allocated substantial amount of resources on CSR activities are only due to satisfy and absorb social and political pressures on the issues of environmental sustainability. To elaborate further, their studies explained that oil and gas companies, mining companies and other related environmental issue companies address the issues of environmental sustainability simply by mobilizing spare resources towards environmental charity program and participating in an environmental corporate volunteer program. By doing this, they at least show their sense of caring towards improving and to ensure continuous environmental sustainability with regards to the nature of the business that they are dealing with. Even though such CSR activities should be encourage, but the activities have not produce any sort of return on the part of business point of view [5].

The literature also gives evidence that CSR have positive relationship with customer loyalty and satisfaction [28]. Various empirical studies among others by [29, 30, 31] found that when organization and business entities continuously engaging with CSR activities, it have a significant positive relationship with customer loyalty and satisfaction in regards to the brands and image of the organization. In contrast, [32] argued that even though CSR activities helps firms to secured loyal customers and enhance customers satisfaction, however, this situation does not gives a significant effects on the firm financial performance. Therefore, considering present competitive economic demands, CSR is somehow viewed as insufficient for organization and business entities to continue with. Thus, organization and business entities must be able to find a new platform and paradigm in dealing with the present economic demands that can fulfils both social and commercial return that can leads to long-term business prosperity and to stay survival. Figure 1 illustrates the conceptual view of CSR towards social benefits.

Fig. 1. Conceptual view of CSR towards social benefits.

Figure 1 provides summary of the conceptual view of CSR towards social benefits. Box 1 in the CSR indicates the resources i.e. spare cash, equipment, human resource; Box 2 signifies the contribution of CSR through charity, physical volunteer activities and philanthropy towards social benefits i.e. social welfare, environmental sustainability and good interaction and relation. The next section explained the new paradigm i.e. corporate social innovation, which is more sufficiently comprehensive as compared to CSR within the era of knowledge and innovation led economy.

3 Review of Corporate Social Innovation

In the literature, corporate social innovation can be defined as new and novel solution embedded into products, processes and services in order to fulfil social needs and to improved quality and quantity of life [33, 15]. Specifically, corporate social innovation includes and integrates social, economic and technological approach in order to meet the demands and pressures of new economic environment [34, 35]. With the integration of social, economic and technological approach, there is a collective dynamic interplay that guarantees organization and business entities a considerable return with regards to these three aspects [36, 15]. As a result, corporate social innovation is seen as the new and outstanding emerging paradigm in order to resolve social, economic and technological issues face by all related parties and attracting the interest and attention of private sectors, policy makers and researches worldwide [6, 37]. According to [38], unlike CSR, corporate social innovation is not necessarily tied up to address specific social purposes but its significant value encompasses wide range of contribution outcomes that includes social, economic and technological benefits. To elaborate further, when organization and business entities implements corporate social innovation in its strategic planning and operational decision making process, it address social and economic issues among others, poor social health, poor education, social income inequality, massive unemployment and poor economic growth [39], this in turn leads to the outcome of profit maximization of private sectors, enhance economic growth, improves living condition of society, better environmental condition and developed and enhance individual's knowledge, skills and competencies [40].

From the above paragraphs, various empirical studies give evidence to justify the outcomes of corporate social innovation. According to [5], corporate social innovation helps to improves societal and economic problems by creating new solution, which refers to knowledge resource embedded into products, processes and services that work to meet pressing social and economic needs and to improve quantity and quality of people's life. Studies by [5, 41] stressed that in the aspects of poor public education, corporate social innovation offers new solutions that comes from knowledge resource that contributes to a better future knowledge workers and at the same time helps specific private sectors gained a substantial amount of market share between competitors. Furthermore, according to [42], the result of deployment of corporate social innovation and knowledge resource on the issues of massive unemployment contributes towards increase in employment among people and also increase in consumption of economic benefits. According to [43, 22], corporate social innovation and knowledge resource leads to the introduction of superior products, processes and services that have a multiplier effects on the economic value in terms of profit maximization, market share monopoly and increase in private performance. A study by [44] also revealed on the positive effects of corporate social innovation and knowledge resource on the issues of social health. Their study found that knowledge resource leads to the establishment of superior medical products that can improves and enhance people's health and subsequently gives a significant return towards profit maximization to the parties involved. Apart from that, [45, 16, 46] describe the new and novel solution which is the main element of corporate social innovation refers to the superior knowledge resource. This superior knowledge resource is embedded into products, processes and services which make them highly innovative and in turn lead to the outcome of improving society well-being that include environmental issues, stimulate economic growth and enhance private needs i.e. profit maximization, market share monopoly and private performance [38]. A recent study by [47] found that a linkage between superior knowledge resource and corporate social innovation is the best possible solution in producing superior products, processes and services towards overcoming social and economic and technological problems. New superior knowledge resource is created through the processes of strategic knowledge management that involves the process of knowledge creation, knowledge transfer and knowledge application [48, 49]. These processes involved the interplay of tacit and explicit knowledge [43, 50]. Figure 2 illustrates the conceptual view of corporate social innovation towards social, economic and technological benefits.

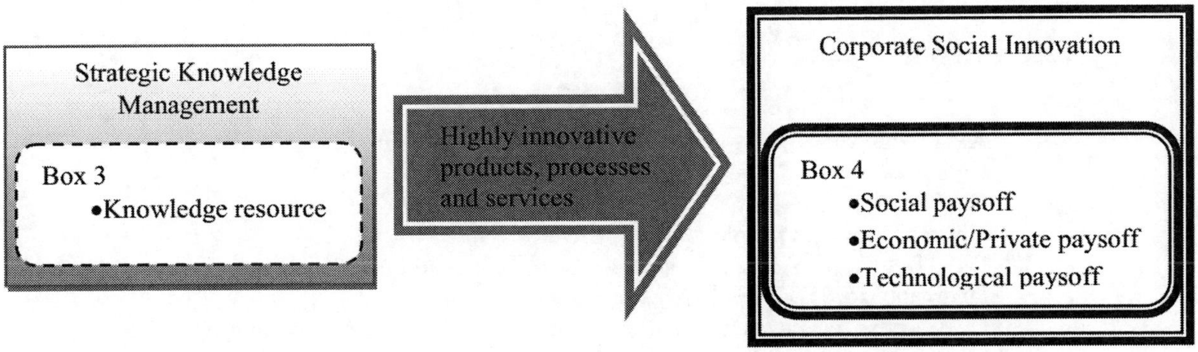

Fig.2. Conceptual view of corporate social innovation towards social, economic and technological benefits

Figure 2 provides summary of conceptual understanding of corporate social innovation and its association with knowledge resource within strategic knowledge management processes. Box 3 in the strategic knowledge management indicates the output i.e. knowledge resource and box 4 signifies the contribution of knowledge resource on corporate social innovation i.e. social, economic and technological payoffs; by producing highly innovative products, processes and services. Therefore, even though CSR approach is still applicable if not entirely out dated, but Corporate Social Innovation offers all parties concern i.e. private and public institutions, a better opportunity in optimizing social, economic and technological payoffs.

4 Conclusions

This paper provides an overview of the existing literature on CSR and corporate social innovation. This paper also provides a brief linkage between knowledge resource and corporate social innovation. The literature highlighted even though CSR can accomplish many social benefits and should be encouraged, but with the demands and pressures of present economic environment, CSR is seen as inadequate and not an ideal solution for organization and business entities to hold on in order to leverage social, economic and technological payoffs concurrently. This paper suggest that corporate social innovation is the new outstanding paradigm that are more sufficient within the era of knowledge and innovation led economy that provide real change in improving the well-being of people's life, enhance economic and environmental growth as well as technological growth. In addition, knowledge resource created through the processes of knowledge creation, knowledge transfer and knowledge application is the vital intangible resource for the success of corporate social innovation. Therefore, there is an urgent need of comprehensive and extensive understanding and analysis on the empirical evidence of corporate social innovation and its relatedness with superior knowledge resource [51]. This new area must be seriously engaged by future research. Thus, this paper hopefully offers an improved understanding and new insight on understanding current states of CSR and new emerging paradigm of corporate social innovation and its linkages with knowledge resource within the context of economic and business environment.

The authors would like to thank the anonymous reviewers. The authors would also like to give special thanks to the Knowledge Transfer Programme (KTP) secretariat for the funding received under the KTP special research programme titled KTP Problem Statements of Industry/ Community Problems in Malaysia and ASEAN Region, SO code 13319.

References

1. A. Lindgreen, V. Swaen, International Journal of Management Reviews. Corporate social responsibility. **12(1)**, 1-7 (2010).
2. C. Bartlett, S. Ghoshal, Sloan Management Review. Building competitive advantage through people. **43(2)**, (2013).
3. R.A. Bettis, M.A. Hitt, Strategic Management Journal. The new competitive landscape. **16(S1)**, 7-19 (1995).
4. H. Aguinis, A. Glavas, Journal of Management. What we know and don't know about corporate social responsibility a review and research agenda. **38(4)**, 932-968 (2012).

5. R.M. Kanter, Harvard Business Review. From spare change to real change: The social sector as beta site for business innovation. **Vol. 77 No. 3,** pp. 122-132 (2015).
6. E.W. Lee, J.M. Restrepo, Policy & Politics. Institutional embeddedness and the scaling-up of collaboration and social innovation: the case of a Hong Kong-based international NGO. **43(3),** 459-471 (2015).
7. J. Cunha, P. Benneworth, Paper presented at the EURA Conference 2013, Enschede, The Netherlands. *Universities contributions to social innovation: towards a theoretical framework.* 3-6 July (2013).
8. M. Lizuka, Working paper series United Nations. Maastricht, The Netherlands Innovation. *Innovation systems framework: still useful in the new global context?.* UNU-MERIT. 005 (2013).
9. A. Cadbury, Twenty-First Century Society. Corporate social responsibility. **1(1),** 5-21 (2006).
10. J. Saul, John Wiley & Sons. Social Innovation, Inc.: *5 strategies for driving business growth through social change* (2010).
11. M. Perkmann, A., Neely, K. Walsh, R&D Management. How should firms evaluate success in university–industry alliances? A performance measurement system. **41(2),** 202-216 (2011).
12. W. Spiess-Knafl, C., Mast, S.A. Jansen, Social Business. On the nature of social business model innovation. **5(2),** 113-130 (2015).
13. M. Perkmann, A. Salter, MIT Sloan Management Review. How to create productive partnerships with universities. **53(4),** 79 (2012).
14. S. Du, V., Swaen, A. Lindgreen, S. Sen, Journal of Business Ethics. The roles of leadership styles in corporate social responsibility. **114(1),** 155-169 (2013).
15. N. Altuna, A. M., Contri, C., Dell Era, F., Frattini, P. Maccarrone, European Journal of Innovation Management. Managing social innovation in for-profit organizations: the case of Intesa Sanpaolo. **18(2),** 258-280 (2015).
16. R. Chiva, P., Ghauri, J. Alegre, British Journal of Management. Organizational learning, innovation and internationalization: A complex system model. **25(4),** 687-705 (2014).
17. E. Garriga, D. Mele, Journal of Business Ethics. Corporate social responsibility theories: Mapping the territory. **53(1-2),** 51-71 (2004).
18. R. Welford, The Journal of Corporate Citizenship. Corporate social responsibility in Europe, North America and Asia: 2004 survey results. **(17),** 33 (2005).
19. A. Dahlsrud, Corporate social responsibility and environmental management. How corporate social responsibility is defined: an analysis of 37 definitions. **15(1),** 1-13 (2008).
20. M. Van Marrewijk, Journal of Business Ethics. Concepts and definitions of CSR and corporate sustainability: Between agency and communion. **44(2-3),** 95-105 (2003).
21. S. Bear, N., Rahman, C. Post, Journal of Business Ethics. The impact of board diversity and gender composition on corporate social responsibility and firm reputation. **97(2),** 207-221 (2010).
22. Y. Yoon, Z., Gurhan-Canli, N. Schwarz, Journal of Consumer Psychology. The effect of corporate social responsibility (CSR) activities on companies with bad reputations. **16(4),** 377-390 (2006).
23. L.M. McDonald, S. Rundle-Thiele, International Journal of Bank Marketing. Corporate social responsibility and bank customer satisfaction: a research agenda. **26(3),** 170-182 (2008).

24. K. Walker, B. Dyck, Business and Society Review. The primary importance of corporate social responsibility and ethicality in corporate reputation: an empirical study. **119(1),** 147-174 (2014).
25. F.M., Edoho, Corporate Social Responsibility and Environmental Management. Oil transnational corporations: Corporate social responsibility and environmental sustainability. **15(4),** 210-222 (2008).
26. K. Babiak, S. Trendafilova, Corporate social responsibility and environmental management. CSR and environmental responsibility: motives and pressures to adopt green management practices. **18(1),** 11-24 (2011).
27. M. Orlitzky, D.S., Siegel, D. A. Waldman, Business & society. Strategic corporate social responsibility and environmental sustainability. **50(1),** 6-27 (2011).
28. P. Martinez, I.R. Del Bosque, International Journal of Hospitality Management. CSR and customer loyalty: The roles of trust, customer identification with the company and satisfaction. **35,** 89-99 (2013).
29. S. Du, C.B., Bhattacharya, S. Sen, International Journal of Research in Marketing. Reaping relational rewards from corporate social responsibility: The role of competitive positioning. **24(3),** 224-241 (2007).
30. S. Pivato, N., Misani, A. Tencati, A European review of Business ethics. The impact of corporate social responsibility on consumer trust: the case of organic food. **17(1),** 3-12 (2008).
31. P.A. Vlachos, N.G., Panagopoulos, A.A. Rapp, Journal of Business Ethics. Feeling good by doing good: Employee CSR-induced attributions, job satisfaction, and the role of charismatic leadership. **118(3),** 577-588 (2013).
32. E. Pol, S. Ville, The Journal of Socio-Economics. Social innovation: Buzz word or enduring term? **38(6),** 878-885 (2009).
33. J.A. Phills, K., Deiglmeier, D.T. Miller, Stanford Social Innovation Review. Rediscovering social innovation. **6(4),** 34-43 (2008).
34. V. Ramaswamy, Strategy & Leadership. Co-creating value through customers' experiences: the Nike case. **36(5),** 9-14 (2008).
35. E. Shaw, A. De Bruin, International Small Business Journal. Reconsidering capitalism: the promise of social innovation and social entrepreneurship? **31(7),** 737-746 (2013).
36. K. Pue, C., Vandergeest, D. Breznitz, Innovation Policy Lab White Paper. Toward a Theory of Social Innovation. 2016-01 (2015).
37. A. Unceta, J., Castro-Spila, J. Garcia Fronti, The European Journal of Social Science Research. Social innovation indicators. Innovation: **1-13** (2016).
38. P. Dawson, L. Daniel, International Journal of Technology Management Understanding social innovation: a provisional framework. **51(1),** 9-21 (2010).
39. P. Harazin, K. Kosi, Social and Management Sciences. Social Challenges: Social Innovation through Social Responsibility. Periodica Polytechnica. **21(1),** 27 (2013).
40. S. Surikova, K., Oganisjana, G. Grinberga-Zalite, Proceedings of the International Scientific Conference. *The Role of Education in Promoting Social Innovation Processes in the Society. In SOCIETY, INTEGRATION, EDUCATION.* Vol. 4, pp. 233-243 (2015).
41. T. Scheuerle, B., Schmitz, W., Spiess-Knafl, R., Schues, S. Richter, International Journal of Social Entrepreneurship and Innovation. Mapping social entrepreneurship in Germany-a quantitative analysis. **3(6),** 484-511 (2015).
42. G. Cajaiba-Santana, Technological Forecasting and Social Change. Social innovation: Moving the field forward. A conceptual framework. **82,** 42-51 (2014).

43. S. El Arifeen, A., Christou, L., Reichenbach, F.A., Osman, K., Azad, K.S., Islam, D.H. Peters, The Lancet. Community-based approaches and partnerships: innovations in health-service delivery in Bangladesh. **382(9909),** 2012-2026 (2013).
44. M. Makimattila, T., Junell, T. Rantala, European Journal of Innovation Management. Developing collaboration structures for university-industry interaction and innovations. **18(4),** 451-470 (2015).
45. S. Battisti, International Journal of Services Technology and Management. Social innovation: the process development of knowledge-intensive companies. **18(3-4),** 224-244 (2012).
46. G. Krlev, E., Bund, E., G. Mildenberger, Information Systems Management. Measuring what matters—Indicators of social innovativeness on the national level. **31(3),** 200-224 (2014).
47. D. Lavie, Academy of Management Review. The competitive advantage of interconnected firms: An extension of the resource-based view. **31(3),** 638-658 (2006).
48. M. Meier, International Journal of Management Reviews. Knowledge management in strategic alliances: A review of empirical evidence. **13(1),** 1-23 (2011).
49. K.L. Turner, M.V. Makhija, Academy of Management Review. The role of organizational controls in managing knowledge. **31(1),** 197-217 (2006).
50. P. Benneworth, J. Cunha, European Journal of Innovation Management. Universities contributions to social innovation: reflections in theory & practice. **18(4),** 508-527 (2015).

Determinants of Internal Whistleblowing Intentions in Public Sector: Evidence from Indonesia

Ari Kuncara W.[1*], *Rizka* Furqorina[2], *Payamta*[3]

[1] Universitas Sebelas Maret, Indonesia
[2] Universitas Sebelas Maret, Indonesia
[3] Universitas Sebelas Maret, Indonesia

Abstract. This research aims to examine the influence of ewuh-pakewuh culture, attitude, perceived behavioral control, personal responsibility for reporting, and personal cost of reporting on the whistleblowing intentions among civil servants in the Directorate General of Taxation and perceived organizational support as moderation variable. This research uses primary data collected through the questionnaire. The data analysis is Structural Equation Model (SEM) with SmartPLS 2.0 software. This study employs a whole population as sample resulting 117 employees of Directorate General of Taxation. This research is divided into two scenarios: less serious corruption case and serious corruption case. The results of this research indicate mixed results. It seems that level of violation might affect the result.

1. Introduction

Indonesian government introduces whistleblowing as a mean to combact corruption. Cases of corruption occurred in Indonesia has became a common phenomenon. Corruption from year to year is increasing, both in terms of quantity and nominal financial losses. Various efforts have been implemented by the government to prevent and combat corruption in Indonesia. Among the efforts, the government has issued Presidential Instruction Republic of Indonesia Number 17 of 2011 about Prevention and Combating of Corruption Action 2012. This instruction contains the enactment of whistleblowing system.

Institution that became a pioneer in the implementation of the whistleblowing system is the Directorate General of Taxation (Direktorat Jendral Pajak or DJP). In Presidential Instruction Republic of Indonesia Number 17 of 2011, DJP is an institution that is prioritized in the implementation of the whistleblowing system. The existence of the whistleblowing system in DJP is expected to be to prevent and combat to any abuses. One of the aims whistleblowing systems in DJP is to rebuild public trust towards the DJP. DJP invites all employees to change the permissive culture into a culture corrective means it will never tolerate the violation. The entire tax officials and the public are expected to contribute actively to become a whistle-blower in the field of taxation.

Prevention efforts with the whistleblowing system have been applied to the DJP, officially, adopted since 2012. During 2012, the DJP recorded of 201 complains from whistle-blowers. This number increased by 63.4% compared to the prior issuance of Regulation of DJP Number: PER-22/PJ/2011. During the year 2013, the number of complains from whistle-blowers was 258, while in 2014, DJP recorded of 228 complaints from whistle-blowers [27]. It seems that the number of complaints from whistle-blowers significantly increased in beginning years of implementation of the system. However, it gradually decreases in 2014. In addition, there was also an employee who abuses their authority. It makes the effectivity of whistleblowing as means of prevention of corruption is questionable.

Based on the extant literature, the intention to report the violations are influenced by several factors such as invidual factors, and organizational factors. Most prior studies, particularly in outside Indonesia, have examined these factors. Meanwhile, Indonesia, particularly Javanese, has unique culture, namely ewuh pakewuh. This culture has not been widely examined in the context of whistleblowing intention.

* Corresponding author: Widagdo2002@yahoo.com

Whistleblowing reporting channels in DJP are included in the internal whistleblowing system. Thus, the internal whistleblowing intention deemed appropriate in the context of this study. The reasons to investigate whistleblowing intentions rather than actions whistleblowing are because of the impossibility and difficulty carrying out investigations of unethical behavior in the workplace by observation [14].

Based on the background of the problems described above, the formulation of the issues to be addressed in this study is as follows. How do ewuh-pakewuh culture, attitude toward whistleblowing, perceived behavioral control, personal responsibility for reporting, personal cost of reporting affect on whistleblowing intention of DJP`s civil servants? How does the perceived organizational moderate relationships between attitude toward whistleblowing, perceived behavioral control, personal responsibility of reporting, and personal cost of reporting and whistleblowing intention of DJP`s civil servants?

2. Literature Review and Hypotheses Development

2.1. Whistleblowing

A 1985 study [26] define whistleblowing as the disclosures made by members or former members of the organization of an illegal, immoral, or otherwise violate the legal legitimacy under the control of their leaders to individuals or organizations that can effect remedial action. Reporting violations (whistleblowing) is disclosure violations or disclosure of acts that are unlawful, unethical or other actions that may be detrimental to the organization and its stakeholders, carried out by a member or leader of the organization to the leadership of other authorized institution follow up action on the breach [22]. Whistleblowing intention referrs to the possibility of respondents report a violation that occurred [1].

Based on Regulation of DJP Number PER-22/PJ/2011, whistleblowing system in the DJP is a system that breach prevention and management reporting violations in the DJP. Whistleblowing system in DJP embraces three principles, namely the principle of prevention, early detection and proper investigation. DJP provides a whistleblowing system as a means for complaints either directly or indirectly. Complaints can be reported directly by way of the reporting coming directly to the Help Desk Directorate for Internal Transformation Decency Apparatus Resources (KITSDA). In addition, complaints also could indirectly report through call center, email, website and writing a confidential letter to head of Unit DJP.

2.2. Short Literatur Review

Various studies have been done on the outside and within Indonesia associated with the factors that influence the intention of whistleblowing [ie. 3, 6, 17, 18, 30]. These factors include cultural factors, demographics, personal cost, responsibility, organizational commitment, locus of control, and the seriousness of the offense.

A 2007 study [16] mentions five factors that influence the actions of whistleblowing: the psychological factor, cultural and ethical factors, structural factors, retaliation, and the type of wrongdoing. These factors will determine how individual actions when he finds violations. When people find their potential infringement damages, then the individual will consider various factors before making reporting. A 2002 study [6] examines the whistleblowing intentions on MBA students in China. The results indicate that locus of control moderates the relationship between votes and whistleblowing intentions judgment. In the study, respondents with an external locus of control are not possible to do whistleblowing. The respondents with an external locus of control will be deemed to uncover the secrets of ethical, but respondents who have an internal locus of control are more likely to report the error in question. A 2008 study [17] proves Confucian culture has a strong influence on whistleblowing intentions on professional accountants. The majority of respondents believe that the moral is the most important factor to encourage whistleblowing. In addition, the research also finds that the culture of guanxi and the fear of retaliation would be discouraged to take action whistleblowing.

A 2010 study [3] proves that attitude, perceived behavioral control, independence commitment, personal responsibility, and personal cost of reporting affect on the whistleblowing intentions public accountant in Barbados. Accountants, who have a positive attitude and high responsibility towards whistleblowing, will tend to do the reporting if he finds violations. The study proves that the organization support will increase the intention to do whistleblowing. Accountants will consider the difficulties in the process of reporting and the threat that would be obtained if the report violations. A 2013 study [18] compares the cultural factors that influence the tendency to report violations in China, Taiwan, and the US. The US has a greater disposition to engage in whistleblowing compared to China and Taiwan. In Taiwan and China, the intention to blow the whistle is affected by large amount of losses and influenced also by the impact of the practice is illegal, immoral or illegal on the company rather than that in the United States. This study reveals guanxi, or personal connections (network) has a great effect on the tendency for students whistleblowing Taiwan and China compared with American students. In another study, using a culture as variable, a study by [30] states that ethnic group influence on employee whistleblowing intentions. The study finds that the intention of doing whistleblowing employees come from the Javanese higher than employees come from non-Java region.

DOI: 10.1051/shsconf/20173401002

Based on the above previous studies, the intention to report the violations are influenced by several factors. Therefore, this study will examine influence attitudes, perceived behavioral control, personal responsibility and personal cost of reporting on the intention of internal whistleblowing intention of employees in DJP. In addition, this study will examine also the moderating role of perceived organizational support. This research is important because, until now, there has been a few of empirical research in Indonesia, which examines the effect of ewuh-pakewuh culture on whistleblowing.

2.3. Hypotheses Development

2.3.1. *Ewuh-Pakewuh* Culture and Whistleblowing

Ewuh-Pakewuh culture is a Javanese culture. According [35], ewuh-pakewuh is a manifestation of the fundamental rules of Javanese life. In addition, according [13], ewuh-pakewuh culture is an attitude and behavior that contains cheekily and worry if the behavior or words will make other people hurt.

According to [19], one of the cultural factors in Indonesia is very influential on the public sector bureaucracy in Indonesia is the Javanese cultural patronage that induces excessive ewuh-pakewuh. It can be seen from the bureaucracy in Indonesia, where the eastern culture in the context of Javanese politeness ewuh pakewuh that is reluctant attitude or awe and uphold respect for the boss or a senior [31].

A 2008 study [17] investigates the influence of Confucian culture on whistleblowing in Taiwan. The result shows that the Confucian culture influences the intention to do whistleblowing of auditing practitioners. In addition, in recent study [18], they find that cultural factors influence the tendency of individuals to take action whistleblowing in the US, China, and Taiwan. In Indonesia, the influence of cultural factors on whistleblowing intention has been studied by [30]. The results show that Indonesian Financial Transaction Reports and Analysis Centre employees who come fom the Javanese have higher internal whistleblowing intention than the intention of employees come from non-Java region. Characters employees come from the Javanese, who tend to have attitudes of pakewuh and not outspoken, tend to avoid conflict. As a result, these Javanese employees do not want to engage in whistleblowing. Ewuh-pakewuh culture will discourage intention of employees to report violations. Based on the above, the proposed hypothesis is:

H1: Culture ewuh-pakewuh has a negative effect on the intention to do whistleblowing.

2.3.2. Attitude toward Whistleblowing and Whistleblowing

Theory of planned behavior (TPB) explains that the intention of the individual to perform a behavior is an indication of the tendency of individuals to perform a behavior and the direct antecedent of the behavior [2]. Intention to perform a behavior can be measured through three main determinants that affect the intention, namely attitude toward the behavior, subjective norm, and perceived behavioral control. Attitude towards behavior is a stage where people evaluate the advantages (favorable) and losses (unfavorable) on an action. Attitude is the individual votes on approval or disapproval of certain behaviors.

Individuals can have confidence that whistleblowing would be beneficial for the organization if it can be uncovered and losses can be prevented. Conversely, individuals also can have confidence in whistleblowing would prejudice the organization, for example, damaging the reputation of the organization. Then, the individual will evaluate the positive or negative side that will be experienced alone, for example, gain reward from the organization on whistleblowing action or the threat of the party reported. Their consideration process is referred to as the process of "attitude factor" experienced by the individual. The tendency of being supportive whistleblowing act logically would increase the interest to take whistleblowing action [34].

A 2009 study [28] finds that attitude has a significant effect on the intention whistleblowing on South Korean police. Othe prior studies [i.e.37, 4] prove that attitudes towards whistleblowing significantly affect the interest of civil servants to take action whistleblowing. Attitude will directly affect whistle=blowing intention, to assess how favorable or unfavorable an individual if he do whistleblowing [3]. Based on the above, the proposed hypothesis is:

H2: Attitude toward whistleblowing has a positive effect on the intention to do whistleblowing.

2.3.4. Perceived Behavioral Control and Whistleblowing

According [5], perceived behavioral control as an individual's belief in a degree of ease or difficulty required to perform certain actions. According to [29], perceived behavioral control refers to the individual's perception of the ease or difficulty in performing a behavior. According to TPB, the individual will perform an action if he is sure to measure the level of difficulty may be obtained. Intention to perform a behavior can be measured through a perceived behavioral control. Perceived behavioral control depends on the presence or absence of opportunities and barriers to behavior.

A 2010 study [32] argues, if people can control themselves in the act that is based on perception, the higher the intention to become a whistle-blower. Perceived behavioral control is influenced by a combination of individual beliefs regarding the factors supporting or inhibiting to perform a behavior by the strength of feeling people will any factors supporting or inhibiting them. DJP has written rules and reporting channel that will support and facilitate the whistle-blower to report violations. Employees will have the intention to commit acts of whistleblowing when they have the perception that such action is easy to do, it is because there are mechanisms that support these actions. Based on the above, the proposed hypothesis is:

H3: Perceived behavioral control has a positive effect on the intention to do whistleblowing.

2.3.5. Personal Responsibility for Reporting and Whistleblowing

A 2010 study [21] identifies personal responsibility for reporting as one of the important variables that drive an individual's decision to report the violation information. According [20], personal responsibility is describes as a commitment to personal responsibility and moral obligation to report an unethical act. According [8], responsibility of reporting fraud actions can be influenced by a sense of social responsibility to co-workers and bosses, attitudes toward the profession, and the values of personal ethics. Individuals may feel obliged to report questionable actions by their moral (moral obligation or social responsibility), or a sense of commitment and loyalty to the organization (responsibility roles).

According [15], a decision to take action of whistleblowing depends on the perception of moral responsibility to report violations that occur within the organization. A 2010 study [3] proves that personal responsibility affects interest to do whistleblowing on internal auditor in Barbados. Prior studies [3] show that personal responsibility for reporting has a positive effect on the interest of the intention to commit acts of whistleblowing. Based on the above explanation, the proposed hypothesis is:

H4: Personal responsibility for reporting has as positive effect on the intention to do whistleblowing.

2.3.6. Personal Cost of Reporting and Whistleblowing

Personal cost of reporting is an individual assessment of the level of discomfort in reporting an unethical act [3]. This discomfort can be pressure, reprisals or retaliation from the offenders. According to [1], members of the institution have a contribution to prevent, deter and detect violations. However, in investigating the behavior of ethical decision makes whistleblowing against disclosure violation gets a lot of obstacles in the reporting process. A 2005 study [23] shows that the threat of retaliation is negatively correlated with interest to do whistleblowing. Thus, the risks to be faced by whistle-blowers must be taken into consideration before doing the reporting.

Individuals with the ability to influence others, a strong position, and authority to replace an employee in the organization tends to have the perception that the personal cost of reporting that will result from the behavior of whistleblowing will be relatively low, so that the individual had the intent higher to take action whistleblowing. Intention employees to report violations to be stronger when the personal cost of reporting is lower rated or personal responsibility to report violations rated is higher [30]. Intention employees to report violations is more powerful when the personal cost reporting is perceived lower or personal responsibility to report violations is perceived higher. Based on the above explanation, the proposed hypothesis is:

H5: Personal cost of reporting for reporting has a negative effect on the intention to do whistleblowing.

2.3.7. Perceived Organizational Support

Perceived organizational support is one form of normative isomorphism. Normative isomorphism aims to increase the professionalism of employees by assigning the most appropriate structure, and perceived organizational support is seen as one of the design of normative isomorphism [9]. According to [3], perceived organizational support is the extent to which an individual perceives that the organization values and supports the actions of individuals. Perceived organizational support encourages employees to adapt to organizational goals.

Based on social exchange theory, employees are more likely to feel obligated to commit to the organization if they get a high degree of organizational support [10]. Thus, employees will feel safe making decisions such as reporting unethical when perceived organizational supports these actions. According [11], a positive relationship, which valued employee perceptions and gain awareness by the organization with awareness in carrying out job responsibilities, affective involvement, and encouraged to innovate.

A 2010 study [3] shows that moderate the relationship of perceived organizational support will increase whistleblowing actions. According [36], protection system moderates the relationship public service motivation and internal whistleblowing intentions. Based on the above explanation, the proposed hypotheses are:

H6a: Perceived organizational support will moderate relationship between attitude toward whistleblowing and intention to do whistleblowing.

H6b: Perceived organizational support will moderate relationship between perceived behavioral control and intention to do whistleblowing.
H6c: Perceived organizational support will moderate relationship between personal responsibility for reporting and intention to do whistleblowing.
H6d: Perceived organizational support will moderate relationship between personal cost of reporting and intention to do whistleblowing.

3. Research Method

This research is a quantitative research by using primary data. The population in this study is employees of Regional Office DJP in Central Java I and II. The sampling technique is census that population is used as a sample. The data collection methode in the study is questionnaires that are adopted from some prior studies. The ewuh pakewuh quetionnaire is adopted from a 2011 study [31]. For variable attitude toward whistleblowing and perceived behavioral control, the questionnaire is adopted from a 2010 study [3]. The questionnaire of personal responsibility for reporting and the personal cost of reporting adopt the questionnaire of 2013 studies [30, 37]. The questionnaire of perceived organizational support is developed based on 2010 study [3].

This study used the intention simulation, which uses vignette. The respondents were asked to evaluate and to make choices based on the information provided. Thevignette used hypothetical cases of occupational fraud in accordance with Regulation of DJP Number: PER-22/PJ/2011, which was developed by the researchers based on the level of seriousness. The first vignette was a vignette to represent the serious corruption, which relates acceptance of bribes. The second vignette was a vignette to represent the less serious corruption, which relates to a mark-up in a procurement project. The validity of the contents of the scenario previously consulted with tax officials, who had the expertise and capability in handling cases of whistleblowing.

Analysis of the data in this study uses the approach of Partial Least Squares (PLS). PLS is one of the statistical methods of Structural Equation Modeling (SEM) based variants. The theoretical framework is presented briefly in the image below:

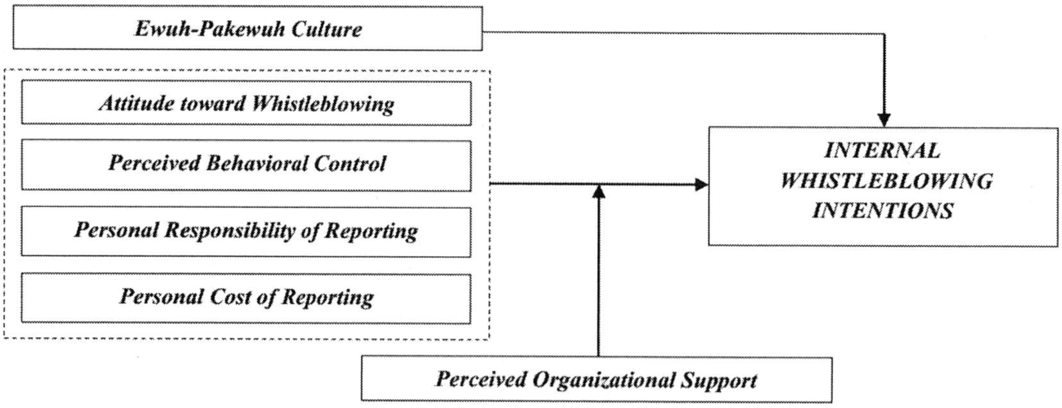

Fig. 1. Theoretical Framework

4. Results and Discussion

4.1. Results

The data in this study is a primary data obtained using questionnaires distributed to tax officials at the Regional Office of Directorate General of Taxation, Central Java Central Java I and II. The number of distributed questionnaires were 255 copies, 135 copies were distributed at the Regional Office of Directorate General of Taxes Central Java I and 120 copies were distributed at the Regional Office of Directorate General of Tax of Central Java II. The survey was conducted in one month, from December 16th, 2015 until January 18th, 2016.

The number of collected questionnaires is 128 copies and as many as of 11 copies is incomplete, so that only 117 copies can be used for this research. The respondents are mostly male (64.96 percent). The respondents hold undergraduate degrees are 70 respondents (59.83 percent). Most respondents are in the age range of 25-35 years (50.43) percent. The respondents mostly occupy executive positions with a number of 99 respondents (84.6 percent) and the percentage of respondents who had experience working in the DJP for >11 years is 50.43 percent.

4.1.1. Case 1: Serious Corruption

This model has a value of goodness of fit of 0.596915, so the value of goodness of fit is categorised as large. Based on the result of the calculation of goodness of fit, it can be concluded that the modification of the model at this stage has met the criteria or fit. . This R square of the model is 54.19 percent.

For a hypothetical two-tailed with alpha testing of 5 percent, the value of the path coefficient indicates an accepted hypothesis if the value of t-statistic is greater than 1.96. Table 1shows that ewuh-pakewuh culture (EP) has t-statistic value of 2.923345. This indicates that the result is significant because the t-statistic value is greater than 1.96. Based on the results SmartPLS, it can be concluded that H1 is empirically proved (H1 accepted). Attitude toward whistleblowing (ATT) has the value of t-statistic of 15.260580, so it can be concluded that the H2 is accepted. Perceived behavioral control (PBC) has a t-statistic value of 14.224069, so the relationship is significant. However, unlike our expectation, the coefficient path is a negative, thus H3 is empirically rejected. Personal responsibility for reporting of (PR) has a t-statistic value of 0.820351, so H4 is rejected. Personal cost of reporting (PC) has a t-statistic value of 0.751600, so that it can be concluded H5 is rejected. In terms of interaction variable, the interaction between perceived organizational support with perceived behavioral control and its interaction with personal cost have a t-statistic value more than 1.96. This indicates that the H6b and H6c are empirically proved. In contrast, H6a and H6d are empirically rejected.

Table 1. Serious Corruption Case

	Original Sample (O)	Sample Mean (M)	Standard Deviation (STDEV)	Standard Error (STERR)	T Statistics (|O/STERR|)
EP -> WBI	-0.066487	-0.070211	0.022744	0.022744	2.923345
ATT -> WBI	0.513034	0.505496	0.033618	0.033618	15.260580
PBC -> WBI	-0.341217	-0.344445	0.023989	0.023989	14.224069
PR -> WBI	0.027178	0.038011	0.033130	0.033130	0.820351
PC -> WBI	0.018276	0.024244	0.024316	0.024316	0.751600
ATT * POS -> WBI	-0.047894	-0.049356	0.050415	0.050415	0.949994
PBC * POS -> WBI	0.143576	0.130231	0.039295	0.039295	3.653767
PR * POS -> WBI	0.196985	0.203299	0.052767	0.052767	3.733132
PC * POS -> WBI	-0.091845	-0.033971	0.080740	0.080740	1.137539

4.1.2. Case 2: Less Serious Corruption

The value of goodness of fit of the model is 0.669780, which is categorized large. Based on the calculation of goodness of fit, it can be concluded that the modification of the model at this stage mets the criteria or fit. This R square of the model is 46.76 percent.

Table 2 shows that ewuh-pakewuh culture (EP) has value of t-statistic of 0.750402. This indicates that H1 is rejected. Attitude toward whistleblowing (ATT) has the value of t-statistic of 6.343492, so it can be concluded that the H2 is accepted. Like the prior result, perceived behavioral control (PBC) has a t-statistic value of 5.714985, but the coefficient path is a negative. Thus, H3 is rejected. The responsibility for reporting of personal (PR) has a t-statistic value of 13.200226, so H4 is accepted. The personal cost of reporting (PR) has t-statistic value of 3.909477, so that it can be concluded H5 is accepted. In terms of the interaction variables, there are no hypotheses that could be proved.

Table 2. Less Serious Corruption Case

	Original Sample (O)	Sample Mean (M)	Standard Deviation (STDEV)	Standard Error (STERR)	T Statistics (|O/STERR|)
EP -> WBI	0.018221	0.025978	0.024282	0.024282	0.750402
ATT -> WBI	0.177189	0.168756	0.027932	0.027932	6.343492
PBC -> WBI	-0.181232	-0.174754	0.031712	0.031712	5.714985
PR -> WBI	0.465643	0.451220	0.035275	0.035275	13.200226
PC -> WBI	-0.117125	-0.124116	0.029959	0.029959	3.909477
ATT * POS -> WBI	-0.103019	-0.062626	0.082852	0.082852	1.243407
PBC * POS -> WBI	-0.132441	-0.122359	0.027802	0.027802	4.763703
PR * POS -> WBI	0.044744	0.029279	0.085277	0.085277	0.524683
PC * POS -> WBI	0.030089	0.007852	0.036724	0.036724	0.819344

4.2. Discussion

The results of hypothesis testing show that H1 indicates mixed result. This hypothesis is accepted in serious case, whereas the hypothesis is rejected and not empirically proven in less serious corruption case. This means that effect of ewuh-pakewuh culture on whistleblowing intention depends on the level of corruption case. In terms of serious corruption case, the employees tend to avoid action that will make hurt other employee, who particularly has managerial position. Rotation system and moving staff regularly might be capable of reducing a sense ewuh-pakewuh in DJP. Staffing system in the DJP in a position is determined by a maximum period of three years. After three years, he will do rotational placement. The rotation of employees is also intended to prevent violations. According [24], organizations/government agencies might face cultural transition, although, some characteristics paternalistic culture in everyday life can still be found.

The results indicate that the hypothesis H2 is accepted on both cases. Attitudes towards whistleblowing affect the intention to commit acts of whistleblowing. Individuals can have a confidence that whistleblowing would have a positive and a negative sides. Then, people would consider a positive or a negative side of his actions. The consideration process is referred to as the process of "attitude" experienced by these employees. The results are consistent with the theory of planned behavior that perceived behavioral control is an aspect that affects an individual's decision to perform an action. The result of this study confirms the results of previous research [28, 3, 32, 37, 4]. Individuals will assess whether the act of whistleblowing will be favorable or unfavorable if he/she does whistleblowing.

Unlike the expectation, the hypothesis H3 is rejected on both cases because the coefficient is a negative. It means that the greater of the perceived behavioral control causes the weaker of the intention to perform the whistleblowing. It is contrary with the theory of planned behavior that if an employee can control themselves in accordance with the perception that is believed to act whistleblowing, it will cause the intention on the employee to take action whistleblowing. The consequence of doing whistleblowing might deter the intention of the employee.

The hipotesis H4 is empirically proved and supported on less serious corruption case, but not in serious corruption. It means that the employee have higher personal responsibility for reporting violation in less serious case than that in serious case. The employees might view that reporting serious case has a negative consequence and they prefer not to do whistleblowing. The results of this study are consistent with prior studies [3, 21] that personal responsibility for reporting a positive effect on the interest of the intention to commit acts of whistleblowing.

Personal cost of reporting empirically has a negative effect on employee intention to commit acts of whistleblowing on less serious corruption case. It seems that the employees feel that reporting of less serious corruption has low personal reporting cost, so it does not have the impact that would harm themself. The result is consistent with previous studies [7, 3, 30, 37, 12, 4] that the personal cost of reporting negatively affect intention to commit acts of whistleblowing.

The results show that perceived organizational support moderates the relation between perceived behavioral control with whistleblowing intentions in both cases. It means that organizational supports might provide a safety and mitigate a negative consequence of doing whistleblowing. Perceived organizational support moderates relation of personal responsibility with the whistleblowing intentions in serious corruption case. This means perceived organizational support will increase whistleblowing intentions. Whistleblowing policies provide an ease in reporting and are able to increase the personal responsibility to report violations.

The interaction perceived organizational support with personal cost of resporting and personal repondsibility of reporting do not significantly affect whistleblowing intentions in less serious case and serious case as well. This may occur because employees feel the support of the organization in action whistleblowing, only felt able to provide ease in conveying violation information. Whistleblowing system in DJP allows the employee to provide information of the violation; however, this support for the whistleblowing does not make employees have a positive attitude towards whistleblowing. Regulation of whistle-blower protection for both the central government and the DJP itself is not able to improve safety of employees to report violations.

5. Conclusion

Whistleblowing is an action taken by members of the organization to uncover violations that occur in organizations. DJP has implemented a whistleblowing system that is internal whistleblowing channels. Whistleblowing system is a system of prevention and identification of fraud/breach that would occur in an organization. With the whistleblowing system, employees will avoid any violation and are expected to participate actively in the participation of preventing violations. Nevertheless, there are still cases were uncovered after implementation of whistleblowing in DJP. This indicates the whistleblowing system functions as a prevention system has not been able to run well.

The results show that effect of the culture ewuh-pakewuh, personal respondsibility of reporting, and personal costs of reporting on the whistleblowing intention provide mixed findings. It seems that the level violation (i.e. level of corruption) affect the result. Similarly, the result of interaction variables are also mixed and it seems depend of type of violation.

This study has several limitations. First, the respondents in this study is only DJP`s civil servants in the Central Java Region I and II, so that research results can not be generalized. Second, this study uses a questionnaire to collect data that the bias is likely to occur. Suggestions for further research can add a more comprehensive method. For example, future study might employ amixed methods by interviewing the informan to clarify the results from quantitative approach. Subsequent research could also examine other factors, such as the effect of educational level as well as rotation and mutation.

References

1. S.A. Ahmad, R.M. Yunos, R.A.R. Ahmad, Z.M. Sanusi. *Procedia,* **164**, 445-450 (2014)
2. I. Ajzen. *OBHDP,* **50**(2), 179–211 (1991)
3. P.A. Alleyne. *Doctoral Diss*. Univ. of Bradford, UK (2010)
4. R. Bagustianto, Nurkholis. *SNA XVIII Medan.* USU (2015)
5. A. Bandura. New Jersey: Prentice Hall (1997)
6. R.K. Chiu. *MAJ,* **17**(9), 581–588 (2002)
7. L.M. Cortina, Magley, J. Vicki. *JOHP,* **8**(4), 247–265 (2003)
8. M. Curtis. *JBE* **68**(2), 191-209 (2006)
9. P.J. DiMaggio, W.W. Powell. *ASM,* **17**, 143–166 (1983)
10. R. Eisenberger, R. Huntington, S. Hutchison, D. Sowa. *JAP,* **71**, 500-507 (1986)
11. R. Eisenberger, P. Fasolo, V. Davis-LaMastro. *JAP,* **75**(*1*), 51-59 (1990)
12. C.K. Eller. *Thesis*. VCU, US (2014)
13. A. Frinaldi, M.A. Embi. *Humanus* **13**(1), 68-75 (2014)
14. N.A. Ghani, J. Galbreath, R. Evans. *AGMS,* **3**(1), 1-18 (2011)
15. J.W. Graham, J. W. *ROB,* **8**,1-52 (1986)
16. H. Hassink, M. de Vries, L. Bollen. *JBE,* **75**(1), 25-44 (2007)
17. D. Hwang, B. Staley, Y.T. Chen, J.S. Lan. *MAJ,* **23**(5), 504-526 (2008)
18. D. Hwang, B. Staley, Y. Tsai, C.L. Chu. *IJAFR,* **3**(2), 202-224 (2013)
19. Ihsan, F. *Ref. Budaya Kerja Birokrasi* (2014)
20. S.E. Kaplan, S.M. Whitecotton. *AAA,* **20**(1), 45-64 (2001)
21. M. Keil, A. Tiwana, R. Sainsbury, Sneha. *Persp. Dec. Sciences,* **41**(4), 787–812 (2010)
22. KNKG (2008)
23. M.J.R. Magnus, C. Viswesvaran. *JBE,* **62**(3), 277-297 (2005)
24. A. Maskun. *JAM,* **7**(1), 162-172 (2008)
25. E.S. Muslim, E. S. *Thesis.* ITB, Indonesia (2000)
26. J.P. Near, M.P. Miceli. *JBE,* **4**(1), 1-16 (1985)
27. DJP Annual Report 2014 (2014)
28. H. Park, J. Blenkinsopp. *JBE,* **85**(4), 545-556 (2009)
29. C.H. Ponnu, K. Naidu, W. Zamri. *IRBRP,* **4**(1), 276-298 (2008)
30. W. Septianti, W. *SNA XVI Manado*. UNSRAT: 1063-1094. (2013)
31. H.I. Soeharjono. *Doctoral Dis.* UGM, Indonesia (2011)
32. A. Sulistomo. *Working Papers*. UNDIP, Indonesia (2012)
33. E. Suryono. *Thesis.* UNDIP, Indonesia (2014)

34. D.H. Tobing.. *Thesis.* UGM, Indonesia (2010)
35. W. Vandenabeele, A.M. Kjeldsen. *EGPA conference 2011*, Romania (2011)
36. R.D. Winardi. *JIEB,* **28**(3), 361. (2013)

SHS Web of Conferences **34**, 01003 (2017) DOI: 10.1051/shsconf/20173401003
FourA 2016

New accountant job market reform by computer algorithm: an experimental study

Yoshitaka Hirose[1,*]

[1]Takasaki University of Commerce Junior College, Department of Business and Administration, 741 Negoyamachi Takasaki Gunma, Japan

Abstract. The purpose of this study is to examine the matching of new accountants with accounting firms in Japan. A notable feature of the present study is that it brings a computer algorithm to the job-hiring task. Job recruitment activities for new accountants in Japan are one-time, short-term struggles. Accordingly, many have searched for new rules to replace the current ones of the process. Job recruitment activities for new accountants in Japan change every year. This study proposes modifying these job recruitment activities by combining computer and human efforts. Furthermore, the study formulates the job recruitment activities by using a model and conducting experiments. As a result, the Deferred Acceptance (DA) algorithm derives a high truth-telling percentage, a stable matching percentage, and greater efficiency compared with the previous approach. This suggests the potential of the Deferred Acceptance algorithm as a replacement for current approaches. In terms of accurate percentage and stability, the DA algorithm is superior to the current methods and should be adopted.

1 Introduction

The purpose of this study is to present an improvement plan to address the problems of the labor market in Japan for Certified Public Accountants (CPA). Japan's successful CPA applicants do their job hunting simultaneously. Job hunting is a short-term, decisive battle. Consequently, accounting firms seek to obtain excellent, successful candidates [1]. The problem with this short-term, decisive battle has been noted previously. The problem has been identified in the United States labor market for medical doctors. However, the labor market of the United States for medical doctors has solved the problem using a computer algorithm. The labor market of Japanese medical doctors uses the same method. This study compares the labor market of these doctors and accountants and considers whether it is possible to use a computer algorithm in the labor market of the accountants.

The approach of this study is an analysis of game theory, which is referred to as matching theory [2]. The important point is the analysis of job hunting in game theory. Game theory can be used to discuss the design of institutions. The research method of this study is experimental research. Multiple alternatives may be compared in the experiment.

[*] Corresponding author: y-hirose@uv.tuc.ac.jp

Carrying out the proposed analysis can result in pre-verification, with the aim of improving institutional design [3].

This paper is organized as follows. First, Section 2 reviews the previous studies and provides an overview of the labor market of the accountants. Next, Section 3 considers an effective algorithm for the solution of this problem. Section 4 then considers the experimental design. The results are shown in Section 5. Section 6 completes the discussion by showing the conclusions and noting future challenges.

2 Previous studies and an overview of the labor market of the new accountants

For new accountants, entering the labor market of Japan involves a short-term, decisive battle. Why has this become a short-term battle? To answer that question, it is necessary to understand the specific circumstances of Japan. First, consider the outline below.

Accounting jobs in Japan are the first jobs for many of the successful candidates. The majority of successful candidates are university students. The Japanese Institute of Certified Public Accountants (JICPA) oversees the new accountants in Japan's labor market. The JICPA conducts self-regulation to ensure that job hunting is carried out in a fair manner. The self-regulation of the JICPA is a fairly strong, compelling force for accounting firms. The JICPA has issued a notice every year for the accounting firms in Japan. The notice defines in detail the start dates for corporation briefings, interviewing, job offer notifications, and the like. Additionally, these rules change each year. The latest rules were published in 2015 [4]. This recent timeline is provided in Figure 1.

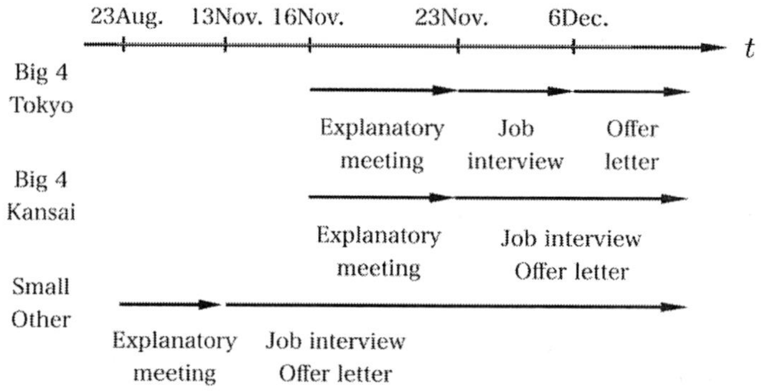

Fig. 1. Timeline of the current rules of job recruitment activities for new accountants.

To understand this figure, one must understand differences in regions and sizes.

The JICPA is divided into three accounting firm groups: (1) Tokyo district of major accounting firms, (2) Kansai district of major accounting firms, and (3) other regions with major accounting firms and small and medium-sized accounting firms. Here, the major accounting firms are the Big 4. More specifically, the Big 4 consists of Ernst & Young, Deloitte, KPMG, and PricewaterhouseCoopers. The Kansai region includes Osaka, Kyoto, and Kobe.

Accounting firms have been requested to comply with the rules of the JICPA. Accordingly, recruitment activities proceed along the schedule of Fig 1. This figure indicates that the recruitment activities of small and medium-sized accounting firms begin first. For accounting firms of small and medium size, it is possible to hold an explanatory meeting immediately after the end of the CPA exam. Accounting firms of small and medium size can also start the job interview process immediately after the CPA exam results are announced, and they can begin sending offer letters. Then the recruiting activities of Kansai and Tokyo's major accounting firms begin. Specifically, Kansai and Tokyo's major accounting firms can hold explanatory meetings beginning on November 16. They can start job interviews on November 23. A major accounting firm in the Kansai region is able to begin sending offer letters on November 23. However, major accounting firms in Tokyo start sending offer letters after December 3. According to the JICPA, accounting firms will not be able to make an offer to a specific person before the "offer letter start." The further details of this rule are as follows:

Type 1. For the major audit firms' Tokyo offices, explanatory meetings start November 16 (Monday) (application acceptance starts November 13 (Friday) 12:00), job interviews start November 23 (Monday) 9:00 (application acceptance starts November 19 (Thursday) 15:00), and offer letters begin after December 3 (Thursday) 9:00 (the joining acceptance period is December 5 (Saturday) 12:00).

Type 2. For the major audit firms' Kansai district offices, explanatory meetings start November 16 (Monday) morning (application acceptance starts November 13 (Friday) 12:00), job interviews start November 23 (Monday) 9:00 (application acceptance starts November 19 (Thursday) 15:00), and offer letters begin any time from the start of the interviews (the joining acceptance period is November 27 (Friday) 17:00).

Type 3. For firms that are not Kansai or Tokyo major accounting firms, including small and medium-sized accounting firms, explanatory meetings can be conducted any time after the end of the CPA exam. However, acceptance of the job interview is carried out after the exam results are announced. The explanatory meeting and offer letter start are determined via consultation in each region.

In this way, the accounting firms may be subject to any of three rules. The application of the rules differs according to firm size and region. A notable aspect of this rule is that the joining acceptance period is established in the Kansai and Tokyo's major accounting firms. Additionally, the joining acceptance deadline has been designated by both date and time. After the joining acceptance deadline of the major accounting firms in the Kansai region passes, the starting date for offer letters of Tokyo's major accounting firms begins. Accordingly, a person who has accepted a job offer form a major accounting firm in the Kansai region cannot possibly consider joining a major accounting firm in Tokyo. Therefore, job hunting for new accountants has become something like a bet. The most typical case is for an accountant to accept a job in Kansai despite wanting to work in Tokyo. Others accept jobs in small and medium-sized firms even though they want to work in Tokyo or Kansai. In other words, Tokyo and Kansai's major accounting firms have missed the opportunity to acquire top talent. The current rules for new accountants and accounting firms are inefficient. However, the accounting firms that are small or medium-sized and those offices of the Big 4 that are not in Tokyo or Kansai might gain excellent employees that could not be obtained without the current rules. In this way, the interests of the audit firms have conflicted. The JICPA has proceeded by trial and error. However, the employment recruitment of new accountants and audit firms remains, basically, an irreversible "one-shot deal."

3 Model: Computational algorithm

Consider the real world to be of the essence, as discussed in the previous section. An offer letter that an audit firm sends out is an exploding offer. It must be answered immediately. The new accountant must choose to accept or reject by the deadline.

In this section, we consider a job recruitment activity that combines computers and humans. First, I consider a valid computer algorithm to solve this problem. Other professionals in the labor market, according to previous studies, are subject to a residency (training physician) matching. The matching of training physicians with hospitals is being analyzed in the field. Such research has been applied to the real world [2].

Previous research identifies three criteria for determining the desirability of matching [2]. They are: (1) truth telling, (2) stability, and (3) efficiency.

1. Truth telling: The truth telling criteria is the requirement that job searchers make honest decisions regarding their preferences. Often, people lie to themselves in the process of job hunting. It is better, however, to job hunt honestly without lying to oneself.

2. Stability: The stability criteria is that everyone in the state that does not have an incentive to the ran missing. Conversely, matching parties that do not have a mutual affection results in employment situations that are not stable.

3. Efficiency: The efficiency criteria are that of Pareto efficiency in welfare economics.

One algorithm that satisfies the truth telling and stability criteria is the Deferred Acceptance (DA) algorithm [5]. This algorithm has been applied to the real world in training physicians matching and school matching. The DA algorithm is as follows.

Step 1: Each new accountant applies for his/her first choice of accounting firms. Each accounting firm accepts new accountants until its capacity is filled. Each accounting firm declines the rest of the new accountants.

Step k: New accountants that were refused in the k-1 step apply for their first choices in the accounting firms that are not otherwise specified. Each accounting firm compares the new entrants and keeps the new accountant. Each accounting firm declines the rest of the new accountants.

End: The algorithm is terminated when the new accountants who were refused are gone. Each accounting firm will have formally adopted new accountants that it will keep.

Actually, though, new accountants do not apply, and accounting firms do not decide to keep them. Essentially, a centralized third party makes the matching determination using a computer. Specifically, JICPA and the Financial Services Agency could cooperate to accomplish this. This is realistic because there is precedent in the cooperation of the Medical Association and the Ministry of Health, Labour and Welfare with a third party organization.

4 Experimental Design

The experimental design was based on previous research referenced [2, 6]. The experiment targeted Doshisha University undergraduates in July 2013 and January 2014. A total of 131 people participated in the following two treatments. The CR (current rule) involved 35 people and the DA (Deferred Acceptance) involved 96 people a subset of 33 people was examined.

The subjects played the role of new accountants' job hunting with three accounting firms, in three-person sets. The hiring capacity of the three accounting firms is one person each. The subjects participate in the experiment for only one of the two conditions (i.e., CR or DA). I conducted a between-subjects design experiment. The experimental procedure consisted of the following steps: First, the experimenter explained the instructions. The experimenter answered any questions. Then the experimenter distributed a document stating the true preference ranking of each subject. The true preference ranking that was distributed to each of the subjects constituted private information. The subjects could not know the true preference order of other subjects. Subjects of the CR treatment were either selected or not selected for entry into each of the accounting firms. For subjects of the DA treatment, three rankings of the accounting firms were selected. Each condition was subject to decision only once. Subjects did not perform repetitive decisions. In addition to having gained the psychological data, the experimenter performed a post-questionnaire of behavioral data for both conditions. Incidentally, the preference order of the accounting firms and the conditions were fixed. In addition, the decision-making of each audit firm was performed by a computer. Each accounting firm of the CR treatment selected higher preference order new accountants. Additionally, each accounting firm of the DA treatment expressed the true preference ranking. Such accounting firms submitted their data without faking the ranking of their desire for new accountants. The true preference rankings of these accounting firms was information that was not known to all subjects. Subjects won three points when they were assigned to the first position of an audit firm. Subjects won two points when they were assigned to the second position of an audit firm. Subjects won one point when they were assigned to the third position of an audit firm. The mean duration of the experiment was about 30 minutes. It should be noted that experiments were carried out context-free. I set the following parameters in the experiment.

Table 1. True preference ranking of the new accountants.

	New accountant 1	**New accountant 2**	**New accountant 3**
First rank	Type 2	Type 1	Type 1
Second rank	Type 1	Type 2	Type 2
Third rank	Type 3	Type 3	Type 3

Table 2. True preference ranking of the accounting firms.

	Accounting firm 1	**Accounting firm 2**	**Accounting firm 3**
First rank	New accountant 1	New accountant 2	New accountant 2
Second rank	New accountant 3	New accountant 1	New accountant 1
Third rank	New accountant 2	New accountant 3	New accountant 3

5 Results

This section shows the experimental results. Three of the results indicate the desirability of the criteria for matching are "truth telling, stability, and efficiency.

Result 1. *The DA treatment has a higher truth-telling percentage than the CR treatment.*

Statistical evidence. I conducted a Welch Two Sample t-test for the difference between the mean values of the two treatments. Under each mechanism, the null hypothesis of equal proportions of truth-telling revelation across the two treatment settings can be rejected at the 1% significance level. $t = -3.125$, $df = 63.959$, p-value = 0.003. The mean of truth telling is listed in the following table.

Result 2. *The DA treatment has a higher stable matching percentage than the CR treatment.*

Statistical evidence. I conducted a Welch Two Sample t-test for the difference between the mean values of the two treatments. Under each mechanism, the null hypothesis of equal proportions of stable matching revelation across the two treatment settings can be rejected at the 1% significance level. $t = -3.796$, $df = 17.101$, p-value = 0.001. The mean of stable matching outcomes is listed in the following table.

Result 3. *The DA treatment has higher efficiency than the CR treatment.*

Statistical evidence. I conducted a Welch Two Sample t-test for the difference between the mean values of the two treatments. Under each mechanism, the null hypothesis of equal proportions of efficiency across the two treatment settings can be rejected at the 5% significance level. $t = -2.725$, $df = 12.079$, p-value = 0.018. The mean of efficiency is listed in the following table.

Table 3. The proportion of truth telling, stability, and efficiency.

	CR	DA
Truth telling	30.3%	66.7%
Stability	9.0%	72.7%
Efficiency	72.7%	87.0%

Therefore, the DA treatment was found to exhibit higher truth telling, stability, and efficiency than the CR treatment.

6 Conclusion

In this paper, I compare the current rules and the DA algorithm. The main results of this study are as follows: 1) The DA treatment has a higher truth-telling percentage than the CR treatment. The way in which current rules tend to apply for a small accounting firm was observed. 2) The DA treatment has a higher stable matching percentage than the CR treatment. 3) The DA treatment has higher efficiency than the CR treatment.

These results lead to the following suggestions. Under the current rules for new accountants, people who engage in job hunting do not pursue their true preference. Stable matching is low under the current rules. The Deferred acceptance alternative to the current rules can solve these problems. It will ensure that new accountants are able to express their true preference rankings. Moreover, it will result in stable matching. This is a good thing for both new accountants and accounting firms. Efficiency is a measure that has been considered desirable in traditional accounting and economics. However, truth telling and stability are measures that are used in market design, matching theory, and game theory. Accordingly, humans need to take advantage of the power of the computer algorithm.

References

1. Japan Auditing Association, *Certified Public Accountants examination system* (Daiichihoki, 1993)
2. A. E. Roth, International Journal of Game Theory, **36**, 3-4 (2008)
3. S. Taguchi, *An experimental Institutional Accounting: A future design for accounting*, (Chuo-Keizai-Sha, 2015)
4. Japanese Institute of Certified Public Accountants, The start time of the recruitment activities of the current fiscal year of the Certified Public Accountant exams, (2015)
5. D. Gale, L. S. Shapley, The American Mathematical Monthly, **69**, 1 (1962)
6. J. Pais, Á. Pintér, Games and Economic Behavior, **64**.1 (2008)

Internal audit function: a comparison between private and public sector in Nigeria

Abdulkadir Madawaki[1,*], *Aidi* Ahmi[2], and *Halimah* @ Nasibah Ahmad[2]

[1]Department of Accounting, Al-Qalam University Katsina, Nigeria
[2]Tunku Puteri Intan Safinaz School of Accountancy, Universiti Utara Malaysia, Malaysia

Abstract. This study compares the internal audit functions between private and the public sector. Features examined include hierarchical rank of internal audit function, internal audit transfer, outsourcing of internal audit services and working relationship of internal audit with external auditor. The study is based on a survey of internal audit managers and chief internal auditors in private and public sector entities in Nigeria. The results revealed that there is no much difference in hierarchical rank of internal audit function in both sectors. While the results showed that differences exist in the reporting status between internal audit in the two sectors with private sector internal audit reporting to audit committee and chief executive officer and public sector internal audit reporting to chief executive officer and chief financial officer. Almost a similar amount of work is outsourced, and private sector entities are more likely to outsource than those in public sector. There is slight difference between internal audit activities and interaction with external auditor in the two sectors with private sector more likely to coordinate in the area of access to audit working papers.

1 Introduction

Over the past two decades, the world has witnessed a series of high profile of corporate collapses, as a result of which has led to various effort to build-up regulatory framework in order to restore investors confidence which internal audit function (IAF) is such a framework within the corporate governance. The expected benefits of strong IAF arise from the view that an effective internal controls make fraud difficult, make fraud discovery likely, provides value-added assurance services and promote effective organizational governance which could results in better financial reporting process and audit function [1]–[4]. The growing role of effective internal audit in enhancing corporate governance has been emphasized through various legislations such as: Sarbanes Oxley Act (SOX), 2002 in the USA, Corporate Law Economic Reform Program Act (CLERAP) 2004 in Australia, Corporate Laws Amendment Act, 2006 in South Africa. In Nigeria Section 61 Sub Section 1, 2 and 3 of the Investment and Securities Act (ISA), 2007 and Nigerian Securities and Exchange Commission Code of Corporate Governance, 2003 places a strong emphasis on private companies (public listed entities) to establish IAF and requires that board of directors and audit committee to evaluate and report on internal controls effectiveness.

* Corresponding author: abdulkadirmadawaki@yahoo.com

While Rule 1701 (2) of Financial Regulations 2009 a Federal Republic of Nigeria Official Gazette addresses the issue of IAF for public companies (Government-owned corporation, and agencies).

Based on the notion that good IAF over financial reporting is an effective corporate governance mechanism and result in higher quality financial reporting in both sectors; this study aims to examine the similarities and differences of IAF in the private sector and that of its counterpart in the public sector. The study is motivated by the growing importance on the task of IAF as a corporate governance device in both sectors and the need to have a greater understanding of IAF and their operations in Nigerian public and private and public sector entities. It has been contended that differences exists between private and public sector internal auditing in relation to their structures and operations [5]. The study extends that of [6], [7] with focus on public sector entities.

The principal interest of corporate governance stem from the separation of ownership and control in public organizations. Agency theory in this setting present the foundation for explaining the functions assigned to IAF since it is the IAF that provides assurance to the management [24]. The shareholders interest has been reinforced through the efforts of professional bodies and government which result in increased demand on the management to ensure that corporations are managed effectively. Much of this demand has been as a result of public expectations in reaction to corporate scandals. The institutional theory postulates that corporation's management and control arrangement have a tendency to go with public expectations. Thus, this study relies on both the agency and institutional theories to compare the similarities and differences between public and private sectors IAF in Nigeria.

2 Literature review

In line with a paradigm of corporate governance developed by the Institute of Internal Auditors (IIA), an effective IAF is one of the four foundations of corporate governance sideways with the management, the board of audit committees and the external auditor. Previous literature for example [8], [9] also uphold the view that IAF is an important corporate governance device that plays a significant role in organizational governance by overseeing the organizational risk, evaluating and assessing control mechanisms.

Prior researches have indicated disparity exists between public and private sector IAF [5] in a comparison of public and private sector IAF in South Africa found that internal controls in national departments are recognized to be effective corporate governance mechanisms but they are not yet well established as those in the public sector.

Studies examining the operational and organizational status of IAF shows that [10] survey the opinions of bank lending officers with regard to the influence of reporting arrangement on internal audit capabilities to avert financial statement fraud. The findings of the study established that IAF that report to senior management are seen as being less able to avert deceitful financial reporting as against those that report exclusively to audit committee.

Similarly, [11] using a case study method looked at five Belgium companies to examine the expectations and opinions of both senior management and internal auditors with regard to the association between the two groups. The study indicated that when internal audit work largely with a management support role there may be lack of perceived independence of the internal audit.

Prior studies have also shown a mixed result with regard to the decision to dismiss the chief internal auditor. For example, [12] study the association between IAF and the audit committee influence on the independence and objectivity of the internal auditors. Their result indicated that 72 percent of the audit committee members were involved in the dismissal of the chief internal auditor. In a similar situation, [13] reported that 52 percent of the Australian and New Zealand audit committee members are involved in the decision to dismiss the chief internal auditor.

Outsourcing of internal audit services has become prevalent in recent years. Prior studies have recognized several incentives for outsourcing internal audit services [14], [15]. A study of [16] investigates the manner and the impact of outsourcing the IAF in the United Kingdom (UK) private and public sector organizations and the impact of outsourcing decision may likely have on auditor independence and quality of audit service. The result of the study indicated that most of the organizations prefer an in-house IAF and that not all respondents are of the opinion that independence may be compromised when internal audit services are outsourced to external audit.

[15] Examine the consequence of SOX as it relates the restriction on outsourcing internal audit services to the external auditor using a sample of 219 responses from 1,000 fortune chief internal auditors. The result of the study indicated that firms with independent, active and expert audit committee are less likely to outsource routine internal audit services to external auditor

Past studies provides proof of the influence of IAF on audit fees and reliance of external auditors on IAF work ([17], [18] all this studies reported that an increased dependence on the work of the IAF by external auditor can translate into lower external audit fees. [18] Find that the use of IAF as a management training ground is positively associated with external audit fees. However, [19] reported an increased in external audit fees as a result of internal control reporting requirements of SOX.

3 Research method

Data were gathered by means of a questionnaire sent to head of internal audits of the sampled organizations. The sample comprises of both private and public sector entities in Nigeria. Private sector entities comprises of 257 listed firms out of which 80 were selected as the final sample. Public sector in Nigeria comprises of three ties of government i.e. the Federal, State and Local Government Councils. The sample in this study comprises of 40 federal public companies excluding federal ministries.

The questionnaires were mailed to 120 head of internal audits and 68 responses were received generating a response rate of 56.66 percent. There were 41 responses from the private sector organizations representing 60.29 percent and 27 responses from public sector representing 39.17 percent. The questionnaire was structure following previous studies with little modification [13], [20]–[22] each question was designed to incorporate responses expressed in multiple choices that enabled the respondents to express their opinion.

In this study, it is presumed that IAF in the private sector to be sound corporate governance device which of its features have to be in line with those of an IAF in the public sector, even though the two settings may follow contrasting contexts this remain an issue of empirical investigation.

4 Results

This section present the results of descriptive analysis performed on data collected through the survey. Table 1 presents the summary of the results from the questionnaire. From the hierarchical rank of IAF, the table shows that 36 internal audit managers in the private sector representing 88 percent indicated that the head of internal audit is located at corporate level compares to 21 respondents representing 78 percent in the public sector. This result suggests that IAF of both sectors was placed at corporate level as opposed to being placed at regional level or state level. This is in line with ISA 2007, SEC Code of Corporate Governance and Financial Regulation Rules of the Federal Government Official Gazette 2009. This finding corroborate with[22] that of who reported that both private and public sector entities IAF in Australia were placed at corporate level.

When asked on the average years do internal auditors stay in a unit before being transferred to another unit within the company, 19 respondents (46.34 percent) from the private sector indicated that on average internal auditors stays between 4 years in the internal audit with 12 percent indicating internal audit stays for 5-6 years. Whereas 22 respondents (81 percent) from the public sector indicated that on average internal auditors stays for more than 8 years in the IAF. A comparison with [22] study indicated that 47 percent of the respondents indicate that on average internal auditors in Australia stay for 2 and 4 years in the IAF. A similar study of [12] indicated that 37 percent (20 respondents) indicated that in IAF such a movement to other units was unlikely for internal auditors in Singapore.

Twenty-three percent of the private sector respondent indicated that audit committee (AC) members are involved in the appointment, evaluation and dismissal of the head of the IAF while 68 percent indicated that AC performs these function together with the chief executive officer (CEO). Whereas, 81.48 percent respondents of the public sector indicated that the CEO are mandated to appoint, evaluate and dismiss the head of IAF. A comparison with Canadian study of [23] reported that 48 percent of the respondents indicated that AC are involved in the dismissal of the IAF. [12] reported 72 percent in Singapore, [20] reported 50 percent of AC are involved in the appointment evaluation and dismissal of internal audit head in Australia.

The table also shows that 71 percent of the private sector IAF report to the AC and CEO while 10 percent indicated that besides AC and CEO they also report to the CEO. The table further shows that 81.5 percent of the public sector respondents indicated that IAF report to CEO while 18.5 percent shows that besides the CEO they also report to the CEO and CFO. In a similar study [22] indicated that 38 percent of the IAF report functionally to the AC, 47 percent shows that besides the AC they also report functionally to CEO and CFO whereas 32 percent of the respondents report administratively to CEO.

There is also slightly significant difference between the two sectors with regard to internal audit service outsourcing. 85.37 percent of the private sector IAF being involved in out sourcing compared to 67 percent of public sector internal audit being involved in outsourcing. Previous studies of [21] indicated that 69.41 percent of Italian firms do not outsourced internal audit services. Whereas, [13] indicated that 68 percent of public sector entities and 65 percent private sector entities engaged in outsourcing internal audit services.

The respondents of both sectors were asked to indicate which of the services were most commonly outsourced. Financial audit rank as the most frequently outsourced service with 46.34 percent of the private sector and 70.37 percent of the public sector outsourcing. Compliance audit encompasses 34.14 percent of private sector and 22.22 percent of public sector. While environmental audit, comprise of 7.32 percent of private sector outsourcing and 0 percent of that in the public sector. In a related study, [22] indicates that information technology and system rank as the most commonly outsourced service with 45 percent of all public sector outsourcing and 50 percent of private sector outsourcing whereas performance audit comprise of 7 percent of public sector outsourcing and 8 percent of that in private sector outsourcing in Australia and New Zealand.

This study also reports responses concerning the coordination of work of IAF and external auditor access to working papers. The results revealed that there is no significant difference between private sector and public sector responses. More than two third of the IAF coordinate in the areas of audit coverage with the external auditor and additionally, in both sectors external auditors have high level of access to internal audit working papers.

Table 1: Summary of the Results

	Public Sector		Private Sector	
	n	**%**	**n**	**%**
Hierarchical Rank of IAF				
Corporate level	21	78.0	36	88.0
Regional level	6	22.0	5	12.0
State level	0	0.0	0	0.0
Average Internal Auditor Stay in a Unit				
Less than 2 years	0	0.0	4	9.76
2 - 4 years	0	0.0	19	46.34
5 – 6 years	0	0.0	12	29.27
7 – 8 years	5	19.0	6	14.63
More than 8 years	22	81.0	0	0.0
IAF recruitment, training and dismissal				
AC	0	0.0	9	23.0
AC/CEO	0	0.0	28	68.0
CEO	22	81.48	3	7.0
CEO/CFO	5	18.52	1	2.0
Day-to-day Reporting				
AC	0	0.0	5	12.0
AC/CEO	0	0.0	29	71.0
CEO	22	81.5	4	10.0
CEO/CFO	5	18.5	3	7.0
Internal audit services outsourcing				
Yes	18	67.0	35	85.37
No	9	33.0	6	14.63
Internal audit services outsourced				
Compliance Audit	6	22.22	14	34.14
Financial Audit	19	70.37	19	46.34
Operational Audit	2	7.41	5	12.20
Environmental Audit	0	0.0	3	7.32
External auditor access to IAF working paper				
Yes	18	67.0	36	89.0
No	9	33.0	5	11.0

5 Conclusion

This study examine the similarities and differences of private sector IAF and that of its counterpart in the public sector using data collected from Nigeria companies. The study is based on a survey of internal audit managers and chief internal auditors in Nigeria. The results revealed that private sector IAF and its counterpart in the public sector have a higher hierarchal ranking where two-third of the internal audit managers and chief internal auditors indicated that IAF was placed at corporate level. Whereas, less than half of the respondents in the private sector indicated that internal auditors stays between 4-5 years in the IAF with more than half of the public sector respondent indicating internal auditors stays for more than 8 years in the unit.

It was also found that more than half of the private sector respondents (68 percent) indicated that both AC and CEO were involved in the appointment, evaluation and dismissal of the head of internal audit unit while 81.48 percent of the public sector indicated that CEO are mandated to appoint, evaluate and dismiss the head of the internal audit unit. The results also showed that 85.37 percent of private sector entities and 67 percent of public sector IAF indicated that they engaged in outsourcing of internal audit services. Whereas, financial audit financial audit risk rank as the most frequently outsourced services in both two sectors.

The study has potential implications for relevant regulating bodies in Nigeria and may find the result useful in assessing the usefulness of IAF rules as stated in the ISA 2007, SEC code of best practices 2003 and the federal government of Nigeria financial regulations rules 2009. The regulations mandated that all private sector and public sector entities to establish IAF. However, with other researchers, this study also had potential limitations which provide avenues for further research. Firstly, consistent with the general trends in questionnaire based research the limited number of respondents could be a hindrance to the generalization of the results and the study does not cover private companies (unlisted) thereby require careful interpretation of the findings. Secondly, the respondents were not asked directly and this may cause misinterpretation of response because the instrument prevents the possibility of explaining and detailing the questions and there is direct control over the actual respondents.

Finally, this study has higher lighted the issue of IAF in private and public sector entities in Nigeria. There is increasing attention towards internal auditing and that companies make different choices with respect to issues in terms of organizational structures and resource management. In the light of these future studies could address the problems of effectiveness of IAF, independence, relationship with AC and management, role of IAF in strengthening financial reporting quality, and if companies are investing increasing efforts in developing IAF. The investigation of these could reveal their actual contribution and provide relevant insights on the issue

References

1. S. Ramamoorti, "Internal Auditing : History , Evolution , and Prospects," *Res. Oppor. Intern. Audit.*, pp. 1–23, (2003).
2. T. Ruud, "The internal audit function: An integral part of organizational governance," *Inst. Intern. Audit. Altamonte Springs, Florida.*, pp. 73–96, (2003).
3. M. L. DeFond and J. R. Francis, "Audit research after Sarbanes-Oxley," *Auditing*, vol. 24, no. SUPPL., pp. 5–30, (2005).
4. E. R. Patterson, J. R. Smith, and R. Patterson, "The Effects of Sarbanes-Oxley on and Internal Control Auditing Strength," *Account. Rev.*, vol. 82, no. 2, pp. 427–455, (2007).
5. K. Barac and M. Van Staden, "Internal auditing as a corporate governance mechanism A comparison between public sector and private sector functions," *African J. Public Policy*, vol. 7, no. 3, pp. 17–31, (2014).
6. M. S. Badara and S. Z. Saidin, "Empirical Evidence of the Moderating Effect of Effective Audit Committee on Audit Experience in the Public Sector: Perception of Internal Auditors," *Mediterr. J. Soc. Sci.*, vol. 5, no. 10, pp. 176–184, (2014).
7. E. E. Onoja and H. Usman, "Internal Audit Techniques and Fraud Prevention: A Case Study of Selected Local Government Councils in Bauch State," *Mediterr. J. Soc. Sci.*, vol. 6, no. 4, pp. 232–244, (2015).
8. L. Paape, J. Scheffe, and P. Snoep, "The relationship between the internal audit function and corporate governance in the EU - a survey," *Int. J. Audit.*, vol. 7, no. October 2002, pp. 247–262, (2003).
9. J. V Carcello, D. R. Hermanson, and K. Raghunandan, "Changes in Internal Auditing During the Time of the Major US Accounting Scandals," *Int. J. Audit.*, vol. 9, no. 2, pp. 117–127, (2005).
10. K. L. James, "The Effects of Internal Audit Structure on Perceived Financial

Statement," *Account. Horizons*, vol. 17, no. 4, pp. 315–327, (2003).

11. G. Sarens and I. De Beelde, "The relationship between internal audit and senior management: A qualitative analysis of expectations and perceptions," *Int. J. Audit.*, vol. 10, no. 3, pp. 219–241, (2006).
12. J. Goodwin and T. Y. Yeo, "Audit Independence and Objectivity : Evidence from Singapore," *Int. J. Audit.*, vol. 125, no. May, pp. 107–125, (2001).
13. J. Goodwin, "A comparison of internal audit in the private and public sectors," *Manag. Audit. J.*, vol. 19, no. 5, pp. 640–650, (2004).
14. D. H. Caplan and M. Kirschenheiter, "Outsourcing and Audit Risk for Internal Audit Services," *Contemp. Account. Res.*, vol. 17, no. 3, pp. 387–428, (2000).
15. L. J. Abbott, S. Parker, G. F. Peters, and D. V. Rama, "Corporate Governance, Audit Quality, and the Sarbanes- Oxley Act: Evidence from Internal Audit Outsourcing," *Account. Rev.*, vol. 82, no. 4, pp. 803–835, (2007).
16. G. Selim and A. Yiannakas, "Outsourcing the Internal Audit Function : A Survey of the UK Public," *Int. J. Audit.*, vol. 226, no. April, pp. 213–226, (2000).
17. W. L. J. Felix, A. A. Gramling, and M. J. Maletta, "The Contribution of Internal Audit as a Determinant of External Audit Fees and Factors Influencing This Contribution," *J. Account. Res.*, vol. 39, no. 3, pp. 513–534, (2001).
18. W. F. Messier, J. K. Reynolds, C. A. Simon, and D. A. Wood, "The effect of using the internal audit function as a management training ground on the external auditor's reliance decision," *Account. Rev.*, vol. 86, no. 6, pp. 2131–2154, (2011).
19. M. D. Beneish, M. B. Billings, and L. D. Hodder, "Internal control weaknesses and information uncertainty," *Account. Rev.*, vol. 83, no. 3, pp. 665–703, (2008).
20. J. Goodwin, "The Relationship Between the Audit Committee and the Internal Audit Function: Evidence from Australia and New Zealand," *Int. J. Audit.*, vol. 7, no. April, pp. 263–278, (2003).
21. M. Arena and G. Azzone, "Internal Audit Departments: Adoption and Characteristics in Italian Companies," *Int. J. Audit.*, vol. 11, pp. 91–114, (2007).
22. J. Christopher, G. Sarens, and P. Leung, "A critical analysis of the independence of the internal audit function : evidence from Australia," *Accounting, Audit. Account. J.*, vol. 22, no. 2, pp. 200–220, (2009).
23. D. Scarbrough, K. Raghunandan, and D. V Rama, "Audit committee composition and interaction with internal auditing canadian evidence," *Account. Horizons*, vol. 12, no. 1, pp. 51–62, (1998).
24. Adams, M.B. "Agency theory and internal audit." *Manag. Audit. J.* vol. 9, no. 8, pp. 8-12, (1994).

SHS Web of Conferences **34**, 01005 (2017) DOI: 10.1051/shsconf/20173401005
FourA 2016

Audit quality in ASEAN region: some efforts and comparisons

Irma Tyasari[1,*], *Nor Zalina* Mohamad Yusof[1†], and *Ku Maisurah* Ku Bahador[1]

[1]Tunku Puteri Intan Safinaz School of Accountancy, Universiti Utara Malaysia, 06010 Sintok, Kedah, Malaysia

Abstract. This paper discusses how Asian economies are responding to the increased global attention on audit quality (AQ) issues. It shows some of the efforts taken by the countries collectively as a region; as well as individual countries' measures. The data is obtained from various reports available in public domain including the information available on various relevant websites. The discussion shows how Asian countries are working through an informal organization they established, called the ASEAN Audit Regulators Group (AARG) in enhancing AQ. Comparison among five ASEAN countries shows variations in term of how much each country has progressed with respect to their efforts in straightening AQ. Some countries are more advanced than the other. This paper contributes to the literature on audit quality by providing insights into how ASEAN as a region responded to the global demand on improving AQ and the gaps in auditing literature on comparative progress of these nations.

1 Introduction

In the year 2002, the US's biggest energy firm, Enron, collapsed; bringing together its auditor, Arthur Andersen & Co. The accounting landscape changed with the establishment of the Sarbanes Oxley Act (SOX) 2002, which has global impact to the profession. Since then, audit quality (AQ) has become a major focus of all relevant parties, be it accounting regulators, practitioners, accounting professional bodies, and accounting academics.

More than a decade later, despite various efforts and measures taken by these parties to enhance AQ, it is still very much a concern. In 2015, for example, a global survey by the International Forum of Independent Audit Regulators (IFIAR) on its 29 members from different jurisdictions finds that audit still unable to provide the level of assurance expected of it by the users of financial statements. More alarmingly, the concern is centered on audit procedures deficiencies with the most deficiencies are also those most fundamental to the quality of financial statements [1]. In the same year, the UK Financial Reporting Council (FRC) also published its report on AQ that indicates that although AQ is improving, but there is still room for improvements.

As concerns mounted, accounting bodies continue to work towards enhancing the AQ. In January 2016, the US Center for Audit Quality (CAQ), which is affiliated with the American Institute of CPA (AICPA), has published their insights learned from a series of roundtable discussions with various stakeholders to gather their feedback on a potential set of audit quality indicators. This is following their release of the "CAQ Approach to Audit

Quality Indicators" (AQI) in 2014 [2]. The AICPA itself has published the "AICPA's Enhancing Audit Quality (EAQ)" initiative in 2014 to address AQ challenges on a holistic, ongoing basis, with the goal of improving audit performance. A year later, they published the "6-Point Plan to Improve AQ" that describes extensive ongoing and new efforts to improve AQ. These initiatives show that AQ is still receiving a significant focus at the global level [2].

Being part of the global economy, ASEAN countries have also been focusing their attention on the issues of AQ. Efforts have been taken although some claimed that the movements could be for legitimacy as much as for efficiency reason [3]. Some argues that such initiatives are due to requirement of international organizations such as the World Bank and the International Monetary Fund (IMF) [4]. Others suggest that the attention is due to practical problems experienced in each country ([4][1], [5][2]). Notwithstanding these, efforts are being taken by these emerging economies, not only at individual country level, but also they are working together collectively at regional level.

In 2011, Malaysia, Singapore, and Thailand, have taken their step to promote better AQ in the region. The Accounting and Corporate Reporting (ACRA) of Singapore, the Securities and Exchange Commission (SEC) of Thailand, and Malaysia's Audit Oversight Board (AOB) – all of which are members of IFIAR, came together and founded the ASEAN Audit Regulators Group (AARG). This informal organization aims to promote AQ in the region through enhanced collaboration between audit regulators. Its "roles extend to elevating audit oversight and quality in the region and being a channel for knowledge sharing and promoting ASEAN capital markets linkage together with cross-bordering offering under equivalent standards of high quality financial reporting" [6].

In 2015, the region has integrated itself economically into the ASEAN Economic Community or AEC. This is a result of the first AEC Blueprint, signed in November 2007 that served as a master plan to chart its establishment in December 2015. This makes ASEAN a huge global market. In 2014, AEC was collectively the third largest economy in Asia and the seventh largest in the world [7]. With the AEC, it is not only that the region needs strong AQ in penetrating the global economy, but each country inside the AEC has to have comparable AQ for being integrated as a single market and production base. However, given these initiatives, comparative studies in ASEAN on this AQ issue are still lacking.

This paper, therefore, aims to discuss what recent initiatives are being taken by ASEAN in relation to AQ. It is hoped that this will provide insight into how much the region has progressed in terms of improving its AQ as well as giving ideas into whether these developing economies are comparable among themselves, and with the developed nations in this regard. The focus is not only on collective efforts in the region, but also on individual country's efforts. This is important particularly with the formation of AEC. However, although the AEC is made up of ten nations in South East Asia, the focus of this study is only on five of its economies, namely Singapore, Thailand, Philippine, Malaysia, and Indonesia. The limited focus is due to availability of data as well as the relative importance of the countries in playing their roles in AEC.

Data is obtained from various reports available in public domain including the information available on various relevant websites. This includes annual reports of accounting regulatory bodies, newspaper articles, and press releases. In this paper, AQ is defined as "the market-assessed joint probability that a given auditor will both (a) discover a breach in the client's accounting system, and (b) report the breach" [8].

---

* Corresponding author: Irmatyasari26@gmail.com

[1] Pusat Pembinaan Profesi Keuangan (Accountant Services Development Centre under the Minister of Finance Department of Indonesia);

[2] Indonesian Institute of Public Accountants

This paper is organized as follow: the next section discusses the efforts taken by ASEAN economies collectively. This is followed by a section that compares the initiatives taken by five of the countries under the AEC as stated above. Comparisons are made in the areas related to their regulations, the number of accountants, the training requirement, and professional bodies' collaboration with academic institutions. Other countries under the AARG are not included due to inability to obtain sufficient data. This paper ends with some concluding remarks.

2 Collective efforts by ASEAN

In ASEAN, there are two regional accountancy organizations, which are, the ASEAN Federation of Accountants (AFA) – established in 1977; and the Confederation of Asian and Pacific Accountants (CAPA), established in 1976. However, the discussion on regional efforts will focus more on the efforts done by AARG. This is because, AFA is not formally recognized and acknowledged by IFAC; while for CAPA, although it is recognized by IFAC, only two ASEAN countries are members of this organization [9]. Furthermore, although AARG is made up of only three ASEAN countries, but the involvement in its efforts come from many other countries and relevant parties including audit firms. In terms of membership, while AARG is consists of accounting regulatory bodies of members' country, AFA and CAPA are consisting of the country members' accounting professional bodies. It is also interesting to focus on AARG as it appears very active since its establishment in 2011, and is affiliated to a world organization, IFIAR.

IFIAR is a global body, established in 2006 by 17 countries; now its membership has reached to 51. The body consists of independent audit regulators from various jurisdictions around the world. In promoting AQ, IFIAR has provided a platform for regulators to share knowledge and practical experience gained from their independent audit regulatory activity. As an affiliation to IFIAR, AARG also has adopted similar approaches to promote AQ, which is through discourse. There are initiatives taken by AARG in relation to enhancement of AQ in the region. The annual meeting and Audit Inspection Workshops mirror those conducted by IFIAR. Working with the World Bank on collaborative efforts to further raising the standard of AQ in ASEAN, and roundtable discussion with other related agencies and Big Four Audit Firms for specific issues concerning audit are other initiatives [1]. The collaboration of AARG has created impact on AQ in the region. The evidence of AARG's contribution is stated in AOB's Annual Report of 2013 [6].

3 Efforts at national level

Discussion and comparisons on efforts at individual country level is focused on four areas, namely, their regulations, the number of accountants, the training requirement, and professional bodies' collaboration with academic institutions [9].

3.1 Enhancements to the statutory and institutional framework

The enhancement to statutory and institutional framework means that the way the regulatory bodies or agencies improve monitoring and enforcement of accounting and auditing mechanism, and how they put place the basic of effective regulations. There should be a consistency between the legal frameworks with the adoption of international standards to create effectiveness. The general impact of the changes caused by the new

DOI: 10.1051/shsconf/20173401005

regulations or new standards is expected to modernize the regulation itself to reflect the adjusted proficient environment that has developed because of emerging worldwide economy [10].

This condition also gave impact to ASEAN countries. There is a progressive move away from self-regulation of auditing by the profession towards an independent regulation within a statutory framework [11]. The regulators have made several efforts in enhancing their statutory and institutional framework concerning AQ due to the changes. One of such effort as evidence is that a country is committed in AQ is by adopting international standards ([3], [12]). However the successful implementation of international standards is very dependent on local conditions. The process of adopting standards and adapting changes has been a challenge among countries regarding the issues of capacities [13].

There is a great deal of variations among ASEAN countries relation their progress and the enhancement of statutory and institutional framework. The main task of regulator is to prepare the formulation of policies, guidance, development and supervision of accountants. In addition, they are also the administrator of registration and licensing of accountants, and imposition of administrative sanctions against the professions. Each regulator has a quality review program which is mandatory because it is one of the requirements of IFAC for its member to demonstrate commitment in quality ([4], [9], [11], [24]). The five countries have established professional associations which have various functions whether as standard setter, supervisor of enforcing compliance, regulatory body, Continuous Professional Development (CPD) and accountants' certification examinations provider.

3.2 Number of professional accountants

As explained in previous paragraph, ASEAN is a large potential market, and strengthening the accountancy profession will ensure adequate number of accountants to meet of its growing economy. Various programs are being offered by the regulators of each country to attract, improve and promote the accountancy profession. Collaborative arrangements between universities and professional associations may support this effort. Besides increasing the number of professional accountants, improving the capacity building of both sides may fill the gap between the interests, and the market's expectation ([15], [16]). An example of such effort is having harmonized international standards (curricula), and giving opportunities for graduates to obtain the appropriate practical experience ([17], [18], [19]).

Indonesia has given a serious consideration regarding the issue above. The country's regulators have created a program to scale up the number of Indonesian accountants because comparing to other ASEAN countries; it only contributes less than 15% of the total accountants in ASEAN in 2015 [9]. Based on the data of AFA, Indonesia has greatly increased the number of its accountants in 2015. This is due to the two programs conducted by IAPI and IAI ([9], [20], [23]). These two programs are similar and comparable to the ones which being conducted in Malaysia, Thailand and Singapore [9]. The table below shows the numbers of accountant's in 5 (five) number ASEAN countries.

Table 1: Numbers of accountant by Year (from 2010- 2015)

Country	2010	2011	2012	2013	2014	2015
Thailand	51,737	61,331	54,319	57,244	63,691	68,777
Singapore	23,626	24,774	25,726	27,397	28,530	24,774
Philippine	21,599	21,939	22,567	22,072	21,586	19,380
Malaysia	27,292	28,148	29,179	30,503	31,454	32,511
Indonesia	8,600	9,624	12,548	17,649	18,507	26,782

(Source: [9])

3.3 Continuous training programs

Another effort which the regulators have committed to meet the challenges of the changes in the auditing profession and may increase the AQ is by having CPD program [22]. The objective is to encourage the culture of continuous learning among professionals [21]. Through this program government can socialize and introduce the updated standards and regulations to the accounting and auditing practitioners. The comparison of CPD requirement for accountants in five ASEAN countries shows that the range of CPD hours is between 18-40 hours per year. This is clearly that a certified accountant must meet the minimum CPD requirements in order to maintain his or her professional status. Each country has its own minimum hour requirement to be fulfilled that has been laid out by its respective professional associations ([9], [24]).

Country	Singapore	Thailand	Philippine	Malaysia	Indonesia
CPD Hours/Year	30	18	20	30	40

(Source: [9], [24], [25])

3.4 Collaboration with higher educational providers

One of the impacts of the global changes is the adoption of new standards which has led professional associations and regulators to collaborate with leading universities to strengthen the accountancy profession. The parties involved are committed to promote and develop the profession ([20], [23]). They play a significant role in the formulation of the curriculum for the accounting degree programs. Among the endeavours involved are providing trainings and workshops to introduce new standards to educators; offering internationally comparable accountancy education to match the contemporary requirements of the profession; revising the curricula and teaching methods based on the new principles of the latest standards, and opening a career path for accounting graduates. There are events of such collaborative programs among the parties involved according to each country, such as ISCA with ICAEW, Ministry of Commerce with FAP, the Commission on Higher Education (CHED) with PICPA, MIA has International Engagement for sharing best practice and lesson learnt and IAI with ICAEW([9], [23]).

4 Conclusion

The discussion in this paper shows how, at regional level, ASEAN is closely following the efforts taken at global level by the IFIAR in improving the audit quality. The Initiatives at this level is more of discursive in nature. The various efforts taken are consistent with its intention to penetrate the global economy through AEC. However at the individual country level, it could be seen that some countries are more ahead then others. More collaborative efforts will be undertaken by the relevant parties to reduce the differing capacity level of these countries.

References

[1] International Forum of Independent Audit Regulator (IFIAR). Retrieved online from https://www.ifiar.org/About-Us.aspx (2015)

[2] Center for Audit Quality (CAQ). Retrieved online from http://www.thecaq.org/reports-and publications/audit-quality-indicators-journey-and-path-ahead (2016)
[3] P. L. Joshi, W. G. Bremser & J. Al-Ajmi. *Perceptions of accounting professionals in the adoption and implementation of a single set of global accounting standards: Evidence from Bahrain.* AA, **24**(1), 41-48 (2008)
[4] Pusat Pembinaan Profesi Keuangan (PPPK). Retrieved online from http://pppk.kemenkeu.go.id/News/20 (2015)
[5] Institut Akuntan Publik Indonesia (IAPI). Retrieved online from http://iapi.or.id/detail/64-Reviu-Mutu (2015)
[6] Audit Oversight Board (AOB). *Annual Report 2013.* Retrieved online from https://www.sc.com.my/general_section/audit-oversight-board/ (2013)
[7] ASEAN Economic Community (AEC). Retrieved online from http://asean.org/asean-economic-community/ (2015)
[8] L. E. DeAngelo. Auditor size and audit quality. JAE, **3**(3), 183-199 (1981)
[9] ASEAN Federation of Accountants (AFA). Retrieved online from http://www.aseanaccountants.org/home (2015)
[10] C. Hayes. *The Ramsay report and the regulation of auditor independence in Australia.* AAR, **12**(2), 3 (2002)
[11] ASEAN Federation of Accountants (AFA). Retrieved online from http://www.aseanaccountants.org/home (2014)
[12] D. Wang, & J. Zhou. *The impact of PCAOB Auditing Standard No. 5 on audit fees and audit quality.* AH, **26**(3), 493-511 (2012)
[13] M. Tokar. *Convergence and the Implementation of a Single Set of Global Standards: The Real-life Challenge.* AE, **2**(1), 47-68 (2005)
[14] Accounting and Corporate Regulatory Authority (ACRA). Retrieved online from https://www.acra.gov.sg/Legislation/Accountants_Act/ (2016)
[15] Nazri Fadzly, Mohamed, and Zauwiyah Ahmad. *Audit expectation gap: The case of Malaysia.* MAJ. **19**.7 (2004)
[16] T.H. Lee, R. Mohamad, N. Z. M. Yusof & M. Ojo. *Practical training and the audit expectations gap: The case of accounting undergraduates of Universiti Utara Malaysia.* JFRA (Emerald Publication) Fourth coming (2007)
[17] J. M. Mula. *Can competency skills for accounting students be internationally harmonised? An Indonesia application.* AEG (SIG5) Symposium, AFAANZ (2007)
[18] M. Saito, K. Hiramatsu, & S. Mayangsari. *Accounting Education for the Implementation of IFRS in Indonesia.* IRB, **12**, 1-21 (2012)
[19] K.I. Dandago, & N. A. B. Shaari. *Effects of focus of accounting research on the quality of accounting education in Malaysian Universities.* AEFR, **3**(10), 1371 (2013)
[20] Ikatan Akuntan Indonesia (IAI). etrieved online from http://iaiglobal.or.id/v03/CA/home (2015)
[21] P. De Lange, B. Jackling, & I. G. Basioudis, *A framework of best practice of continuing professional development for the accounting profession.* AE, **22**(5), 494-497 (2013)
[22] Y.S. Cheng, Y.P. Liu, & C.Y. Chien. *The association between auditor quality and human capital.* MAJ, **24**(6), 523-541 (2009)
[23] Institut Akuntan Publik Indonesia (IAPI). Retrieved online from http://iapi.or.id/assets/collections/doc/prp.pdf (2015)
[24] Institut Akuntan Publik Indonesia (IAPI). Retrieved online from http://iapi.or.id/peraturan/ (2015)
[25] Pusat Pembinaan Profesi Keuangan (PPPK). Retrieved online from http://pppk.kemenkeu.go.id/News/20http://pppk.kemenkeu.go.id/News/Details/29 (2015)

SHS Web of Conferences **34**, 01006 (2017) DOI: 10.1051/shsconf/20173401006
FourA 2016

Ownership patterns and control of top 100 Malaysian listed companies

Ayoib B Che-Ahmad[1], and Aree S Mustafa[2]

[1,2]Tunku Puteri Intan Safinaz, School of Accountancy, Universiti Utara Malaysia, Malaysia

Abstract. The incentive of this study stems from the significance of protecting minority shareholders' rights. It analyses the most dominant ownership structure practices within listed firms in Bursa Malaysia (BM), in which controlling shareholders (CSs) hold control on many firms despite relatively small portion of investment. Malaysian firms are characterised by high levels of concentrated ownership with families owning more than 70% of the listed firms in BM [9]. This study finds that the separation between cash flow rights (real investment) and control rights (voting rights) for Top 100 Malaysian listed firms is accomplished through pyramidal ownership and golden share arrangements. Hence, the divergence between cash flow rights (CFRs) and voting rights (VRs) exacerbates Type II Agency problem. This study calls for future studies in the area of accounting and finance to introduce and operationalise new measurement of Type II Agency Cost in order to better understand the agency conflicts within this unique market. It also better explains the state of the minority shareholders in the BM.

1. Introduction

Previous literature has examined various firms' ownership structure in many countries. Many of previous studies compared between dispersed ownership and concentrated ownership, in which individual or families represent controlling shareholders (CSs) through ownership on almost all equities [2]. In comparison, this study examines an ownership structure in which individuals practise control on many affiliated firms with less stake of real investment. In Malaysia, there are two predominant methods to separate between CFRs and control rights (CRs) or voting right (VRs). They are pyramidal ownership and golden share [20]. These kinds of ownership structure permit CSs to maintain their control on many firms with relative small portion of equities. The separation between CFRs and VRs entrenches the CSs position and maintains corporate control in their hand far from market competition with relative small stake of equities [9]. Despite of the fact that this separation reduces agency conflicts between agent and principle that is known as Type I Agency Problem, Type II Agency Problem would emerge [15]. The separation between CFRs and CRs also enhances majority shareholders incentives to expropriate minority shareholders wealth [3]. CFRs in this study are defined as CSs rights to claim in the firm's liquidity and profits. Meanwhile, CRs indicates CSs capability to influence firms' strategic decisions for instance, in terms of board of directors' selection. These types of ownership structures distort CSs motivations to place more attention on governance mechanisms, extending from legal protection of minority shareholders to reputational boundaries on CSs. The same reason encourages political and market scrutiny in numerous countries to focus on problems to innate pyramidal ownership and golden share. Therefore, this paper addresss ownership structure practices within Malaysian listed firms. To the best of our knowledge, there is no study that addressed the sectorial concentration of wedge using pyramidal ownership and golden shares in BM. The study's main objective is to explain the mechanisms of divergence between CFRs and CRs practiced in BM market, which provides a useful ground for future studies to investigate Type II Agency Problems in Malaysian business environment. With respect to theoretical contribution, this paper extends agency theory literature by analyzing deeper the agencies conflicts and provides measurement of such conflict in the unique Malaysian market. Previous study documents the ownership-control structure as the most significant elements that exacerbate Type II Agency Problem [2]. The practical contribution of this study is that it would offer strong foundation for policy makers to suggest an optimal policy to improve the corporate governance mechanisms and provide protected and conducive environment for minority shareholders.

2. Mechanisms for Separating Cash Flow and Control

This study addresses Malaysian listed firms in term of ownership-control structure mechanisms are pyramidal-ownership structure and golden share [6].The following section addresses pyramidal ownership structure and golden shares.

2.1 Pyramids

This section allocates to illustrate the nature of CFRs and CRs and Pyramidal structure. Firms involve with pyramidal ownership structure have divergence between CFRs and CRs that has a negative implementation on minority shareholders protections. This is a consequence of high incentives of Ultimate Ownership (UO) to expropriate minority shareholders wealth [22]. CFRs are a real ownership of investors through buying shares. That is because ownership is a consequence of investment and CFRs represent as an indicator of investors investment in the corporation [19]. While, CRs indicates to UO voting rights [9]. Reasonably, CFRs shall be equal to CRs in corporation as indicators of shareholders real investment. Notwithstanding that pyramid structure leads to divergence between CFRs and CRs. Pyramidal structure is defined as "a situation in which the same entrepreneur, through a chain of control relations, controls many firms" [7]. This means that UO own the majority of ownership of one firm that in turn owns the majority ownership of another firm. The clearest example of pyramidal structure is that of Halim bin Saad, an entrepreneur from Malaysia. Halim owns about 28.3 real stake ownership of Renong Berhad as shows in figure (1). Accordingly, Halim represents the UO of the Renong Berhad. At the same time, Renong Berhad possesses about 32.5 of share ownership of United Engineers Malaysia (UEM). Consequently, Renong Berhad represents a UO of UEM. It can be seen that Halim controls Renong Berhad that in turn hold a major control on UEM. This indicates that Halim directly controls Renong Berhad in the first level of the pyramidal structure and in the second level Halim controls UEM indirectly. In addition, Halim controls through both of Renong and UEM on Kinta Kelas indirectly. Halim ownership in Kinta Kelas arises from his real ownership in Renong Berhad and UEM. As a result of this pyramidal ownership structure Halim own real ownership about 5.73 in Kinta Kelas. The ownership value is finds out using the following method:

Halim's CFR in Kinta Kelas = 28.3% x 32.5% x 62.4% = 0.05739 ~ 5.73%

Meanwhile, the theoretical ownership comes through individual investment in the corporation. According to [18] the ownership arises as a consequence of investment. If Halim stake of ownership is about 5.73 per cent, this indicates that his investment stake in Kinta Kelas is 5.73 per cent. The following example illustrates Halim stake of ownership in Kinta Kelas in dollar value. For instance, Kinta Kelas worth is about 100,000,000

Malaysian Ringgit (MR). Therefore, Halim controls Kinta Kelas worth that is about (100,000,000) with an investment about 5,730,000 (5.53 per cent × 100,000,000). Halim has indirect control rights in Kinta Kelas. This is achieved through Halim's direct control on Renong Berhad at first level that in turn own major control on UEM at the second level of the pyramidal, and at the final level Halim controls Kinta Kelas through UEM. Based on the definition provided by [12] CRs represent the weakest link in the chain of control relationship. Accordingly, Halim has CRs about 28.3 per cent and this consider as the weakest link in the chains of control relationship through the three levels of the pyramidal structure. Obviously, Halim with 5.73 per cent of CFRs or 5,730,000 RM worth of investment hold about 28.3 per cent of CRs in Kinta Kelas with worth about 100,000,000 RM. This align with [13] propose that pyramidal ownership structure creates sharp divergence between ownership and CRs that deviates from the traditional view of one share-one vote. Importantly, UO incentive to expropriate shareholders wealth increased as a result of the separation between CRs and CFRs [9]. The divergence between CRs and CFRs find out following [14] whether via the ratio of cash flow rights to control rights or the difference between cash flow rights and control rights.

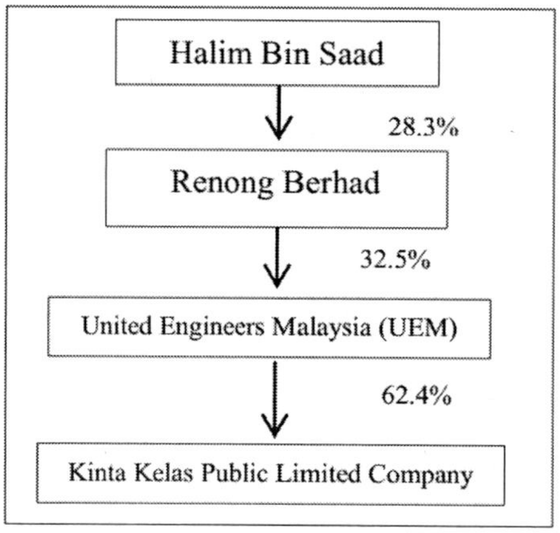

Figure 1: Pyramidal ownership structure
Source : [16]

The divergence between CFRs and CRs for Kinta Kelas finds out using the following two methods:

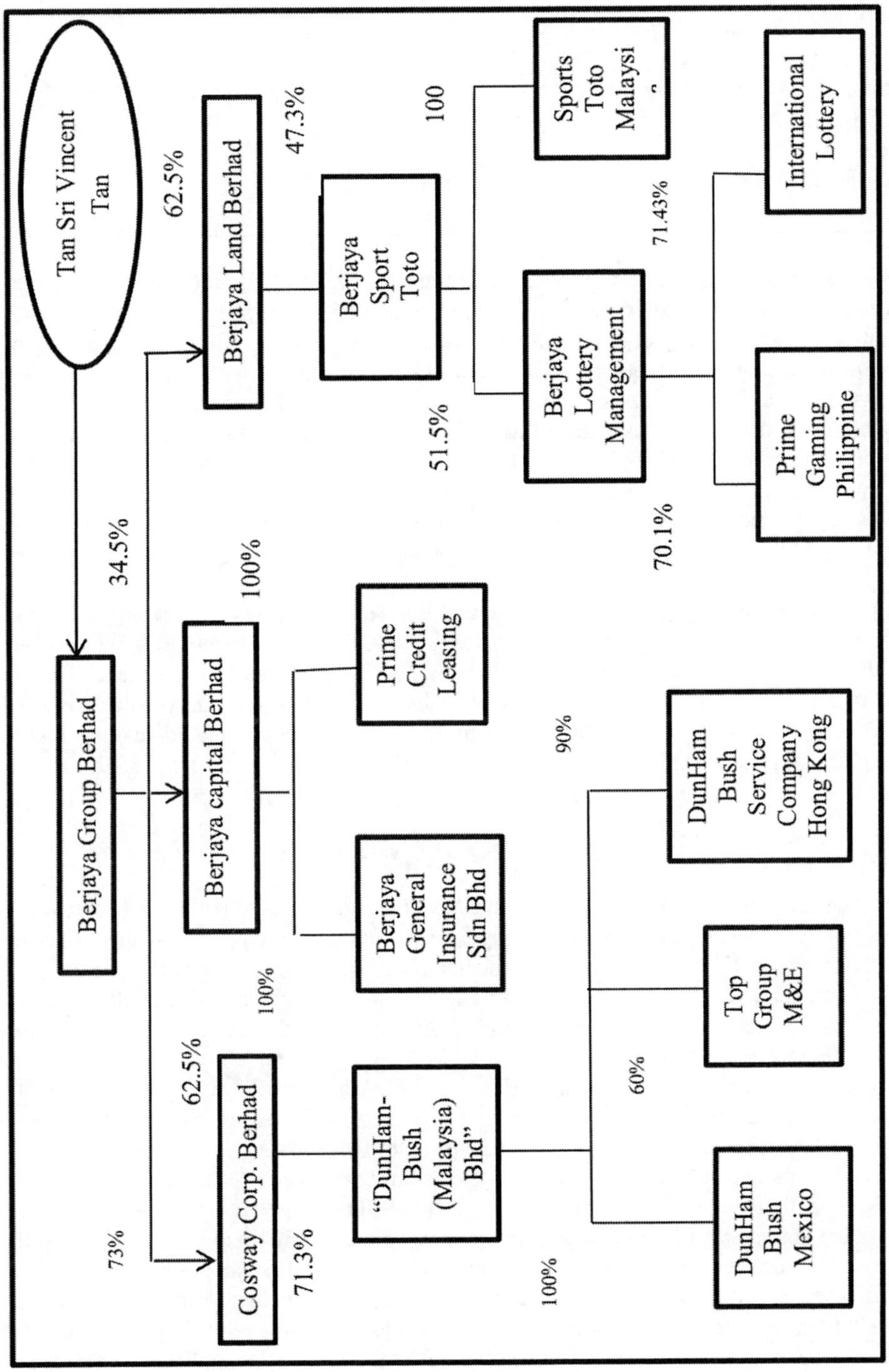

Source: [6]
Figure 2. DunHam-Bush (Malaysia) Bhd

1-Ratio of cash flow to voting rights = Halim's cash flow rights/ Halim's voting rights = 5.73%/ 28.3% = 0.2024	2- the variance between cash flow rights and voting rights = Halim's cash flow rights - Halim's voting rights = 5.73% - 28.3% = 22.57%

According to previous techniques, the larger divergence between CFRs and CRs reflects the sharper differences between investment and voting rights and verse versa. The second case is DunHam-Bush (Malaysia) Berhad Figure 2. The UO of this firm is Cosway Corporation Berhad with ownership stake about 71.3%. The UO of Cosway Corporation Bhd on the other side is Berjaya Group Bhd with about 73% of CFRs. Tan Sri Vincent Tan is a UO with stake about 34.05% of Berjaya Group Bhd. To figure out the CFRs and CRs of Tan Sri Vincent Tan in DunHam-Bush (Malaysia) Bhd, this work first considered the holding of Cosway Corporation Bhd in DunHam-Bush (Malaysia) Bhd, the holding of Berjaya Group Bhd in Cosway Corporation Bhd and finally Tan Sri Vincent Tan's holding in Berjaya Group Bhd.

According to this ownership structure, Tan Sri Vincent Tan's cash flow rights in DunHam-Bush (Malaysia) Bhd is about 17.723%, (=71.3%×73%×34.05%) whereas his CFRs is about 34.05%. The CFRs is determined according to the weakest link in the chain of controlled rule [14]. The ratio of CFRs is about 52.05%. This value shows that there is violation of one share–one vote policy in DunHam-Bush (Malaysia) Bhd and it might enhance controlling shareholders incentive to expropriate the interests of minority shareholders.

2.2 Golden shares

The Malaysian government holds a single unit "golden share" which uses it to veto on essential decisions that is relate to national interest consequences [16]. Golden share define as a set of Government's special authority and legal constrictions on privatize firms [8]. This kind of control-ownership is dominant amongst major privatized utilities, for instance Tenaga Nasional Berhad and Malaysian Airlines [1]. The Malaysian government is able to influence on the strategic industries of some Government Linked Corporations (GLCs) using Golden share [12]. This share is directs by the Ministry of Finance's investment organization (MOF Inc.). Golden share are typically own by government, given it veto power in major decision of the company or privilege to select board of director, right to speak at general assembly meeting and to oppose any decision made by the board of directors that conflicts with the government policies [21] . The same view is documented by [11] that golden share uses in order to improve control on many affiliated firms with small stake of real investment. For instance, Tenaga Nasional Berhad issue golden share with privilege to select board of director of the company, but not more than 6 directors in each time. In addition, the holder of special shares in Telekom Malaysia Berhad have the concession to issue additional shares which carry more than one voting rights or substantial disposal of firm assets and the decisions relate to merger and takeover as shows in the figure 3.

DOI: 10.1051/shsconf/20173401006

Class D Non-Convertible Redeemable Preference Shares of RM1.00 (sub-note (b)) At 1 January and 31 December	*	*	*	*
TOTAL ISSUED AND FULLY PAID-UP SHARE CAPITAL	3,758.0	2,630.6	3,719.4	2,603.6

* Amount less than RM0.1 million

(a) Special Rights Redeemable Preference Share (Special Share)

The Special Share of RM1.00 would enable the Government through the Minister of Finance to ensure that certain major decisions affecting the operations of the Company are consistent with the Government's policy. The Special Shareholder, which may only be the Government or any representative or person acting on its behalf, is entitled to receive notices of meetings but does not carry any right to vote at such meetings of the Company. However, the Special Shareholder is entitled to attend and speak at such meetings.

Certain matters, in particular, the alteration of the Articles of Association of the Company relating to the rights of the Special Shareholder, the dissolution of the Company, any substantial acquisitions and disposal of assets, amalgamation, merger and takeover, require the prior consent of the Special Shareholder.

The Special Shareholder has the right to require the Company to redeem the Special Share at par at any time. In a distribution of capital in a winding up of the Company, the Special Shareholder is entitled to the repayment of the capital paid up on the Special Share in priority to any repayment of capital to any other member. The Special Share does not confer any right to participate in the capital or profits of the Company.

Figure 3: Wedge using golden shares
Source: Annual report of Telekom Malaysia Berhad 2015

3. Research design

3.1 Sample and data collection

The population of this study is limited to the top 100 firms listed in BM as obtained from the Bursa website as at 31 December 2015. The period covered is 2015 including financial and non-financial firms. This study uses the data stream in order to collect firms' total assets. In order to determine the top 100 firms this study arrange firms from bigger to smaller size of firms total assets. This work sought to illustrates the extent to which the top 100 firms listed in BM for the year-end 2015 used wedge. In order to get wedge data this study uses BM website and firms' annual reports. Financial institution for instance financial intermediaries, insurance companies and banks are about 21% of the top 100 firms listed in BM. firms classified within trade and services sector are about 30% per cent. Plantation, consumer products and IPC are about (8%, 7%, 2%) respectively. Construction firms are about 14% of the top 100 firms listed in BM. The remaining industry product, construction and REITS are about (7%, 5%, 6%) respectively.

4. Data analysis and results

Figure 4 shows the statistic descriptive of ownership structure of the top 100 firms listed in BM. The Malaysian pyramidal firms are about 85% of the top 100 firms listed in BM. They are using wedge to deviate between CFRs and CRs. Firms using golden shares are about 5% of the top 100 firms listed in BM. While non-pyramidal firms listed within the top 100 firms in BM are about 10%. Non-pyramidal firms dose not practice any kinds of separation between CFRs and CRs. The bar graph in Figure 5 display the sectorial concentration of wedge (pyramidal, Non-pyramidal and golden shares) practiced by the top 100 firms listed in BM in 2015. The prevalent ownership structure to divergence between CFRs and CRs is pyramidal ownership, with 21%, 22%, 13% of firms working in the finance, trade and services and properties respectively. While Non-pyramidal ownership structure is about 3%, 2%, 2% of firms working in the trade and services, consumer products and construction respectively. Firms using golden shares own by government are about 5% of the top 100 firms listed in BM and that classified within trade and services sector.

 DOI: 10.1051/shsconf/20173401006

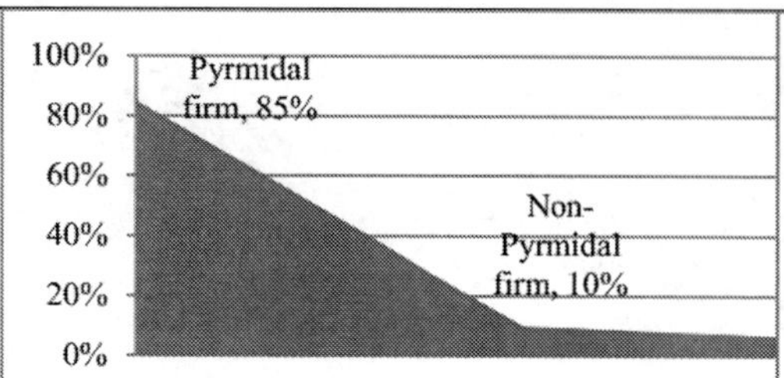

Figure 4 Control-ownership patterns in the top 100 listed firms in BI
Source: authorial computation

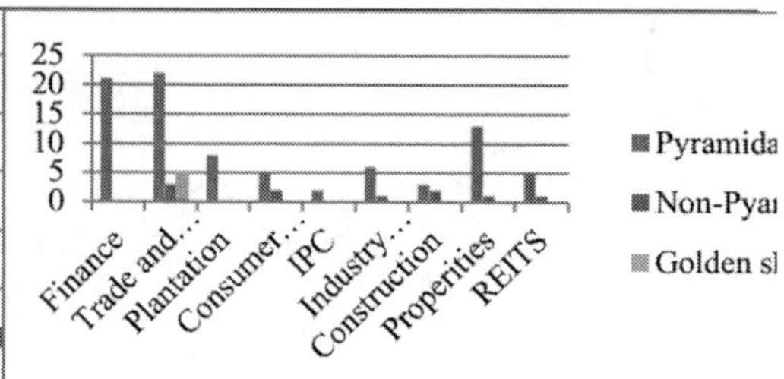

Figure 5 sectorial distribution of wedge
Source: authorial computation

5. Conclusion

This paper illustrates the ownership-control structure of the top 100 firms listed in BM using pyramidal structure and golden shares. The majority of listed firms are family own firms, these families control on many affiliated firms using pyramidal ownership structure. Thus, the insider system is the most common structure in the Malaysian capital market, and this increase insider wealth. The controlling shareholders utilize wedge mechanism to separate between CFRs and CRs. This assists controlling shareholders to control on many affiliated firms with small stake of real investment. Whereas, golden shares allow government to direct firms' strategic decisions for instance, nominate the board directors. Previous literature reports that wedge improve controller incentive and ability to maximize their interest at the expenses of minority shareholders and this kind of agency conflicts known as Type II Agency Problem [5, 10]. This study recommends regulator and policy maker to operationalize and introduce new measurement of corporate governance in order to offer a protected environment for minority shareholders. For instance blocking minority position through some policies that allow minority shareholders to recall for general assembly meeting (challenging management manipulation), participate in firm's decision making particularly that's relate to dividend policy and selecting independent directors.

References

1. Ab Razak, N. H., Ahmad, R., & Aliahmed, H. J. *Government ownership and performance: An analysis of listed companies in Malaysia*. *COC*, ***6***(2), 434–442 (2008)
2. Amran, N. A., & Ahmad, A. C. *Corporate governance mechanisms and performance: Analysis of Malaysian family and non-family controlled companies*. *JMAA*, ***6***(2), 1(2010)
3. Ararat, M., Aksu, M., & Tansel Cetin, A. *How board diversity affects firm performance in emerging markets: Evidence on channels in controlled firms*. *CGIR*, ***23***(2), 83–103(2015)
4. Ararat, M., Black, B. S., & Yurtoglu, B. B. *Corporate Governance , Business Groups , and Market Value : Time-Series Evidence from Turkey Corporate Governance , Business Groups , and Market Value : Time-Series Evidence from Turkey*, LER (13) (2014)
5. Ararat, M., Süel, E., Aytekin, B., & Alkan, S. *Women on board Turkey: 2014 (2nd Annual Report: Independent Women Directors Project)*. BizEd (**Vol. 3**) (2014)
6. Bany-Ariffin, A. N., Nor, F. M., & McGowan, C. B.. *Pyramidal structure, firm capital structure exploitation and ultimate owners' dominance*. IRFA, ***19***(3), 151–164(2010).
7. Morten Balling, Elizabeth Hennessy, R. O. (1998). *Corporate Governance,*

 DOI: 10.1051/shsconf/20173401006

Financial Markets and Global Convergence.

8. BORTOLOTTI, B., & FACCIO, M. *Reluctant privatization, Boston Meetings Paper,* ECGI, **40** (2006)
9. Claessens, S., Djankov, S., & Lang, L. H. P. *The separation of ownership and control in East Asian corporations*. JFE, ***58***(1), 81–112 (2000)
10. Demirag, I., & Serter, M. *Ownership patterns and control in Turkish listed companies. Corporate Governance:* IR, **11**(1), 40–51(2003)
11. Faccio, M., & Lang, L. H. P. *The ultimate ownership of Western European corporations*. JFE, **65**(3), 365–395(2002)
12. Gomez, E. T., & Jomo, K. S. *Malaysia's Political Economy: Power, Profits, Patronage*. Cambridge: Cambridge University Press (1999)
13. Grossman, S. J., & Hart, O. D. *One share-one vote and the market for corporate control*. JFE, **20**, 175–202(1988)
14. La Porta, R., Lopez-de-Silanes, F., Shleifer, A., & Vishny, R. *The quality of government*. JLEO, **15**(1), 222–279 (1999)
15. La Porta, R., Shleifer, A., & Vishny, R. W. Lόpez-De-Silanes F. Law and finance//Journal of Political Economy (1998)
16. Lee, C.. *Telecommunications reforms in Malaysia*. APCE, **73**(4), 521–540 (2002)
17. Lemmon, M. L., & Lins, K. V. *Ownership structure, corporate governance, and firm value: Evidence from the East Asian financial crisis*. JF, **58**(4), 1445–1468 (2003)
18. Morck, R., Wolfenzon, D., & Yeung, B. *Corporate governance, economic entrenchment and growth.* NBER, **43**(3,Sep), 655-720 (2004)
19. Morck, R., & Yeung, B. Y. *Special issues relating to corporate governance and family control,* WBPRWP, **3406** (2004)
20. Muhamed, A. B. I. N., Strätling, R., & Salama, A. *The Impact Of Government Investment Organizations In Malaysia On The Performance Of Their Portfolio Companies.* APCE, ***85***(3), 453–473(2014)
21. Sun, Q., Tong, W., & Yu, Q. *Determinants of foreign direct investment across China*. JIMF, **21**(1), 79–113(2002)
22. Y. Li, E., Cho, C.-C., & Wu, C.-H. *Role of auditor in agency conflict and corporate governance: empirical analyses of Taiwanese firms*. CMS, **8**(3), 333–353(2014)

Cloud Computing Adoption in Organisations: Review of Empirical Literature

Haslinda Hassan[1,*], *Mohd Herry* Mohd Nasir[1], *Norhaiza* Khairudin[1]

[1]Tunku Puteri Intan Safinaz School of Accountancy, Universiti Utara Malaysia, 06010 Sintok, Kedah, Malaysia

Abstract. This study reviews literature on cloud computing adoption in organisations to identify its influential factors and its operationalisation in prior literature. We classify the factors that influence the cloud computing adoption using the three contexts suggested by the Technology-Organisation-Environment (TOE) framework, namely, technology, organisation, and environment. The finding suggests that the influences of these factors vary across studies and most of the studies have operationalised cloud computing adoption using intention to adopt cloud computing or binary variable, rather than the actual use of the technology.

1 Introduction

Cloud computing technology is becoming ubiquitous and exploited increasingly by organisations to leverage the information technology (IT) opportunities [1] and to stay competitive and innovative [2]. This technology has been claimed as being able to provide economic advantages, such as reducing computing costs, scalability, and flexibility. These advantages are becoming the major drivers for the increasing use of cloud computing in business organisations [1, 3].

Cloud computing refers to the use of shared computing resources over the Internet [4]. The National Institute of Standards and Technology (NIST) defines cloud computing as "a model for enabling convenient, on-demand network access to a shared pool of configurable computing resources (e.g. networks, servers, storage, applications, and services) that can be rapidly provisioned and released with minimal management effort or service provider interaction" [5]. This technology evolved through the recent advancements in hardware, virtualisation technology, distributed computing, and service delivery over the Internet [6]. The cloud computing services provide an alternative to the use of local servers' applications [4].

An increasing number of studies are taking place in determining factors that influence cloud computing adoption in business organisations. The studies have explored the importance of technological, organisational, and environmental factors to cloud computing adoption in the organisations, however, their influences vary across studies. Hence, there is a need to analyse the influential factors in different studies to acquire a better understanding of the cloud computing adoption in the organisations. Specifically, this paper attempts to

* Corresponding author: lynn@uum.edu.my

DOI: 10.1051/shsconf/20173402001

identify the factors that influence the cloud computing adoption in the organisations and to determine how the existing studies operationalise the cloud computing adoption.

The next section presents literature review on cloud computing, focusing specifically on understanding cloud computing, factors influencing the cloud computing adoption in organisations, and the operationalisation of cloud computing adoption. The final section concludes with the study's limitation.

2 Literature Review

2.1 Understanding cloud computing

The term "cloud" is derived from the idea where users are able to access applications from anywhere in the world on demand [7]. Cloud computing uses ubiquitous resources that can be shared by business users over the Internet, allowing them to communicate with many servers at the same time [2].

There are three types of cloud computing services, namely, Infrastructure-as-a-System (IaaS), Platform-as-a-Service (PaaS), and Software-as-a-Service (SaaS). IaaS is the most basic service level that offers customers the infrastructure services over the Internet, including storage, network, and software. The customer mainly manages the operation of the infrastructure while the provider generally maintains the operations of the data centre [4]. PaaS offers services, such as application design, development, testing, deployment, and programming language hosting tools, in which the access to these services are provided online to customers who desire to build an application for their organisation [3]. Hence, customers who use these PaaS services do not have to buy and manage the underlying infrastructure to develop the application as they are available online. In the SaaS model, users can utilise a hosted set of software that they do not own, but pay for the elements of the utilisation [4]. Customers can install the software according to the types of exclusive and non-exclusive method, including public, private, hybrid, and community clouds [5] and the applications can be used at anytime and anywhere over the Internet [3].

2.2 Factors influencing cloud computing adoption in organisations

This study reviews literature on cloud computing adoption in organisations to identify factors that influence the cloud computing adoption in the organisations and to determine how the existing studies operationalise the cloud computing adoption. The search of literature on Google Scholar was conducted on 11th March 2016 where "cloud computing adoption" was used as the keyword. The search was, however, limited to the title of the articles only.

As of the date, we found that there were 352 articles with "cloud computing adoption" at the title of the articles. Nonetheless, several articles with the "adoption of cloud computing" at the title of the articles were also included in the list and were, therefore, considered for review. Many of the studies were, however, excluded from our review due to the studies being exploratory, qualitative in nature, individual used as a unit of analysis, conference papers, lecture notes, and press releases. Articles that did not fall within these criteria are deemed relevant for review. Of the 352, only nine (9) studies focused on the factors that influence cloud computing adoption in organisations and, therefore, selected for review (Table 1). This data indicates that in spite of the increasing attention of researchers to cloud computing, the empirical studies on the factors that influence cloud computing adoption in organisations, are however, limited. It is likely that cloud computing is a relatively recent area of research in the IS discipline [6].

 DOI: 10.1051/shsconf/20173402001

Most of the studies were carried out in the Asia-Pacific region [see, for example, 3, 9], Europe [1, 6], and North America [10]. Many of them focused on both small and large companies in more than one industry. Only one study emphasised on the SMEs [11] and on a single industry [2].

Technology-organisation-environment (TOE) framework has been widely used in prior studies to examine factors influencing cloud computing adoption in organisations [see, for example, 12]. TOE framework suggests that the process by which an organisation adopts and implements technological innovations is influenced by factors from three contexts: technology (T)—technologies that are relevant to organisations; organisation (O)—the organisation's size, scope, and the amount of slack resources available in the organisations; and environment (E)—external environment in which an organisation conducts its business [8]. Other studies have used a combination of technology acceptance model (TAM)-TOE framework [3], diffusion of innovation (DOI)-TOE [6], institutional theory [9], transaction cost theory [13], and information processing theory [10]. One study, however, did not use any specific theory [11]. TAM model generally explains the individual's behavioural intention to use IT, hence, we argue that the model is more appropriate to be used to measure technology acceptance at the individual level rather than at the organisation. Although mimetic pressure, coercive pressure, and normative pressure (as proposed by the institutional theory) were covered in one study [9], the effects of these factors on cloud computing adoption, however, have not been tested.

Recognising the importance of the TOE framework in prior studies, this framework is used to present the various technological, organisational, and environmental factors found to influence the cloud computing adoption in organisations (see Table 1 for details). Most studies have explored the importance of these factors on cloud computing adoption in their studies, however, their influences on the adoption vary across different industry contexts. We also found that not all studies covered factors from all the three contexts [9, 11, 13].

From the technological context, three common factors tested in prior studies were relative advantage, compatibility, and complexity. These three factors are derived from Roger's DOI theory [14]. Relative advantage was covered by most of the studies and was found significant. The significant importance of relative advantage indicates that the organisations recognise the relative advantages of cloud computing, such as in reducing IT costs, creating the competitive advantage, and enabling easier access to information. Nonetheless, organisations in high-tech industry have perceived relative advantage as significant but negatively influencing cloud computing adoption. These organisations may have realised the relative advantage of using cloud computing, however, they may have a lower level of cloud computing knowledge [2]. Compatibility was seen as not an important factor to consider cloud computing in all the three studies [1, 2, 6]. Conversely, the significant finding of complexity to cloud computing adoption [1, 3, 6, 11] implies that the organisations have a level of fear and concern regarding the adoption of cloud computing services [1].

From the organisational context, a number of studies on cloud computing adoption have witnessed significant role of top management support on cloud computing adoption in Taiwan [2], Australia [9], and Portugal [6]. The result suggests that top management plays the decision process by demonstrating support in the form of committing financial and organisational resources and engaging in the process [6]. Although the adoption of cloud computing is not limited to large organisations, organisation size was found to be a facilitator of cloud computing adoption [2, 6]. Larger organisations with more resources were argued to have more tendencies to adopt cloud computing than smaller organisations. Evidence from the study conducted by Oliveira et al. indicates that firms, with an established technology infrastructure and competent workforce, are better suited for cloud integration [6].

DOI: 10.1051/shsconf/20173402001

Table 1. Factors influencing cloud computing adoption in organisations.

Study	Country	Theory	Technological Context[a]							Organisational Context[b]				Environmental Context[c]					
			t1	t2	t3	t4	t5	t6	t7	o1	o2	o3	o4	e1	e2	e3	e4	e5	e6
[1]	UK	TOE	√	√	√*					√	√*	√		√*					
[2]	Taiwan	TOE	√*	√	√					√*	√	√*		√*					
[3]	India	TAM & TOE	√*		√*									√*					√*
[6]	Portugal	DOI & TOE	√*	√	√*					√*	√*	√*		√	√				
[9]	Australia	Institutional theory								√*									
[10]	USA	Information processing theory							√				√*					√*	
[11]	APAC region	—	√*		√*	√	√	√											
[12]	Taiwan	TOE	√*			√*					√*			√					
[13]	Australia	Transaction cost theory				√*										√*	√		

Note. Only empirical studies that have focused on the factors influencing cloud computing adoption in organisations (hence, organisation as the unit of analysis) are covered and listed in Table 1. *Factor found significant to cloud computing adoption.

[a]Technological context: t1-relative advantage (perceived benefits); t2-compatibility; t3-complexity; t4-business concerns (e.g., confidentiality); t5-sharing and collaboration; t6-reliability; t7-task uncertainty.

[b]Organisational context: o1-top management support; o2-technology readiness; o3-firm size; o4-inter-organisational uncertainty.

[c]Environmental context: e1-external pressure; e2-regulatory support; e3-perceived vendor opportunism; e4-perceived legal uncertainty; e5-environmental uncertainty; e6-trading partner support.

From the environmental context, external pressure was mostly tested and found significant [1, 2, 3]. Complex architecture of cloud computing requires tighter integration of customers, business partners, and suppliers and is influenced by the adoption status of its trading partner [3]. The finding implies that when competitors implement cloud computing as a competitive instrument, other organisations face strong competition, hence, feel pressure to adopt cloud to maintain a competitive edge [3]. Both regulatory support and perceived legal uncertainty were found to be not significant to cloud-computing adoption. This does not necessarily mean that the organisations disregard prevailing standards and regulations, but rather that existing legislation protecting the use of cloud computing has not been earnestly embraced by the organisational decision makers [6].

2.3 Operationalisation of cloud computing adoption in organisations

Overall, most of the studies have operationalised cloud computing adoption using adoption intention [see, for example, 3, 10, 12], or binary variable [1, 2] (see Table 2). For example, cloud computing adoption was operationalised using intention to increase the level of computing adoption [9, 13] and the summation of the degree of SaaS, PaaS, and IaaS adoption, where three response options (has not adopted; planning to adopt within 12 months; already adopted) were used for each of the three service models [12]. The choice of intention rather than the actual adoption of the technology was due to the relative newness of cloud computing as a business tool [10].

Table 2. Operationalisation of cloud computing adoption.

Study	Operationalisation
[1]	Binary variable
[2]	Binary variable
[3]	Intention to adopt cloud computing
[6]	(a) stage of cloud computing adoption; and (b) Intention to adopt cloud
[9]	Intention to increase the level of cloud computing adoption
[10]	Intention to adopt cloud computing
[11]	(a) Its ease of use; (b) its reliability; and (c) security and privacy
[12]	Intention to adopt cloud computing
[13]	Intention to increase the level of cloud computing adoption

Intention to use refers to the degree to which an individual intents to perform a particular behaviour or act. It is, however, argued that the intention to use does not necessary lead to adoption [15]. In addition, measuring technology adoption using binary measures, such as adopt or not adopt cloud computing services, is deemed inadequate as it does not completely capture the reach and richness of the use of IT innovations and cannot inform post-adoptive variability [16].

3 Conclusion

This study reviews literature on cloud computing adoption in organisations, focusing specifically on its influential factors and its operationalisation in prior literature. The factors (technology, organisation, and environment) are classified based on the TOE framework. We find that most studies have explored the importance of these factors to cloud computing adoption, nonetheless, their influences vary across studies. Most studies have operationalised cloud computing adoption using intention to adopt or binary variable (adopt or not adopt). The literature chosen for the review is, however, limited to the quantitative studies of cloud computing adoption in the organisations only.

The authors would like to thank Universiti Utara Malaysia and the Ministry of Higher Education Malaysia for financially supporting this research (FRGS/1/2014/SS05/UUM/02/29).

References

1. A. Gutierrez, E. Boukrami, R. Lumsden. Technological, organisational and environmental factors influencing managers' decision to adopt cloud computing in the UK. Journal of Enterprise Information Management, 28(6), 788-807 (2015).
2. C. Low, Y. Chen, M. Wu. Understanding the determinants of cloud computing adoption. Industrial Management & Data Systems, 111(7), 1006-1023 (2011).
3. H. Gangwar, H. Date, R. Ramaswamy. Developing a cloud-computing adoption framework. Global Business Review, 16(4), 632-651 (2015).
4. S. Bhardwaj, L. Jain, S. Jain. Cloud computing: A study of infrastructure as a service (IAAS). International Journal of Engineering and Information Technology, 2(1), 60-63 (2010).
5. P. Mell, T. Grance. The NIST definition of cloud computing. National Institute of Standards and Technology, 53(6), 50 (2009).
6. T. Oliveira, M. Thomas, M. Espadanal. Assessing the determinants of cloud computing adoption: An analysis of the manufacturing and services sectors. Information & Management, 51(5), 497-510 (2014).
7. N. Sultan. Cloud computing for education: A new dawn? International Journal of Information Management, 30(2), 109-116 (2010).
8. L.G. Tornatzky, M. Fleischer. The process of technological innovation. Lexington, MA: Lexington Books (1990).
9. O.M. Yigitbasioglu. The role of institutional pressures and top management support in the intention to adopt cloud computing solutions. Journal of Enterprise Information Management, 28(4), 579-594 (2015).
10. C.G. Cegielski, L.A. Jones-Farmer, Y. Wu, B.T. Hazen. Adoption of cloud computing technologies in supply chains: An organisational information processing theory approach. The International Journal of Logistics Management, 23(2), 184-211 (2012).
11. P. Gupta, A. Seetharaman, J.R. Raj. The usage and adoption of cloud computing by small and medium businesses. International Journal of Information Management, 33(5), 861-874 (2013).
12. P.F. Hsu, S. Ray, Y.Y. Li-Hsieh. Examining cloud computing adoption intention, pricing mechanism, and deployment model. International Journal of Information Management, 34(4), 474-488 (2014).
13. O. Yigitbasioglu. Modelling the intention to adopt cloud computing services: A transaction cost theory perspective. Australasian Journal of Information Systems, 18(3), 193-210 (2014).
14. E.M. Rogers. Diffusion of innovations (5th Ed.). New York, NY: Free Press (2003).
15. W. Money, A. Turner. Assessing knowledge management system user acceptance with the technology acceptance model. International Journal of Knowledge Management, 1(1), 8-26 (2005).
16. K. Zhu, S. Dong, S.X. Xu, K.L. Kraemer. Innovation diffusion in global contexts: Determinants of post-adoption digital transformation of European companies. European Journal of Information Systems, 15(6), 601-616 (2006).

SHS Web of Conferences **34**, 02002 (2017) DOI: 10.1051/shsconf/20173402002
FourA 2016

The development of state/region owned goods management's monitoring instrument design

Yogy Ikhwanto[1], and *Irwan* Taufiq Ritonga[2]

[1]Fellowship of Magister of Accounting, Universitas Gadjah Mada, Yogyakarta, Indonesia
[2]Lecturer at Magister of Accounting, Universitas Gadjah Mada, Yogyakarta, Indonesia

Abstract. The problems in state/region owned goods in Indonesian state and local governments are suspected to occur because of weak monitoring programs, according to many studies. A tool or instrument in implementing this monitoring program is expected to address this problem. Such tool currently doesn't exist yet. This research aims to fill that gap by developing a monitoring instrument design for state/region owned goods by using *Daerah Istimewa Yogyakarta* (DIY) Local Government as a research context in order to take valuable inputs for the design. This research is using developmental research method. Government Regulation were used for normative reference and Friedman's results-based accountability quadrat were used in developing good indicators for the instrument. This research is succeeded in formulating the indicators that made up the instrument. Indicators compiled are divided into compliance-based indicators and results-based indicators. Indicators are formulated based on the validation and inputs from employees of DIY's Assets Management Agency and experts from academia. This instrument still has some limitations that need improvement through further research.

1 Introduction

Problematic management of state or region owned goods (SOG/ROG) is still being a major issue for many governmental entities in Indonesia. The impact of the problem can be seen in the financial statement's opinion obtained because of the ties between the SOG/ROG's management and accounting. Proven as an example, Biannually Inspection Summary I of Year 2015 of Republic of Indonesia's Audit Board (BPK RI) indicated as many as 230 financial statements of local governments obtained qualified/disclaimer of opinion from fixed asset accounts and/or other assets, as well as 37 from the inventory account [1].

Several studies have tried to examine the causes of the problems in managing SOG/ROG in the context of different local governments. Fatana [2] conducted a research of ROG problem in Lombok Barat regency and found that weaknesses in the administration of ROG were caused in part by the lack of ROG management monitoring program. Other research that focused on the management of the fixed assets as a form of ROG found positive and significant relationship between monitoring and controlling program with fixed asset optimality [3-4].

If SOG/ROG monitoring program is so important, then how to make sure that the monitoring program as mandated by Government Regulation No. 27 Year 2014 about State/Local Government's Goods Management effectively improve the quality SOG/ROG management? Based on good practices, a monitoring and/or evaluation program are not only a means to look at the suitability of the activities with the regulations, but also to measure the achievement of results through performance indicators [5]. To achieve these objectives, an instrument that contains a list of questions or indicators of performance is commonly used as practiced by various local and international entities. The existence of such instruments is believed by many organizations to be effectively assisting the implementation of monitoring and evaluation of programs.

As best as the author searched, such monitoring instrument doesn't exist yet for SOG/ROG management, whether published officially or developed academically (no previous research). Knowing the importance of monitoring and evaluation as well as the condition of vacant monitoring instruments related to the SOG/ROG management as stated above, then it is required to conduct a developmental research to produce an instrument or a tool to implement the mandate of monitoring the SOG/ROG management stipulated in Government Regulation No. 27 of 2014 which is capable of monitoring the implementation, performance, and the results of SOG/ROG management. Hopefully such instrument can improve the effectiveness of the monitoring program. To achieve this objective, a model or framework that is practical, but also results-based in its development is needed. Therefore, the development of SOG/ROG management's monitoring instrument design that is practical and results-based will be the goal of this research.

2 Literature review

2.1 The management of SOG/ROG

The legal basis for the SOG/ROG management is Government Regulation No. 27 Year 2014 on the Management of State and Region Owned Goods. Based on Government Regulation No. 27 Year 2014, the management of SOG/ROG includes the management stages as follows: needs planning, procurement, usage, utilization, security and maintenance, valuation, deletion, transfer, extermination, administration, guidance, supervision, and control.

Author could not find a specific previous research similar to that conducted in this research because this research is using developmental research method that is rarely used. A similar research method with a different topic conducted by Muktiali [6] who developed the instrument of monitoring and evaluation of development outcomes in *Semarang* City.

SOG/ROG management also has close links with the accounting. Issuance of Government Regulation No. 27 of 2014 set the various provisions related to the management of SOG/ROG that contribute to government accounting [7]. Determination of the status of use by a decree is a basis for accounting records. Administration in general gives evidence for the accounting records like proof of purchase, proof of transfer, and evidence of removal. In addition, the regulation also concerns the assessment of land and buildings in case of utilization and transfer to get fair value which is in accordance with good accounting practices internationally.

In addition, there is also an obligation in applying accounting policies depreciation of fixed assets in 2015. This obligation provides an opportunity to measure the performance of government fixed asset management using depreciation account. In particular, when the entity applying good management of fixed assets, the entity will report accumulated depreciation that is lower compared with entities that do not manage their fixed assets as well [8]. This is because the entity maintained its fixed assets in good condition through maintenance. Moreover, because of the nature of the depreciation of fixed assets itself that provides an indicator of the power serviceability fixed assets to the government or the public [9].

2.2 Monitoring and evaluation

Monitoring is an internal activity that is designed to provide regular feedback on the progress of a program, the problems encountered, the efficiency of program implementation and achievement of program outcomes [10]. This definition is consistent with the intent of monitoring in Government Regulation No. 39 of 2006 Article 1 paragraph 2. Meanwhile, the evaluation is an activity to systematically and objectively measure how well a program is achieving the desired objectives and how far the changes can be attributed to the program [11].

There are many monitoring and evalution frameworks well-known in the world. For present study, the author chose to use results-based accountability (RBA) framework developed by Friedman [12]. The RBA is more appropriately used as a monitoring and evaluation framework in administrative activities such as SOG/ROG management compared to other framework. The framework can be seen in the figure below.

	Quantity	**Quality**
Effort	**(1) What did we do (#)**	**(2) How well did we do it (%)**
Effect	**(3) Is anything better off? (# *output*)**	**(4) Is anything better off (% *outcomes*)**

Figure 1. Friedman's results-based accountability framework

Friedman grouped all performance indicators in the world into four categories pictured in quadrants as shown in the figure above. Most effective performance indicator that can drive real change and outcomes are in quadrant number 2 or 2nd quadrant, which are performance quality indicators, and quadrant number 4 or 4th quadrant which are outcome quality indicators. Indicators for the instrument designed must be formulated to meet the criteris in this two categories.

3 Research method

The authors chose the Local Government Special Region of Yogyakarta (DIY) as a research context to undertake preliminary research and validation test of the monitoring instrument developed.

The author used a developmental research approach for this research. Developmental research can easily be interpreted as a study that aims to produce or develop a product [13]. However, a more complete definition of a developmental research is also a research that provides a way to apply theory, validate the good practice that is already running, and develop a procedure, technique, or a new instrument based on the analysis of a particular case [14]. The research's process framework can be pictured in the figure below.

DOI: 10.1051/shsconf/20173402002

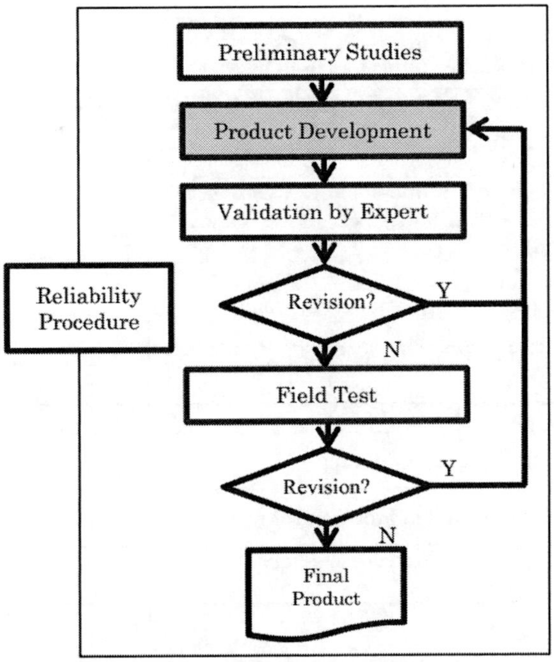

Figure 2. Research's Process Framework

In general, the series of activities in developmental research can be explained as follows:

a. *Preliminary studies*, consisting of literature studies and field studies. Literature studies aim to find a model that suits procedural formulation of the problem. The field study is conducted to assess the needs, understand the process, and understand the contextual conditions in which the product will be implemented. The field study will be conducted with a qualitative approach through in-depth interviews and focus group discussion.
b. *The product development stage*. Friedman's result-based quadrant [15] is used to help to determine the class of the the performance indicators developed in the instruments.
c. *Expert's Validation stage*, namely the involvement stage of experts associated with the products developed to determine whether the products developed by the research were ready to be applied in the field.
d. *The pilot phase*, which is a stage to evaluate the usability of the product on a limited basis in the context of the research (Government of DIY). Limited test will be done by monitoring the performance of ROG management at the governmental unit (satuan kerja perangkat daerah) in DIY using instruments that have been developed.
e. *Procedure of reliability,* which is to document all the procedures described above well enough so that if other researchers perform the same procedure on the same case it will get the same findings and conclusions, as described by Yin [16] concerning the reliability in qualitative research.

4 Results and discussion

This research succeeded in formulating a set of indicators that make up the instrument in monitoring ROG management in local government of DIY through the development stage by the author and validation by experts. The instrument developed also have passed the pilot phase to determine the usability of the indicators in the context of research with the available data.

The Summary of compiled performance indicators that make up the ROG management's monitoring instruments can be presented in the following table.

Table 1. Summary of Performance Indicators in ROG Management's Monitoring Instrument

No	Performance Indicators	Scoring
Compliance Monitoring Indicators		
Stage: Needs and budget planning		
1.	Timely submission of proposals of ROG needs (including procurement, maintenance, usage, removal, and extermination)	Timely= 1 Late= 0
2.	The suitability of the proposed ROG needs to the standard (including items standards and price standards)	Conformed= 1 Not conformed= 0
3.	Completeness and correctness of information in ROG needs proposal	Σ conformed criteria/ Σ criteria
Stage: Procurement		
4.	Timely submission on procurement report	Timely= 1 Late= 0
5.	Completeness of information in procurement report	Σ conformed criteria / Σ criteria
6.	Compliance of realization in types and quantities of procurement against it's budget	Conformed= 1 Not conformed= 0
7.	Timely procurement based on cash flow predictions	91%-100% Conformed= 1 71%-90% Conformed= 0,75 51-70% Conformed= 0,50 21-50% Conformed= 0,25 <20% Conformed= 0
	Availability of acceptable excuse for untimely procurement	Acceptable: 1 – (score above) Not Acceptable: 0
Stage: Security and Maintenance		
8.	Timely submission of maintenance report	Timely= 1 Late= 0
9.	Compliance of budget and performance target in maintenance realization	91%-100% Conformed= 1 71%-90% Conformed= 0,75 51-70% Conformed= 0,50 21-50% Conformed= 0,25 <20% Conformed= 0
	Availability of acceptable excuse for unconformed maintenance realization	Acceptable: 1 – (score above) Not Acceptable: 0
10.	Quality of physical, administrative, and legal security of assets	Σ conformed criteria / Σ criteria
Stage: Administration		
11.	Conformity of bookkeeping with the status decalred for the ROG	Conformed= 1 Not conformed= 0
12.	Conformity of room and unit inventory book with the main inventory book	Conformed= 1 Not conformed= 0
13.	Conformity of five-yearly inventarization	Conformed= 1 Not conformed= 0
14.	Conformity of yearly inventarization for supplies and inventories	Conformed= 1 Not conformed= 0
15.	Conformity of installation of ownership sign	Conformed= 1 Not conformed= 0
16.	Timely reconciliation of ROG	Conformed= 1 Not conformed= 0
17.	Conformity of depreciation calculation with the accounting policy	Conformed= 1 Not conformed= 0

No	Performance Indicators	Scoring
Stage: General condition		
18.	The ratio of active inventory items used to total inventory	According to the results
19.	The ratio of the value of inventory items in good condition to total value of inventory	According to the results
20.	The ratio of maintenance expenditures to depreciation expense	According to the results

Source: Developed.

Based on its characteristics, the author divided the indicators developed into two categories, namely compliance-based indicators that belong to 2nd quadrant of Friedman's RBA and indicators of the quality of ROG management work in general that belong to 4th quadrant of results-based indicators according to Friedman's. Limited pilot phase of the indicators above has also been conducted in The Government of DIY. However, the evaluation of indicators of general condition of ROG management cannot be done at the moment because of unavailability of rule of thumb value and therefore become the limitation of this instrument.

Indicators that measure the ROG management conditions in general require a separate study to determine the rules of thumb value that can be empirically proven. Studies can be done by measuring this indicator within a certain time on a variety of government entities that are considered good in managing SOG and/or ROG based on the indicators of compliance offered in the instrument. With proper statistical procedure, a particular rule of thumb value can be obtained.

The current instrument have not set the weight of each indicator in order to do the sums to get the value of the final performance indicators. This weighting can be highly subjective because of different perceptions of individuals. To determine the weight in empirical basis, it is necessary to study the perception of many people involved in the management of SOG and/or ROG.

The use of depreciation account of fixed assets as a component of performance indicators still leaves a note for the application of the depreciation policy of fixed assets in Indonesia is still not in good practice. Most government entities are still using the straight-line method to calculate the depreciation of fixed assets in all types of assets while the Government Accounting Standard No. 7 paragraph 57 allows the use of other methods. This could result in the record of depreciation that is not conformed with the actual assets' value decrease. A study that aims to determine the appropriate depreciation method that is conformed with the characteristics of each type of government's fixed assets is expected to help address this deficiency.

5 Conclusions and suggestions

This research succeeded in formulating a set of indicators that make up the instrument in monitoring ROG management in local government of DIY through the development stage by the author, validation by experts, and limited pilot phase. Based on its characteristics, the author divided the indicators developed into two categories, namely compliance-based indicators and indicators of the quality of ROG management work in general that belong to results-based indicators. Based on the discussion presented in the previous chapter, we can conclude some limitations in the research as well as the advice for conducting the next research as follows.

a. Indicators that measure the ROG management conditions in general require a separate

 DOI: 10.1051/shsconf/20173402002

study to determine the rules of thumb value that can be empirically proven. Studies can be done by measuring this indicator within a certain time on a variety of government entities that are considered good in managing SOG and/or ROG based on the indicators of compliance offered in the instrument. With proper statistical procedure, a particular rule of thumb value can be obtained.

b. The current instrument have not set the weight of each indicator in order to do the sums to get the value of the final performance indicators. This weighting can be highly subjective because of different perceptions of individuals. To determine the weight in empirical basis, it is necessary to study the perception of many people involved in the management of SOG and/or ROG.

c. The use of depreciation account of fixed assets as a component of performance indicators still leaves a note for the application of the depreciation policy of fixed assets in Indonesia is still not in good practice. A study that aims to determine the appropriate depreciation method that is conformed with the characteristics of each type of government's fixed assets is expected to help address this deficiency.

References

1. Badan Pemeriksa Keuangan RI, *Ikhtisar Hasil Pemeriksaan Semester I BPK RI Tahun 2015,* (2015)
2. D. Fatana, *Evaluasi Sistem Pengendalian Internal Pengelolaan Barang Milik Daerah* (Unpublished Thesis, 2015)
3. E.B. Bles, *Analisis Faktor-Faktor Yang Mempengaruhi Pengelolaan Aset Tetap (Tanah Dan Bangunan): Studi Pada Pemerintah Daerah Kota Jayapura* (Unpublished Thesis, 2015)
4. Jusmin, *Pengaruh Manajemen Aset Terhadap Tingkat Optimalitas Aset Tetap (Tanah Dan Bangunan) Pemerintah Kota Baubau* (Unpublished Thesis, 2015)
5. UNDP, *Handbook on Monitoring and Evaluating for Results* (2002)
6. M. Muktiali, Jur. RIPTEK Bap. Kot. Sem., *Penyusunan Instrumen Monitoring Dan Evaluasi Manfaat Program Pembangunan Di Kota Semarang,* **3.2,** 11-20 (2009)
7. J. Hoesada, M. Ling, *Barang Milik Negara/Daerah* (2014)
8. D. Edgerton, *Valuation and Depreciation: A Guide for Not-for-Profit and Public Sector under Accrual Based Accounting Standards* (2013)
9. J. Hoesada, *Dua Puluh Lima Alasan Penyusutan Aset Tetap Dalam Akuntansi Pemerintahan* (2007)
10. M. Bamberger, E. Hewitt, *Monitoring and Evaluating Urban Development Programs* (1986)
11. N. Frankel, A. Gage, *M&E Fundamentals A Self Guided Minicourse* (2007)
12. M. Friedman, *Trying Hard Is Not Good Enough: How to Produce Measurable Improvements for Customers and Communities,* **4th** (2005)
13. S. Kantun, *Hakikat Dan Prosedur Penelitian Pengembangan* (2013)
14. R.C. Richey, J.D. Klein, Jou. Of. Comp. In Hig. Ed., *Developmental Research Methods: Creating Knowledge from Instructional Design and Development Practice,* **16.2**, 23-38 (2005)
15. M. Friedman, *Trying Hard Is Not Good Enough: How to Produce Measurable Improvements for Customers and Communities,* **4th** (2005)
16. R.K. Yin, *Case Study Research: Design and Methods,* **4th** (2014)

Socio-technical analysis of Indonesian government e-procurement system implementation: barriers to enhance information transparency and accountability

Fitri, Amalia

Accounting Department, Faculty of Economics and Business, Universitas Gadjah Mada, Indonesia

Abstract: E-governance has become increasingly important to deliver better public services, and increase public trust. With a specific focus on e-procurement and the issues of transparency and accountability this research investigates the role and barriers of information technology in enhancing information transparency and accountability to the public, using the actor-network theory and notion of delegation approach. This research concludes that information technology was delegated to automate the procurement process to increase transparency, accountability and prevent fraud. However, barriers of e-literacy, lack of leadership, reluctance of implementation, and lack of infrastructure created obstacles to attain the goals. This infers that social and technical aspects are interrelated and empower each other to support the technology in enhancing information transparency and accountability. Enhanced collaborative approach between the developers and users in the application development and implementation should be taken on board to enhance e-procurement system implementation quality.

Keywords: e-procurement, transparency, accountability, actor-network theory, delegation, role, barriers

1. Introduction

1.1. Research Background

The Asian financial crisis in the late 1990s enforced Indonesian government to increase their public sector's transparency and accountability and gain a stable economic condition (1, 2). IT in public sector open new ways of public management through e-government to deliver better public services, enhance communication between citizens and government, and improve citizen participation which eventually increase citizen trust (3).

However, there are challenges specifically for developing countries, which lack of capability and integrity to conduct good government governance. Based on the Seventh Annual Globe e-Government Survey 2002, Indonesia was ranked 94th amongst the 197 countries assessed (4),

 DOI: 10.1051/shsconf/20173402003

due to the lack of infrastructure, political will, and competent human resources. This condition initiated Indonesia's government commitment to develop and implement e-government policy to "improve its relationship with citizens and the business sector through enhanced, cost-effective, and efficient delivery of services, information and knowledge using IT" (5, p. 483). One of Indonesia's e-government reform initiatives was to improve public spending efficiency through public e-procurement system (1).

1.2. Research Justification and Research Questions

Audit practice through check and balance becomes important to ensure social transparency and accountability, and also build public trust due to the increase of fraud, and corporate failures which intensified the demand for financial audit to establish good governance (6). Indonesia's ineffective public procurement system which incurs higher cost, leads to poor project performance. Moreover, the decentralization reform which gives the local government rights to disperse their decision making closer to the citizens, enhance the complexity in monitoring local procurement projects, which can create an uncompetitive business environment (7). The Indonesian government together with donors such as United States Agency for International Development (USAID) initiated the use of IT to facilitate e-procurement to enhance public service and access to Indonesia's rural areas. Despite the reform, Indonesia's public e-procurement has not been very successful due to socio-economic problems and digital divide (1, 7).

As IT represents a high risk in developing countries, its implementation to bridge the power distance between civil servants and the public must be monitored with higher norms and integrity (8, 9). Therefore, the relationship and network developed between the actors, social, and technical aspects involved in the government IT reform must be analyzed to understand the technology roles and challenges in accommodating the social demand and functions of information transparency and accountability. This research **aims** to analyze the role of IT in increasing e-procurement transparency and accountability using the actor network theory approach, to capture insights from technology developers and users and find out in the role of technology in fulfilling the e-procurement reform goals. In accordance with the research aims, below are the **research questions:**

1. How is the adoption of technology meant to enhance transparency and accountability for the e-procurement system's users?
2. What are the developer's insights about the barriers to transparency and accountability of the e-procurement system?
3. What are the users' insights about the barriers to the transparency and accountability of the e-procurement system?

2. Literature Review

2.1. Information Transparency, Accountability, and Barriers in Indonesian e-Procurement

As part of the national reform, in December 2004 World Bank and Organization for Economic Co-operation and Development-Development Assistance Committee (OECD DAC) together with developing countries (including Indonesia) and donors initiated the development of a set of tools and standards which guides the improvement and assessment of the national procurement systems (7). As part of the seven Flagship of The National Information Technology Board Republic of Indonesia, the Institution of Government Procurement Policy (IGPP) was established in 6 December 2007 under the Presidential Decree 106/2007 to create an efficient and effective spending of state budget, increase the human resource competency in e-procurement, and increase the capacity of IGPP as a policy and decision maker (10). In the first years of its implementation, 44 EPSU were established in 2008 and 2009, and in December 2010 this increased to 128 EPSU, which served 244 governmental institutions in 25 provinces across Indonesia (10).

IT has claimed to reduce the cost of collecting, distributing, and accessing government information thus provide government, citizens, and businesses greater access to information (11). In e-government, transparency is delivered through government proactive dissemination to release of government materials, public meeting and protection for whistleblowers (11, 12). In terms of accountability, IT enables greater citizen participation in politics and decision-making, increasing government responsibility to accommodate citizen aspiration and represent their views and interest in governing the nation (13). The effort to enforce transparency and accountability can be seen in India where e-government has facilitated online rural property records which increased the speed of record access and prevent bribery (14). The delivery of transparent data, the security of networks, confidentiality of information, and reliability and verifiability of information facilitates government accountability to the citizen (15). Citizen participation in e-government increases and is supposed to enhance process-based (responsiveness and information accessibility) and institutional-based trust (information transparency and government responsibility) (16). However, previous findings state that IT utilization does not guarantee the success of anti-corruption in transparency and accountability but may open new opportunities for corrupt behaviors instead (11). Internal resistance to transparency, creation of new means of corruption through IT, literacy rate, technology infrastructure, and digital divide are amongst the common barriers to e-government reform (2, 3).

2.2. Actor Network Theory

Actor Network Theory (ANT) has recently gained significant use in the field of information system research, emphasizing on the social-human-technology interaction (17, 18). Actors in forms of texts, technical artifacts, human skills, and institutions play a centre role in the networks and define the relationships between each other (19). Scientific and technological innovation, social patterns and organizational arrangements are to be stabilized through the network and interaction of human and non-human where conflicts and controversies can be overcome through translation of actors' interests, persuasion, and the enrolment of network and the mobilization of resources (18, 20, 21). In this way, the scientific and technological innovation reflects the societies as they are shaped by the interaction of actants within the networks (22, 23). To establish a strong network and relation of actants, translation process involves power to stabilize social asymmetry and disputes between actors to align interests. The interaction process generates power, which is translated into a command resulting from each agent whom translates the power into their own desired outcome (20). Power persuades of translation through deformation of concepts and acts, which differ from diffusion as networks become more concentrated through iteration of actor-network relation. Society is held together by heterogeneous means of natural, technological, textual and topographical factors (23). Thus,

 DOI: 10.1051/shsconf/20173402003

it is true that organizations, social relations and technology are interrelated and often assumed able to support organization development (24) and possess political strength to stabilize and translate organization processes (25). The notion of delegation refers to transferring roles, values, intensions, and rules to technological artifacts or machines (25, 26). The roles and functions of humans are delegated to technology to create social durability (27) through the development of machines and technology, which embeds social and organizational role previously performed by human actors. This delegation was initiated to create social durability and discipline the society by tying humans to moral laws and build social morality.

2.3. Linking ANT to Indonesian e-Procurement Transparency and Accountability

The e-procurement system results from the interaction and collaboration of various stakeholders with different interests, which are aligned through the translation process. The technology and features translated within it were delegated to ensure safe and reliable procurement process and prevent any deviations, which were prone to occur in the manual procurement system. Referring to the politics of technology, (24) stated that politics are exercised based on the dominant society interest and represents the use of information in contextual terms, which denies the concept of technology neutrality. Thus, there may be unintended outcomes from the implementation such as users' resistant and users who do not confirm with the intended script inscribed in the technology, which questions the moral order of technology presupposed. Related to this, this research will refer to the notion of delegation and actor network theory to study the role of IT in Indonesian e-procurement system to establish transparent and accountable process, and also study the barriers in enforcing transparency and accountability for the public procurement.

3. Research Methods

Case study method allows researcher to focus on a specific situation to study various activities, which can combine data collection methods of interviews, archives, questionnaires and observations (28). Semi-structured interviews has been conducted with the developers and management of Indonesian Government e-Procurement System which are the e-Procurement Directory staff in the Institution of Government Procurement Policy (IGPP), and also with the users of e-procurement system: EPSU West Java Province and EPSU Central Kalimantan Province. Two staff from each organization was interviewed. The author gathered preliminary information from published materials from IGPP and EPSU website to understand about the project and research contexts. The researcher adopted the contact summary sheet method by (29) for the semi-structured interviews to validate the categorization of results based on the actor network theory used in the analysis. This enables the author to focus on the relevant information, which confirms the theoretical propositions used to design the case study and research scope (30).

4. Results and Discussion

4.1. Actors and Networks

To study the roles and barriers of technology in the e-procurement system, it is first important to identify the actors and their relationships in the e-procurement application development and implementation. In the global network, IGPP consist of policy and strategy development division, monitoring-evaluation and information system development division, human resources development division and law division, which collaborates to manage and develop the application. These divisions constantly collaborate and are in close contact with the E-Procurement Services Unit (EPSU) in the local network to enhance coordination of implementation. IGPP also collaborates with donors and international organizations such as United States Agency for International Development (USAID), World Bank and Organisation for Economic Co-operation and Development (OECD) to develop the regulation and application of e-procurement (31). From 2007 until 2009 USAID through the Millennium Challenge Corporation-Indonesia Control of Corruption Project rewarded a grant to build the first five regional EPSU in five provinces in Indonesia (32). The collaborations and interactions within actors in the global network also occur in in the monitoring and evaluation of the e-procurement. Based on the interviews, IGPP periodically conduct meetings to gather feedbacks from users and related parties, and discuss about further application development. Some of the feedbacks suggest that the application should also provide services for e-catalogue, e-purchasing, e-contract management, e-payment, and e-budgeting.

4.2. Translation and Power

The translation process involves all entities, which are enrolled into their identities and stabilized through project management, monitoring, and evaluation which control transactions between local and global network (20). The translation process begins with problematization in which the actors explain the problematical situation to other actors in the network to align interest through interessement and make them convinced to accept the idea of the situation, thus enrolling them into power and identities, also mobilising the actors which stabilize the network through the obligatory point of passage (20). In this project, IGPP communicate the urgency of e-procurement to governmental organization.

4.3. Delegation and Durability: The adoption of Technology to Enhance Transparency and Accountability for the e-Procurement System

The roles of human actors in procurement process are delegated to technology through the development of e-procurement application which embeds social and organizational role starting from procurement registration until result announcement. The interview results explain the developer's views that the role of IT is very important to make information more accessible and transparent throughout the whole e-procurement process. Previously, the announcements can only be accessed in the formal offices for a limited of time and accessible to limited people. Using IT, documents are uploaded into the system, encrypted and are only allowed to be uploaded in the specified time so it is difficult for the committee and participants to modify them. This e-procurement system also increases the independency of procurement committee from the head of institution and external parties, because sometimes there are pressures to pronounce a particular winner due to conflict of interest. The transparency of this system protects the procurement committee from such deviation. The users confirmed with the developers' insights on the importance of IT. IGPP also developed document security and e-audit application to increase public trust, document security, privacy through increased

DOI: 10.1051/shsconf/20173402003

transparency and accountability. In this system, every user has to register online to be given an ID and password. E-audit and document security application plays a significant role to ensure the credibility and reliability of information disclosed to the public. Data security application secures and encrypts documents submitted by the vendors, which are secured by password and can only be open by the procurement committees and other verified users. This degree of document security and transparency enable immediate identification of irregularities in the system, and also immediate tracing of fraud. To support IT implementation, all EPSU personnel are certified, trained and officially employed to maintain their performance and integrity.

4.4. Barriers to Transparency and Accountability of The e-Procurement System

Despite of the importance of the applications, barriers of information transparency and accountability have prevented the full utility of the application. IGPP believe that the users view that technology is hard to implement which makes them reluctant to use this application. Thus, building trust through the transparency and accountability of the system will take great effort of training, and monitoring. From the developers' perspectives, there are no technical barriers, which prevent the application implementation to enhance transparency and accountability, but more of problems concerning human resources competency, leadership, and willingness to put some efforts in the improvement. Problems also occur in the e-audit implementation. The developers view that the progress of e-audit implementation has been slow due to the transition process and also in the human resource competency and infrastructure. Moreover, auditors lack of understanding of the e-procurement process and many of them are more convenient using the manual audit process. The problem of digital divide which leaves the majority of rural areas in Indonesia untouched by IT also create reluctance for users to implement this application. Moreover, since e-procurement is not yet compulsory, users tend to delay the implementation because they doubt whether this application will give them flexibility as in the manual system. Another problem is the lack of data interoperability and integration between the e-procurement application database with the institutions which issues legal documents required such as the tax office and law and attorney offices. Without interoperability the verification process is difficult and has to be done manually and is more prone to deviation.

4.5. Discussion

(27) argued that society is made durable; nothing is purely technological or social, as society is held together by heterogeneous means of natural and technological factors. Based on the case study, we can infer that the success of technology implementation does not only depend on the technical infrastructures and machines in the application, but also require the collaboration of competent and determined human actors to manage and support the technological roles in the organizations. Technology was delegated to substitute the role of human actors in the manual procurement process to prevent irregularities and fraud. Specifically, document security application and e-audit application were developed to increase information transparency, accountability and integrity of the application to increase public trust. IGPP have expressed the problematical procurement situation (problematization) to governmental organizations, aligned actor's interest and translated them into roles to achieve the desired outcome of the project. From the analysis, it is interesting to hear from the application developers and users that the main problems in the application implementation more relates to the organizational aspects of

the users and staff such as lack of leadership to enforce implementation in the governmental organization, e-literacy, infrastructure and low level of public trust. These problems which are embedded in digital divide in developing countries (11) must be tackled with the deployment of technology, which provides access as well as training and engage users to participate in the application. Moreover, as greater connections between users, empowerment of citizens to participate in e-procurement and collaboration to promote the culture of transparency will enhance the social benefits of transparency (11). Regarding the heterogeneous actors involved in this project, we can infer that global network possess political interest and power over the local network. Even though governmental organizations are encouraged to be independent in the application management and implementation, problem of digital divide, and e-literacy increase the networks interdependence. These problematical situation shows lack of collaboration and restraint between actors in the global network, mostly the willingness of related parties to fully support the program.

Despite that the role of human actors in the manual procurement process were delegated into technology, the combination of organizational aspects and technical aspects remains crucial in the achievement of the goals. We can see that the application may not have hitches, but the implementation will struggle without the support of management and users' competency, leadership and trust. Lack of collaboration between human and non-human actors will create barriers to reach the desired outcome of technology delegation. What must be focused on are the social factors of user's acceptance, user's literacy, sufficient infrastructure, trust and leadership, which are the foundations of the application's success. This confirms previous research by (11, 33, 34) that social determinants of users' acceptance, culture, education and trust determine the success of technology implementation.

5. Conclusion

This research identified the actors and networks involved in Indonesian e-procurement which assisted to study the role and barriers of information technology to enhance information transparency and accountability. The data was obtained from six interviews with e-procurement application developers and users and also supported by documentary data. Briefly explained, human and policy intervention, and different perceptions on the role of technology between developers and users creates obstacles in the application development and implementation; barriers of information transparency and accountability. In the translation process on e-procurement application, the lack of leadership from governmental organizations, reluctance to implementation, e-literacy of users, lack of infrastructures and lack of public trust create obstacles in the alignment of the actors' interest and enrolment into the project to implement the application. The role of information technology specifically the e-audit and document security application becomes crucial in preventing fraud, and maintaining the independence of procurement committees, also enhance transparency and accountability. The users and developers were overall satisfied with the e-procurement application and emphasized more in the obstacles of governmental organization leadership, users' e-literacy, lack of public trust, lack of interoperability of data, and lack of infrastructures. Due to the limitation of this research, there are some suggestions for further works. First, further research should also explore other IT reforms in the Indonesia ICT Flagship Program such as e-education, and e-budgeting. Moreover, a wider scope of research, which involves other actors such as donors, experts, academicians, and suppliers, will present richer findings to compare and contrast different insight.

References

1. World Bank. (2001). Indonesia country procurement assessment report: Reforming the public procurement system. Retrieved March 21, 2011, from http://siteresources.worldbank.org/INTINDONESIA/Resources/Procurement_Assessment.pdf.
2. Heeks, R. (2003). *Reinventing government in the information age: International practice in IT-enabled public sector reform*. New York: Routledge.
3. Tolbert, C., and Mossberger, K. (2006). The effects of e-government on trust and confidence in government. *Public Administration Review*, May-June.
4. Wahid, F. (2004) Lessons from e-government initiatives in Indonesia. *Media Informatika*, **2(2)**, 13-21.
5. Mirchadani, D., Johnson Jr, J., & Joshi, K. (2008). Perspectives of citizens towards e-government in Thailand and Indonesia: A multigroup analysis. *Inf Sys Front*, **10**, 483-497.
6. Power, M. (1997). *The audit society: Rituals of verification*. New York, Oxford university press.
7. OECD. (2007). Snapshot assessment of Indonesia's public procurement system. Retrieved March 21, 2011 from http://www.oecd.org/dataoecd/27/47/39254688.pdf
8. Ciborra, C. (2005). Interpreting e-government and development: Efficiency, transparency or governance at a distance? *Inf Tech & People*, **18(3)**, 260-279.
9. Allen, A., Juillet, L., Paquet, G. and Roy, J. (2001). E-governance and government online in Canada: Partnerships, People and Prospects, *Gov Inf Quarterly,* **18**, 93-104.
10. Institution of Government Procurement Policy (IGPP). (2010) Profil LKPP. Retreived May 26 2011 from http://www.lkpp.go.id/v2/profil_lkpp.pdf.
11. Bertot, J., Jaeger, P., & Grimes, J. (2010). Using ICTs to create a culture of transparency: E-government and social media as openness and anti-corruption tools for societies. *Gov Inf Quarterly*, **27**, 264-271.
12. Shim, D., and Eom, T. (2008). E-government and anti-corruption: Empirical analysis of international data. *Int Journal of Public Adm*, **31**, 298-316.
13. Kumar, A. (2003). E-government and efficiency, accountability and transparency. *EJISDZ*, **12**, 2, 1-5.
14. Bhatnagar, S. (2003). E-government and access to information. Global Corruption Report. Retreived June 6 2011 from http://unpan1.un.org/intradoc/groups/public/documents/apcity/unpan008435.pdf.
15. Welsch, E., and Hinnant, C. (2002). Internet use, transparency, and interactivity effects on trust in government. *Proceedings of the 36th Hawaii International Conference on System Sciences (HICSS'03).*
16. Welsch, E., Hinnant, C., and Moon, M. (2005). Linking citizen satisfaction with e-government with trust in e-government. *Journal of Public Adm Research and Theory,* **15(1)**, 37-58.
17. Elbanna, B. (2003). The implications of the local configuration of a standard e-procurement system on the organisation power circuits. *Presented in the 16th Bled e-Commerce Conference e-Transformation, Bled, Slovenis, June 9-11.*
18. Latour, B. (1987) *Science in Action*. Cambridge: Harvard University Press.

19. Stanforth, C. (2007). Using actor-network theory to analyze e-government implementation in developing countries. *The Massachusetts Institute of Technology Information Technologies and International Development,* **3(3)**, 35-60.
20. Callon, M. (1986). The sociology of an actor–network: the case of the electric vehicle. In: M, Callon., J, Law., & A. Rip. (Eds.), *Mapping the Dynamics of Science and Technology, Sociology of Science in the Real World* (pp. 19-34). London: The Macmillan Press.
21. Law, J. (1992). Notes on a theory of actor–network: ordering, strategy and heterogeneity. *Sys Practice,* **5(4)**, 379–393.
22. Bijker, W. E., and J. Law, eds. (1992). *Shaping technology/Building society: Studies in sociotechnical change.* Cambridge, MA, MIT Press.
23. Law, J. (1999) After ANT: Complexity, Naming and Topology. In J, Law., & J, Hassard. *Actor Network Theory and After* (pp. 1–14). Oxford: Blackwell.
24. Bloomfield, B.P. (1995). Power, machines and social relations: Delegating to information technology in the National Health Service. *Organization,* **2**, (3-4), 489-518.
25. Akrich, M. (1992). The De-Scription of technical Objects, in W. Bijker and J. Law (eds) *Shaping technology/Building Society*, 205-224, Cambridge, MA, MIT Press.
26. Latour, B. (1992). Where Are the Missing Massess?, in W. Bijker and J. Law (eds) *Shaping Technology/Building Society*, 225-258. Cambridge, MA, MIT Press.
27. Latour, B. (1991). Technology is Society Made Durable, in J. Law (ed.) *A Sociology of Monsters: Essay on Power, Technology and Domination*, 103-131.
28. Eisendhardt, K.M. (1989). Building theories from case study research. *The Academy of Management Review,* **14(4)**, 532-550.
29. Miles, M.B., & Huberman, M. (1994). *Qualitative Data Analysis: An Expanded Sourcebook (2nd ed).* Beverly Hills, CA: Sage Publishing.
30. Yin, R. (1994) *Case study research: Design and methods* (2nd ed.). Beverly Hills, CA, Sage Publishing.
31. Yulianto, A. (2010) *Konsep Fungsi dan Implementasi Portal Pengadaan Nasional.* Retreived August 6 2011 from http://www.lkpp.go.id/v2/files/content/file/29112010161242Andik%20Yulianto-Konsep%20Portal%20Pengadaan%20Nasional-V1-3.pdf.
32. Arifiyadi, T. (2010) Bab III Konsep pengadaan secara elektronik (e-procurement) instansi pemerintah. Retreived August 6 2011 from http://www.lontar.ui.ac.id/login.jsp?requester=file?file=digital/135548.Analisis%20hukum.pdf. [Accessed August 6 2011].
33. Jaeger, P. T., & Matteson, M. (2009). E-government and technology acceptance: The implementation of Section 508 guidelines for e-government websites. *Electronic Journal of E-Government*, **7(1)**, 87–98.
34. Heeks, R. (2005) E-government as a carrier of context. *Journal of Public Policy*, **25**, 51–74.

Exploratory Study Based on Stakeholder Theory in the Development of Accounting Information Systems in the Catholic Church: A Case Study in the Archdiocese of Semarang, Indonesia

Fransiscus Asisi Joko Siswanto[1], *Francisca Reni Retno* Anggraini[1], and *Bernardinus Sri* Widodo[2]

[1]Accounting Department, Sanata Dharma University, Mrican Tromol Pos 29, Yogyakarta, Indonesia
[2]Sanata Dharma Mecatronics, Paingan Maguwoharjo Sleman, Yogyakarta, Indonesia

Abstract. This study aims to find a strategy in the development of computer-based accounting information system in the church. With exploratory study on the theory of stakeholders, this study identifies the needs of financial information for the purposes of making a decision required by the parish priest, the parish treasurer, and a team of economists at the archdiocese of Semarang (AS). This research was conducted by using qualitative and quantitative approach. Qualitative method is conducted by applying a focus group discussion with economist team in AS (the users who have major influence in the development of the system). In addition to that, quantitative method is also applied to the parish treasurer (the users who have great interest in the system development). The results showed that the parish treasurer has high perceived usefulness, perceived ease of use, perceived of relevance, and the self-efficacy toward the accounting information system (AIS) for the parish. This study provides an answer on the benefits of a bottom-up strategy based on the stakeholder analysis in the development of AIS in the area of the Catholic Church AS.

1 Introduction

Research on the usefulness of accounting information systems at religious organizations is still an interesting issue. [1] says that many issues arise in the implementation of computer-based information system on a voluntary organization. Such issues are cultures, small budgets, users who work voluntarily, and the complexity of the organization purposes, etc. Therefore, a specific strategy is needed in the development of information systems within the organization.

The Catholic Church has a specific organizational structure. It is the sacred papal throne at the top, the dioceses, and the parishes at the lowest. The lower institutions will be responsible to the upper institutions. In 2008, the Archdiocese of Semarang (AS) has made accounting information system for the parish as outlined in the Technical Guidelines of

Parish Finance and Accounting. It was designed with the aim to materialize the credibility of the Church in terms of finance and accounting that are expected to enhance the visionary and credibility of the ecclesiastical property management. After being applied for almost seven years, some obstacles have been encountered. There have been discrepancies in the practice between the guidelines that are over-complex and the software of General Ledger (GL) which is not easily applied. This resulted in the expected goals which have not been fully achieved. This condition is supported by the research conducted by [2] who found that there is high user resistance. The results of their research also showed that the resistance is influenced by the low performance which is far from expectancy.

Based on the preliminary observations, the development of parish accounting information systems is applied by top down policy so that it may cause the inability of the system to be implemented properly. The weakness of the top down strategy is the lack of ability to accommodate the needs and abilities of the user so that the user considers the system difficult to run. This might be the cause of the low performance which is far from expectancy. Based on this condition, the research was conducted with the exploratory study based on the stakeholder theory to discover an effective strategy in both the preparation and implementation of the computer-based accounting information systems for the church. The exploratory study identifies the needs of financial information for the purpose of making a decision required by the parish priest, the parish treasurer, and the team of economists at the archdiocese of Semarang (AS).

This study aims to find the right strategy in the development of accounting information system (AIS) of the parish. In this case the study is more focused on the lowest organizational unit of the church, the parish. The research question in this study is whether the AIS which is developed from the stakeholder analysis can increase the perceived usefulness, perceived ease of use, perceived of relevance, and the user's self-efficacy.

This research is based on the stakeholder theory to identify and explore the needs of the users of accounting information system of the parish. The theory of stakeholders is still rarely used in the research on non-profit organizations [3]. The main user of accounting information is the parish priest, the parish treasurer, and the bishop of the diocese (represented by the economist team). The parish priest and the parish treasurer require accounting information to take a decision on the level of the parish, to analyze the performance of the parish and the units under the parish, while the bishop uses it to determine the subsidies provided to the parishes and to examine the financial performance of it.

The results of this study show that the parish treasurer of AIS for parish recognizes the high perceived usefulness, the perceived ease of use, the perceived of relevance, and the user's self-efficacy. This research gives contribution to AIS development theory that stakeholder analysis is beneficial in the development of AIS, especially in religious organization. This research also gives contribution to Archdiocese of Semarang in developing AIS of the parish.

2 Literature Review

2.1. Stakeholder Theory

As presented by [3], the stakeholder theory is first proposed by Freeman (1984) and the theoretical development has been done by Donaldson and Preston (1995) and Mitchell et al. (1997). Donaldson and Preston (1995), in [3], states that the theory of stakeholders can be explained with three views: (1) descriptive (stakeholders related to the organization operations), (2) the normative (related to the goal of developing the moral guidelines for determining the managerial response to the stakeholder), (3) instrumental (exploring the connection between the stakeholder management and the objectives of the company). Cooper (2014) states that the stakeholders can be identified based on the power to influence, the legitimacy of the relationship, and the urgency of the claims of the stakeholders. Nuseibeh and Easterbrook (2000) in [4] defines stakeholders as individuals or organizations that benefit or suffer losses due to the success or failure of a system. In designing a system, the stakeholders are managers, designers, and users of the system. Further [4] states that because the stakeholders are all people who are affected by or who receive the impact on the system designing, they need to be considered in the design. Their perspectives are sometimes contradictory therefore the system designers should accommodate their different opinions. Therefore [4] suggests that the system designer shall pay greater attention to the users of the system implemented earlier, since they have more experience in using the earlier system. Their opinions can be used as a basis to meet their needs. However, considering the users of the system only would not be enough to achieve success in the design and implementation of the system. The designers should also identify the needs of other stakeholders. Freeman (1984) in [5] states that in the non-profit organization recognizing the stakeholders is an important factor in the face of a changing environment and involving stakeholders in the management is the most effective strategies for achieving organization success.

[3] states that the recent stakeholder theory is assumed to be able to explain the various conflicts of interest in non-profit organizations. Benjamin (2008), [3], states that the stakeholders in non-profit organizations are donators, the party receiving the benefit (beneficiaries), the regulator, and the organization partners. Accountability in a non-profit organization includes upward (corresponding to funders or donators), lateral (related to staff members) or / and downward (to the client). Speckbacher (2008) in [3] states that non-profit organization's stakeholders can be divided into internal stakeholders (composing of managers and volunteers) and external (composing of funders). Stakeholders in the Catholic churches include the economist team in archdiocese (middle managers), the parish priest (lower level managers), the system user (parish treasurer) and the donators (parishioners and other funders). Economist team in archdiocese establishes policies related to finance which will be implemented in all the parishes in the Archdiocese. The parish priest is making policies at the parish level. Parish treasurer manages and reports the financial administration to the parish priest and economist team in archdiocese. Parishioners and other donators need financial reports to assess the accountability of the church.

2.2. Stakeholder Analysis

The term of stakeholder analysis is introduced by Freeman (1984) in his book Strategic Management in [4]. Nuseibeh and Easterbrook (2000), in [4], state that the requirement of system software starts with basic assumption that there is a wide range of stakeholders who have different desires and may be conflicting. A solution that meets the objectives of all stakeholders and also eliminates (reduces) the conflicts is found by systematic exploration.

The steps in stakeholder analysis are: (1) Identifying the organization stakeholders and the ability to choose the key stakeholders of the organization is the most important thing because key stakeholders are people or organizations that are concerned with the system designed, (2) Making a priority of stakeholders, stakeholders can be grouped into four groups, namely: (a) high influence, high interest, (b) low influence, high interest, (c) high influence, low interest, (d) low influence, low interest, (3) Understanding the perspective of the stakeholders, the best way to understand the stakeholders is to conduct semi-structured interview by asking open-ended questions to start a discussion, (4) Accommodating the stakeholders' perspectives into the design of the system, it is the most difficult stage because the system designer must accommodate all the interests that are sometimes contradictory.

2.3. *Technology Acceptance Model* (TAM)

TAM is a model which is introduced by [6]. TAM argues that individual acceptance toward information technology (IT) is influenced by perceived usefulness (PU) and perceived ease of use (PEU). [7] found that user's self-efficacy (SE) influences PEU and perceived of relevance (PR) influence PU. It means PU, PEU, PR, and SE influences the individual acceptance toward IT. PU is the extent to which a person believes that using a technology will enhance her or his performance. PEU is the extent to which a person believes that using a technology will be free of effort. PR is the extent to which a person believes that the capability of a system to enhance the individual's job performance. Then, SE is the belief that the IT has capability to perform a particular behavior.

3 Research Method

Research design of this study is sequential exploratory so it is conducted using mixed method that is qualitative and qualitative approach. Qualitative method is done by focus group discussion (FGD) with economist team in AS (users has major influence in the development of the system) in order to develop AIS of the parish. Furthermore, quantitative method is applied by conducting the survey to the parish treasurer in order to examine their PU, PEU, PR, and SE.

This research was conducted with a case study on AS, which currently has a need to establish a new AIS that replaces the old one. Sequential exploratory research is conducted for problem solving. The research is started by setting the area in which the problem is found, then literature review is used to guide researcher collecting the data and data analysis. The data and data analysis are used to state the hypothesis.

4 Analysis

4.1. Preliminary Study

The results of the research conducted by [2] showed the users' reluctance of using the accounting software for parish based GL, because it is difficult to use. It may cause the low expectancy to GL. Based on FGD with a team of economists AS, we obtained information that AS needs parish's financial information to decide the parish which should get subsidy and how much it is.

Furthermore, [2] states there is low expectancy on the GL software for the parish treasury. It is caused by the nature of their work which is voluntarily and the GL software which is not user friendly. They just want their work can be completed easily and in a simple way. We suspect that the low expectancy on the GL software may not be caused only by the nature of their work which is voluntarily but also that they have little knowledge and computer skills.

4.2 Stakeholder Analysis

4.2.1. Identifying organizational stakeholders

Based on the results of early studies, researchers suspect that the source of the failure of the application of the accounting system for the parishes is due to the lack of ability of the parish treasurer to use computers therefore they want software that is easy to use, yet can produce financial reports quickly. The stakeholders of AIS of the parish are a team of economists of diocese, parish priest, parish treasurer, parishioners, and other donators.

4.2.2. Making a priority stakeholders

After identifying the stakeholders, the researchers then grouped the stakeholders as follows. Stakeholders with high influence and high interest is a team of economists at the diocese, stakeholders having low influence and high interest is the parish treasurer, stakeholders having high influence but low interest is parish priest and stakeholders who have low influence and low interest are parishioners and other donators.

4.2.3. Understanding the perspective of stakeholders

Each parish treasurer was given a questionnaire obtaining about their experience with the financial information system program or the program which is still in use, and their expectations of the new system program. Based on the results the majority of respondents did not know about the General Ledger program that had been used by the parish. The majority of the respondents expect that there is a new program used to process financial transaction data in their parishes. A team of the economist diocese wants an easy to use accounting software so that the financial information for the diocese will be accepted quickly, and the decisions related to finances can be done quickly. Parish priest also wants accounting software to be easily used, that he can take decision quickly. Parishioners and other donators do not require reports quickly but they require transparent and accountable report.

4.2.4. Accommodating the stakeholders' perspective into the design of the system

Intensive discussion is conducted to a team of the economist diocese that has the authority to make policies relating to the financial administration of the parish. Then, from the discussions, it is discovered that the financial transactions in a church or parish are frequently repetitive transactions. So the accounting system must make special accounting journal to record the transactions, while the transactions which do not occur repeatedly will be processed through the general journal.

4.3 Survey about user acceptance

After the parish treasurers are given the training, they are requested to fill the questionnaire. It is conducted to examine the level of user acceptance toward IAS of the parish. The level of user acceptance is determined by PU, PEU, PR, and SE.

5 Conclusion

The Catholic Church organization is specific that it is structured from the highest to the lowest, the sacred papal throne, dioceses, and parishes. The Parishes are administered voluntarily by the people who live in it. To achieve its vision and mission, the Catholic Church aspires to provide financial information that is transparent and accountable. In 2008, the Archdiocese of Semarang (AS) has implemented an accounting information system for the parish as outlined in the Parish Technical Guidelines for Finance and Accounting. It is prepared with the aim to materialize the credibility of the Church in terms of finance and accounting Parish. It is expected to enhance/accelerate the visionary and credibility of the ecclesiastical property management.

We identify that the stakeholder of the parish accounting information system is a team of economists at AS, the parish priest, the parish treasurer, and parishioners and other donator. The stakeholders with high influence and high interest is a team of economists at the diocese, while the stakeholders who have low influence and high interest is the parish treasurer, and the stakeholders who have high influence but low interest is the parish priest and the stakeholders who have low influence and low interest are parishioners and donators. The parish treasurers are the person who is the most accountable to prepare and provide parish financial statements. But, they work voluntarily by using/sharing their spare time along with their main job. Therefore, they need tools in order to present financial reports quickly and accurately.

Based on the survey, we found that PU, PEU, PR, and SE of the users toward the IAS are high. It means that the stakeholder analysis can be used as the strategy in developing of AIS in the parish. Although the study found a new strategy of development of accounting information systems, but this study has limitations, for example, the results of this study are highly specialized and only suitable for the organization of the Catholic Church. In addition, the research is more qualitative than quantitative therefore it is more in the form of subjective interpretations of the object studied.

References

1. G.G. Morgan. J. of Strat. Inf. Syst., 4,3 (1995)
2. A.D. Kuntara, I. I.H. Suryandari, dan A.F. Prasetyawan. *Working Paper* (2015)
3. K.R. Cooper. *Dissertation* (2014)
4. J. Bauntelle. http://boxesandarrows.com/understanding-organizational- stakeholders-for-design-success/ (2004)
5. M. Shimaoka. The 2nd International conf. on Gov. Perf. Mgt. and Leadership, *Portland State University* (2011)
6. Davis, F.D., Bagozzi, R.P., and Warshaw, P.R. Mgt. Sc., 35,8 (1989)
7. Venkatesh, V., and Davis, FD. Mgt. Sc.,46,2 (2000)

Social proof in social media shopping: An experimental design research

Yurita Yakimin Abdul Talib[1,*] and *Rafeah* Mat Saat[1]

[1]Tunku Puteri Intan Safinaz School of Accountancy, Universiti Utara Malaysia, Malaysia

Abstract. The practices of social proof techniques to attract consumers to shop on social media have proliferated over time and been used extensively. The reviewed literature uncovers that social community recommendation, customers' ratings and reviews, celebrity's endorser and numbers of likes, affect consumers' purchasing decisions. However, the effect of different types of social proof techniques on purchasing intention is unknown. This study empirically compares the effect of number of followers, celebrity endorser and social community recommendation on consumers' purchasing intention. An experiment has been conducted and the results reveal that the consumers' purchasing intention differs between groups. Further analysis discovers that the impact on consumers' purchasing intention is different between high number of followers and low number of followers, and between having social community recommendation and not having social community recommendation. Though, the impact of these two techniques is equal wherein no technique is superior to other. In order to gain purchasing engagement and boost online sales, online businesses on social media are encouraged to use the power of social proof technique, either by increasing the number of followers or providing more social community recommendations.

1. INTRODUCTION

The evolution of the Internet and its related technology has created new forms of interactions for acquisition of goods and services through social networking platforms, such as Facebook, Instagram, Twitter, WhatsApp, and Telegram. Online shopping has migrated from the web sites shopping experience to social media shopping experience. An online shopping engagement via social media is created by means of customers' reviews, social community recommendations, referrals and customers' ratings. The rise of social media allows users to share and access information within the social community [1]. Online buyers' reliance on social community shared information, a kind of social proof techniques, is generally useful and effective at increasing consumer trust [2]. Trust formation from trusted sources is the key to a successful online purchasing engagement [3,4].

Social proof techniques are generally effective at increasing consumer trust on online purchase. Amblee and Bui [2] discovered that electronic word-of-month (eWOM) among social community, such as online ratings and reviews, is a useful and effective signal of product reputation for a shopper yet to have a purchasing experience. Social interaction of consumers in the social media leads to increased trust to the online sellers and intention to buy [5]. Social media enables sellers to interact directly with their potential buyers and help them to understand their customers' social habits in order to engage conversion with them, which ultimately will help to increase sales. Social media also allows customers to review and evaluate their purchased product, and thus provides information to other people in their social community [6]. As a number of positive reviews and recommendations grow, consumers have more favorable attitude toward a product [7]. Hence, firms are encouraged to use various social proof techniques offered by social media to drive consumers' purchase [8].

Although prior studies have uncovered the benefits of social proof techniques, there is no yet study on comparing the techniques and determining the most effective technique. This study seeks to extend prior studies by exploring the

* Corresponding author: yurita@uum.edu.my

following research question: Are different types of social proof techniques are equally effective in influence consumers' purchase intention in a social commerce site? The answer to the question will have practical implications for online business in general and social commerce business specifically.

2. LITERATURE REVIEW

2.1 Social Media

Social media is defined as a group of internet-based applications that build on Web 2.0 technology that allows for the 0creation and exchange of User Generated Content [9]. Similarly, Parr [10] defines social media as a use of Internet or Web-based technologies to share and discuss information with users across the globe. The Web 2.0 technology enables users from various platforms to create, discuss and share content within the community to create social networks. Hence, the fundamental elements of social media are Web 2.0, content (information) and community. Other elements of social media are openness, accessibility and speed of information. Openness indicates that anyone can join almost all social media platforms for free [11]. The contents on social media platforms are shared instantaneously and accessed by anyone in the social community anywhere and anytime [12].

According to Mayfield [11], social media can be classified into seven types, such as social networks sites (SNSs), forums, blogs, podcasts, wikis, content communities and microbloggings. SNSs are websites that allows people to connect virtually with their real world friends and families to form their social community [11]. Facebook and Instagram are examples of SNSs. Blogs are online journals (normally personal) that contain text, images, audio and video that are shared in the Internet and owned by bloggers. Blogs allow the bloggers and readers to communicate online and form a community [11]. Another type of social media is Wikis, a website that allows users to add, modify and delete contents. Wikipedia, an online encyclopedia, is the most popular Wiki website. The Internet forum starts with a group of people share and discusses a same topic or interest. The discussion on the forum, known as thread, can be triggered by any members of the forum community [11]. Unlike blogs, there is no ownership for the forums.

The emergence of social media has gained tremendous popularity and is widely accepted as a new form of interactions. Across the world, there were 21.6 billion users of the social media in 2015, and it is expected to grow [13]. Accordingly, social networking was the second most popular activities among Malaysian internet users [14].

2.2 The Evolution of Social Commerce

Online shopping or e-commerce is a common trend in society nowadays. Online shopping is described as purchasing products or services via the Internet. In 2014, one-quarter of 300,000 global respondents were users of e-commerce and more than half (55%) are willing to use it in the future [15]. Malaysians are ranked among the most passionate online shoppers. By 2015, Malaysia's online shopping market is expected to increase to RM5 billion (US$1.52 billion) [15]. Variety of products could be purchased online easily including tickets, clothing items, home accessories, electronic gadgets, foods, and many more. Shopping activities over the internet has been part of the lifestyle due to the advantages offered by online services. For instance, individuals are attracted to shop online due to wider products selection, cost and time saving, as well as convenient shopping experience [16].

One of the major developments in the e-commerce industry recently is social media shopping or social commerce that is expected to evolve as a widely acceptable practice. "Social commerce is no longer media hype or a business fad. It becomes an established practice, although the specifics of this practice continue to emerge and evolve" [17](Wang and Zhang, 2012, p. 121). Social media shopping has received a tremendous role in changing the way people shop nowadays. Shopping via social media platforms such as Facebook and Instagram has taken over in-store shopping experience. The rise of social media shopping is contributed by the increased of social media users. Across the world, there were 21.6 billion users of the social media in 2015, and it is expected to grow [13]. Accordingly, social networking was the second most popular activities among Malaysian internet users [14].

As a subset of e-commerce, social commerce involves commercial activities, such as marketing, selling and buying products and services, using social media technologies to support online interactions and user contributions [18]. The interactions among the social media users may result in discovering products, aggregating and sharing product information, and collaboratively making shopping decisions [19]. In essence, the two important attributes of social commerce are commercial activities and social networks. The capability of social networks that allows for users' content participation, such as reviews and recommendations, for example, represents the major advantages of social commerce [18]. These social proof techniques increased consumers trust towards the online sellers and their products, and hence, increased online shopping engagement [2,3,4].

2.3 Social Proof

Social proof or informational social cue is a psychological phenomenon where people assume the actions of others are the correct behaviors to follow .The actions of others around us are important guides in a situation when decision makers are unclear about the value of the decision to be made [20]. Unconsciously, the collective information of the society or social community influences and contributes to our decisions. Regardless of how independent were think we are, our decisions to do or not to do certain things are conditioned by others around us. The force that influences our day to day decisions is known as social proof.

Social proof is much noticeable in social network platforms such as Facebook, Instagram and Twitter. When making a buying decision, many people need proof (authority) that a seller and its product are legitimate. In online shopping via networks, the proof can be obtained from reviews and recommendations by others. Social networks connect online shoppers directly with others to form a social community. The immediacy provided by social networks allows online shoppers to obtain and share products-related and sellers-related information within their social community. A market survey by Nielsen [15] reported that 92% of respondents worldwide trust recommendations by friends and families, and 70% of them ranked online consumer reviews as the second most trusted source of information. Amblee and Bui [2] discovered that electronic word-of-month (eWOM) among social community, such as online ratings and reviews, is a useful and effective signal of product reputation for a shopper yet to have a purchasing experience. The trust developed among the social community might possibly impact one's purchasing decision. Social media characteristics that facilitate the social interaction of consumers lead to increased trust and intention to buy [21,22]. Online businesses should consider using eWOM technologies as they are effective and primary source of social proof [2].

Nowadays, it's a trend for online sellers in social networks, such as Instagram, to promote their Instagram Shop (InstaShop) through reviews by famous celebrities, bloggers, or other reputable sellers. Famous celebrities' reviews were used by sellers to attract online shoppers to purchase their product [23]. Celebrity's endorsement is a signal to the consumers the seller and its product is trusted and has good reputation. Dzisah and Ocloo [24] found that social proof by means of celebrity's endorsement does not only influence on marketing promotions, but also on consumers' purchase or re-buying behaviors. A quasi-experiment research that examined Pepsi Cola commercials with and without celebrity endorsement discovered that celebrity endorsed advert has a significant influence on students' purchase intention [25]. Jain [26] found that celebrity endorsement in India helps in brand promotion, and thus, impact on sales. Therefore, businesses are recommended to grab the opportunity offers by celebrity endorsed adverts in order to expand their operations and promote their product.

In addition, social media platforms provide a 'liking' function. For example, Facebook 'like' button and Instagram 'follower' button indicate numbers of people following the business or sellers' online page. The amount of followers, fans, views and likes that offered by social networks are among important and effective tools to affect others' perception about the sellers. It is believe that the higher the number of 'like' or 'follower' of sellers, the higher the trust of shoppers on the legitimacy of particular sellers. A user on Twitter with a million followers is perceived as more trustworthy and reputable than a similar user with a thousand followers, resulting in faster growth of followers and higher engagement [27]. If a person 'like' social media seller page, they are more likely to buy the product, recommend the product to others and share the product content. For that reason, a ghost followers industry exists for the sole purpose of increasing social proof on social media [27]. However, Sitaram Asur et al. [28] empirically discovered that number of followers do not contribute to trend creation. Similarly, Thackeray et al. [29] found that 'liking' the page does not result in purchasing the product. They further explained that the insignificant relationship may be due to mismatch between the shared content and users' preference.

Based on the preceding literature discussion, we propose the following hypotheses for testing the effect on social proof techniques on buyers' intention to buy.

H_1 –Consumers' purchase intention will be higher when number of followers is high rather than no followers

H_2 –Consumers' purchase intention will be higher when there is celebrity endorsement rather than no celebrity endorsement

H_3 –Consumers' purchase intention will be higher when there is social community recommendation rather than no social community recommendation

H_4 – There is significant difference of the impact of social proof techniques on consumers' purchase intention

DOI: 10.1051/shsconf/20173402005

3. METHODOLOGY

3.1 Design and Measure

Experimental designs are highly functional, when seeking the associations between cause and effect in consumer research [30]. In this current study, an experimental design consisting of four between-subject conditions was applied. The manipulated independent variable is social proof (number of followers, endorsement by celebrities, recommendation by social community, and no information of social proof). The intention to purchase on social media is a measured variable. Treatment 1 showed the highest number of followers. Celebrity endorses the smart-phone in the treatment 2 and treatment 3 showed the condition where friends or recognized community tagging others in the Instagram. Treatment 4 is a control treatment which contains photos of smart-phone without any number of followers and endorsement from friends or celebrity. Participants were randomly assigned to one of the four conditions.

The fictitious company was depicted in each scenario under the name of "Gadjet Budget Enterprise" that sells cell phones on-line via Instagram. A fictitious company is created to control for any predisposition participants may have related to real companies and to minimize subject bias [31]. The selection of smart phones is explained by the fact that consumers especially university students are familiar with smartphones. The respondents' perceived on purchase intention were assessed with three questions based on the scale adapted from Chiu et al. [32]. A five-point Likert scale (from strongly disagree to strongly agree) was used for the questions. The Cronbach alpha of these three items was .88.

3.2 Research procedure and respondents

The subjects in this research are undergraduate students enrolled in information systems and accounting courses in one public University in Malaysia. Convenience sampling of university students is chosen to represent gender and the group of consumers of smart phone. The researcher contacted the instructors of the courses and asked their permission for students' participation in this study. Upon obtaining the consent, the respondents are randomly assigned to one of the four conditions of the experiment. First of all, researchers gave a brief explanation about the study. Respondents are given a sufficient time to read the scenario. In the scenario, the respondents have to imagine that they need to buy a smartphone for school work and leisure purposes. Their parents are giving them enough money for that purpose. The snapshot of Instagram seller and its product was provided to the respondents.

After reading the scenarios and the snapshot of the Instagram seller, respondents were asked to respond to manipulation checks regarding the scenario. Only those that answered the manipulation question right that count in this study. Then, participants were required to answer questions designed to assess their likelihood to buy from the Instagram seller. Finally, the respondents answered demographic questions, such as age and gender. Each scenario had around 26 to 33 number of respondents. Total of respondents was 122, which 19.7 percent male and 80.3 percent female. In terms of age groups, 89 percent of the respondents were between the ages of 18 and 22, 11 per cent in the 23-25 age groups.

4. RESULTS

A series of analysis of variance (ANOVA) were conducted in order to explore the impact of Instagram's social proof functions on purchase intentions (Table 1). Participants were divided into four groups according to the social proof functions in Instagram (Group 1: number of followers; Group 2: celebrity endorser; Group 3: social community recommendations and Group 4: control group). ANOVA tests revealed that there is a statistically significant difference in the mean of the Instagram social proof functions between the four treatment groups towards purchase intention ($F= 3.301$, $p = .05$). The results indicate that participants' intention to buy from Instagram seller is higher when there is recommendation from their social community, than where there is high number of followers or celebrity endorser or no social proof at all.

Table 1: ANOVA test

Variables	Mean	F	Sig
Purchase Intention		3.301	.023
Number of followers	7.687		
Celebrity endorser	7.344		
Community recommendation	9.461		
Control group	7.314		

A Tukey post-hoc test reveals that the number of follower condition was significantly different from no number of follower condition. The number of follower condition was not significantly different from celebrity endorser and social community recommendation. The celebrity endorser condition was not significantly different from any celebrity endorser condition and community recommendation. The community recommendation condition was significantly different from no community recommendation. In addition, the results indicate that there is no significant difference among these three social proof conditions towards consumers' purchase intention.

Table 2: Post-hoc test

Group Experiment	Group Experiment	Sig
Number of followers	Celebrity endorser	.954
	Social community recommendations	1.000
	Control group	.029
Celebrity endorser	Number of followers	.954
	Social community recommendations	.969
	Control group	.108
Community recommendation	Celebrity endorser	.969
	Number of followers	1.000
	Control group	.043

5. DISCUSSION AND CONCLUSION

Social proof is prominent on social networks such as Instagram and Facebook; hence many online businesses have taken advantage by promoting and selling their products on social networks. The results of the experiment show that social proof techniques, specifically high number of followers and social community recommendations, does indeed play significant factors that impact on buyers' intention to buy from social media sellers. The results are consistent with prior studies that confirms eWOM (i.e., social community recommendation) and high number of followers are effective marketing tool to enhance product evaluation and hence, purchasing intention [2]. High number of followers is a signal to social media community that the particular sellers are legitimate and trusted although the supporters (i.e., followers) may be passively experienced. Similarly, positive recommendations help social community to learn about and have more favorable attitude towards unfamiliar products and sellers. Both techniques are equally effective in influencing consumers' purchasing intention, and hence, online sellers can choose either to increase the number of followers or to have more community recommendations.

On the other hand, having celebrity endorsement does not significantly different from not having celebrity endorsement. This indicates that celebrity endorsement is not relevant in the context of our study, unlike other studies (e.g., [24,25]). There are few possible reasons. First, the respondents were young people aged between 18 to 23 years old; who may not familiar with the chosen celebrities. Second, there is a possibility of mismatch between the product (i.e., mobile phone) and the celebrities' image. Therefore, it is suggested to online sellers to ensure the selected celebrities have the value and image that match with their brand and product.

Our study has some limitations. The use of students' data in one University only lacks of diversity and limits its generalizability. In addition, an experimental design research create artificial situation that do not always represent real-life situations. We shall replicate this study into real world social media shoppers to gauge holistic and realistic view of the influences of different social proof techniques on consumers' purchasing intention. Future research can also extend current study to investigate whether there is interaction effect among the social proof techniques.

References

1. J. Chen, H. Xu, & A.B. Whinston, Moderated online communities and quality of user-generated content, *Journal of Management Information Systems*, 28, 2, pp. 237-268 (2011)
2. N. Amblee, and T. Bui, Harnessing the influence of social proof in online shopping: The effect of electronic word of mouth on sales of digital microproducts, *International Journal of Electronic Commerce*, 16, 2, 91–113 (2012)
3. P. Beatty, I. Reay, S.Dick, & J. Miller, Consumer trust in e-commerce web sites, *ACM Computing Surveys*, 43(3), 1–46 (2011)
4. M. Zhou, & D. Tian, An integrated model of influential antecedents of online shopping initial trust: Empirical evidence in a low-trust environment, *Journal of International Consumer Marketing*, 22(2), 147–167(2010)
5. M. Hajli, A study of the impact of social media on consumers, *International Journal of Market Research*, 56(3), 387-404 (2014)

6. Y. Chen, S. Fay, Q.Wang, The Role of Marketing in Social Media: How Online Customer Reviews Evolve, *Journal of Interactive Marketing* 25(2), 85-94 (2011)
7. D.H. Park, & J. Lee, eWOM overload and its effect on consumer behavioral intention depending on consumer involvement, *Electronic Commerce Research and Applications*, 7, 4, 386–398 (2009)
8. N.A. Hashim, S. Mohd Nor & H. Janor, Riding the waves of social commerce: An empirical study of Malaysian entrepreneurs, *Malaysian Journal of Society and Space* 12 issue 2,83 – 94 (2016)
9. A.M. Kaplan & M. Haenlein, Users of the world, unite! The challenges and opportunities of social media, *Business Horizons* 53(1), 59-68 (2010)
10. B. Parr, (2008). Ben Parr's Entrepreneurial Musings. Retrieved from http://benparr.com/2008/08/itshttp://benparr.com/2008/08/its-time-we-defined-social-media-no-more-arguing-heres-the-definition
11. Mayfield (2008). What is social media, Retreived from http://www.icrossing.com/icrossing-what-is-social-media
12. V. Tapria & P. Kanwar, Understanding Social Media. Varinder Tapria & Priya Kanwar & Ventus Publishing ApS, (2012)
13. Statista (2015). Retrieved from http://www.statista.com/statistics/278414/number-of-worldwide-social-network-users/
14. Internet Users Survey 2014 (2014). Retrieved from http://www.skmm.gov.my/skmmgovmy/media/General/pdf/Internet-Users-Survey-2014.pdf
15. Nielsen. (2014). E-commerce: evolution or revolution in the fast-moving consumer goods world? Retrieved from http://ir.nielsen.com/files/doc_financials/Nielsen-Global-E-commerce-Report-August-2014.pdf.
16. J. H.Wu & S. C. Wang, What drives mobile commerce?: An empirical evaluation of the revised technology acceptance model. *Information & Management*, 42(5), 719-729, (2005)
17. C. Wang & P. Zhang, The evolution of social commerce: The people, management, technology, and information dimensions. *Communications of the Association for Information Systems*, 31(5), 1-23 (2012)
18. T. P. Liang & E. Turban, Introduction to the special issue social commerce: a research framework for social commerce. *International Journal of Electronic Commerce*, 16(2), 5-14 (2011)
19. J. Shen, Social comparison, social presence, and enjoyment in the acceptance of social shopping websites. Journal of Electronic Commerce Research, 13(3), 198-212 (2012)
20. H. Rao, H.R. Greve and G.F. Davis, Fool's gold: Social proof in the initiation and abandonment of coverage by Wall Street analysts. *Administrative Science Quarterly*, 46, 3 (2001), 502–526 (2001)
21. T. P. Liang, Y. T. Ho, Y. W. Li & E. Turban, What drives social commerce: The role of social support and relationship quality. International Journal of Electronic Commerce, 16(2), 69-90 (2011)
22. M. Hajli, A research framework for social commerce adoption. *Information Management & Computer Security*, 21(3), 144-154 (2013)
23. R. Rahim, R. Aravinthan & N.S. Hooi, Man dupes online shoppers. The Star, pp. 40 (2015, October 2)
24. W.E. Dzisah and C.E. Ocloo, Celebrity Endorsement and Consumer Buying Behaviour; Enhancing the Promotion Function of Marketing in the Central Business Area of Accra, Ghana. *European Journal of Business and Management*. Vol.5, No.25, 2013 (2013)
25. A. Apejoye, Influence of Celebrity Endorsement of Advertisement on Students' Purchase Intention. *J Mass Communicat Journalism* 3:152 (2013)
26. V. Jain, Celebrity Endorsement And Its Impact On Sales: A Research Analysis Carried Out In India. Global *Journal of Management and Business Research*. Volume 11 Issue 4 (2011)
27. Devumi (2016). Retrieved from https://devumi.com/social-proof-in-digital-success/
28. S. Asur, B. A. Huberman, G. Szabo, and C. Wang, Trends in social media - persistence and decay. In 5th International AAAI Conference on Weblogs and SocialMedia, (2012).
29. R. Thackeray, B.L Neiger, A.K. Smith & S.B. Van Wagenen, Adoption and use of social media among public health departments. *BMC Public Health* 12:242 (2012)
30. I. Assiouras, O. Ozgen & G. Skourtis, The impact of corporate social responsibility in food industry in product-harm crises. *British Food Journal*, *115*(1, SI), 108–123 (2013)
31. C.H. Cho, J. R. Phillips, A.M. Hageman & D.M. Patten, Media richness, user trust, and perceptions of corporate social responsibility: An experimental investigation of visual web site disclosures. *Accounting, Auditing & Accountability Journal*, *22*(6), 933–952 (2009)
32. C.M.Chiu, C.C. Chang, H.L Cheng & Y.H. Fang, Determinants of customer repurchase intention in online shopping. *Online Information Review*, Vol. 33 Iss 4 pp. 761 – 784 (2009)

SHS Web of Conferences **34**, 02006 (2017) DOI: 10.1051/shsconf/20173402006
FourA 2016

Strategic IS Planning Practices: A Case of Medium Manufacturing Company in Malaysia

Raja Haslinda Raja Mohd. Ali [*], *Rosli* Mohamad, *Yurita Yakimin* Abdul Talib, *Akilah* Abdullah

Tunku Puteri Intan Safinaz School of Accountancy (TISSA-UUM), Universiti Utara Malaysia

Abstract. Deployment of Information Technology/Information Systems (IT/IS) has been recognized as one of the potential means to supply information for timely business decision making. Despite growing importance of IT/IS in business, Small Medium Enterprises (SMEs) embrace constant challenges to deploy IS effectively, and thus, hardly to optimize the IT/IS strategic values. In such situation, IS planning is getting paramount to promote more effective IT/IS management of the firms. This study reports IS planning practices of a manufacturing-based SME. The face to face interview has been carried out to explore IS planning activities in a chosen firm. The qualitative analysis suggests that the selected company did undergo the four phases of strategic IS planning process namely, planning, analysing, selection, and implementation. The owner act as the champion of the strategic IS planning process. Further, the owner and the employees plays an important role as a decision maker. Specifically, the owner decides mainly on acquiring a new technology related to the firm's core business process whereas the users responsible on the selection of relevant application (s) to be used in their respective departments. Apart from that, the owner also plays a role as a researcher for the new technology acquired whilst the users provide an input on the applications to be acquired.

1 Background and Objective of Study

Greater reliance on Information Technology/Information Systems (IT/IS) on daily operations and strategic needs urges today's organisations to emphasise on the IS planning. Appropriate IS planning practice enables organisations to narrow the gap between IT/IS and the strategic objectives of the organisations. IS alignment then leads to a greater competitiveness and value derived from IS-related investment [1].

Increased importance of IT/IS in organisation has attracted considerable research to understand the impact of IT/IS on firm performance. Specifically, IS planning emerges as one of the primary focuses as long as IT/IS deployment is concerned. IS planning refers to a process of specifying an organization's selection of computer-based applications and technologies that correspond to the organisations strategic objectives [2].

[*] Corresponding author: rj.linda@uum.edu.my

In contrast to considerable focus on earlier studies of IS planning amongst large organisations, studies involving smaller firms (SMEs) are somewhat limited [3, 4]. Taking into consideration considerable contribution of SMEs in most nations' economy, how these firms embrace IT/IS deserve special attention. More importantly, the firms are facing somewhat unique challenges while embracing IT/IS in optimising IT/IS related values [5]. Among possible explanations are the lack of strategic orientation in their IT/IS deployment [6] and limited resources capabilities, particularly knowledge and skills to plan and manage IS-related projects [7]. Apart from that, management and administration of the firms are more likely on ad-hoc basis and informal in nature [8]. Thus, how the firms plan their IT/IS worth further investigation. In response, this study tries to understand IS planning practices in the context of SMEs with a specific focuses on the role of top management and system user.

This study adapted SISP success model [2] that has been extensively used for large organisation context to establish in-depth understanding of top management role and user involvement in IS planning practice of SMEs. The framework underpins most of the strategic IS planning research. The framework illustrates the strategic IS planning as input-process-output model. The input for strategic IS plan comprises of seven components that include environmental factors (both internal and external as well as planning resources), planning process, information plan, plan implementation, alignment (between internal environment and the organisations' strategic goals). Meanwhile IS strategic plan represents the output of the planning process.

2. Research Method

Researchers employed qualitative research strategy based on a single case study. A manufacturing firm located in Northern region of Malaysia was identified for this study. A screening process had taken place during the phone call to ensure the firms had employed certain types of IS/IT related applications. Data were obtained via semi-structured themed interview with the owner/manager of the respective firms. Based on the owner/manager consent, another series of interview was carried out involving the system user. An open-ended interview method allows respondents to express the key issues using their own terms [9]. This enables researchers to systematically accumulate detailed and richer data from the respondents [10].

Data were collected on four main thematic focus; (i) description of IS planning stages and activities that have been applied in a firm based on [11, 12] framework, and (ii) roles of top management and users in IS planning activities. Each interview sessions had taken place for about 1 ½ hours. Researchers presented the interview questions in English and digitally recorded the interview session upon obtaining interviewees' permission.

Authors followed [13] guidelines in analysing the qualitative results. The researcher initially listened to the recordings and had them transcribed. The data was divided into units and a meaningful code was assigned to data units one by one. The coding of all units (including the ones set aside in the "miscellaneous" stack) was repeated several times, with codes merged, renamed, and split where appropriate. The second researcher then analysed the data independently to strengthen the credibility of the outcomes. The next section reports data analysis results and discussion of the outcomes.

3. Findings and Discussion

Company A is a medium sized firm that was bought over from the in 1995. As time goes by, this factory has grown to become one of the successful medium firm with about 50

DOI: 10.1051/shsconf/20173402006

permanent employees. The main business process for Company A is paddy processing and it also acts as a rice supplier. This firm has five interrelated functional departments of purchasing, sales, drying, packaging, and paddy processing. Administrative tasks in the firm mainly involve purchasing and sales as well as production such as processing, drying, packaging, and storage. Some of the systems and machines used in Company A were already there when the company was bought over suggesting more than 20 years of age. However, there are few technologies that were bought within the last five years. Some of the business processes are computerised but some are not. The decision on whether it needs to be computerised depends on the criticality of the business process. For Company A, accounting and manufacturing are two most critical functions. For this company, 50 employees are easily to track and manage, so manual system is still reasonable to be used.

3.1 Stages in strategic IS planning

The owner started the process by identifying the issues that they are facing currently. The mentioned issue related to hiring young workers who are willingly to work in a rice factory. Most of their workers are elderly. He stated that,

> *"It's so difficult to find a worker nowadays. That is why, when you see in my company most of them are elderly, 76 years old, 69 years old also we have."*

Further in this stage, the company also aims to reduce the amount of wages. Decision on acquiring a new technology (robotic) is made due to the assumption that the numbers of workers in the factory can be reduced. Before the robotic was acquired, three employees need to be hired to do the task. After the robotic has been acquired, only an employee is needed to do the same job.

Company A, however, did not have a specific period for the planning. When the new technology or systems is needed, then they start the planning process. In addition, the owner believed that there is no need to prepare a planning earlier as the technology growth rapidly.

Next, the owner analysed the external business environment during this stage especially on the supplier credibility and history company background. He mentioned,

> *"Yes, we do look at their company's objective. We visited the factory...We have to be really sure about the brand and the plant."*

Then, the top management selected the specific technology that is considered the best solution to their knowledge. Once the proposed project is implemented, the business process was changed from manual to fully computerised. For example, the work that was once conducted by three employees, with the implementation of the new technology, only one person is needed by a click of a button. In sum, Company A completed four stages of strategic IS planning process as presented in Figure 1 below.

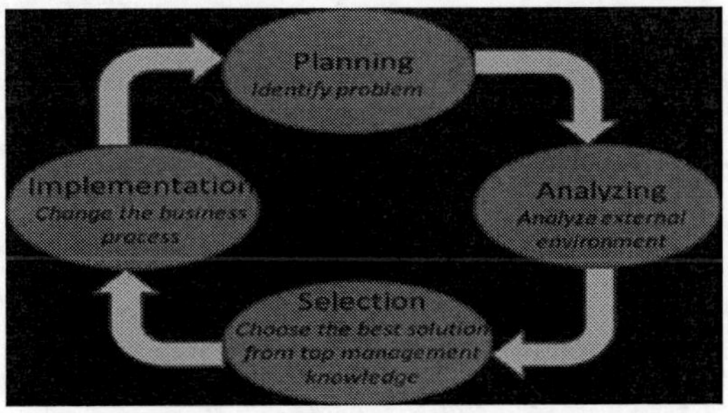

Figure 1: SISP process for Company A

3.2 Champion

The champion of the strategic IS planning process in Company A is the owner, as the owner decides everything related to acquiring the new information system and technology. The owner asserted,

> *"Yes, I decide everything, include what to buy. The directors also do not know. I have many directors. What I want to purchase, I need to table everything in the meeting."*

The owner decides the technology that the company needed and presented to the board of directors for their approval. This is important to ensure that the board of directors know about the investment to be made.

3.3 Roles of Top Management in Strategic IS Planning

This section discusses on the roles of top management in strategic IS planning development. Findings suggested that top management plays an important role in making decision and studying the technology to be acquired.

3.3.1 Decision maker

All decision with regards to acquiring the new technology will be made by the owner. When asking whether the owner discussed with his employees about the proposed technology, he asserted,

> *"The most important is I'm the one who makes the decision. Then I tabled to the board of directors."*

However, the employees do assist him on some parts of the decision making process as preparing the proposal and budget. The owner further urged,

> *"Of course, I get the help from my employee [the accounts executive]. We discussed. I need to use a system to prepare a budget. I don't know. So they [the accounts executive] need to assist me. I do not know how to prepare the paperwork. I only have ideas."*

3.3.2 Researcher

The owner did some studies about the proposed technology including went for a site visit at the supplier factory. One of the reasons is that he could not rely on his employees as for him the employees do not have knowledge about the proposed technology. He asserted,

> *"Ooo..they do not know either [about the proposed technology]. Further, during the site visit, only one person can go. As I make the decision, I had to go. Most of my employees, they only know about their daily tasks. I am the one who needs to know everything."*

In sum, the top management in Company A plays important role in decision making and collected information with regards to the proposed technology.

3.4 Roles of Users in Strategic IS Planning

This section discusses on the roles of users in strategic IS planning development. Findings suggested that users also play important roles in making decision and providing input about the applications to be used or acquired in their respective department.

3.4.1 Decision maker

The decision with regards to the business units will be decided by the business units itself and not the owner. For example, the application used in the accounting department was decided by the Accounts Manager. The owner stated that,

> *"For example accounts department, they will make their own decision. I do not know about their applications....the HR manager's refused to use any applications for her department, so I said OK...They know better about their department".*

Further, when respondents were asked about the decision made on applications to be used in their department, one of the users assert,

> *"We make our own decision."*

3.4.2 Input Provider

The users also provide inputs during the planning stage to make sure that the applications acquired can assist them to do their daily tasks. One of the users explained,

> *"We are a group of accounts executive that report to a chief accountant at the Head Office... So, we will discuss on what systems that we want to use. Our company (Company A) has its own procedure and its own accounting system. I will assist them [by giving information] on any issues related to the systems such as upgrading the systems..."*

In sum, the users play an important role in strategic IS planning. However, their role is limited to the application that is used in their own department only. They did not make any decision nor provide input for the major technology acquired by their organization.

4. Conclusion

Every organization will have different phases to develop their strategic IS planning. Some organizations may have completed or undergone all the phases as suggested by the literature. However, some organizations may complete some phases only because it is applicable to the needs and nature of their organization. Company A, a medium-sized manufacturing company only undergone four phases, namely planning, analysing, selection, and implementation. As and when this company face any issues related to the IS and business process in their organization, they will prepare the planning. Therefore, conceiving phase is not been included as they already decided the proposed technology and applications to be acquired.

Further, the results aligned with previous studies on SMEs that the top management is the champion of the strategic IS planning process in which they have the power to make the decision related to the overall organization [14,15]. The users will only decide on the specific applications to be used in their respective department. In this study, the owner of Company A did all the studies related to the proposed technology without assistant from his employees. The owner's reluctant to gain assistance from his employees might be due to the reason that, as most of the employees in his company are elderly, they might not have knowledge about the newest technology available in the market.

In conclusion, this study provides a clear understanding on how medium-sized organization practiced IS planning with a specific focuses on the role of top management and system user.

We would like to thank Universiti Utara Malaysia for providing funding on this research project.

References

1. V. Basu, E. Hartono, A. L. Lederer, V. Sethi. Inform. Manage. *The impact of organisational commitment, senior management involvment, and team involvement on strategic information systems planning.* **39**, 513-524 (2002)
2. A. L. Lederer, H. Salmela. J. Strategic Inf. Syst. *Toward a theory of strategic information systems planning.* **5** (3), 237-253 (1996)
3. P. Powell, M. Woerndl. Eur. J. Inform. Syst. *Time to stop searching the important things?*. **17**, 174-178 (2008)
4. A Garg, D. P. Goyal. J. Adv. Manage. Res. *Striving towards strategic alignment in SMEs: An empirical analysis.* **9** (1), 77-95 (2012)
5. T. Fuller. Int. Small Bus. J. *Fulfilling IT needs in small businesses: A recursive learning mode*l. **14** (4), 25-45 (1996)
6. C. Hagmann, C. McCahon. *Inform. Manage.* Strategic *information systems and competitiveness.* **25**, 183–192 (1993)
7. S. Pavic, S.C.L. Koh, M. Simson, J. Padmore. Benchmarking: *An Int. J. Could e-business create competitive advantages in UK's SMEs?*. **14** (3), 320 – 351 (2007)
8. P. Bharati, A. Chaudhury. Int. J. E-Bus. Res. *SMEs and competitiveness: The role of information systems.* **5** (1), i - ix (2009)
9. R. S. Y. Berry. Paper presented at the British Educational Research Association Annual Conference, University of Sussex at Brighton. *Collecting data by in-depth interviewing.* September 2-5 (1999). Retrieved July 1, 2013 from http://www.leeds.ac.uk/educol/documents/00000l172.htm
10. J. F. Gubrium, J. A. Holstein (Eds). *Handbook of interview research.* Thousand Oaks, CA: Sage (2002)
11. H. E. Newkirk, A. L. Lederer, A. M. Johnson. Eur. J. Inf. Syst. *Rapid business and IT change: drivers for strategic information systems planning.* **17** (3), 198-218 (2008)
12. A Cassidy. *A Practical Guide to Information Systems Strategic Planning.* Auerbach Publications. NW: Taylor & Francis Group (2006)
13. D. A. Erlandson, E. L. Harris, B. L. Skipper, S. D. Allen. *Doing Naturalistic Inquiry: A Guide to Methods.* California: Sage publications (1993)
14. P. Bharati, Chaudhury, A. Commun. ACM. *Studying the current status of technology adoption.* **49** (10), 88-93 (2006)
15. Chia-An Chao, Aruna Chandra. JSBED. *Impact of owner's knowledge of information technology (IT) on strategic alignment and IT adoption in US small firms.* **19** (1), 114-131 (2012)

Do Financial Advisors Live Up to Their Reputation: The Case of Major Assets Restructurings of Chinese Listed Companies

Cen Wu[1,*], *Qingquan Tang*[1], *and Jiali Jenna Tang*[2]

[1]Sun Yat-sen Business School, Sun Yat-sen University, Guangzhou, China

[2]University of Massachusetts Lowell, Lowell, MA, USA

Abstract We investigate whether top-tier advisors provide superior services by examining the relationship between reputation (measured by whether it is a top-10 advisor ranked on deal value) of financial advisors on the bidder side and stock market-based/accounting-based performance of bidders. Using Chinese listed companies with major assets reorganizations (MARs, M&As with large-scale target), we find top-tier advisors are associated with higher excess returns (CARs), implying that reputation generates a verification effect on investors. But we find no significant relationship between the advisor reputation and bidders' accounting-based performance post-MARs. The findings indicate that although advisor reputation can attract M&A business and sends positive signal to the market, it does not lead to stronger financial performance in the long run. That is, the so-called top-tier financial advisors fail to live up to their reputation. We also find that payment premium is an intervening variable between advisor reputation and the long-term accounting-based performance of bidders, suggesting that top-tier advisors fail in their duties to help clients achieve greater share of synergy gains.

1 Introduction

After going through the phase of extensive growth, Chinese firms face many obstacles in continuous development, such as excess capacity and rising costs. In order to achieve transformation and gain core competitive advantages, firms need to effectively adjust their assets structures, capital structures, and business models. Such objectives are difficult to meet with the traditional mergers and acquisitions (M&As), major asset reorganizations (MARs) are thought to be more efficient means, which are usually associated with positive market response. Based on China Securities Regulatory Commission (the CSRC), an M&A is identified as an MAR if it meets one of the following three standards: assets being purchased or sold are no less than 50% of the bidder's assets, or relating to no less than 50% of the bidder's sales, or no less than 50% of the bidder's net assets listed in the audited consolidated financial statement of the most recent fiscal year. Due to the complexity of conducting MARs and strict requirements of timely information disclosure, the CSRC requires bidders to employ independent[1] financial advisors to enhance the effectiveness of information exchange in MARs.

Theoretically, top-tier advisors are more capable in promoting information exchange and reducing transaction cost (Golubov et al., 2012) [1], because reputation is a proxy for ability, suggested by the classical work of Klein and Leffler (1981)[2] and Shapiro (1983)[3], and their models are applicable to the case of investment banks' services (Chemmanur and Fulghieri, 1994[4]). But empirical results are mixed. While some find top-tier advisors perform superior services for their clients in return for premium fees (Golubov et al., 2012[1]), some report a negative or at best insignificant relationship between bidder reputation and bidder returns (Servaes and Zenner, 1996[5]; Rau, 2000[6]; Hunter and Jagtiani, 2003[7]; Ismail, 2010[8]).

While the current literature has extensively discussed M&As in China, very few focuses on the role of financial advisors in enhancing firm performance during MARs. Top-tier advisors can send signals to the stock market that both the bidder firms and M&As are of high quality, and top-tier advisors are more capable of selecting more premium targets and offering constructive advice. However, they take no responsibility for investors[2] and may fail to do their

[1] Financial advisors hired by bidders in MARs should have no benefit connection with the bidder or the seller.

[2] The disclaimer is written in the first few pages of an independent financial advisor report.

duties in due diligence, which is revealed by Chinese media[3]. These lead to our primary research questions: Does top-tier advisors lead to superior stock market-based performance? And are they behave as expected and generate strong accounting-based performance in the long run? Our results show that top-tier advisors are associated with better bidder[4] CARs but not better long-term accounting-based performance of bidders, it is consistent with the notion that financial advisors do not live up to their reputation. To explain above results, we find that top-tier advisors are associated with higher payment premium, resulting in wealth being transferred from bidders to sellers. We conclude that top-tier advisors care more about deal completion than helping clients negotiate favorable terms or achieve greater share of synergy gains.

Although China is not the only setting in which we can investigate reputation of financial advisors, the Chinese setting presents some advantages over that of developed countries. First, the bidder of an MAR must hire a financial advisor under the CSRC requirement. So by restricting our sample to MARs, we are able to remove samples of in-house acquisitions and avoid nonrandom sample selection. Besides, due to the larger scale of targets, financial advisors should play a more important role in MARs. Second, financial advisors are unlikely to focus on long-term interest in an unstable market, which is a reason that they do not live up to their reputation. Chinese M&A market is now under market reform, which provide a perfect setting to investigate what will financial advisors do to maximize short-run profit.

This study makes important contributions to the M&A and financial advisor literature. First, to our knowledge, this is the first study to investigate the outcomes of hiring a top-tier advisor in China. With the current M&A trend, many studies have investigated M&As in China's market. However, these studies ignore one crucial factor in enhancing M&A synergy, that is, the role of financial advisors in facilitating M&A process. We fill this gap by examining the effects of advisor reputation. Second, results show a mismatch between stock market-based and accounting-based performance of bidders. We argue that advisors' pursuit of deal completion and contingent fees can explain it. This paper extends the studies on the difference between short-run and long-term performance of bidders in M&As (Chi et al., 2011[9]; Dutta and Joginitial, 2009[10]). Third, payment premium is found to be an intervening variable between reputation and financial performance. It gives direct evidence that top-tier advisors fail in their duties and do not live up to their reputation.

We organize the rest of the paper as follows. Section 2 discusses the related literature and presents the hypotheses. Section 3 presents the data and methodology. Section 4 presents the empirical results, and Section 5 concludes the paper.

2 Related literature and hypotheses development

2.1 Stock market-based performance: expectation of the reputation

Professional firms play a role as rubber stamp and demonstrate themselves as reliable trustees of the management (Meyer, 1994[11]; Pfeffer, 1981[12]). In this section, we argue that hiring top-tier advisors has identification effect, because it sends signals that inform investors of at least two things: the quality of listed company and the quality of M&As.

First, it provides verification on quality of the bidder. If the bidder suffers from poor corporate governance or low information transparency, the M&A tends to have smaller chances of being approved by the regulatory parties (e.g. the CSRC). Besides, M&As by such bidders may very likely be driven by the management's or the board's pursuit of private interests. Consequently, such bidders are less likely to generate financial returns from M&A, which leads to reputation loss for the affiliated financial advisors, because reputation is strengthened by strong word-of-mouth. Top-tier investment banks are more cautious in choosing clients (Fang, 2005)[13], and they also have more experience in accessing conditions of clients. Thus, top-tier advisors prefer to provide service to firms of high quality.

Second, it provides verification on quality of the M&A. High reputation is accumulated from past experience, which can be proxy for high ability. Bao and Edmans (2011)[14] argue that, financial advisors share different levels of responsibility in M&As, depending on whether they initially propose the deal. If the financial advisor proposes the deal, top-tier advisors are more capable to select appropriate targets and increase completion possibility since they have accumulated more experience and network resources from past service. If financial advisors are merely involved in execution, top-tier advisors with more negotiation experience can play a better role to reduce transaction cost and negotiate favorable terms. Thus, regardless of what role they play, top-tier advisors help ensure better deal outcomes. Derrien and Dessaint (2014)[15] show that the quality of financial advisor's service is affected by its market share ranking, which is usually the proxy for reputation. Francis et al. (2014)[16] also find that financial advisors value their reputation in the local market, confirming that reputation can play a verification role.

In conclusion, investors will expect top-tier advisors to enhance MAR performance, which leads to our first hypothesis:

[3] Securities Daily, March 27, 2015, the first C01 version. (in Chinese)

[4] Normally, only the bidders are listed companies and they acquire shares or assets from unlisted companies in China, which lead to the fact that relevant literature usually only investigates returns of bidders rather than both (Chi et al., 2011)[9].

H1: Higher reputation of financial advisors is associated with higher bidder CARs.

2.2 Accounting-based performance: truth of the reputation

Different from the stock market-based performance that reflects what investors expect from top-tier advisors, the accounting-based performance of bidders reflects the quality of service provided by top-tier advisors. So the former can be considered as the "expectation" of top-tier advisors' service and the latter "truth". We investigate the "truth" in this section.

In a market where reputation mechanism perfectly functions, top-tier advisors will live up to their reputation, because of the following two reasons. For one thing, they have more experience in selecting appropriate targets, negotiation, controlling deal risk and completing deals. So top-tier advisors can lead to better outcomes. For another, high reputation often attracts clients of higher quality. And advisors with a lot influence are more powerful in political lobbying, through which they can maximize their own interest. As a result, top-tier advisors are more cautious in selecting and carrying on deals, to avoid bad outcomes damaging reputation (Golubov et al., 2012)[1]. In conclusion, top-tier advisors are associated with better accounting-based performance of bidders.

However, when reputation mechanism does not perfectly function, the above scenario will not hold and reputation will not be proxy for ability. We offer three reasons that top-tier advisors may not provide superior service in China. First, political connections of Chinese companies can accelerate the pace of IPO activity (Piotroski and Zhang, 2014)[17]. It is reasonable to expect that if a financial advisor has political connections, it may also be more capable to guarantee or accelerate deal completion. So in order to attract clients, financial advisors may put more resources to build political connections than improve service quality. In such cases, financial advisors accumulate their reputation by political factors other than superior service or more experience.

Second, financial advisors will seek long-term development only if they have stable expectations in M&A market and relevant policies. However, China is in the middle of M&A market reform. In 2013, the CSRC released a quality appraisal system of financial advisors that will affect whether and how long should M&As of listed companies be authorized. There is a high probability that more laws or policies will be released in the near future. Uncertain expectation will drive advisors to pay little attention in building reputation by superior services or word of mouth. In order to maximize short-term profit, they care more about deal completion, because it determines advisory fees (Rau, 2000[6]; Kolasinski and Kothari, 2008[18]). Besides, by accelerating deal completion, financial advisors can increase their client base and future sales. In consequence, they might sacrifice due diligence and service quality to some extent. In such cases, financial advisors accumulate their reputation by expanding market share recklessly.

Third, long-term MAR performance is associated with a lot of complex influence factors, so performance lower than expected can not be easily attributed to the failure of advisory service, especially when deal completion is also the client's priority. It is quite possible in China because MARs are often drive by private interest of the controlling shareholders. They need a completed deal to transfer wealth. In such cases, financial advisors share less responsibility and caring less.

Based on the above argument, we hypothesize the following hypothesis:

H2: Higher reputation of financial advisors is not associated with higher long-term performance of bidders.

3 Data and methodology

We obtain the MARs data from the China Stock Market Trading Database (CSMAR), during the sample period ranging from 2006 to 2013. 2006 is when MARs started to be common, continuously taking place in every following year. In 2013, the CSRC released a quality appraisal system of financial advisors and the results were announced in 2014. It is a more direct proxy for advisors' ability, investor's cognition on top-tier advisors may be affected. We obtain the financial advisor data from each company's MAR announcement.

We download financial advisor league table from Mergermarket database. If a financial advisor is among the top-10 according to the value of deals on which they advised for a sample of M&A transactions by Chinese listed or unlisted companies during the period January 1, 1998 to December 31, 2012, it is a top-tier advisor[5]. Accumulation period for advisor reputation starts in 1998, because it is when Mergermarket database starts to disclosure league table and accumulation of reputation needs some time before having effects. The period ends in 2012, because in 2013 financial advisors' market share may be affected by the quality appraisal system.

Table 1 presents descriptive statistics. We employ the standard market model to calculate CAR. ***CAR*** is in the 3-day event window (-1, +1). The estimation period is (-150, -30). ***ΔROA*** is difference in ROA of year t+1 and t-1, t is the year of MAR announcement. ROA is adjusted by subtracting industry-year median. ***Premium*** is equal to difference between payment price and net asset of target divided by net asset of target. ***Rep*** is equal to 1 if a financial advisor is

5 Ten top-tier advisors are China International Capital Corporation Limited, CITIC Securities Co Ltd, Huatai United Securities Co Ltd, Haitong Securities Co Ltd, GF Securities Co Ltd, China Merchants Securities Co Ltd, Guotai Junan Securities Co Ltd, Southwest Securities Co Ltd, China Galaxy Securities Co Ltd, Bank of China International Holdings Limited. No foreign banks are included because they are not allowed to be the independent financial advisor in MARs.

among the top-tier, and 0 otherwise. ***Exp*** is equal to period expense of year t-1 divided by total asset at the end of year t-1. ***Sale*** is equal to natural log of sale of year t-1. ***Lev*** is financial leverage of year t-1. ***JR*** is equal to 1 if the president of the board is also the general manager, and 0 otherwise. ***JR*** controls how powerful the manager can be. ***Share*** is equal to percentage of controlling shareholder's share in year t-1. ***Exe_share*** is equal to 1 if the management hold shares in year t-1, and 0 otherwise. ***Soe*** is equal to 1 if the company is a SOE in year t-1, and 0 otherwise. ***Compare*** is book value of target divided by total asset at the end of year t-1. ***Bookvalue*** is equal to natural log of book value of target. ***Pay*** is equal to natural log of payment price.

Table 1 Descriptive Statistics

Variables	N	Mean	S.D.	Median	Minimum	Maximum
CAR3	300	0.24	0.88	0.14	-0.2	9.39
CAR5	300	0.28	0.88	0.19	-0.3	9.45
ΔROA	300	0.03	0.12	0.02	-0.4	0.39
Premium	300	2.24	3.68	1.07	-0.33	25.86
Rep	300	0.54	0.5	1	0	1
Exp	300	0.26	0.41	0.17	0.01	2.53
Sale	300	20.51	1.64	20.48	16.36	24.8
Lev	300	0.55	0.33	0.52	0.05	1.68[6]
JR	300	1.79	0.41	2	1	2
Share	300	35.85	14.67	34.44	8.93	75.19
Exe_share	300	0.91	0.29	1	0	1
Soe	300	0.59	0.49	1	0	1
Compare	300	1.96	9.11	0.34	0.01	141.33
Bookvalue	300	20.22	1.27	20.25	17.08	23.61
Pay	300	21.08	1.16	21.22	17.7	24.09

4 Empirical results

4.1 Outcomes of hiring top-tier advisors

We run the following model to test H1 and H2. *Performance* stands for ***CAR*** or ***ΔROA***.

$$Performance = Rep + ControlVariables + \xi \quad (1)$$

Panel A of Table 2 presents regression results. The second column shows that the coefficient of ***Rep*** is significantly positive, while the third and fourth columns show that the coefficients of ***Rep*** are both insignificantly. There results are consistent with H1 and H2. There exists mismatch between market-based and accounting-based performance of bidders associates with hiring top-tier advisors.

4.2 Discussion: Why not accounting-based performance?

A natural question that arises from above findings is why top-tier advisors can only enhance short-run bidder CARs but not accounting-based performance of bidders. Mismatch between short-run and long-term performance of bidders is also found by some literature (Dutta and Joginitial, 2009[10]; Chi et al., 2011[9]), arguing that investors are likely to be shortsighted at the beginning, due to lack of sufficient information (Delong and Deyoung, 2007)[20]. Such matters may be more pronounced in China, because Chinese investors are deficient in investment knowledge and skills (Wang et al., 2006)[21]. Being more immature and inexperienced than investors in developed countries, they are more likely to be misled by reputation, ignoring the "truth" that reputation may not be proxy for better advisory services.

We give further empirical evidence that reputation is not proxy for better service. As discussed earlier, top-tier advisors are more capable of negotiating favorable terms for bidders, among which payment price is a key term. For bidders, accepting a higher payment price leads to less share of synergy gains, which will destroy long-term financial performance. However, the sellers will accept deal terms more quickly given a higher payment price. Because top-tier advisors may care more about deal completion (Rau, 2000)[6] and contingent fees (if any) are positively related to payment price, we argue that top-tier advisors are associated with higher payment premium, which is negatively related to long-term accounting-based performance of bidders.

We run the following models to test mediating effect. *Performance* stands for ***ΔROA***.

$$Performance = Premium + ControlVariables + \xi \quad (2)$$

$$Premium = Rep + ControlVariables + \xi \quad (3)$$

Panel B of Table 2 presents the results. The first column shows that the coefficient of ***Premium*** is significantly negative. The second column shows that ***Rep*** is significantly and positively related to ***Premium***. Sobel test is significant

6 Samples with a leverage larger than 1 are not excluded because MARs are often motivated by serious financial risks.

at 10% level (z=-1.684). These results suggest that ***Premium*** is an intervening variable between ***Rep*** and ***ΔROA***[7]. Wealth is transferred from bidders to sellers through high payment price. Because MARs in China are often driven by private interest of the controlling shareholders, this result implies that top-tier advisors may accumulate their client base by supporting tunneling.

Table 2 Empirical Results

	Panel A: Main empirical results			Panel B: Mediating Effect	
	CAR	*ΔROA*	*ΔROA*	*ΔROA*	*Premium*
Rep	0.211**	0.018	0.020		0.622**
	(2.04)	(1.23)	(1.34)		(2.26)
Premium				-0.008**	
				(-2.30)	
Exp			-0.049**	-0.053**	-0.810*
			(-2.02)	(-2.19)	(-1.78)
Sale	-0.132***	-0.026***	-0.032***	-0.031***	-0.062
	(-3.55)	(-4.80)	(-5.18)	(-5.10)	(-0.54)
Lev	1.191***	0.035	0.051**	0.053**	0.165
	(7.08)	(1.44)	(2.02)	(2.11)	(0.35)
JR	0.089	0.021	0.022	0.016	-0.388
	(0.69)	(1.15)	(1.22)	(0.86)	(-1.14)
Share	0.001	0.000	0.000	0.000	0.001
	(0.30)	(0.26)	(0.44)	(0.45)	(0.10)
Exe_share	0.272	0.100***	0.093***	0.092***	-0.148
	(1.51)	(3.85)	(3.55)	(3.56)	(-0.31)
Soe	-0.048	0.030*	0.031*	0.030*	0.060
	(-0.41)	(1.81)	(1.89)	(1.80)	(0.19)
Compare	-0.020***	-0.004***	-0.003***	-0.003***	-0.006
	(-3.26)	(-4.02)	(-3.56)	(-3.59)	(-0.38)
Bookvalue	-0.051	-0.009	-0.010	-0.040**	-4.158***
	(-0.68)	(-0.81)	(-0.90)	(-2.32)	(-20.80)
Pay	0.051	0.029***	0.032***	0.064***	4.202***
	(0.66)	(2.62)	(2.90)	(3.65)	(20.44)
Constant	1.299	0.046	0.102	0.047	-1.741
	(0.93)	(0.23)	(0.50)	(0.24)	(-0.46)
Adj.R^2	0.151	0.113	0.123	0.135	0.659
F	2.521***	2.091***	2.170***	2.296***	17.047***
N	300	300	300	300	300

* Estimated significance at the 10% level, ** Estimated significance at the 5% level, *** Estimated significance at the 1% level.

4.3 Robustness tests

4.3.1 Two-stage Heckman procedure

We consider potential endogeneity of bidder-advisor matching. For example, top-tier advisors are more likely to be hired by large companies because of high advisory fees. OLS estimates are therefore biased. To address this concern, we adopt the two-stage Heckman (1979)[22] procedure. In the spirit of Golubov et al. (2012)[1], we construct a new variable ***Rep_before***, which is equal to 1 if in the 5 years prior to the deal, the bidder has employed a top-tier advisor at least once for an M&A, related party transaction, equity issue, or stock option incentive plan[8], and 0 otherwise. ***Rep_before*** effects the choice of financial advisor but not the M&A outcome. Unreported results show that in the first-stage equation, the coefficient of ***Rep_before*** is significantly positive, suggesting that if bidders have employed a top-tier advisor before, they are likely to hire a top-tier advisor again in MAR. From the first-stage equation, we construct an ***IMR*** (Inverse Mills Ratio) that reflects self-selection. After controlling ***IMR*** in the second-stage equation, the coefficient of ***Rep*** is still significantly positive in the ***CAR*** equation, and remains insignificant in the ***ΔROA*** equation.

4.3.2 Other sensitivity tests

We use two alternative event windows to calculate CAR, (-2,+2) and (-4.+4), the latter is the longest continuous period that CARs are significantly different from zero.

We also use an alternative definition of long-term financial performance. Starting from the first announcement, an M&A deal usually takes longer than a year to complete. So we replace ***ΔROA*** with difference in ROA one year before

[7] This kind of mediating effect is actually a "suppressing effect", where X is insignificantly associated with Y because two intermediary effects (X to M and M to Y) are of similar size but opposite in direction (Shrout and Bolger, 2002)[23].

[8] In any of these cases, a listed company should hire a financial advisor under the CSRC requirement.

and after *deal completion*, which is the date when an official permission paper is received by the bidder from the CSRC. None of these variations changes our results.

5 Conclusion

Using MARs of Chinese listed companies from 2006 to 2013, this paper investigates how financial advisors' reputation affects market-based/accounting-based performance of bidders. Contrary to prior literature that mainly focus on the U.S. advisory market, we find reputation has verification effect on Chinese investors, but there is no evidence that higher reputation of financial advisors is associated with better accounting-based performance of bidders. It suggests that investors are misled by the reputation of top-tier advisors, although it is not proxy for better advisory services. In addition, we discuss why financial advisors do not live up to their reputation. Investors may be shortsighted and payment premium is found to be an intervening variable between advisors' reputation and long-term accounting-based performance. These results suggest that reputation mechanism needs improvement in Chinese M&A market. While hiring a financial advisor, firms should pay more attention to the quality of advisory services rather than the ranking of market share.

References

1. A. Golubov, D. Petmezas, N. G. Travlos, J. Finance **67**, 271-311 (2012)
2. B. Klein, K. B. Leffler, J. Polit. Econ. **89**, 615-641 (1981)
3. C. Shapiro, Q. J. Econ. **98**, 659-679 (1983)
4. T. J. Chemmanur, P. Fulghieri, J. Finance **49**, 57-79 (1994)
5. H. Servaes, M. Zenner, Rev. Financ. Stud. **9**, 787-815 (1996)
6. P. R. Rau, J. Financ. Econ. **56**, 293-324 (2000)
7. W. C. Hunter, J. Jagtiani, Rev. Financ. Econ. **12**, 65-81 (2003)
8. A. Ismail, Rev. Quant. Finance Account. **35**, 411-429 (2010)
9. J. Chi, Q. Sun, M. Young, Emerg. Markets Rev. **12**, 152-170 (2011)
10. S. Dutta, V. Jog, J. Bank Financ. **33**, 1400-1412 (2009)
11. J. W. Meyer, B. Rowan, Am. J. Sociol. **83**, 340-363 (1977)
12. J. Pfeffer, *Power in Organizations*. (Ballinger, Cambridge, MA, 1981)
13. L. H. Fang, J. Finance **60**, 2729-2761 (2005)
14. J. Bao, A. Edmans, Rev. Financ. Stud. **24**, 2286-2315 (2011)
15. F. Derrien, O. Dessaint, Available at SSRN 1988928 (2014)
16. B. B. Francis, I. Hasan, X. Sun, M. Waisman, J. Corp. Finance **25**, 202-215 (2014)
17. J. D. Piotroski, T. Zhang, J. Financ. Econ. **111**, 111-136 (2014)
18. A. C. Kolasinski, S. Kothari, J. Financ. Quant. Anal. **43**, 817-842 (2008)
19. W. Wang, A. M. Whyte, J. Bank Financ. **34**, 44-54 (2010)
20. G. DeLong, R. DeYoung, J. Finance **62**, 181-216 (2007)
21. X. Wang, K. Shi, H. Fan, J. Econ. Psychol. **27**, 762-780 (2006)
22. J. Heckman, Econometrica **47**, 153-161 (1979)
23. P. E. Shrout, N. Bolger N, Psychol. Methods **7**, 422-445 (2002)

Improved Accuracy of Ratio Multiple Valuation

Julianto Agung Saputro[1,*] and *Jogiyanto* Hartono[2]

[1]Accounting Faculty of Economics and Business, Universitas Gadjah Mada, Indonesia
[2]Universitas Gadjah Mada, Indonesia

Abstract. Multiple valuation is widely used by investors and practitioners but its accuracy is questionable. Multiple valuation inaccuracies are due to unreliability of information used in valuation, inaccuracies comparison group selection, and use of individual multiple values. This study investigated the accuracy of valuation to examine factors that can increase the accuracy of the valuation of multiple ratios, that are discretionary accruals, the comparison group, and the composite of multiple valuation. These results indicate that multiple value adjustment method with discretionary accruals provides better accuracy, the industry comparator group method combined with the size and growth of companies also provide better accuracy. Composite of individual multiple valuation gives the best accuracy. If all of these factors combined, the accuracy of valuation of multiple ratios will give the best results.

1 Introduction

Many studies have been conducted to obtain a valuation model of the firm's most accurate and closer to the real situation. Firm value is a variable that can't be observed [1] and so can't be measured directly, therefore we need a model to get the value appropriately and accurately. Which valuation model is the best for measuring the value of the firms, is one of the important questions in investing. Many studies have conducted in-depth and wide to make a valuation model that can measure the value of the firms.

Valuation can be done with absolute or multiple ratios approaches. The survey results show the ratio approach is more widely used by investors, practitioners, and other users [2]. Multiple ratios are used in making investment decisions, valuation, and analysis of the company's specific interests, such as the Initial Public Offering (IPO) [3, 4]. Multiple ratio valuation is a model that is simple and easy to use [5, 6]. Practitioners more use of P/E, P/B, and P/CF multiple [7, 8]. Analysts relied P/E in the analysis [9, 10].

Although multiple ratios are widely used, the accuracy of this multiple is still doubtful [11]. Multiple accuracy ratio becomes an important issue because of its inexact will have an impact on the error analysis and decision making. Many factors determine the accuracy, a value driver factor that determined by the reliability of financial reporting information [11], the selection of a comparison group [11, 12, 13], compositing multiple ratio [14,15], and measurement error models [16]. This study aims to examine three factors that can improve the accuracy of multiple valuation, the reliability of financial reporting information, group comparison selection, and compositing individual multiple.

* Corresponding author: julianto.agung@mail.ugm.ac.id

This study contributes to the development of valuation models, by inserting discretionary accruals adjustment factor, selecting comparison group, and compositing multiple factor. These factors have an impact on the accuracy of multiple valuation. This study also contributes to investors and practitioners with better alternative valuation models.

2 Hypothesis Development

Multiple valuation can be grouped into three, namely accrual, book value, and cash flow multiple [17]. Accruals multiple is a multiple that uses accounting accrual data. Multiple accruals used to compare stock prices to accounting accrual, which included in this category are the P/E, P/SALES, P/GROSS, P/EBITDA, and P/EBIT[1] [17]. Multiple P/E is widely used by market participants because it is easy to use and P/E data is available in a lot of media [18]. P/E model is predominantly used by analysts, as the basis of valuation techniques before performing further analysis [19, 10, 9, 20, 21]. Book value multiple uses the book value data on the balance sheet. This category consists of P/B, P/ASSET, and P/INV[2]. P/B has an advantage in an analysis of the time and using the same accounting standards among companies, making it easier to make comparisons over time between the companies. Conservative investors often prefer P/B, because it offers more real measure of the firm value. P/CF is the ratio of stock price to operating cash flow per share outstanding. Which includes cash flow multiples are P/CF and P/DIV[3] [17]. These multiples use of operating cash flow of shares, non-cash income such as depreciation and amortization added to net income. P/CF is more widely used than the P/DIV [20].

Earnings information, as a value driver, has an important factor in the accuracy of multiple valuation. P/E has a weakness, because earnings numbers were heavily skewed [22]. This bias is because earnings determined by the selection of accounting accrual methods and is often used to earnings manipulation [23, 24].

H1: P/E after adjusting discretionary accruals provides better results than the multiple P/E without adjustment discretionary accruals.

A comparison group selection can be done by the same industry or the industry combined with specific factors such as size and growth. Industry comparison provides better accuracy than the company as a whole [25]. Groups of different industries will give different risk levels, so the industry group can be a good comparison group [26]. The accuracy of the comparison group selection will improve the valuation accuracy [27]. Valuation by industry combined with growth (ROE) gives the best results compared with the industry and market, and industry combined with the size of the companies [12].

H2: Comparison group using industry combined with the growth and size provides better accuracy than the comparasion group and market without the combined.

Compositing multiple valuation from individual multiple valuation can be the best way to improve the valuation accuracy [22]. Combining additional information simultaneously from each of multiple valuation is useful to improve the valuation accuracy [15]. Compositing individual valuation can provide the complete information of each

---

[1] P/E is a price multiple divided by earnings per share, P/GROSS is a price multiple divided by the gross profit of the outstanding shares, P/EBITDA is a price multiple divided by earnings before interest, taxes, depreciation and amortization of outstanding shares, and P/EBIT is a price multiple divided by earnings before interest and taxes of outstanding shares.
[2] P/B is a price multiple divided by book value of the outstanding shares, P/ASSET is a price multiple divided by total assets of the outstanding shares, and the P/INV is a price multiple divided by invested capital of the outstanding shares.
[3] P/CF is a price multiple of operating cash flows divided by the price of the outstanding shares, P/DIV is a price multiple divided by common stock dividend of shares outstanding.

multiple valuation, so it does not just rely on valuation and ignore the others. Composite may provide better accuracy than individual multiple valuation [31].
H3: Composite multiple generates a better value than individual multiple valuation.

3 Method

Samples are taken from data compiled by Bloomberg. The population used is a listed company on the Indonesian Stock Exchange in 2000 through 2014. The total sample used was 2,278 firms. Multiple valuations are an expression of market value relative to the value driver [28, 29]. Values driver for multiple valuation can use accrual multiple approaches, book value multiple, and cash flow multiple [17]. Multiple value of an equity multiple $\lambda_{i,t}$ of firm i in year t shown in equation (1). $P_{i,t}$ is the current stock price when financial statements published and $X_{i,t}$ is the value driver.

$$\lambda_{i,t} = \frac{P_{i,t}}{X_{i,t}} \quad (1)$$

Discretionary accruals can reduce the reliability of the earnings number, therefore accruals multiple should be adjusted [30]. Calculations of discretionary accruals use earnings management models developed by Kothari [31]. Discretionary accruals is used as adjustment factor for accrual multiple value. Equation (2) is used to compute multiple value adjustment with discretionary accruals ($\lambda_{i,t}^{adj}$). P/E that has been adjusted is called the P/E Adj.

$$\lambda_{i,t}^{adj} = \lambda_{i,t} - DA_{i,t} \quad (2)$$

Estimated value of equity used harmonic mean of the values of the ratio multiple comparison company multiplied by the value driver. Harmonic mean is one measure of central tendency, which is not sensitive to outliers than the arithmetic mean [2]. Harmonic mean giving less weight to multiple ratio is relatively high [23]. Harmonic mean gives better results in minimizing price error [11, 26, 32, 33]. Harmonic mean has the smallest bias if the method of measurement error used is the percentage error [16]. Because of the difference in performance is relatively small, harmonic mean is a statistical measure that is suitable for multiple estimates.

$$\hat{\lambda}_a = \frac{1}{\frac{1}{n}\sum_{i=1}^{n}\frac{1}{\lambda_i}} \quad (3)$$

Multiple aggregate value comparison companies ($\hat{\lambda}_a$) is calculated using harmonic mean, n is the number of comparison companies and λ (λ^{adj}) is a multiple value (accrual multiple) comparison companies. Estimated value of equity can be obtained by multiplying the value of multiple aggregate with a value driver of the target company (4).

$$\hat{P}_{i,t} = \hat{\lambda}_a \,.X_{i,t} \quad (4)$$

Multiple valuation accuracy was measured using the magnitude of price error or percentage error. The best accuracy value indicated by the smallest of the price error value, the value of zero or close to zero. Valuation accuracy of the estimated equity value was evaluated by calculating the absolute scale of price error (5).

$$\left|\frac{e_{i,t}}{P_{i,t}}\right| = \left|\frac{\hat{P}_{i,t}-P_{i,t}}{P_{i,t}}\right| \quad (5)$$

Benchmarking for performance accuracy of multiple valuation uses the size dispersion of the
distribution of absolute scale pricing error $\left|\frac{e_{i,t}}{P_{i,t}}\right|$, that are mean, median, interquartile, and standard deviation.

Comparative group conducted to test the accuracy by grouping companies according to industry groups and combinations using the size and growth. Company size is proxied by total assets (TA) and growth is proxied by ROE [12]. There are 4 comparator groups used:

1. MARKET: all the companies sampled unless the target company.

2. INDUSTRY: using the Industry Classification Benchmark (ICB).
3. INDUSTRY + TA: industrial group with a combined total assets.
4. INDUSTRY + ROE: ROE combined with the industry group.

Compositing multiple is done using valuation results based on some individual multiple valuation, as input, by giving equal weight to each of multiple ratios. The same weighting is used as a third source of financial statements is equally important and complementary.

$$\hat{P}_{i,t} = \sum_{i=1}^{n} \beta_i \cdot \hat{\lambda}_a \cdot X_{i,t} \quad (6)$$

Equation (6) shows the estimated value of the equity to composite, β_i are weights for each individual multiple, each individuals multiple are given a weighting of one third since the composite was used three individuals multiple.

4 Results and Discussions

The result in Table 1 suggest that P/E multiple has the smallest absolute price error value which indicates the best accuracy of all accruals multiple (*). Multiple P/B is also a multiple with smallest absolute price error. P/CF is better than P/DIV in term of smallest error.

Table 1. Absolute Price Error Distribution

Multiple		**Mean**	**Median**	**75-25**	**90-10**	**STD**
Accrual	P/E	0.6937*	0.5227*	0.5124	0.9452*	0.8902*
	P/SALES	0.9521	0.7252	0.4683	0.9762	1.9907
	P/GROSS	0.7428	0.5961	0.4817	1.0244	1.3706
	P/EBITDA	0.7185	0.5656	0.4553*	1.0135	0.9294
	P/EBIT	0.7134	0.5586	0.4847	1.0018	0.9383
Book Value	P/B	0.6801*	0.5561*	0.4994	0.9266	0.9222*
	P/ASSET	0.8771	0.6343	0.4752*	0.8315*	2.3249
	P/INV	0.7654	0.6043	0.5071	1.0858	1.0759
Cash Flow	P/CF	0.8344*	0.6593*	0.5015*	1.1296*	1.0785*
	P/DIV	1.0804	1.0000	0.5174	1.2182	2.0775

*The smallest absolute value of the price error in multiple groups.

Paired sample t test (Table 2) showed significant difference between the P/E before and after adjusted for discretionary accruals. It shows that the discretionary accruals adjustments give effect to the accrual multiple P/E. Table 3 shows that the multiple P/E Adj have better accuracy than the P/E, it can be seen from the relative performance value (RP) of the mean, interquartile range, 90-10 percentiles, and standard deviations were negative. Thus hypothesis 1 is supported.

Table 2. Paired Sample t Test

	Mean	Std. Dev	Std. Err Mean	Lower*	Upper*	T	df	Sig. (2-tailed)
P/E - P/E Adj	-.00097	.01754	.00037	-.00169	-.00025	-2.628	2277	.009

*95% confidence interval of the difference. P/E Adj is the P/E after adjusting discretionary accruals.

The test of hypothesis 2 result (Table 4), indicate that the industry with a combination of TA or ROE gives better results than the industry and market. This shows that hypothesis 2 is supported. This result indicates that multiple P/E Adj and P/B with a comparator group companies that based on the industry combined with the ROE are the best. Multiple combination is better than multiple P/E Adj or P/B. Multiple combined showed the best accuracy of all the comparison groups.

Table 3. Price Error Comparison P/E Adj and P/E

Multiple	Mean	75-25	90-10	STD
P/E Adj	0.6909	0.5112	0.9267	0.8805
P/E	0.6937	0.5124	0.9452	0.8902
Relative Performance (RP)	**-0.0041**	**-0.0022**	**-0.0199**	**-0.0111**

RP (relative performance) calculates the difference performance of the two multiple by comparing the pricing error value of two multiple ratio (MR) of a distribution of dispersion. RP is a percentage of the difference between the relative pricing error of MR1 and MR2, RP = [(MR1-MR2) / MR1] x 100%. MR1 is the P/E Adj and MR2 is the P/E.

Table 4. Distribution Absolute Price Error Individual and Combined Multiple

	Comparison	P/E Adj	P/B	P/CF	Composite
Mean	MARKET	0.6909	0.6801	0.8344	**0.6302**
	INDUSTRI	0.6862	0.6532	0.7737	0.5863
	INDUSTRI+TA	0.6648	0.6489	**0.7293***	**0.5854***
	INDUSTRI+ROE	**0.6141***	**0.6290***	0.7916	0.6131
Median	MARKET	0.5239	0.5561	0.6593	**0.4957**
	INDUSTRI	0.5545	0.5391	0.5973	0.4790
	INDUSTRI+TA	0.5139	0.4954	**0.5606***	**0.4572***
	INDUSTRI+ROE	**0.4759***	**0.4888***	0.5653	0.4714
Interquartile	MARKET	0.5112	0.4994	0.5015	**0.4568**
	INDUSTRI	0.5293	0.5060	0.5437	0.4369
	INDUSTRI+TA	0.5479	**0.4831***	**0.4762***	**0.4247*****
	INDUSTRI+ROE	**0.5009***	0.5155	0.5149	**0.4007****

*Indicates the absolute value of the smallest price error. Symbols ** and *** show the first and second best of valuation accuracy.

The accuracy test results between individual multiple and combined (Table 4) show the greatest accuracy is multiple combination, for the mean (0.6302), median (0.4957), and interquartile (0.4568). These results indicate that the hypothesis 3 is supported, meaning composite multiple ratio has a better accuracy than that with individual multiple ratios.

5 Conclusion

This study shows that the discretionary accruals have an impact on the accuracy of multiple valuation. P/E adjusted discretionary accruals can improve accuracy although not too high elevation. Improved accuracy is not very high due to the possibility of capital market in Indonesia is not yet efficient. It is not because of the inability of ratios multiple in measuring the value of the firms, but may be investors or market participants who have not been able to absorb the accounting information properly.

Industry groups provide better accuracy than all comparators in the market. Industry groups showed similar patterns for the company's operations. Better accuracy occurs because the comparison used is similar in operation and risk. Industry groups combined with the size and growth showed better accuracy than any industry groups. Compositing multiple valuation in P/E Adj, P/B, and P/CF give better accuracy than individual multiple valuation. Compositing multiple provides better accuracy because it can provide more complete information. Composite results showed the best accuracy by using a comparison group industry combined with the company's growth.

This study is limited to the Indonesia Stock Exchange data only. The next research can test the various stock exchanges around the world to test the consistency. This study is also limited to a multiple valuation using the equity valuation. The next research can test the accuracy by comparing the impact of the equity and the entity valuation, and combine both those valuations. The next research can also be developed by combining multiple scoring models with intrinsic valuation.

References

1. A. Curtis, The Accounting Review **87,** 121-148 (2012).
2. H. Brase, C.P. Brase, *Understanding basic statistics* (Cengage Learning, 2011).
3. M. Kim, J.R. Ritter, Journal of Accounting and Economics **53**, 409 (1999).
4. J.D. Brushwood, //arizona.openrepository.com/arizona/handle/10150/556239 (2015).
5. E. Lie, H.J. Lie, Financial Analyst Journal **58**, 44 (2002).
6. L.E. DeAngelo, The Accounting Review **65** 93 (1990).
7. S.C. Tasker, http://lists.exeter.ac.uk/items/6D34285B-0349-9734-CB98-A94E59ADB93A.html (1998).
8. R. Barker, The European Accounting Review **8,** 195 (1999).
9. L.D. Brown, C.C. Andrew, B.C. Michael, Y.S. Nathan, JAR **53**, 1 (2015).
10. M.T. Bradshaw, The Accounting Review **79,** 25-50 (2004).
11. I.A. Cooper, N. Lambertides, http://ssrn.com/abstract=2291869 (2014).
12. C.S.A. Cheng, R. McNamara, Review of Quantitative Finance and Acc **15**, 349 (2000).
13. S. Bhojraj, C.M.C. Lee, D.K. Oler, Journal of Accounting Research **41,** 745 (2003).
14. K.K. Yee, Financial Analyst Journal **60**, 23 (2004).
15. Y.K. Yoo, Review of Accounting and Finance **5**, 108 (2006).
16. I. Dittman, E. Maug, //papers.ssrn.com/sol3/papers.cfm?abstract_id=884337 (2008).
17. A. Schreiner, www1.unisg.ch/www/edis.nsf/SysLkpByIdentifier/3313/$FILE/dis3313.pdf (2007).
18. S. Block, Journal of Investing **19**, 7 (2010).
19. E.G. Demikaros, N.C. Strong, M. Walker, Accounting Horizons **18**, 221 (2004).
20. J.E. Pinto, T.R. Robinson, J.D. Stowe, http://papers.ssrn.com/sol3/papers.cfm?abstract_id=2657717 (2015).
21. W.P. Dukes, Z.J. Peng, P.C. English II, Journal of Investing **15**, 90 (2006).
22. S.N. Kaplan, R.S. Ruback, The Journal of Finance **50**, 1059 (1995).
23. J. Liu, D. Nissim, J.K. Thomas, Financial Analyst Journal **63**, 56 (2007).
24. J.A. Ohlson, China Journal of Accounting Research **7**, 65 (2014).
25. J. Liu, D. Nissim, J.K. Thomas, Journal of Accounting Research **40**, 135 (2002).
26. N. Abuaf, Journal Of Financial Applied **2**, 123 (2011).
27. S. Young, Y. Zeng, The Accounting Review **90**, 2571(2015).
28. S. Bhojraj, C.M.C. Lee, Journal of Accounting Research **40**, 407 (2002).
29. S.H. Penman, *Financial Statement Analysis& Security Valuation* (McGraw-Hill, 2013).
30. K. Schipper, L. Vincent, Accounting Horizons **17** (2003).
31. S.P. Kothari, A.J. Leone, C.E. Wasley, Journal of Acc and Economics **39**,163 (2005).
32. R.P. Beatty, S.M. Riffe, R. Thompson, Accounting Horizons **13**, 177 (1999).
33. J. Liu, D. Nissim, J.K. Thomas, //faculty.som.yale.edu/jakethomas/papers/multiples.pdf (2002).

Design and Analysis of Financial Condition Local Government Java and Bali (2013-2014)

Nur Dewi Natrini[1] and *Irwan* Taufiq Ritonga[2]

[1]Fellowship of Magister of Accounting, Gadjah Mada University, Yogyakarta, Indonesia
[2]Lecturer of Magister of Accounting, Gadjah Mada University, Yogyakarta, Indonesia

Abstract This research aims to identify financial condition of local government in Java and Bali year 2013-2014. It is due to government financial condition, according to several researchers, provides an image on the ability of a government in fulfilling their obligations whether in the form of debt or service fulfillment in timely manner. According to assessment upon financial condition, local government is able to identify how to fulfill public needs, how to utilize resources and how to proceed resources so that it can be more productive. As for the measurement method of financial condition itself, the standard method cannot be determined. Therefore, indicator used for measuring local government financial condition is Brown's (1993)[2] indicator development adjusted to Indonesian government. In order to develop the indicator, this research employs qualitative method by comparing GASB No.34, SAP Government Regulation ("*Peraturan Pemerintah*" - PP) 71 Year 2010, SAP PP 24 year 2005 and literature studies and expert validation. In order to obtain a balanced comparison, this research also employs clusters developed by Baidori (2015)[1] for government in Java and Bali. Results of this research showed that among 7 analyzed clusters, there are variations of results, even though each cluster has similar socioeconomic condition to each other. This variation upon Indonesian local government financial condition is caused by regional autonomy.
Key words: Financial condition analysis, indicator, cluster, regional autonomy

1 Introduction

Law Number 23 year 2014 on Local Government defines hat regional autonomy refers to rights, authorities and obligations of a region to govern and manage their government and community interest in their local region. Implementation of regional autonomy leads to a quite enormous flow of funds from central government to local government (Syahruddin, 2006)[21]. In 2006, there was an increase of fund transfer to local government as many as Rp647 quintillion compared to fund transfer in 2014 that was Rp596.5 quintillion (Directorate General of Budget, 2015). Transfer of State Budget ("*Anggaran Pendapatan Belanja Negara*" - APBN) fund is used to optimize the role of local government in guiding and coordinating governmental implementation, including to manage their finance, whether between province and vertical instances, inter-vertical instances and between regions or cities (PP No. 19 Year 2010)[13].

However, central government only makes available of principles of local financial management without detailed provisions about its management (Ritonga et al., 2012a). As a result, each region has its own programs and activities that different from each other, according to their own local needs and potentials (Prita, 2015). Differences of programs and activities done in each region will lead to difference in allocating budget (Ritonga et al., 2012a , 2012b). Incidence of the difference in budget allocation will lead to a different financial condition among local governments (Ritonga et al., 2012a , 2012b)[15,16]

Financial condition describes the ability of a government in fulfilling their obligations whether in the form of debt or service fulfillment in timely manner (Wang et al., 2007 and Kioko, 2013)[22,7]. According to assessment upon financial condition, local government is able to identify how to fulfill public needs, how to utilize resources and how to proceed resources so that it can be more productive (Williams, 2003)[24]. The importance of financial condition measurement was illustrated by Wang and Liou (2009)[23] as a checking upon human health, financial condition health of an organization which is considered as a complex and a multidimensional matter so that every change in any financial condition will affect other parts.

This research developed a new indicator using Brown (1993)[2] indicator analysis tool which was adjusted to the condition in Indonesia. The result of those indicators was used to measure financial condition of governments in

[1] Corresponding author: dewi.natrini@gmail.com

Java and Bali. However, in order to generate a balanced financial condition analysis, this research utilized cluster technique as developed by Baidori (2015)[1]. This goes along with Zafra-Gómez (2009)[25] who stated that if financial condition measurement model is equipped with the implementation of cluster analysis, i.e., by categorizing local government in a similar socioeconomic characteristic, the result of the financial condition evaluation will be much more effective. The new development of financial condition indicators was based on assessment upon financial condition which has no system that can be generally implemented (Wang et al., 2007)[22].

2. Literature Review

2.1 Definition of Local Government Financial Condition

Generally, financial condition is defined as an ability of a government in fulfilling their obligations whether in the form of debt or service fulfillment in continuous and timely manner (Kioko, 2013 dan Wang et al., 2007)[7 dan 22]. The same issues was expressed by Hruza (2015)[6] that provision of service and materials needed by community is an irreplaceable role of government in modern democracy era. Several researchers have different terms of financial condition, among others are

Table 1. Terms of Financial Condition

Researchers	Terms
Crosby and Robbin (2013)[27]	*Fiscal Health,* refers to a primary determiner of government ability in fulfilling desire and needs of their community.
Cabaleiro et al. (2012) and Cuadro-Ballestros et al. (2013) [28,29]	*Financial Health,* is described as the main requirement in fulfilling the aim of any institutions.
Kloha et al. (2005) and Trussel and Patrick (2009)[30,31]	*Fiscal Distress,* is described as a condition threatening local government to serve their citizen and to maintain public functions that are considered as essential functions
Brusca et al. (2015)[26]	*Financial Sustainability,* is defined as an ability in managing the expected revenue and predicting long term financial risks without affecting the decrease of revenue or redundancy of expense.

The most acceptable definition of local government financial condition is referring to the ability of a government in fulfilling their obligations in finance in a timely manner and the ability to maintain services provided for community (Ritonga, 2014)[17].

2.2 Brown's *10-point test* Financial Measurement Model

Brown (1993)[2] conducted a quick financial condition measurement using 10-point test for financial ratio of local government of which the population is less that 100,000 people. The tools of the corresponding test are established specifically for providing a more focused measurement tool regarding financial condition upon a smaller local government (Rivenbark and Roenigk, 2011)[18]. Ten ratios utilized by Brown (1993)[2] consist of 4 financial fundamental factors including *revenue* (ratio 1-3), *expenditures* (ratio 4), *operation position* (ratio 5-7), and *debt structure* (ratio 8-10) (Ritonga, 2014; Maher and Nollenberger, 2009)[17,8]. The aim of establishment of 10-point test measurement tools by Brown (1993)[2] is due to the needs of measurement tools of local government financial condition that shall be quick and effective, and also to improve comparative data availability of the city provided from financial indicators.

The strength of Brown's (1993)[2] 10-point test model is on its benchmarking. That comparison is seen from the ranking conducted by Brown from local government starting from the worst to the best aspects according to each category (Rivenbark and Roenigk, 2011; Rivenbark et al., 2010)[18,19]. Brown's financial condition Measurement comprises of 3 steps, among others are: a. Calculating 10 financial ratios based on data in annual financial statements, b. Comparing the ratios between each local government with another according to the same population magnitude. Brown (1993)[2] classified the city by forming the following four categories: (1) a city of which total population is between 50,000-100,000 people, (2) a city of which total population is between 30,000-50,000 people, (3) a city of which total population is between 15,000-30,000 people, and (4) a city of which total population is less than 15.000 people, c. Conducting a ranking upon local financial condition according to the classification specified on the second step, starting from the best local government to the worst local government.

DOI: 10.1051/shsconf/20173403003

2.3 Clustering Results by Baidori (2015) in Indonesian context

Local government financial condition analysis will be more meaningful and useful if the local governments involved are classified into some categories with the same characteristics (Priyambodo and Ritonga, 2014)[32]. Data in the process of cluster forming developed by Baidori (2015)[1] includes eight variables of General Allocation Fund calculation components, i.e. total population; local area; Human Development Index; Construction Cost Index; Gross Regional Domestic Product ("*Produk Domestik Regional Bruto*" - PDRB) per capita; Own-Source Revenue ("*Pendapatan Asli Daerah*" - PAD); Tax Revenue Sharing Fund (*Dana Bagi Hasil Penerimaan Pajak*) and Natural Resources Revenue Sharing Fund (*Dana Bagi Hasil Penerimaan Sumber Daya Alam*). According to clustering results from Baidori (2015)[1], there are 5 clusters in city group and 6 clusters in regional group formed. The following are translation of each cluster of both city and regional groups.

Table 2 City Cluster

Cluster 1	Batu	**Cluster 2**	Blitar
	Mojokerto		Cilegon
	Pasuruan		Denpasar
	Probolinggo		Madiun
	Semarang		Magelang
	Serang		Pekalongan
	Tasikmalaya		Salatiga
Cluster 4	Bandung		Surakarta
	Banjar		Tangerang
	Bekasi		Tangerang Selatan
	Bogor		Tegal
	Cimahi		Yogyakarta
	Cirebon	**Cluster 3**	Surabaya
	Depok	**Cluster 5**	Kediri
	Malang		
	Sukabumi		

Table 3 Region Cluster

Cluster 1	Bandung Barat	Kendal	Purbalingga	**Cluster 2**	Bandung
	Bangli	Klaten	Purwakarta		Banyuwangi
	Banjarnegara	Klungkung	Purworejo		Cianjur
	Bantul	Kudus	Rembang		Garut
	Banyumas	Kulon Progo	Semarang		Jember
	Batang	Kuningan	Serang		Malang
	Blitar	Lamongan	Situbondo		Sukabumi
	Blora	Lebak	Sleman		Tasikmalaya
	Bondowoso	Lumajang	Sragen	**Cluster 3**	Badung
	Boyolali	Madiun	Subang	**Cluster 4**	Bangkalan
	Brebes	Magelang	Sukoharjo		Pamekasan
	Buleleng	Magetan	Sumedang		Sampang
	Ciamis	Majalengka	Tabanan		Sumenep
	Cirebon	Mojokerto	Tegal	**Cluster 5**	Bekasi
	Demak	Nganjuk	Temanggung		Bogor
	Grobogan	Ngawi	Trenggalek		Cilacap
	Gunung Kidul	Pacitan	Tuban		Gresik
	Jembrana	Pandeglang	Tulungagung		Indramayu
	Jepara	Pati	Wonogiri		Karawang
	Jombang	Pasuruan	Wonosobo		Sidoarjo
	Karanganyar	Pekalongan			Tangerang
	Karangasem	Pemalangan		**Cluster 6**	Bojonegoro
	Kebumen	Ponorogo			
	Kediri	Probolinggo			

3. Research Method

This research employed qualitative method since the researcher aims to explore deeper the financial condition of local government by developing regional financial condition indicators in the first place. Hence, this research utilized development approach model. It was described by Sugiyono (2015)[20] that research and development method is a research method that is used for examining in order to develop an existing product (innovation) or to create a new tested product. Data collection in this research used documentation method. According to Cresswell (2014)[3] during a research process, researcher may gather qualitative documents in the form of public documents (such as newspaper, papers, agency reports) or private documents (such as journal, diary, letters, e-mails). Data of this research were taken from secondary data, i.e., Examination Report ("*Laporan Hasil Pemeriksaan*" - LHP) of BPK RI upon Student Worksheets ("*Lembar Kerja Peserta Didik*" - LKPD) in Java and Bali on 2012-2014 LKPD, publication data from Central Bureau of Statistics "*Badan Pusat Statistik*" - BPS), Regional Medium-term Development Plan ("*Rencana Pembangunan Jangka Menengah Daerah*" - RPJMD), Annual Local Government Working Plan ("*Rencana Kerja Pemerintah Daerah*" - RKPD) in 121 local governments.

Translation of this research steps are as follow:

1. Analysis of financial condition measurement method used by Brown (1993)[2] by looking for account equations existing on financial statement in Indonesia on the basis of literature review. According to the corresponding step, a new financial condition measurement is obtained and it will be validated by the experts.
2. Calculating ratio according to annual financial statement and other supporting data
3. Comparing data of inter-local government according to Baidori's (2015) clusters. By comparing the ratio, each local government in each cluster will be classified into 4 quartiles as follow:
   a. Quartile 1 (0-25%) shows that local government in particular cluster has a bad ratio according to their rank.
   b. Quartile 2 (25%-50%) shows that local government in particular cluster has a good ratio according to their rank.
   c. Quartile 3 (50%-70%) shows that local government in particular cluster has a better ratio according to their rank.
   d. Quartile 4 (70%-100%) shows that local government in particular cluster has the best ratio according to their rank.
4. Conducting grading or ranking upon regional financial condition according to the scale proposed by Brown (1993)[2]. That scale was designed to enable local government of which the ratio is above 50% (quartile 3 and above) to obtain a positive total score. It means that local government of which its ratio majority is above 50% has a better financial condition compared to the local government of which its ratio majority is below 50% (Brown, 1993). The local government of which ratio is between 50%-70% (quartile 3) will receive a total score of 9 points out of 9 financial condition indicators. Meanwhile, the local government of which ratio is between 25%-50% (quartile 2) will receive a total score of zero (0) point, and he local government of which ratio is between 0=25% (quartile 1) will receive a negative total score (-9) points out of the new financial condition indicators.
5. The determination of financial condition upon local government financial condition existing in each cluster. Brown (1993)[2] concluded that local government financial condition is defined according to assessment scale detailed as follow:

Table 4 Assessment Scale of Financial Condition

Total score	Relative assessment scale upon entire local governments
10 or more	Between the best
5 up to 9	Better than the most cities
1 up to 4	Average
0 up to -4	Worse than the most cities
-5 or less	Between the worst

4. Research Findings and Discussion

4.1 Indicators of Financial Condition according to the version of Indonesian Government

Analysis of local government financial condition in Java and Bali area is measured using modified indicators that were referring to measurement tools developed by Brown (1993)[2]. The modification of financial condition indicator aims to adjust the measurement tool so that it may run in accordance with financial condition in Indonesia. If Brown (1993)[2], in developing these indicators, used American financial statement under GASB no 34, this research used SPAP of PP 71 year 010 and PP 24 year 2005 and the related laws and regulations.

The following is translation of 10 financial condition indicators developed to measure local government financial condition in Java and Bali areas:

Table 5 New Indicator Translation

Brown Ratio (1993)	Financial Statement of America	Financial Statement of Indonesia	New Ratio Indonesian version
Ratio 1: Total Revenues / Population	Total Revenue is a total of all revenues stated in Governmental funds statement. The revenues in the corresponding statement/report are divided into 4 groups, i.e., General Fund, Capital Projects Fund, General Debt Service Fund, Non-major Governmental Fund	PP 24 Year 2005 and PP 71 Year 2010 divide revenues into type or source of revenue, they are Own-Source Revenue ("*Pendapatan Asli Daerah*" - PAD), Transfered Revenue (sharing fund (*dana bagi hasil*), general allocation fund (*dana alokasi umum*), special allocation fund (*dana alokasi khusus*), and Other Authorized Revenue (*Lain-lain Pendapatan yang Sah*).	**Ratio 1:** (Total Revenue - Special Allocation Fund (*Dana Alokasi Khusus*) - DAK)) / Population **Interpretation:** High ratio, showing a bigger ability to obtain supplementary revenue
Ratio 2: Total General Funds Revenue from own sources / Total General Revenue	General fund form own resources account according to American Government statement/report is: Taxes (Real estate, Sales and use, Personal income, Other) General revenue account according to American Government statement/report consists of: Taxes (Real estate, Sales and use, Personal income, Other), Federal, State, and another categorical aid, Charges for services, Investment income, Other Revenues	Law No 33 Year 2004 defines that PAD a revenue obtained by local/region collected under local regulations and laws and regulations. General revenue in American version is translated in Indonesian Government financial statement into PAD, Sharing Fund, provincial tax sharing and other revenues and DAU.	**Ratio 2** : PAD / (PAD+DBH+DAU+Provincial Tax Sharing Revenue) **Interpretation:** High ratio, showing that government does not rely on external governmental organization
Ratio 3: General Fund sources from other funds / Total general fund sources	General Fund Sources Account from other funds consists of: Principal Amount bond issued, Transfer from General debt service fund, and Transfer From Non-major Debt Services Fund	Regional revenue out of own-source revenue according to PP 24 year 2005 consists of Regional Fiscal Balance Funding (*Dana Perimbangan*) (DBH, DAU, DAK) and Other Authorized Revenue.	**Ratio 3 :** (DBH+DAU+Provincial Tax Sharing Revenue+Other Authorized Revenue) / (PAD+DBH+DAU+Provincial Tax Sharing Revenue) **Interpretation:** Low ratio, showing that government is not in operational transfer to fund general government operational
Ratio 4: Operating expenditures / Total Expenditures	Operating expenditures account consists of: General Government, Public safety and Judicial, Education, City University, Social Services, Environmental protection, Transportation Services, Parks, recreation, and cultural activities Total expenditures account consists of: General Fund, Capital Projects Fund, General Debt Services Fund.	PP 71 year 2010 describes operational expenditure as budget expense for daily activities of local government that are beneficial for short terms. Expenditure, according to PP 71 Year 2010, is classified into three classifications, they are: Operational expenditure, Capital Expenditure and Unforeseen Expenditure.	**Ratio 4:** Operational expenditure / Total Expenditure **Interpretation:** Low ratio, showing that the infrastructure is well-maintained

Brown Ratio (1993)	Financial Statement of America	Financial Statement of Indonesia	New Ratio Indonesian version
Ratio 5: Total Revenues / Total Expenditures	Total Revenues and Total Expenditures account consist of: General Fund, Capital Projects Fund, General Debt Services Fund.	Revenue account in Indonesian Government financial statement consists of: PAD, Transfered Revenue and other authorized revenue.	**Ratio 5:** (Total Revenue – DAK) / (Total Expenditure-DAK) **Interpretation:** High ratio, showing that local government experiences positive equity between periods
Ratio 6: Unreserved general fund balance / Total general General Revenue	Formula for unreserved general fund balance in American Version is Revenue – Expenditure – Other Financing Uses – Other Financing Sources + Fund Balance at the Beginning Year (Only for accounts in general fund)	Unreserved Fund Balance account in Indonesian version is: PAD + DBH+ DAU+ Provincial Tax Sharing and other revenues - Operating Costs + Unreserved Fund Balance in Indonesian version of the previous year.	PAD + DBH+ DAU+ Provincial Tax Sharing and other revenue - Operating Costs + Unreserved Fund Balance in Indonesian version of the previous year / (PAD + DBH + DAU + Sharing Revenue + Other authorized revenue). **Interpretation:** High ratio, showing that there is a resource that can be utilized to overcome temporary shortcomings of short term liabilities
Ratio 7: Total general fund cash and investments / Total General Fund Liabilities	General Fund Account– Balance Sheet: Cash and Cash equivalent, and Investment. Total General Fund Liabilities: Accounts payable and accrued liabilities, Accrued tax refunds, Accrued judgments and claims, Due to component units, Estimated dis-allowanced of federal state and other aid, and Other liabilities.	Account in financial statement balance sheet in Indonesia: Cash and cash equivalent, short-term investment, Third Party Calculation debt, domestics total assets part, other short-term debt, other long-term total assets.	**Ratio 7:** (Cash + Short-term Investment) / Short-term liabilities. **Interpretation:** High ratio, showing sufficiency of cash and cash equivalent that can be used to pay off short-term liabilities
Ratio 8: General Fund Liabilities / Total General Fund Revenues	Total General fund liabilities account is: Accounts payable and accrued liabilities, Accrued tax refunds, Accrued judgments and claims, Due to component units, Estimated dis-allowanced of federal state and other aid General fund revenue account: Taxes (Real estate, Sales and use, Personal income, Other), Federal, State, and another categorical aid, Charges for services, Investment income, Other Revenue	Current liabilities account: Third Party Calculation Debt, Debt Interest, domestics total assets part, other short-term debts, other long-term total assets General revenue account: PAD+DBH+DAU+Provincial Tax Sharing and other Revenues	**Ratio 8:** Short-term Liabilities/ (PAD+DBH+DAU+Provincial Tax Sharing Revenue) **Interpretation:** Low ratio, showing that short-term liabilities can be only served by normal flow of annual revenue
Ratio 9: Debt Services / Total Revenues	General debt services funds account is Administrative and other and Debt services (Interest and Redemptions). Total revenue account consists of General Fund, Capital Projects Fund, General Debt Services Fund.	According to SAP, interest expenditure in financial statement of Indonesian Government is a part of operational expenditures. Total revenue account consists of Own-Source Revenue ("*Pendapatan Asli Daerah*" - PAD), Transfered Revenue (sharing fund (*dana bagi hasil*), general allocation fund (*dana alokasi umum*), special allocation fund (*dana alokasi khusus*), and Other Authorized Revenue (*Lain-lain Pendapatan yang Sah*)	**Ratio 9:** Interest Expenditure / (Total Revenue - DAK) **Interpretation:** Low ratio, showing that government is able to pay off debt requirements on due date

Brown Ratio (1993)	Financial Statement of America	Financial Statement of Indonesia	New Ratio Indonesian version
Ratio 10: Direct Long Term Debt / Population	Non-current Liabilities account consist of Bonds and Note Payable Due within one year and Bonds and Note Payable. Long-term Debt account in ratio 10 is presented in net position balance sheet in American financial statement	Account in Indonesian Government financial statement is long-term liabilities (foreign debt, banking sector-domestic debt, obligation domestic debt, Obligation Premium (discount), other long-term debt)	**Ratio 10:** Long-term Debt / Population **Interpretation:** Low ratio, sowing that government possesses an ability to pay off long-term debt

4.2 Calculation of each ratio and scoring for Local Government Financial Condition.

Scoring is the third step after calculation of each ratio in all local governments in each cluster. Scoring value is obtained from the result of financial indicators classification in the four quartiles specified according to Brown's (1993)[2] steps.

Table 6 Example of Score calculation of Pamekasan Region (*Kabupaten Pamekasan*) year 2014

No	Ratio	Points for each quartile				Score
		Q1 (0-25%)	Q2 (25%-50%)	Q3 (50%-75%)	Q4 (75%-100%)	
		-1	0	1	2	
1	Ratio 1				2	2
2	Ratio 2				2	2
3	Ratio 3				2	2
4	Ratio 4	-1				-1
5	Ratio 5				2	2
6	Ratio 6		0			0
7	Ratio 7		0			0
8	Ratio 8	-1				-1
9	Ratio 9				2	2
10	Ratio 10				2	2
Total Score						**10**

According to table 6, Pamekasan Region has total score of 10. The total score is interpreted according to the rank of financial condition developed by Brown (1993). Pamekasan Region obtains a total score of 10, so that it can be assumed that the financial condition of Pamekasan Region in 2014 is considered as the best condition. Score determination to describe financial condition applies for city/municipal and regional government in each cluster.

4.3 Analysis of City Government Financial Condition Year 2013-2014

The next step is to analyze financial government of local government in each cluster. The following is one of the analysis on regional government year 2013-2014 in cluster 2.

Table 7 Score for Financial Condition of Regional Government Year 2013 and 2014(Cluster 2)

No	Regional Government	Score	
		2013	2014
1	Bandung	11	8
2	Banyuwangi	11	12
3	Cianjur	6	6
4	Garut	1	1
5	Jember	8	11
6	Malang	6	8
7	Sukabumi	5	0
8	Tasikmalaya	8	10

Description:

	The Best
	Better
	Average
	Worse
	The Worst

According to Table 7, it can be identified that the region of which the financial condition is consistent is Banyuwangi Region, with the best financial condition, and Cianjur Region, with a better financial condition. Sukabumi Region experiences a decrease of two ranks in those classifications, from the previous financial condition that shows a better result into a worse financial condition. It is due to the decrease on ratio 1, ratio 2, ratio 3, and ratio 5 of which almost all of them show score of -1 except for ratio 1 of which the score shown is 0. Not only significantly decreasing, Sukabumi Region also secures the lowest financial condition in cluster 2. The decrease in ratio 2 and 3 can be defined as that there is a dependence on source from external organization in funding operational of local government compared to the regions in cluster 2.

5. Conclusion

According to the 7 analyzed clusters, there are variations of results, even though each cluster has similar socioeconomic condition to each other. Variation of local government financial condition in Indonesia is caused by regional autonomy as specified in Law Number 23 Year 2014 that defines that regional autonomy refers to rights, authorities and obligations of a region to govern and manage their government and community interest in their local region. It goes along with the research conducted by Ritonga et al. (2012a and 2012b)[15,16] that the differences of programs and activities done in each region will lead to difference in allocating budget. Incidence of the difference in budget allocation will lead to a different financial condition among local governments (Ritonga et al., 2012a and 2012b)[15,16]. Shortcomings and limitations of this research, and also as suggestions for the research in the future, are described as follow:

a. This research was only conducted for two years in analyzing the financial condition of local government in Java and Bali. Researches in the future should be able to utilize financial information for more that two years to complete the analysis of financial condition of local government.

b. This research is only limited to Local Governments in Java and Bali that only represent 20% of all local governments in Indonesia. All of the shortcomings may be used as consideration for the next researches in the future in order to expand research object to enrich the analysis of financial condition of local government in Indonesia.

References

1. Baidori 2015, 'Pengklasteran Pemerintah Daerah di Jawa dan Bali Berdasarkan Variabel Sosio ekonomi (Komparasi Laporan keuangan Daerah)', *Thesis*. Yogyakarta: Universitas Gadjah Mada.
2. Brown, KW 1993, The 10-Point Test of Financial Condition: Toward an Esay-to-Use Assessment Tool for Smaller Cities. *Government Finance Review*, vol. 9, no. 6, pp. 21-26.
3. Creswell, JW 2014, Research Design: Qualitative, Quantitative, and Mixed Methods Approaches, **4th** Edition, *Sage Publications*, California.
4. DJPK 2013, Analisis Realisasi APBD Tahun Anggaran 2012. Direktorat Jenderal Perimbangan Keuangan-Kementerian Keuangan.
5. Direktorat Jenderal Angggaran (DJPK) 2015, Budget In Brief APBN 2015. Direkotrat Penyusunan Anggaran Republik Indonesia
6. Hruza, F 2015, Public Sector Organization Financial Ratios Recent Development as a Metter of Financial Innovation. Investment Management and Financial Innovations, vol.12, no.2, pp. 88-93.
7. Kioko, SN 2013, Reporting on the Financial Condition of the States 2002-2010. *Jurnal of Public Budgeting, Accounting, and Financial Management*, vo.25, no.1, pp165-198.
8. Maher, CS, dan Nollenberger, K 2009. Revisiting Kenneth Brown's "10-Point Test." *Government Finance Review*, vol. 25, no. 5, pp. 61-66.
9. Maher, CS 2013, Measuring Financial Condition: An Essential Element of Management During Periods of Fiscal Stress, *The Journal of Government Financial Management*, pp. 20.
10. Prita, AD 2015. 'Analisis Kondisi Keuangan Berdasarkan Model Brown'. *Skripsi*. Yogyakarta: Universitas Gadjah Mada.
11. Republik Indonesia 2005, Peraturan Pemerintah Nomor 24 Tahun 2005 tentang Standar Akuntansi Pemerintahan.
12. Republik Indonesia 2010, Peraturan Pemerintah Nomor 71 Tahun 2010 tentang Standar Akuntansi Pemerintahan.

13. Republik Indonesia 2010, Peraturan Pemerintah Nomor 19 Tahun 2010 tentang Tata Cara Pelaksanaan Tugas dan Wewenang serta Kedudukan Keuangan Gubernur Sebagai Wakil Pemerintah di Wilayah Provinsi
14. Republik Indonesia 2014, Undang-Undang Nomor 23 Tahun 2014 tentang Pemerintahan Daerah.
15. Ritonga, I, Clark, C, dan Wickremasinghe, G 2012a. Assessing Financial Condition of Local Government in Indonsia: An Exploration. *Municipal and Public Finance,* vol. 1, no. 2, pp. 37-50.
16. ____2012b, "*Factors Affecting Financial Condition of Local Government in Indonesia*": http://www.apira2013.org/proceedings/pdfs/K159.pdf.
17. Ritonga, I 2014, Analisis Laporan Keuangan Pemerintah Daerah, Lembaga Kajian manajemen Pemerintah Daerah, Yogyakarta.
18. Rivenbark, WC, Roenigk, DJ, Allison, GS 2010, Conceptualizing Financial Condition in Local Government, *Jurnal of Public Budgeting, Accounting, and Financial Management*, vol.22, no. 2, pp. 149-177
19. Rivenbark, WC, dan Roenigk, DJ 2011, Implementation of Financial Condition Analysis in Local Government, *Public Administration Quarterly,* vol.35, no.2, pp. 241-267
20. Sugiyono 2015, Metode Penelitian dan Pengembangan (Research and Development), Alfabeta: Bandung.
21. Syahrudin 2006, Desentralisasi Fiskal: Perlu Penyempurnaan Kebijakan dan Implementasi
22. Wang, X, Dennis, L, dan Tu, YSJ 2007, Measuring financial condition: A Study of US states, *Public Budgeting & Finance*, vol. 27, no. 2, pp. 1-21.
23. Wang, X dan Liou, KT 2009, Assesing the Change in Financial Condition: An Anlaysis of Panel Data From U.S.States, *Public Budgeting and Finance,* vol.21, no.2, pp.165-197
24. Williams,DW 2003, Measuring Government in the Early Twentieth Century. *Public Administration Review,* vol.63, no.6, pp.643-659.
25. Zafra-Gomez, JL, Lopez-Hernandez, AM, Hernandez-Bastida, A 2009, Evaluating Service Quality and Minimizing the Effects of the Socioeconomic Environment: An Aplication to Spanish Municipalities. *The American Review of Public Administration*, vol.39, no.4, pp.425-449.
26. Brusca, I, Rossi, FM, Aversano, N 2015, Drivers for the Financial Condition of Local Government: A Comparative Study Between Italy and Spain. *Lex Localis- Journal of Local Self Government*, vol.13, no.2, pp.161-184
27. Crosby, A, dan Robbins, D 2013, Mission Impossible: Monitoring Municipal Fiscal Sustainability and Stress in Michigan. J.of Public Budgeting, Accounting, and Financial Management, vol.23, no.3, pp.522-555.
28. Cabaleiro, R, Buch, B, Vaamonde, A 2012, Developing a Method to Assessing the Municipal Financial Health. *American Review of Public Administration*, vol.43, no.6, pp.729-751.
29. Cuadrado-Ballesteros, B, Mordan, N, Garcia-Sanchez, IM 2014, Is Local Financial Health Associated with Citizens Quality of Life? Soc Indic Res, vol.119, pp.559-580.
30. Kloha, P, Carol, SW, dan Kleine, R 2005, Developing and Testing a Composite Model to Predict Local Fiscal Distress, *Public Administration Review,* vol.65, no.3, pp.313-323.
31. Trussel, JM, dan Patrick, PA 2009, A Predictive Model of Fiscal Distress in Local Government, *J.of Public Budgeting, Accounting, and Financial Management*, vol.21, no.2, pp.578-616
32. Priyambodo, VK dan Ritonga, I 2014, Pengklasteran Pemerintah Daerah untuk Memaksimalkan Analisis Kondisi Keuangan Pemerintah Daerah, *Simposium Nasional Akuntansi XVII*. Mataram

Value Relevance of available-for-sale financial instruments (AFS) and revaluation surplus of PPE (REV) components of other comprehensive income

Maryam Yousefi Nejad[1,*], *Azlina* Ahmad[2]

[1]FEP Faculty economics and managements, Universiti Kebangsaan Malaysia
[2]FEP Faculty economics and managements, Universiti Kebangsaan Malaysia

Abstract. Value relevance studies have thus far provided mixed evidence. Studies also found that fair value reporting has a significant impact on the value relevance of other comprehensive income. Additionally, most studies on value relevance of other comprehensive income focused on developed countries where the capital market is more efficient. In these settings due to the existence of active market, fair valuation of assets may not pose a major problem. In Malaysia the mandatory reporting of comprehensive income by all listed firms based on the requirement of FRS101, coincides with the adoption of FRS139 the standard for financial instrument and could be an interesting setting to investigate. This study hypothesized that other comprehensive income and its components are associated with share price. The results indicate that other comprehensive income and its components are value relevant. The results indicate that the move towards a more comprehensive income reporting through the preparation of the Statement of Comprehensive Income results in more informative financial reporting.

1 Introduction

One of important concept related to other comprehensive income reporting is the concept of fair value (e.g. [1][2]. Fair values of assets and liabilities change according to the changes in market value [3]. These changes give rise to the unrealized gains and losses which should be report in other comprehensive income. In great majority of countries, the market are less efficient to provide the necessary and reliable data for fair value measurement. Firms in these countries would have to use price of similar assets and liabilities or use projection of future benefits, and combine it in a model with all reasonable assumptions that are available in the market [3]; [4]. This situation may result in estimated values that are less reliable and may not truly be comparable between firms and countries [3]; [4].

In developed countries, market for most assets exist and pose no problem of reliability on the fair value measurement. However, in developing countries, some assets do not have available market, therefore, pose problem with fair value measurement, hence,

* Corresponding author: maryam_youssefinejad@yahoo.com

DOI: 10.1051/shsconf/20173403004

reliability is questioned. Most studies on value relevance of comprehensive income, however are done in developed countries such as the USA [5] and Europe [6], where the market is efficient and fair valuation of assets is reliable. However, very little studies [7] have been carried out in developing countries, which provide a different setting due to differences in economic environments, rules and regulation as well as accounting standards. For example, in Malaysia as a developing country, majority of firms are family firm, and very conservatives compare to other developing countries. Thus, can findings of past studies be generalized to countries where market efficiency is far less than advanced countries and using of fair value may not be well accepted by investors due to the unreliability of the information for valuing assets?

In Malaysia the preparation of statement of comprehensive income was made mandatory beginning 1st January, 2010 under FRS 1011. The statement required public listed companies to separately report comprehensive income in their financial statements [8]. Adoption of FRS 101 which coincide with FRS 139 in Malaysia, provide a good opportunity to empirically investigate whether the other comprehensive income and its components are value relevant. Among components of other comprehensive income this study examines the value relevance of unrealized changes on available-for-sale financial instruments (AFS) and revaluation surplus of Property, Plant and Equipment (REV). The AFS and REV are components of other comprehensive income that are affected by fair value measurement. Moreover, AFS is influenced by FRS 139 coincides with the mandatory adoption of FRS 101 in 2010 and REV is influenced by FRS 116 in 2006.

This study contributes to the literature by providing empirical evidence on the value relevance of other comprehensive income. Despite many studies on value relevance of comprehensive income, there are limited empirical evidence on value relevance of other comprehensive income and its components after mandatory reporting of other comprehensive income in 2010. This study also contributes to the literature in a way that examines value relevance of other comprehensive income in a different market setting. Moreover, this study provides the empirical evidence on value relevance of two components of other comprehensive income which are affected from asset market, namely unrealized changes on available-for-sale financial instruments (AFS) and revaluation surplus of PPE (REV). The findings would help to better understand the value implications of components of other comprehensive income. This study provides empirical evidence on value relevance of other comprehensive income in a developing country. The result could be useful for policy maker in developing countries that have not mandatory reporting of comprehensive income and contemplating to do so. Additionally, this study addresses the concept of value relevance which is of interest to standard setters and policy makers of countries with developing economics that try to grab attention of investors to make more investment, and accordingly, accelerate economic growth.

2 Literature review

An assessment of the literature shows that the value relevance of comprehensive income, other comprehensive income and its components varies across countries and industries, and has also changed over time. In other words the results conducted so far by studies show the mixed results. Therefore, it is relatively difficult to compare the results of the past available studies in different countries. There are several possible explanations why these studies could not provide consistent results and further research is required to fill those gaps.

First, the IASB introduced the mandatory adoption of IFRS for listed companies with financial years starting on or after January 1, 2005. Moreover, the revision of IAS 1 and reporting of other comprehensive income components is only mandatorily applied for financial years starting on or after January 1, 2010. Therefore, comparable accounting data for doing empirical research has only been available since 2011. Due to this limitation, as a result, there are limited studies which their data are collected after effective date of IAS 1. The results obtained from the period after mandatory of comprehensive income reporting may lead to find results which they are more appropriate for assessing whether the capital market better understands the value implications of components of other comprehensive income.

Secondly, some of the prior studies suffered from potential measurement error. They had to calculate the other comprehensive income amounts (as-if data) due to unavailability of data instead of using actually reported values (as-reported data). For example [9],[10], [11], [12], [13] and [14] in their analysis used as if data. Despite of them for example [2] [15] and [6] use actual data of comprehensive income and other comprehensive income, however findings of them on value relevance of comprehensive income, other comprehensive income and its components, show mixed results.

Third, one of the reasons for the mixed result in the studies of value relevance of other comprehensive income, could be acceptance of IFRS, and its consequence, fair value. Early, the FASB required firms to report comprehensive income as the sum of net income and other comprehensive income. This requirement raises debates about the value relevance of other comprehensive income components as most important part of comprehensive income (e.g., [16] [5][10][17]. Then, the reporting of other comprehensive income is getting more attention because of the convergence of U.S. Generally Accepted Accounting Principles (GAAP) to International Financial Reporting Standards (IFRS) [18]. After introduction and adoption of IFRS and fair value accounting comprehensive income, and specifically other comprehensive income, become increasingly important for either practitioners or researchers [1]. This is because unrealized gains and losses in fair value of assets and liabilities are disclosed in other comprehensive income. The information provided by fair value could provide financial statement users with more complete, relevant, and representationally faithful information that should result in an improved basis for decision making [19].

Additionally, most studies carried up in developed countries where there are active market and obtaining a reliable measure for fair value is not a problem. However, there are very few studies on developing countries. One major issue is because in countries with developing economy asset market are less efficient to provide necessary and reliable data for fair value measurement. Firms in these countries would have to use price of similar assets and liabilities. This situation may result in estimated values that are less reliable.

Therefore, with respect to above, there is need to investigate value relevance of other comprehensive income with actual components of other comprehensive income instead of calculated data. Examining the value relevance of other comprehensive income also seems necessary after mandatory of other comprehensive income. Additionally the value relevance of other comprehensive income should be tested between different market efficiency and between different environments after convergence of IFRS.

3 Hypotheses development

This study identifies Efficient Market Hypothesis (EMH) as the relevant underlying theory. As discussed above, an information is said to be value relevant if it affect investors' decision and is reflected in the share price. This is in line with the efficient market hypothesis which states that an efficient market is a market in which its share prices always fully reflect all available information. The reporting of other comprehensive income and its components is aimed at enhancing financial reporting transparency. Therefore in line with the Efficient Market Hypothesis, other comprehensive income and its components information would have an effect on share price as such information could positively affect the ability of investors to make more effective decisions.

Therefore, based on the objectives of this study and the assertion of semi-strong form market efficiency this study hypothesize that:

H1: Other comprehensive income is positively associated with share price.

H1a: The unrealized changes on available-for-sale financial instruments component of other comprehensive income is positively associated with share price.

H1b: The revaluation surplus of Property, Plant and Equipment component of other comprehensive income is positively associated with share price.

4 Methodology

This study takes a quantitative approach and applies panel data design. The research samples were obtained from firms listed on the Main Market of Bursa Malaysia from 2011 to 2013. Data for this research was obtained from the Thomson Reuters DataStream Professional (DataStream) and Annual Reports of companies from 2011 to 2013.

In order to statistically examine the hypotheses, this study applies Ohlson's valuation model. In Ohlson's model, book value of equity and earnings are the explanatory variables of a firm's share price as follows:

$$P_{it} = \alpha_0 + \alpha_1 BVE_{it} + \alpha_2 NI_{it} + \alpha_3 v_t$$

Where:

P_{it} = Share price

BVE_{it} = Book value of equity per share

NI_{it} = Net income per share

v_t = error term

This study develops the Ohlson's model and specifies Model 1 for testing first hypothesis (H_1, H_{1a} and H_{1b}) that investigates whether total other comprehensive income and its components are value relevant. The Model 1, 2 and 3 are expressed as follows:

Models	Models' number
$PRICE_{it} = \alpha_0 + \alpha_1 BVE_{it} + \alpha_2 NI_{it} + \alpha_3 OCI_{it} + \alpha_4 Size_{it} + \alpha_5 Lev_{it} + \varepsilon_{it}$	(1)
$PRICE_{it} = \alpha_0 + \alpha_1 BVE_{it} + \alpha_2 NI_{it} + \alpha_3 AFS_{it} + \alpha_4 Others_{it} + \alpha_5 Size_{it} + \alpha_6 Lev_{it} + \varepsilon_{it}$	(2)
$PRICE_{it} = \alpha_0 + \alpha_1 BVE_{it} + \alpha_2 NI_{it} + \alpha_3 REV_{it} + \alpha_4 Others_{it} + \alpha_5 Size_{it} + \alpha_6 Lev_{it} + \varepsilon_{it}$	(3)

DOI: 10.1051/shsconf/20173403004

5 Findings

The regression analyses result of the first hypothesis (H1) reveals that total other comprehensive income is positively associated with share price. This means that total other comprehensive income is value relevant. The regression results of testing hypotheses H1a and H1b indicate that the unrealized changes in available for sale financial instruments (AFS) and revaluation surplus of property, plant and equipment (REV) are significantly and positively associated with share price. In other words, unrealized changes in available for sale financial instruments (AFS) and revaluation surplus of property, plant and equipment (REV), are value relevant. In general it can be concluded that information on other comprehensive income and its component is informative and useful for investors' decision making.

References

[1] M. Bertoni and B. De Rosa, "COMPREHENSIVE INCOME , FAIR VALUE , AND CONSERVATISM : A CONCEPTUAL FRAMEWORK FOR," pp. 1–14, 2013.

[2] K. Kanagaretnam, R. Mathieu, and M. Shehata, "Usefulness of comprehensive income reporting in Canada," *J. Account. Public Policy*, vol. 28, no. 4, pp. 349–365, Jul. 2009.

[3] R. Ball, "(IFRS): pros and cons for investors International Financial Reporting Standards (IFRS): pros and cons for investors," no. June 2013, pp. 37–41, 2006.

[4] M. Hoogendoorn, "International Accounting Regulation and IFRS Implementation in Europe and Beyond - Experiences with Eirst- time Adoption in Europe," vol. 3, 2006.

[5] D. Chambers, T. J. Linsmeier, C. Shakespeare, and T. Sougiannis, "An evaluation of SFAS No. 130 comprehensive income disclosures," *Rev. Account. Stud.*, vol. 12, no. 4, pp. 557–593, May 2007.

[6] A. Mechelli and R. Cimini, "Is Comprehensive Income Value Relevant and Does Location Matter? A European Study," *Account. Eur.*, 2014.

[7] Q. Huang, J. Ye, and G. Du, "Value Relevance of Comprehensive Income in Small and Medium-size Enterprises," 2014.

[8] A. Abdul Rahman and M. D. H. Hmadan, "Reporting Comprehensive Income: Evidence from Malaysian ACE Market Companies," *SSRN Electron. J.*, pp. 1–26, 2012.

[9] C. S. A. Cheng, J. K. Cheung, and V. Gopalakrishnan, "On the usefulness of operating income, net income and comprehensive income in explaining security returns," *Account. Bus. Res.*, 1993.

[10] D. Dhaliwal, K. R. Subramanyam, and R. Trezevant, "Is comprehensive income superior to net income as a measure of firm performance?," *J. Account. Econ.*, vol. 26, no. 1–3, pp. 43–67, Jan. 1999.

[11] M. Brimble and A. Hodgson, "The Value Relevance of Comprehensive Income and Components for Industrial Firms," *Work. Pap. Griffith Univ. Bus. Sch. (Universiteit Amsterdam)*, 2004.

[12] J. Ernstberger, "The value relevance of comprehensive income under IFRS and US GAAP: empirical evidence from Germany.," *Int. J. Accounting, Audit. Perform. Eval.*, vol. 5, no. 1, 2008.

[13] I. Goncharov and A. Hodgson, "Measuring and Reporting Income in Europe," *J. Int. Account. Res.*, vol. 10, no. 1, pp. 27–59, Jan. 2011.

[14] H. S. Deol and J. A. Nazari, "The Decision Usefulness of Comprehensive Income Reporting: Evidence from Canada," 2013.

[15] H. Zülch and P. Pronobis, "The Predictive Power of Comprehensive Income and Its Individual Components under IFRS" no. 95, 2010.

[16] G. C. Biddle and J.-H. Choi, "Is Comprehensive Income Useful?," *J. Contemp. Account. Econ.*, vol. 2, no. 1, pp. 1–32, Jun. 2006.

[17] D. a. Jones and K. J. Smith, “Comparing the Value Relevance, Predictive Value, and Persistence of Other Comprehensive Income and Special Items,” *Account. Rev.*, vol. 86, no. 6, pp. 2047–2073, Nov. 2011.

[18] C. Lee and M. S. Park, “Subjectivity in fair-value estimates, audit quality, and informativeness of other comprehensive income,” *Adv. Account.*, Jun. 2013.

[19] J. P. Jones and S. D. Stanwick, “Fair value accounting: A guide to understanding the current standards,” *J. Corp. Account. Financ.*, vol. 11, no. 1, pp. 103–108, 1999.

DOI: 10.1051/shsconf/20173404001

The Quality of Accounting Earnings and Change in Political Power Map: Evidence from Indonesia

Iman Harymawan[1,*], *Dian* Agustia[2] , and *Khusnul* Prasetyo[3]

[1, 2, 3] Airlangga University, Indonesia

Abstract. This study investigates the earnings quality of politically connected firms listed in Indonesian Stock Exchange during 2006-2010. This study compare two definition (Soeharto Dependency Index (SDI) and Politically Exposed Person (PEP)) of political connections to capture the effect of the changes of political power map on the earnings quality of politically connected firms in Indonesia. The finding shows that the quality of accounting earnings of politically connected firms which were formerly closely related to President Soeharto in 1998 are not significantly different to other firms. Furthermore, when this study employ politically exposed person as a proxy of political connections, the findings shows that connected firms have significantly lower quality accounting earnings. Consistent with prior findings, this study shows that change in political power map in Indonesia affect the level of earnings quality of politically connected firms.

1. INTRODUCTION

In the annual survey of corruption perceptions, Transparency International rates Indonesia as one of the most corrupt countries on earth. Political power is one of the influential factors in many aspects in this country. Despite a number of recent media articles on the topic, prior studies have not examined the quality of accounting earnings for politically connected firms in Indonesia. Previous studies of connected firms in Indonesia look at stock returns following rumors about Soeharto's (former Indonesia President) health [1]. This study shows that firms that were close to Suharto suffered negative returns when bad news about the President's health hit the stock market. Leuz and Oberholzer-Gee [2] analyze how politically connected firms in Indonesia make strategic financing decisions. Their results suggest that foreign securities and close political connections are substitutes. They find that firms that close to Suharto were significantly less likely to have publicly traded securities abroad.

The results of international studies suggest that firms with political connections have lower quality accounting earnings [3]. Firms with political ties dislike the transparency and scrutiny because high levels of public scrutiny can be difficult to reconcile with political favors of often dubious legality [2]. However, whether this is the case for Indonesia and whether the quality of accounting earnings in Indonesia is influenced by the

* Corresponding author: harymawan.iman@feb.unair.ac.id

change of structure of political power map is the focus of this paper. This study also further analyzes the quality of accounting earnings among politically connected firms (non-State Owned Enterprises (SOEs)), State-Owned Enterprises and non-politically connected firms for listed companies in Indonesia.

The first part of this paper examine the link between political connections and the quality of accounting earnings in Indonesia's listed companies using the Soeharto Dependency Index [1]. Leuz and Oberholzer-Gee [2] find that for firms which closely related to Soeharto tend to access to foreign securities due to the difficulty re-establishing connections with a new government when their patrons fall from power. Firms with foreign securities attract the attention of foreign analysts and the international press coverage [[4] ; [5]]. Due to the increase of public scrutiny, they also have to increase their transparency. The result of this paper support this view which shows that the quality of accounting earnings of firms which were formerly closely related to Soeharto is not significantly different to other firms during 2005-2010 due to change in political structure map. In respect of relation, the political connections dummy variable has negative relation to accounting quality variable. It is indicate that the quality of accounting earnings for firms which formerly closely related to Soeharto is relatively better than other firms.

In the second part, this study employ *politically exposed person* to define politically connected firms. This study find that firm with political connections is associated with a lower quality of accounting earnings. Interestingly, when we include State-owned Enterprise (SOE) as politically connected firms, we find that the quality of accounting earnings of politically connected firms is become higher. This finding shows that SOE in Indonesia have better quality of accounting earnings than firms which have *politically exposed person* as a board of directors. Overall, consistent with prior findings, these findings give us the notion that the changes of political power structure influence the quality of accounting earnings of politically connected firms. This study contributes to literature on the comparison of the earnings quality of politically connected firms using two different definitions of political connections. This paper also introduces the new proxy of political connections using *politically exposed person.* The rest of this paper is organized as follows. The second section will discuss literature review. Third section describes the financial and political connection data. Finally, we present the result and conclude.

2. HYPOTHESIS DEVELOPMENT

Until the end of Suharto reign, Indonesia is highly centralized and stable political structure [1]. A year after Soeharto stepped down Indonesia has had a multi-party system. Before 1999, there are only 3 parties participated in the elections. However, in the two legislative elections since the fall of the Soeharto regime, no political party has managed to win an overall majority of seats, resulting in coalition governments. Greater democracy in the post-Suharto era has introduced new forms of corruption. There are now many more political parties and these also need money. Soeharto's dramatic fall in May 1998 created political vacuum, with weak institutions and lack of popular leaders. With the collapse of rigid centralized control, the state became increasingly disjointed and self-serving. Corruption, which elaborately well organized under Suharto, grew more random and unpredictable. The public clamored for "reformasi", which generally stood for freedom, justice, and prosperity.

In 2004, the faltering transition from President Suharto's three decades of authoritarian rule experienced a breakthrough with the decisive popular election of President Susilo Bambang Yudoyono. For the first time in its 59-year history, Indonesia's executive is directly accountable to the people. For the era of "reformasi", this innovation marks a new beginning after a long series of false starts. However, how is the improvement of Indonesia transparency (in general) and how is the quality of accounting earnings (transparency) for listed firms (in particular) politically connected firms in reformation era for Indonesia is still remain a question.

Prior studies have examined political connections as they are expected to provide benefits to firms. Faccio et al., [6] find that politically connected firms are more likely to be bailed out during economic crises. Several studies across different markets also show that firms with political connections gain preferential access to credit [[7], [8], [9], [10], [11]] Chaney, et al. [3] in their international study find that firms with political connections is associated with a lower quality of accounting earnings. Furthermore, they document that political connections have incremental explanatory power beyond country, regulatory and firm-specific ownership characteristics. Leuz and Oberholzer-Gee [2] find that for firms which closely related to Soeharto tend to access to foreign securities due to the difficulty re-establishing connections with a new government when their patrons fall from power. Firms with foreign securities attract the attention of foreign analysts and the international press coverage [[4], [5]]. This facts leads to the first research question examined in this study – *How is the quality of accounting earnings for firms which formerly closely related to Soeharto in reformation era?*. This study expect that connected firms (SDI) will have better (or no difference) earnings quality than non-connected ones.

H1: Firms with political connections (SDI) will have better (or not significantly different) earnings quality than non-connected ones

Goldman, So, and Rocholl [12] find that in the country with strong legal system, companies that are connected to the winning party are significantly more likely to experience higher procurement contracts. Then, when the losing party regains their political power in the next period, in turn, they will experience an increase in procurement contracts significantly. This study suggests that the value of current political power is matter to their connection's benefit. Leus and Oberholzer-Gee [2] explain three reasons that well-connected firms are less likely to have publicly traded debt or equity securities abroad. First, well-connected firms have access to preferential financing at home and therefore do not need to access foreign capital markets. Second, firms with political ties dislike the transparency and scrutiny that come with publicly traded securities. Third, foreign securities make it more difficult for insiders to extract private control benefits. They also argue that Firms with political ties dislike the transparency and scrutiny because high levels of public scrutiny can be difficult to reconcile with political favors of often dubious legality. This leads to the second research question examined in this study – *do firms with politically exposed person as board director is associated with lower quality of accounting earnings?*. This study expects to find that firms with connection (PEP) will have lower earnings quality than non-connected ones.

H2: Firms with political connections (PEP) will have lower earnings quality than non-connected ones.

3. DATA

3.1. *Sample*

For initial board members data in Indonesia Stock Exchange (IDX) obtained from Indonesia Capital Market Data (ICMD) directory. The first definition of political connection data (SDI) is from Soeharto Dependency Index which used in Fisman [1] and Leuz and Oberholzer-Gee [2] to define firms which closely related to Soeharto. The second definition of political connection data is from updated version of *politically exposed person* (PEP) list data from Tjondro [13]. In this data, they define politically connected firms in two ways. First, if one (or more) board of directors in the firm is categorized as *politically exposed person* based on Bank of Indonesia definition. Second is State-owned Enterprise (SOE). This approach (board connections) has been previously used in cross-country studies (Faccio et al., 2006) and studies in other countries [[14], [15], [16]].We update this list by adding some additional companies which not included in their list. Finally, we obtain financial data from Osiris database which provide financial report information for all listed companies in Indonesia Stock Exchange (IDX). My sample period covers year 2006 to 2010.

3.2. *Variables*

All variables definition used in this analysis are provided in this section. Dummy variables are used to identify firms with and without connected firms (SDI and PEP), SOE, and non-connected firms. To measure earnings quality, I follow Chaney, et al., [3] which used performance adjusted current accruals measure (REDCA) to measure the quality of financial reporting. This method is developed by Ashbaugh, LaFond, and Mahyew [17]. REDCA is computed as the difference between total current accruals (TCA) and expected performance (i.e.,ROA) adjusted total current accruals (EPTCA). We also use sales growth, year, and industry (2 digits SIC code) as control variables. If the politically connected firm has positive relation with REDCA, it means that the accrual quality is high or in other words the quality of accounting earnings is low.

4. EMPIRICAL ANALYSIS

4.1. *Soeharto Dependency Index and quality of accounting earnings*

In this section, we use cross sectional samples to examine the quality of accounting earnings for firm which were formerly closely related to Soeharto during 2006-2010. To examine this issue, we employ ordinary least square (OLS) regression to see how quality of accounting earnings varies on the listed companies in Indonesia.

$$REDCA = \beta_0 + \beta_1 connected + \beta_2 sales\ growth + \beta_3 year + \beta_4 industry + \varepsilon$$

Table 1 shows the result of the quality of accounting earnings for firm which formerly closely related to Soeharto in 2006-2010. The result shows the political connections dummy has negative relation to the quality of accounting earnings. The firms which were formerly closely related to Soeharto have better accounting earnings relatively to other firms in the reformation era. However, it is insignificant.

This findings consistent with Leuz and Oberholzer-Gee [2] which find firms which closely related to Soeharto tend to access to foreign securities due to the difficulty re-establishing connections with a new government when their patrons fall from power. Therefore they need to access foreign securities which requires higher of public scrutiny (higher transparency).

Table 1. Soeharto Dependency Index and quality of accounting earnings.

The dependent variable to measure accounting information quality is the standard deviation of the firm's discretionary accruals REDCA. Connected is dummy variable set to 1 if the firms formerly closely connected to Soeharto and 0 otherwise.

	1	2	3
Connected firms (SDI)	-3.6578 (-0.35)	-3.3250 (-0.40)	-3.6576 (-0.35)
Sales growth	-0.0015 (-1.39)	-0.0015 (-1.42)	0.0015 (-1.38)
Industry fixed effect	Yes	No	Yes
Year fixed effect	No	Yes	Yes
Intercept	0.8042 (0.03)	-2.3583 (-0.41)	0.2.6267 (-0.10)
Number of obs.	836	836	836
Adjusted R2	-0.0369	-0.0021	-0.0395

4.2. *Politically exposed person and quality of accounting earnings*

In the second section, we examine the quality of accounting earning using new set of political connections data based on *politically exposed person* who serve as board director in firms as a proxy. The objective of this test is to examine whether the quality of accounting earnings of politically connected firms in reformation era is become better or even worst. As we discussed earlier, in the political level, Soeharto's dramatic fall in May 1998 created political vacuum, with weak institutions and lack of popular leaders. With the collapse of rigid centralized control, the state became increasingly disjointed and self-serving. Corruption, which elaborately well organized under Suharto, grew more random and unpredictable.

Table 2 shows the relations between quality of accounting earnings and firms which have one or more *politically exposed person.* Surprisingly, after controlling some variables, we find that politically connected firms have significantly positive association with the quality of accounting earnings. This finding shows that politically connected firms are significantly have higher accrual quality or in other words is associated with lower quality of accounting earnings. This finding consistent with Goldman, So, and Rocholl [12] which find that in the country with strong legal system, firms that are connected to the winning party are significantly more benefit to their connection. It is also consistent with the explanation from Leus and Oberholzer-Gee [2] on why well-connected firms are less likely to have publicly traded debt or equity securities abroad. First, well-connected firms have access to preferential financing at home and therefore do not need to access foreign capital markets. Second, firms with political ties dislike the transparency and scrutiny that come with publicly traded securities. Third, foreign securities make it more difficult for insiders to extract private control benefits.

Table 2. Politically exposed person and quality of accounting earnings.

The dependent variable to measure accounting information quality is the standard deviation of the firm's discretionary accruals REDCA. Connected is dummy variable set to 1 if the firm has one (or more) politically exposed person as board of directors and 0 otherwise.

	1	2	3
Connected firms (PEP)	14.7739 (2,06)	10.4604 (1.71)	14.7963 (2.06)
Sales growth	-0.0015 (-1.38)	-0.0015 (-1.37)	-0.0015 (-1.37)
Industry fixed effect	Yes	No	Yes
Year fixed effect	No	Yes	Yes
Intercept	-16.4077 (-.065)	-5.0818 (-0.81)	-18.3816 (-0.71)
Number of obs.	800	800	800
Adjusted R2	-0.0326	0.0012	-0.0353

As additional test, we construct dummy variable for state-owned enterprise in the sample. Interestingly, when we include State-owned Enterprise (SOE) as politically connected firms, we find that the relations between SOE and the quality of accounting earnings still positive but insignificant. This finding show that SOE in Indonesia have better quality of accounting earnings than firms which have politically exposed person as a board of directors. Overall, consistent with prior findings, these findings give us the notion that the changes of political power structure also influence the quality of accounting earnings of politically connected firms.

5. CONCLUSION

This paper examines the effect of changes in political power map to the quality of accounting earnings of politically connected firms in Indonesia in the reformation era (2006-2010). We employ two definitions as a proxy of political connections. First is Soeharto Dependency Index which classifying the firms which closely related to President Soeharto (old definition). Second is *politically exposed person* as a proxy of political connections (new definition). This study finds no difference between politically connected firms which formerly closely related to President Soeharto in 1998 and non-connected firms. Surprisingly, when we employ politically exposed person as board of directors as an indicator, we find that politically connected firms have significantly lower quality of accounting earnings. Additionally, this study also find interesting result that state-owned enterprise (SOE) have better quality of accounting earnings rather than firms with politically exposed person as a boards. Overall, consistent with prior findings, these findings give us the notion that the changes of political power structure also influence the quality of accounting earnings of politically connected firms. There are several limitations in this research. First due to time constrain of data collection, there are some others control variables such as market capitalization and operating cycle that have not been added into the model so far. Second, there are some missing variables in loans (which a part of REDCA calculations) data that might be influence the result. Third, there are still several other settings that might be developed as robustness checks such as the relation of number *politically exposed person* and the dependency index to the extent of financial opacity and employ some alternative measures the quality of accounting earnings.

REFERENCES

[1] Fisman, R. (2001). Estimating the Value of Political Connections. *The American Economic Review, **91***(4), 1095-1102.

[2] Leuz, C., & Oberholzer-Gee, F. (2006). Political relationships, global financing, and corporate transparency: Evidence from Indonesia. *Journal of Financial Economics, **81***(2), 411-439.

[3] Chaney, P. K., Faccio, M., & Parsley, D. (2011). The quality of accounting information in politically connected firms. *Journal of Accounting and Economics, **51***(1-2), 58-76.

[4] Baker, H. K., Nofsinger, J. R., & Weaver, D. G. (2002). International Cross-Listing and Visibility. *Journal of Financial & Quantitative Analysis, **37***, 495-521.

[5] Lang, M. H., Lins, K. V., & Miller, D. P. (Writers). (2003). ADRs, Analysts, and Accuracy: Does Cross Listing in the United States Improve a Firm's Information Environment and Increase Market Value? [Article], *Journal of Accounting Research*

[6] Faccio, M., Masulis, R. W., & McConnell, J. J. (2006). Political Connections and Corporate Bailouts. *The Journal of Finance, **61***(6), 2597-2635.

[7] Backman, M. (2001). *Asian Eclipse: Exposing the Dark Side of Business in Asia* (Revised edition ed.): John Wiley & Sons.

[8] Claessens, S., Feijen, E., & Laeven, L. (2008). Political connections and preferential access to finance: The role of campaign contributions. *Journal of Financial Economics, **88***(3), 554-580.

[9] Cull, R., & Xu, L. C. (2005). Institutions, ownership, and finance: the determinants of profit reinvestment among Chinese firms. *Journal of Financial Economics, 77*(1), 117-146.

[10] Johnson, S., & Mitton, T. (2003). Cronyism and capital controls: evidence from Malaysia. *Journal of Financial Economics, 67*(2), 351-382.

[11] Khwaja, A. I., & Mian, A. (2005). Do Lenders Favor Politically Connected Firms? Rent Provision in an Emerging Financial Market*. *Quarterly Journal of Economics,* ***120***(4), 1371-1411. doi: 10.1162/003355305775097524
[12] Goldman, E., Rocholl, J., & So, J. (2013). Politically Connected Boards of Directors and The Allocation of Procurement Contracts. *Review of Finance*, ***17***(5), 1617-1648.
[13] Tjondro, E. (2012). *Effect of Financial Leverage, Profitability, and Type of Bank Lending Decision using Political Tie as Moderating Variables.* Master thesis, Airlangga University, Surabaya.
[14] Gray, S., Harymawan, I., & Nowland, J. (2016). Political and government connections on corporate boards in Australia: Good for business?. *Australian Journal of Management,* ***41***(1), 3-26.
[15] Harymawan, Iman, and Nowland, J. "Political connections and earnings quality: how do connected firms respond to changes in political stability and government effectiveness?." *International Journal of Accounting & Information Management* 24.4 (2016).
[16] Hillman, A. J. (2005). Politicians on the board of directors: Do connections affect the bottom line?. *Journal of Management*, 31(3), 464-481.
[17] Ashbaugh, H., LaFond, R., & Mayhew, B. W. (2003). Do nonaudit services compromise auditor independence? Further evidence. *The Accounting Review,* ***78***(3), 611-639.

DOI: 10.1051/shsconf/20173404002

The mediating effect of leverage and dividend policy on the influence of corporate governance towards firm value

Noorlailie Soewarno[1,*] and *Stephanie* Yulia Arifin[1] *Bambang* Tjahjadi[1]
[1] Universitas Airlangga, Indonesia

Abstract. This study examines the mediating effect of leverage and dividend policy on the influence of corporate governance towards firm value. This is a quantitative study used secondary data of 181 companies listed on the Indonesian Stock Exchange (IDX) in the year of 2014. The results showed that leverage did not mediate corporate governance-firm value relationship, and devidend policy partially mediated corporate governance-firm value relationship

1. Introduction

In 2002, the importance of corporate governance became more obvious as a series of corporate disasters such as frauds and meltdowns led to a series of bankruptcy filings after the 1998 economic crisis, included Enron and WorldCom)[1].
When a company is required to improve their performance quality through implementation of corporate governance, many steps have to be taken as form of responsibility to investors and creditors. Some of them are dividend distribution, controlling level of debt and repayment of principal loan amount and interest expense. The more payment for loan means lower return for shareholders. It is parallel with the findings of Hastori et al. [2], which said that there was a significantly negative relation between dividend payout policy and agency costs while leverage had significantly positive influence towards agency cost. Agency problem may occur when dividend paid is smaller because of higher payment for loan or higher amount of retained earnings. Many studies have been conducted to examine the relationship between corporate governance, leverage, dividend policy and firm value. Ionescu [3] argued that conducting corporate governance practices could transmit credible signal to investors. Appropriate corporate governance implementation could reduce the risk of managers' misbehaviors. Therefore, it can be assumed that there was significant positive connection between quality of corporate governance and firm value.

When investors feel ensured that their investment is profitable and information given is transparent, firms can be indicated as having high level of credibility. The trust of investors, then, can increase the attractiveness of a firm in stock market which leads to increase in firm value. Furthermore, there were still inconsistency in the results of previous studies. Therefore the influences are still required to be reexamined. Also, there was no research that tried to combine corporate governance, leverage, dividend policy and firm value all

* Corresponding author noorlailie@gmail.com

together. Therefore, the author is interested to study those relationships in Indonesian case. The purpose of this research was to investigate the existence of a significant direct and/or indirect influence of corporate governance, leverage and dividend policy towards firm value in companies listed in Indonesia Stock Exchange (IDX).

2. Literature review

In the agency theory, agents who tend to fulfill their own interest, tend to act recklessly by investing in low or even negative present value project using excessive external funding. This harmful activity can cause asymmetry information between agents and principals in terms of segregation of functions between ownership and management. This agency problem indicates that firm value will increase if principals regulate agent's behaviors in managing resources. Several previous studies found that the relationships between corporate governance and firm value varied. Board size and institutional shareholders were found that they influenced firm value positively while other indicators did negatively [4]. If the indicators are combined, corporate governance still has significant and positive connection with firm value [5]. Corporate governance can transmit strong signal to convince investors about companies' credibility, meaning that corporate governance contributes to increasing firm value [3] [6].

Previous studies investigating the relationship between corporate governance and leverage found that corporate governance was related negatively with the level of leverage [7] [8] [9]. This happened because if the company improved its corporate governance practices, management would eventually consider debt as disciplining device. Corporate governance will direct management only to take appropriate possible actions [10]. In the other hand, it was also found that corporate governance could increase the level of leverage because it expanded its credit-access [11]. Companies which have more independent directors has more long-term debt [12]. There is a positive and significant relationship between debt and firm value in countries with double taxation system but not in countries with imputation system [13]. The difference in those relationships depends on corporate tax rate and personal tax rate which reduces the tax advantage coming from corporate borrowing [14]. On the contrary, leverage also can influence firm value negatively [15]. The company which wants to increase firm value should only take leverage up to the optimum point because after that point, leverage will bring more cost than benefit [16].

H1: Leverage mediates the influence of corporate governance towards firm value

The relationship between corporate governance and dividend policy may differ depending upon the access of interest-bearing debt [17]. In South Africa, Kenya and Ghana, it was found that there was positive relationship between corporate governance and dividend payout as the other researches [18][19], while in Nigeria, corporate governance was linked negatively to dividend payout. Companies in Nigeria tended to reduce the cost of debt [17]. Regarding dividend policy and firm value, dividend per share and dividend yield are both factors in determining share prices [20]. Dividend per share has significant and positive relation with firm value while dividend yield has significant and negative relationship with firm value [20]. Dividends have positive information that earnings and other variable do not [15]. Investors in countries with poor investor protection will prefer dividends more because private benefits are more likely to happen [21]. Indonesia itself is considered as having poor legality, shareholder rights, and judicial efficiency compared with 13 other countries with emerging markets [7]. Conversely, it was found that dividend yield influenced shares prices negatively but dividend payout sometimes gave positive influence [22]. Dividend policy may have clientele effect on investors [20].

H2: Devidend policy mediates the influence of corporate governance towards firm value

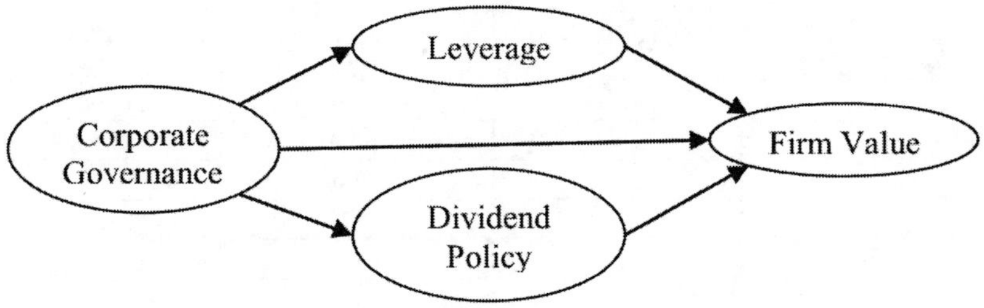

Figure 1 Conceptual Framework

3. Research method

The sample used in this study are companies listed in the Indonesia Stock Exchange (IDX) in the year of 2014, which are amounted to 181 companies. The data used is the secondary data which is financial data from financial statements of companies listed in IDX for the period which ended on December 31, 2014, dividends paid during 2014 and shares prices on April 30, 2015. The data was processed using the Partial Least Square (PLS) Warp. The following table elaborates indicators and operational definition of variables:

Table 1 Operational Definition of Variables

Variables	Symbol	Indicators	Definition
Corporate Governance (CG)	BS	Board size	The number of members in board of commissioners
	AC	Audit committee	The number of members in audit committee of a company
	AQ	Audit quality	Whether a company using big four public company or not, measured by a scale of 1 for companies using big-four public accounting firm and 0 for companies using non-big-four public accounting firm
Firm value (FV)	Q	Tobin's Q	(Market value of equity + debt)/Assets
	PBV	Price to book value	(Market value per share)/Book value per share
Leverage (Lev)	D/E	Debt to equity ratio	Total debt/Total equity
	D/A	Debt to assets ratio	Total debt/Total assets
Dividend Policy (Div)	DPR	Dividend payout ratio	Dividend per share/ Earnings per share
	DY	Dividend yield	Dividend per share/ Price per share

4. Results and discussion

4.1 Direct influences

Table 2 Direct Influences

Direct Relationship	B-value	P-value	Conclusion
CG → Lev	0.26	< 0.01	Accepted
CG → Div	0.28	< 0.01	Accepted
CG → FV	0.18	< 0.01	Accepted
Lev→ FV	-0.17	< 0.01	Accepted
Div → FV	0.21	< 0.01	Accepted

Corporate governance is found to have positive and significant impact on leverage. Corporate governance can increase the creditors' trust so that the credit access will be higher, the level of leverage increases and cost of debt decreases [11]. This can happen because higher level of CG make a company able to have more long-term financing [12]. By reducing agency problems, corporate governance, then, will increase the trust of creditors about the optimal condition of the company which lead to increase in leverage.

Corporate governance positively and significantly correlate to dividend policy. By improving corporate governance, a firm can increase the minorities' right which led to increase in dividend policy. The empirical finding shows that corporate governance influences dividend policy positively. Corporate governance can protect minorities' right in dividend policy [18]. Thus, higher level of corporate governance means higher dividend policy.

Corporate governance also has positive impact on firm value. Companies in countries with weak legal protection system, including Indonesia, can increase their credibility by improving their Corporate governance because it signals about investor protection and asymmetric information reduction [3][6]. Corporate governance can solve agency problems. Having more transparent information and good relationship can assure the investors about the credibility of a firm.

This study showed that leverage affects firm value negatively, and supported study the previous study by [15] [16]. Leverage can increase firm value until certain point [16]. After it is reached, the increase of leverage will influence negatively towards firm value. The increase experienced by firm value until the maximum point is caused by the existence of leverage which brings more tax benefit than cost of debt. When the cost of debt is higher than the benefit, firm value will decline because the investors feel that the investment is too risky. Thus, firm value will increase until certain point then there is a decline.

In the agency theory, it is explained that the interest of shareholders is to gain more return while the interest of management is to retain or have private benefit. By sharing part of profits as dividends, investors can consider that the firm has minorities' right which lead to increase in firm value. This research also found that dividend policy positively influence firm value. Investors in countries with poor investor protection (such as Indonesia) will prefer dividends more because private benefits are more likely to happen [21].

4.2 The mediating effects of leverage on corporate governance-firm value relationship

Figure 2 showed the empirical result of the direct effect of corporate governance on firm value. It showed that p-value <.01with the beta value of 0.18.

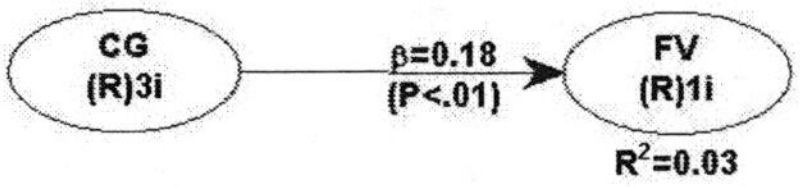

Figure 2. Direct Effect of Corporate Governance on Firm Value

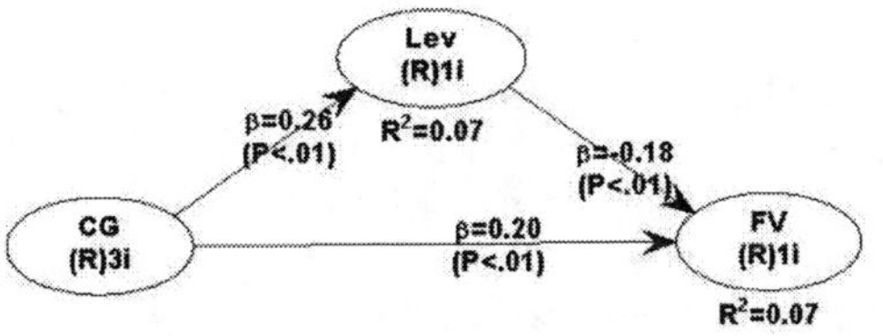

Figure 3. The Mediating Effect of Leverage on Corporate Governance-Firm Value Relationship

Figure 3 showed that the direct effect was still significant (p-value <.01), but the beta value increased from 0.18 to 0.20. This was not a sign for the existence of mediating effect. Therefore the hypothesis 1 stating that leverage mediates the influence of corporate governance towards firm value cannot be accepted. Ensuring that management takes the best action to create long-term and sustainable value is a main topic in agency theory. By having corporate governance, a firm can signal to investors about its credibility. Besides, corporate governance also can increase the transparency of a firm. Transparency can reduce the asymmetric information between management and investors. Corporate governance has already influenced firm value, whether the companies have high or lower leverage. Corporate governance can signal to investors about investors' protection and reduction of asymmetric information through fulfilling their transparency obligations [3] [6]. Corporate governance influences leverage positively [11] [12]. By implementing and improving corporate governance, a company can increase the confidence of creditors, hence increasing the leverage level too. Higher leverage means higher risk, and at the same time can potentially lower firm value. Decreasing firm value is caused by uneasy investors of higher risk. Even though debt brings tax benefit; it also brings higher risks for investors, such as bankruptcy costs [15].

4.3 The mediating effects of devidend policy on corporate governance-firm value relationship

Figure 4 showed that the direct effect was still significant (p-value is more than 1%, but still less than 5 %), and the beta value decreased from 0.18 to 0.14. This was a sign for the existence of partial mediation effect. Therefore the hypothesis 2 stating that dividend policy mediates the influence of corporate governance towards firm value can be accepted.

DOI: 10.1051/shsconf/20173404002

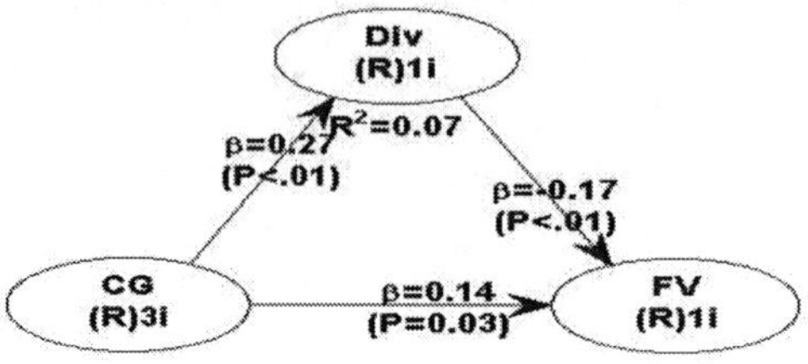

Figure 4. The Mediating Effect of Devidend Policy on Corporate Governance-Firm Value Relationship

In countries with weak legal protection system, corporate governance could enhance credible investor protection signal and reduce asymmetric information [5] [16]. Corporate governance could increase the right of minorities to obtain their dividend which meant higher corporate governance showed higher dividend policy [19]. The investors will feel safer about their dividend rights of minorities if companies have corporate governance. The trust of investors, then eventually will increase firm value.

5. Conclusion and suggestions

The focus of this study is to examine the mediating effects of leverage and dividend policy on corporate governance-firm value relationship in the Indonesian stock Exchange (IDX). Empirical findings showed the following results. Firstly, leverage did not mediate the influence of corporate governance towards firm value. Secondly, dividend policy mediated the relationship between corporate governance and firm value. Further research should expand into industrial sectors and/or using a longer period and using different indicator for each variable. For the practitioners, it is better to improve corporate governance, keep the leverage low and watching the effects of dividend policy towards firm value.

References

1. Monks, Robert A. G. and Nell Minow. *Corporate Governance*. New Jersey: Wiley-Blackwell (2004)
2. Hastori, Hermanto Siregar, et al. Agency Costs, Corporate Governance and Ownership Concentration: The Case of Agro-industrial Companies in Indonesia. *Asian Social Science*, 11(18): 311-319 (2015)
3. Ionescu, Luminita. Effects of Corporate Governance on Firm Value. *Economics, Management, and Financial Markets*, **7**(4), 215-220 (2012)
4. Marouan, Kouki and Dabboussi Moez. Impact of Corporate Governance on Shareholder Value Creation: Evidence from Tunisian Context. *International Business Research*, **8**(5) (2015)
5. Siahaan, Fadjar O.P. The Effect of Good Corporate Governance Mechanism, Leverage, and Firm Size on Firm Value. *GTSF International Journal on Business Review (GBR)*, **2**(4):137-142 (2013)
6. Klapper, Leora F. and Inessa Love. Corporate Governance, Investor Protection, and Performance in Emerging Markets. *World Bank Policy Research Working Paper* (2002)

7. Bunkanwanicha et al., (2008)
8. Arping, Stefan and Zacharias Sautner. Corporate Governance and Leverage: Evidence from a Natural Experiment. *Tinbergen Institute Discussion Paper*, 10-019/2 (2010)
9. Richardson, Grant, et al. Corporate Tax Aggressiveness, Outside Directors and Debt Policy: An empirical Analysis. *Journal of Corporate Finance*, **25**:107-121 (2014)
10. Saad, Noriza Mohd. Corporate Governance Compliance and the Effects to Capital Structure in Malaysia. *International Journal of Economic and Finance*, **2**(1): 105-114 (2010)
11. Funchal, Bruno and Daniel Gottlieb. Corporate Governance, Bankruptcy Law and Firms' Debt Financing Under Uncertainty. *Corporate Ownership and Control*, **6**(2):47-50 (2008)
12. Alves, Paulo, E.B. Couto, and P.M. Francisco. Board of Director Composition and Capital Stucture. Research in International Business and Finance, **35**, 1-32 (2015)
13. Jiang, Yuanyuan. *Taxes, Debt, and Firm Value: New Evidence*. Yale University, ProQuest Dissertations Publishing (2004)
14. Myers, Steward C. Determinats of Corporate Borrowing. *Journal of Financial Economics*, **5**: 147-175 (1977)
15. Fama, Eugene F. and Kenneth R. French. Taxes, Financing decisions, and Firm Value, (1977). (Online), (https:/ssrn.com/abstract=1871 accessed on Oct. 13 (2015)
16. Modigliani, Franco and Merton H. Miller. Corporate Income Taxes and the Cost of Capital: A Correction. *The American Economic Review*, **53**(3) (1963)
17. Abor, Joshua and Vera Viador. Does Corporate Governance Explain Dividend Policy in Sub-Saharan Africa? *International Journal of Law and Management*, **55**(3):201-225 (2013)
18. Kowalewski, Oskar, I. Stetsyuk, and O. Talavera. *Corporate Governance and Dividend Policy in Poland.* German Istitute of Economic Research, 702 (July).*German Institutte for Economic research*, 702 (July) (2007)
19. Schellenger, Michael H., D. D. Wood, and A. Tashakori. Board of Director Composition, Shareholder Wealth and Dividend Policy. *Journal of Management*, **15**(3):457-467 (1989)
20. Sharif, Taimur, H. Purohit, and R. Pillai. Analysis of Factors Affecting Share Prices: The Case of Bahrain Stock Exchange. *International Journal of Economics and Finance*, **7**(3):207-216 (2015)
21. Pinkowitz, Lee, R. Stulz, and R. Williamson. Does the Contribution of Corporate Cash Holdings and Dividends to Firm Value Depend on Governance: A Cross-country Analysis. *The Journal of Finance*, **61**(6):2725-2751 (2006)
22. Okafor, C. A. and C. O. Mgbame. Dividend Policy and Share Price Volatility in Nigeria. *Jorind* (9)1 June 2011:202-210 (2011)

 DOI: 10.1051/shsconf/20173404003

Corporate Governance Quality and Audit Quality in Malaysia

Ho Wai Kee[1,*], *Oo* Yu Hock[2], and *Kwong* Chee Kueng[3]

[1]UCSI University, Kuala Lumpur, Malaysia
[2]Asia E University, Kuala Lumpur, Malaysia
[3]University Putra Malaysia, Selangor, Malaysia

Abstract. This paper examines the impact of corporate governance quality on audit quality in Malaysia. The sampling frame is 457 Malaysian non-finance listed companies, over the periods 2003 to 2007 (pre-2007 Code period) and 2008 to 2012 (post -2007 Code period), consisting of 2,285 observations for each period. This study uses pooled ordinary least square (OLS) to test the research hypotheses and model. The results show that the effectiveness of the audit committee (AC) has no significant influence on audit fees in the pre- and post-2007 Code period, and the effectiveness of the board has no significant influence on audit fees in the pre-2007 Code period, although it has significant influence on audit fees in the post-2007 Code period. The results suggest that the existing corporate governance framework in relation to AC has limitation in its governance role on audit process. Our study contributes to existing literature conducted in the US, the UK and Australia where their institutional settings are different from that of Malaysia. In addition, our study is based on the 2007 Code's recommendation which contributes to the previous research conducted in Malaysia and provides an insightful evidence to the regulator on the corporate governance regime in Malaysia.

1 Introduction

In Malaysia, the board of directors (board) of public-listed companies have a fiduciary duty to act in the interest of the company. The board is required to establish an independent audit committee (AC) to implement and support the oversight function of the board. The governance role of the AC is to ensure that the interest of the shareholders are properly protected through the oversight of financial reporting and external audit processes. The governance role of the board and AC are governed by the Listing Requirements of Bursa Malaysia Securities Berhad (previously known as Listing Requirements of Kuala Lumpur Stock Exchange (KLSE)) [1, 2], [3-5] and Malaysian Code on Corporate Governance [6, 7].

In the past two decades, high-profile financial cases pertaining to corporate governance scandals continuously plagued in Malaysia, such as Perwaja Steel Bhd, Aokam Perdana Bhd, Idris Hydraulic (M) Bhd, Transmile Group Berhad, Axis Incorporation Bhd, Megan Media Holdings Berhad, Linear Corp Bhd, Silver Bird Group Bhd and Malaysia Airline

[*] Corresponding author: howaikee1002@gmail.com

Systems [8, 9]. These cases have caused a loss of public confidence in the role of auditors in preventing fraud. In addition, investors had lost confidence in the Malaysia market during the 1997/1998 Asian Financial Crisis [10]. Consequently, the corporate governance framework in Malaysia was reformed. The regulators, the public and investors are now closely scrutinised on corporate governance in Malaysia.

A number of studies have examined the relationship between corporate governance quality and audit quality, particularly on the governance role of the board and AC, in relation to their oversight of the external audit process. These studies are based on settings in the US [11, 12, 13], the UK [14, 15, 16], Australia [17] and Malaysia [18, 19, 20].

Previous studies conducted in the US, the UK and Australia did not provide conclusive evidence on the relationship between audit fees and the effectiveness of the board and AC. [15] and [12] found that audit committee independence has a significant and positive relationship with audit fees. However, this finding contrasted that of [11, 14, 17]. The results found in [12] on the relationship between audit committee expertise and audit fees also contrasted with the findings of [13, 16, 15, 17]. Similarly, [15] and [17] found that the frequency of audit committee meetings has a significant and positive relationship with audit fees did not support the findings of [11] and [12]. In the study of [12], there was no evidence to show that board independence has a significant relationship with audit fees; this result was not supported by [13, 15, 17]. The differences opinions in the US, the UK and Australia could be due to the country variations of corporate governance requirements.

Previous studies in Malaysia have shown that the effectiveness of the board and AC variables were tested separately [18, 19, 20], with data prior to the implementation of the 2007 Code [18, 19] and the study period covering a short time frame, i.e. three years or less [18, 19, 20]. Furthermore, the findings of these studies on the relationships between the audit fees and the effectiveness of the board and AC were inconclusive. [18] found that board independence, audit committee expertise and the frequency of audit committee meetings are significant and positively related to audit fees. This findings contrasted that of [19] and [20] who found board independence and frequency of audit committee meetings have no significant relationship with audit fees.

Our study examines the relationship between corporate governance quality and audit quality. We use the effectiveness of the board and AC as proxies of corporate governance quality and audit fees paid to the external auditor as the proxy of audit quality. The effectiveness of the board and AC are the best practices in corporate governance of the Code, the 2007 Code and corporate governance guide. The sample consists of 457 Malaysian non-finance public-listed companies which had traded their shares on the main board of Bursa Malaysia from 2003 to 2012. We divided the ten-year data into two periods, i.e., pre-2007 Code period and post-2007 Code period, to test the relationship between the audit fees and the effectiveness of the board and AC before and after the introduction of the 2007 Code.

Our study shows that the effectiveness of the board and AC have no significant influence on audit fees in the pre-2007 Code period. In the post-2007 Code period, the effectiveness of the board is significant and positively related to audit fees; however, there is no evidence to show that the effectiveness of AC has significant influence on audit fees in the post-2007 Code period. This suggests that the existing corporate governance framework particularly on the board has an influence on the quality of audit process, but corporate governance framework in relation to audit committee has limitation in its governance role on audit process. Our results partially support the regulatory initiatives intention at enhancing role and responsibilities of the board and AC in order to improve the audit process.

Nevertheless, this study contributes to the literature in two ways. Firstly, our data is based on the recommendations of the Code, the 2007 Code and corporate governance guide with a longer time-frame study period from 2003 to 2012 which enabled us to provide further empirical evidence that lending support to the latest corporate governance principles and best practices in enhancing the audit process. And secondly, our findings provide an insight evidence to the regulators, policy-makers, industry players and investors which shows that the principles and best practices of 2007 Code have its limitations in enhancing the audit process, alerting that such limitations need to be addressed in order to provide and sustain a sound corporate governance mechanism.

The next section of this paper provides the hypotheses development followed by the research methods, results and discussion. The last session of this paper provides a conclusion of the study.

2 Hypotheses development

Audit committee is viewed as a monitoring mechanism and serves to support the oversight function of the board. There are studies based on agency theory which suggested that the board that emphasis a high quality of monitoring will demand a higher quality audit resulting in greater audit effort by the auditor and resulting higher fees [15, 27]. This study used the effectiveness of audit committee and the effectiveness of board as proxies for corporate governance quality.

2.1 The effectiveness of audit committees (ACE) and audit fees (AF)

A large number of international and national studies in previous years have addressed whether ACs are effective in terms of providing better quality of audit or improved internal control structure. Empirical evidence documented conflicting results on the impact of the effectiveness of the board and AC on the audit fees and the direction of the association between these variables. Consistent with previous studies [18, 15, 20], and based on the best practices in corporate governance of the Code, the 2007 Code and the corporate governance guide, we claim that audit committee independence, frequency of audit committee meetings, audit committee with financial expertise and audit committee size contribute to the effectiveness of AC.

An independent AC demands a wider scope of audit from the external auditors and willing to support external auditors during scope intervention with management [12]. Audit committee with financial expertise has a better understanding on the auditing issues, risks, and the audit procedures that auditors propose to address auditing issues and risks [21, 22, 12]. ACs who meet regularly have a better understanding of current auditing issues and have a positive influence on the scope of audit at different stages of the audit [12]. Company with larger AC is able to enhance its status and increased resources which may increase its effectiveness in fulfilling its monitoring role [15]. An effective AC in assuring the quality of audit process demands additional audit procedures from the external auditor, therefore increasing audit fees [14, 12, 15]. Based on these observations, we posit that an effective AC demands wider scope of audit from the external auditor, leading to a higher audit fees paid to the external auditor. Therefore, we offer our first hypothesis as follows:
H_{1a}. Companies with an effective AC are likely to be charged with higher audit fees in pre-2007 Code period.
H_{1b}. Companies with an effective AC are likely to be charged with higher audit fees in post-2007 Code period.

2.2 The effectiveness of the board (BOD_E) and AF

 DOI: 10.1051/shsconf/20173404003

To be truly effective, a board needs more independent [11, 23] and diligent [11] directors on the board. Previous studies have found that a higher percentage of independent directors on the board helps to reduce the frequency of fraudulent financial reporting [24] and lower the occurrence of earnings overstatement [25]. [26] suggested that an effective board meets frequently. The board that meets frequently shows better diligence in discharging its responsibilities and improving its governance role of monitoring the financial reporting process [11]. Consistent with the best practices in corporate governance that recommended in the 2007 Code, we also included the board duality as one of the measurement for an effective board [7]. Board duality requires higher control of risk and higher audit effort and leading higher audit fees paid to external auditor [15].

Our study is based on the previous studies [11, 23], the best practices in corporate governance of the Code, the 2007 Code and the corporate governance guide to determine the effectiveness of the board (Table 2). Following the discussion above, we posit that an effective board will demand additional audit procedures from the external auditor to improve the financial reporting process, thus increasing the amount of audit fees paid to the external auditors. Therefore, we tender the following hypothesis:

H_{2a}. Companies with an effective board are likely to be charged with higher audit fees in pre-2007 Code period.

H_{2b}. Companies with an effective board are likely to be charged with higher audit fees in post-2007 Code period.

3 Research methods

3.1 Sample

The sample consists of the Malaysian non-finance listed companies that had traded their shares on the main board of Bursa Malaysia for the years from 2003 to 2012. These companies are selected because they have gone through the changes of Malaysia corporate governance framework in 2000 (The Code) and 2007 (2007 Code). We split the study periods into two, i.e., 2003 to 2007 (pre-2007 Code period) and 2008 to 2012 (post-2007 Code period). In the earlier period, corporate governance practices in Malaysia were based on principles and best practices of the Code and in the latter period were based on the principles and best practices of the 2007 Code. Consistent with previous studies, we have excluded finance listed companies in this study as they are governed by a different regulatory body [27, 28]. All data were hand collected from the annual financial statements available in the website of Bursa Malaysia. The sample size is 457 companies after further eliminating the companies with missing data. Table 1 provides the sample size and number of firm-year observations by industry.

Table 1. Sample Size and Number of Observations

Industry	Number of Sample (n)	Number of Observations: Pre-2007 Code Period	Number of Observations: Post-2007 Code Period
Construction	31	155	155
Consumer	84	420	420
Industrial	143	715	715
Plantation	29	145	145
Property	62	310	310
Technology	14	70	70
Trading/Services	94	470	470
Overall	457	2,285	2,285

DOI: 10.1051/shsconf/20173404003

3.2 Measurement of variables

3.2.1 Dependent variable

Consistent with recent studies on audit fees [13, 19, 15], our dependent variable is the statutory audit fees paid to external auditors, measured as the natural logarithm of audit fees (lnAF).

3.2.2 Independent variables

ACE and BOD_E are the main independent variables in this study. ACE and BOD_E are measured based on the best practices in corporate governance of the Code, the 2007 Code and the corporate governance guide. The best practices in corporate governance in relation to AC include audit committee independence, frequency of audit committee meetings, audit committee with financial expertise and audit committee size. ACE is a dichotomous variable with 1 denoting (a) all AC members are non-executive directors with majority of independent directors, (b) at least one member is a member of accounting association or body, (c) at least three members in AC [6, 7], and (d) AC members meet at least four times per year [29] and 0 otherwise.

The best practices in corporate governance pertaining to the board consist of board duality, board independence and frequency of board meetings. BOD_E is a dichotomous variable equal to 1 when (a) the roles of chairman and chief executive officer (CEO) are not combined, (b) at least one-third of the board membership are independent non-executive directors [6, 7], and (c) board members meet at least five times per year [29] and 0 otherwise.

3.2.3 Control variables

In addition to the two independent variables, we control for the effects of other variables on audit fees that were found in previous studies [15, 30]. The control variables include LOSS, lnTA, LEV, BIG4, lnSUBs, NSH5_S. LOSS is an indicator variable equal to 1 if the company is making loss during the year and 0 otherwise. lnTA is measured as the natural logarithm of total assets. LEV is the leverage level of the company, measured by the ratio of long-term debts to total assets. BIG4 is an indicator variable equal to 1 if the company is audited by big-four auditor and 0 otherwise. lnSUBs represents the complexity of the company, measured as the natural logarithm of number of subsidiaries. NSH5_S is the total number of shareholders holding 5% or more shares in the company. We also control for the director's remuneration (DR) on audit fees as the 2007 Code recommends that public-listed companies pay sufficient remuneration to attract and retain directors [7]. Hence, we expect director's remuneration as one of the determinants of audit fees. DR is measured as the natural logarithm of remuneration paid to the directors. Consistent with previous studies [28, 15], we include year and industry dummies to control for the unobserved effects during the sample period and the industry effect respectively.

3.3 Regression model

The regression model use to test the hypotheses in this study is as follows:

$$lnAF_{it} = \beta_0 + \beta_1 ACE_{it} + \beta_2 BOD\_E_{it} + \beta_3 lnDR_{it} + \beta_4 LOSS_{it} + \beta_5 lnTA_{it} + \beta_6 LEV_{it} + \beta_7 BIG4_{it} + \beta_8 lnSUBs_{it} + \beta_9 NSH5\_S_{it} + YEAR\ dummies + INDUSTRY\ dummies + e_{it}$$

Table 2 displays the definition of variables in the regression model. We use the ordinary least squares (OLS) regression model to explore the relationship between the audit fees and the effectiveness of the board and AC.

Table 2. Definition of Variables

No.	Variables	Definition
1	LnAF	Natural logarithm of audit fees (RM'000).
2	ACE	Audit committee effectiveness is a dichotomous variable; 1 if (a) all members of the AC are non-executive directors with majority of independence directors, (b) at least one member is a member of accounting association or body, (c) at least three members in AC and (d) AC meeting minimum four times per year and 0 otherwise.
3	BOD_E	Board effectiveness is a dichotomous variable; 1 if (a) the roles of chairman and CEO are not combined, (b) at least one-third of the board membership are independent non-executive directors and (c) board meeting minimum five times per year and 0 otherwise.
4	lnDR	Natural logarithm of director's remuneration (RM'000).
5	LOSS	A dichotomous variable, 1 if the company making loss during the year and 0 otherwise.
6	lnTA	Natural logarithm of total assets (RM'mil).
7	LEV	The ratio of total long-term debts to total assets.
8	BIG4	A dichotomous variable, 1 if the company is audited by big four auditor and 0 otherwise.
9	lnSUBs	Natural logarithm of number of subsidiaries.
10	NSH5_S	Total number of shareholder holding 5% or more shares in the company.
11	YEAR	Year dummies. Dummy variables for each year from 2003 to 2012.
12	INDUSTRY	Industry dummies. Dummy variables for each industry, i.e., construction, consumer, industrial, plantation, property, technology and trading/services.

3.4 Correlations

Panel A and Panel B of Table 3 report on the correlation matrix for the relevant variables for the pre-2007 Code period and post-2007 Code period respectively. There is no multicollinearity problem with other experimental variables and control variables as the correlations are all below 0.7 [31]. We also calculate the variance inflation factor (vif) for all variables and the results show that all vif values are less than 10.0 suggesting that multicollinearity is not a substantive issue in this study [31, 32].

DOI: 10.1051/shsconf/20173404003

Table 3. Pearson Correlation Matrix

Variable	(1)	(2)	(3)	(4)	(5)	(6)	(7)	(8)	(9)
Panel A: 2003-2007 (n=457, 2,285 observations)									
(1) lnAF	1.00								
(2) ACE	0.05	1.00							
(3) BOD_E	0.03	0.08	1.00						
(4) lnDR	0.48	0.01	-0.03	1.00					
(5) LOSS	-0.07	-0.01	0.06	-0.22	1.00				
(6) lnTA	0.77	0.06	0.01	0.45	-0.17	1.00			
(7) LEV	0.19	0.04	0.06	0.07	0.03	0.14	1.00		
(8) BIG4	0.09	0.03	-0.05	0.09	-0.05	0.14	0.00	1.00	
(9) lnSUBs	0.73	0.06	0.05	0.36	0.03	0.56	0.16	-0.04	1.00
(10) NSH5_S	-0.04	-0.04	0.02	0.00	-0.06	-0.04	0.02	-0.04	0.00
Panel B: 2008-2012 (n=457, 2,285 observations)									
(1) lnAF	1.00								
(2) ACE	0.07	1.00							
(3) BOD_E	0.01	0.08	1.00						
(4) lnDR	0.53	0.11	-0.04	1.00					
(5) LOSS	-0.08	0.03	0.11	-0.19	1.00				
(6) lnTA	0.79	0.03	-0.02	0.55	-0.17	1.00			
(7) LEV	0.27	0.03	0.10	0.17	0.04	0.31	1.00		
(8) BIG4	0.24	-0.02	-0.07	0.17	-0.14	0.29	0.06	1.00	
(9) lnSUBs	0.71	0.13	0.06	0.43	-0.02	0.58	0.30	0.04	1.00
(10) NSH5_S	0.02	0.01	-0.02	0.01	-0.03	0.03	0.05	-0.06	0.10

4 Results and discussion

4.1 Regression results

Table 4 presents the OLS regression result of lnAF on ACE, BOD_E and control variables. The results show significant explanatory power with R^2 of 0.762 in the pre-2007 Code period and 0.764 in the post-2007 Code period. Table 4 shows that in the pre-2007 Code period and post-2007 Code period, ACE is not significantly related to lnAF. This result is in contrast with our prediction that an effective AC demands wider scope of audit from the external auditor, leading to higher audit fees paid by the company. Thus, we have to reject the hypothesis H_{1a} and H_{1b}. The result shows that ACE has no influence on lnAF. BOD_E is not significantly related to lnAF in the pre-2007 Code period. Therefore, we have to reject hypothesis H_{2a}. BOD_E is significantly and positively related to lnAF at 5 percent level in the post-2007 Code period. This result in the post-2007 Code period is consistent with our prediction that an effective board demands higher quality of audit services to ensure reliability of financial statements, to complement its own monitoring managerial behavior and resulting in higher audit fees paid by the company. We do not reject the hypothesis H_{2b} in post-2007 Code period. This suggests that only BOD_E has a significant influence on lnAF in the post-2007 Code period.

The OLS regression result of lnAF on control variables shows that, in the pre-2007 Code period and post 2007 Code period, lnDR, lnTA, lnSUBs are significant and positively related to audit fees at 1 percent level. LEV shows a statistically significant positive relationship with audit fees at 10 percent level in the pre-2007 Code period. In the post-2007 Code period, LOSS and BIG4 are significant and positively related to audit fees at 1 percent level. The result in this study also shows that there are both industry effect and time effect in the research model.

Table 4. Regression of lnAF on ACE, BOD_E and Control Variables

	2003-2007			2008-2012		
	Coefficient	t-statistic		Coefficient	t-statistic	
Constant	2.785	6.330		2.785	7.810	***
ACE	-0.024	-0.530		0.022	0.610	
BOD_E	0.026	0.860		0.004	0.110	**
lnDR	0.080	3.540	***	0.061	3.100	***
LOSS	0.059	1.560		0.119	3.320	***
lnTA	0.344	16.530	***	0.367	18.160	***
LEV	0.275	1.850	*	-0.147	-0.840	
BIG4	0.055	1.390		0.149	3.680	***
lnSUBs	0.413	14.090	***	0.402	14.180	***
NSH5_S	-0.009	-1.160		-0.009	-0.840	
R^2	0.762			0.764		
F-Statistic	113.000	***		120.7	***	
Industry effect	Yes			Yes		
Year effect	Yes			Yes		

Note: Variables are as defined in Table 2; *, **, *** significant at 10, 5 and 1 percent levels.

4.2 Discussion

In this study, the results show that ACE has no significant influence on lnAF in the pre- and post-2007 Code period is in contrast with the result of [14, 12, 15]. These studies had argued that an effective AC demands addition audit procedures in order to assure the quality of audit process, leading to an increase in audit fees, is not supported by the results of our study. The results also show that BOD_E has no significant impact on lnAF in the pre-2007 Code period. The finding of BOD_E has a significant influence on lnAF in the post-2007 Code is consistent with the US study of [11, 13] and the UK study of [15] who found that an effective board helps to reduce the frequency of fraudulent financial reporting [24], lower the occurrence of earnings overstatement [25], improve its governance role of monitoring the financial reporting process [11] and enhance the quality of audit and financial reporting process [23] and therefore will demand more audit procedures from the auditors, leading to an increase in audit fees.

In this study, we also find that lnTA, lnSUBs, LEV have a significant influence on lnAF in the pre-2007 Code period, lending support to the findings by [19, 15, 18, 15] had argued that companies with large total assets have more complex operations; companies with more subsidiaries have complex group transactions and weak internal control; companies with higher leverage will demand higher audit quality to protect themselves from business and finance risk and therefore increase the audit fees paid to the external auditor. In the post-2007 Code period, lnTA, lnSUBs, LOSS and BIG4 have significant influenced on lnAF.

These results also provide support to the findings by [19, 15]. Consistent with our expectation, lnDR has a statistically significant and positive relationship with lnAF in the pre- and post-2007 Code. This suggests that lnDR is an important determinant of audit fees.

5 Conclusion

Our study has provided an overview of the impact of the corporate governance quality on audit quality in Malaysia during the period from 2003 to 2012 (i.e., pre-and post-2007 Code period), using the effectiveness of the board and AC to proxy for corporate governance quality and audit fees to proxy for audit quality. The timeframe is interesting and appropriate as the Code was revised in 2007 and the 2007 Code was aimed at strengthening the roles and responsibilities of the board and AC. The sample has focused on 457 non-financial companies that had listed and traded their shares on the main board of Bursa Malaysia from 2003 to 2012. There is no evidence to show that the effectiveness of AC has a significant influence on audit fee in the pre- and post-2007 Code period. The results also show that the effectiveness of the board does not has a significant influence on audit fees in the pre-2007 Code period but it has significant influenced on the audit fees in the post-2007 Code period. This suggests that the existing corporate governance framework particularly on the board has an influence on the quality of audit process, but the corporate governance framework in relation to audit committee has limitation in its governance role on audit process. Our results partially support the regulatory initiatives intention at enhancing role and responsibilities of the board and AC in order to improve the audit process.

This study is subject to a number of limitations that need to be considered in future research. Firstly, the use quantitative research method and secondary data to explore the effectiveness of the board and AC may not reflect the true performance of the board and AC. We suggest to use qualitative research method to explore the performance of the board and AC on governance outcomes by including other attributes such as the board and AC's individual behavior, informal networks of the board and AC, the relationship among the board, AC and external auditors and the supervisory and compliance structure of the company. Secondly, our study did not make allowance for the political connections and ethnicity documented in previous studies conducted in Malaysia; this should be considered for future study.

References

1 Kuala Lumpur Stock Exchange (KLSE), Main Board Listing Requirements of Kuala Lumpur Stock Exchange, Kuala Lumpur Stock Exchange, Kuala Lumpur (1993).
2 Kuala Lumpur Stock Exchange (KLSE), Listing Requirements of Kuala Lumpur Stock Exchange, Kuala Lumpur Stock Exchange, Kuala Lumpur (2001).
3 Bursa Malaysia Securities Berhad, Listing Requirements of Bursa Malaysia Securities Berhad, Bursa Malaysia Securities Berhad, Kuala Lumpur (2005).
4 Bursa Malaysia Securities Berhad, Listing Requirements of Bursa Malaysia Securities Berhad, Bursa Malaysia Securities Berhad, Kuala Lumpur (2006).
5 Bursa Malaysia Securities Berhad, Listing Requirements of Bursa Malaysia Securities Berhad, Bursa Malaysia Securities Berhad, Kuala Lumpur (2009).
6 Financial Committee on Corporate Governance, Malaysian Code on Corporate Governance March 2000, Securities Commission, Kuala Lumpur (2000).
7 Securities Commission, Malaysian Code on Corporate Governance (Revised 2007), Securities Commission, Kuala Lumpur (2007).
8 N.A. Zainal Abidin, H. Nasibah Ahmad, AAMJ, **12(1)**, 23–34 (2007).

9 Securities Commission Malaysia, Enforcement related press releases, available at: http://www.sc.com.my/enforcement/enforcement-related-press-releases/(accessed 11 September 2015) (2015).
10 Securities Commission Malaysia, Malaysian Code on Corporate Governance 2012, Securities Commission Malaysia, Kuala Lumpur (2012).
11 J.V. Carcello, D.R. Hermanson, T.L. Neal, JR., R.A. Riley, Contemp. Account. Res., **19(3)**, 365-384 (2002).
12 L.J. Abbott, S. Parker, G.F. Peter, K. Raghunandan, Auditing-J Pract TH, **22(2)**, 17-32 (2003).
13 A.M.Y. Chan, G. Liu, J. Sun, Account. Financ., **53(4)**, 1129-1147 (2013).
14 P. Collier, A. Gregory, Eur. Account. Rev., **5(2)**, 177-198 (1996).
15 M. Zaman, M. Hudaib, R. Haniffa, J. Bus. Financ. Account., **38(1 & 2)**, 165-197 (2011).
16 I. Adelopo, K. Jallow, P. Scott, J. Appl. Account. Res., **13(2)**, 100-121 (2012).
17 J. Goodwin-Stewart, P. Kent, Account. Financ., **46(3)**, 387-404 (2006).
18 P. Yatim, P. Kent, P. Clarkson, MAJ, **21(7)**, 757-782 (2006).
19 S. Johl, N. Subramaniam, and M. Mat Zain, Int. J. Account., **47(3)**, 302-332 (2012).
20 A.I. Husnin, A. Nawawi, A.S.A.P. Salin, ASS, **9(5)**, 1911-2025 (2013).
21 F.T. DeZoort, Account ORG SOC, **23(1)**, 1-21 (1998).
22 F.T. DeZoort, S. Salterio, Auditing-J Pract TH, **20(2)**, 31-47 (2001).
23 H.Y. Lee, Appl. Finan. Econ., **18(8)**, 629-638 (2008).
24 M.S. Beasley, Account. Rev., **71(4)**, 443-465 (1996).
25 P.M. Dechow, R.G. Sloan, A.P. Sweeney, Contemp. Account. Res., **13(1)**, 1-36 (1996).
26 N. Vafeas, J Financ Econ, **53(1)**, 113-142 (1999).
27 M.A. Bliss, F.A. Gul, A. Majid, JCAE, **7(2)**, 82-98 (2011).
28 E.A. Abdul Wahab, M. Mat Zain, Manag. Audit. J., **28(8)**, 735-754 (2013).
29 Bursa Malaysia Berhad, Corporate Governance Guide, 2nd Ed., Bursa Malaysia Berhad, Kuala Lumpur (2012).
30 D.C. Hay, W.R. Knechel, N. Wong, Contemp. Account. Res., **23(1)**, 141-191 (2006).
31 J.F. Hair, W.C. Black, A.L. Babin, R.E. Anderson, *Multivariate data analysis. (7th ed.)*, Pearson, Upper Saddle River, NJ (2010).
32 J.M. Wooldridge, *Introductory econometrics: a modern approach (5th international ed.)*, South-Western Cengage Learning, Singapore (2013).

 DOI: 10.1051/shsconf/20173404004

The association between corporate social responsibility disclosure of cigarette company and company's financial performance

Amelia Desy Ratna Yuwita[1] and *Devi* Sulistyo Kalanjati[1,*]
[1]Universitas Airlangga, Indonesia

Abstract. This research examines the association between Corporate Social Responsibility (CSR) disclosure and financial performances-Return on Assets (ROA), Return on Equity (ROE), and Stock Return-within the cigarette companies listed on Indonesian Stock Exchange. This research used 3 cigarettes companies; PT Gudang Garam Tbk., PT Hanjaya Mandala Sampoerna Tbk., and PT Bentoel Internasional Investama. Simple linear regression is used to examine the association between CSR disclosure and the cigarettes companies' financial performance. The study reveals that the disclosure of CSR only has positive influences toward Return on Assets; yet, it does not correlate with the Return on Equity and Stock Return.

Keywords: *Corporate Social Responsibility (CSR), Return on Assets, Return on Equity, Stock Return*

1 Introduction

According to Act No. 40 2007 on Limited Liability Company (Perseroan Terbatas-PT) in Article 74 par (1), a company which conducts activities in the field and/or related to the natural resources is required to implement the Social and Environmental Responsibility. The company as a legal entity has to have a corporate social responsibility (CSR). The studies about CSR has growing rapidly caused by many cases where the companies did not make a positive contribution to the community and giving them a direct negative impact instead.

This research focuses on the CSR disclosure in the cigarette companies in Indonesia. Tobacco industry manages and utilizes the natural resources such as tobacco plants as main raw material. The tobacco industry in Indonesia, the fifth largest tobacco market in the world, has been established by the Indonesian government as one of the 10 major industries in the country which implied by the high number of employee recruitments and the contribution of industries in the country's income. According to [1], the tobacco sector in Indonesia has more than 4 million employees, including farmers, manufacturing sector employees, the sellers, and the distributors. In 2013, tobacco industry donated 95.79% of 155.82 trillion rupiah of the customs income to the country, or roughly 10.9% from the total of the country's income. They become one of the main pillars and driving forces of the Indonesian economy and their roles in increasing the growth of the national economy are parts of its positive contributions for the nation. Their CSR activities are also well known although at the same time, unfortunately, their main operating activities has a significant

* Corresponding author: Devi Sulistyo Kalanjati. devi.kalanjati@yahoo.com

DOI: 10.1051/shsconf/20173404004

social impact on the health of human resources in Indonesia.

This study examines the relationship of CSR disclosure and corporate performances; ROA, ROE, and Stock Returns in companies engaged in the tobacco sector which also listed on the Indonesia Stock Exchange during 2007-2014. In 2007, the policy regarding the CSR disclosure in Indonesia was released; therefore the research results could be more legitimate and relevant in the present time. Meanwhile cigarettes companies are choosen because they have been criticized a lot by the community in general, because of the direct impact of their products are considered to be adverse to the human health. In this case, whether the disclosure of CSR has association with ROA, ROE and the stock return is important to be analyzed.

2 Literature Review

[2] in [3] define the CSR as actions that appear to further some social good, beyond the interest of the firm and that which is required by law. In short, the social responsibility is simply the reciprocal/payback of the company to the community and the surrounding environment, since the company has also taken the advantage of the community and the surrounding environment, which unfortunately the process of taking advantage from the company often causes damages to the environment and other social impacts.

According to the Financial Reporting Council [4], the disclosure of CSR activity can be viewed from various perspectives; level of disclosure, the theme of the disclosure, location or where such disclosure is made in the annual report, and also the type of disclosure. The level of disclosure range means that the amount of items disclosed are compared with the total amount of the items that should be disclosed. [5] mentions the themes included in the accounting of CSR are: social, employment, products, customer, and also the environment. The disclosure of CSR can be found in various ways, e.g. narrative qualitative, non-monetary quantitative and monetary quantitative [6]. According to [7] there are some sections in the annual report such as in chairperson's report, operational review, corporate governance and CSR section. [7] also revealed that the most section used for social disclosure is in the CSR section.

[8] stated that the financial statement is designed for the decision makers, especially those outside of the company, regarding to the financial positions and the results of operation. The Ratio analysis shows the relationship between the elements of financial statements. expressed in the form of a simple mathematical [9]. [10] explains that the profitability ratio measures the company's ability to generate profits by using the resources of the company itself e.g. assets, capital, or company's sales. [10] also describes some of the ratios be used to calculate profitability among others are ROA and ROE. ROA is the ability to use assets to generate earnings after taxes. It measures the effectiveness and efficiency of the company's management in managing the assets. ROE shows the company's ability to generate earnings after tax by using the company's capital. ROE is a micro factor used by the investors to consider before taking the investment decisions.

Stock return is the income expressed as a percentage of the initial capital investment [11]. Stock returns is also an important factor that a prospective investor should consider before making a stock investment decisions. According to [12], the stock return can be divided into two; the stock realized and expected return.

The Act No 40 year 2007 in Section 74 about Limited Liability Company (PT) [13] states that: (1) Companies that have been in business related to the natural resources, are required to implement CSR. (2) The CSR referred in paragraph (1) is a mandatory that are budgeted and accounted as an expenses which its implementations are done in regard to decency and fairness. (3) The company that does not carry out the obligations referred in paragraph (1) will be sanctioned according to the established law. (4) Further provisions on CSR responsibility are regulated by the Government.

The CSR activities are part of good corporate governance. The activities are expected to raise the financial performance of the company because of their alignments to the public. The public is not only able to choose a good product by its value but also through its corporate governance. Logically thinking, if the people who become customers will have a positive assessment of the company, then they will be loyal to the products produced, which also would raise the image of the company and will be reflected through the increase in company's performance [14].

The companies of the cigarette industry are playing two competing roles. Being those whose main operating activities have a negative impact to social health but on the other side, their CSR activities are quite significant and well known in Indonesia. Therefore, the main focus in this research is to examine the associations between the CSR activities and the company's financial performance.

3 Methodology

This research used 3 cigarettes companies, namely PT Gudang Garam Tbk., PT Hanjaya Mandala Sampoerna Tbk., and PT. Bentoel Internasional Investama, as the populations for period of 8 years from 2007- 2014. The Author used simple linear regression to examine the association between CSR disclosure and financial performance.

The analysis of the model used panel data regression analysis which combines the time series and cross section data. The purpose of this test is to determine the association between independent variable and dependent variable by using Eviews 6.0 software for Windows. The independent variable in this research is CSR Index (CSDI), while the dependent variables used are ROA, ROE, and Stock Return (SR).

First: The instrument used in this variable is the CSR disclosure checklist which consists of 79 items. This checklist was created based on Global Reporting Initiatives (GRI).

Determining the value of disclosure index checklist of CSR were performed by using the dichotomy approach, means that each disclosure items of CSR in the instrument is given a score of one (1) for each item disclosed, and a score of zero (0) for each item that was not disclosed. So, the CSR disclosure index is calculated by the formula: the total items disclosed divided by the total score of items disclosed added with the total items that are not disclosed.

Second: ROA is calculated by dividing the earnings after taxes to the total assets, while ROE also divides the earnings after taxes to the total equity. On the other hand, the stock return is calculated by dividing the difference between the stock price index of individual companies and the year before, with individual company stock price index in the previous year.

Third: To determine which approach the researcher used in panel data regression, the Chow (F-Statistics) and Hausman tests were performed.

If the F-Statistics or F-Chow value is higher than F-table, then H0 is rejected, it means that fixed effect approach used followed by the Hausman test. But, if F-Chow value is smaller than F-table, then it fails to reject H0, it means that the pooled least square or summon effect approach is chosen. When PLS is used, the Hausman test does not need to be conducted.

If the result of F-Chow test stated that fixed effect approach were used, then the next test was Hausman Test, in order to determine whether the fixed effect approach or random effect approach that would be used. This test use Chi-square value, if the probability value is smaller than the significant test result (α), and then H0 is rejected, which means that fixed effect approach is used. Otherwise, if the probability of Hausman is higher than α,

DOI: 10.1051/shsconf/20173404004

then fail to reject H0, which means random effect approach is chosen.

Fourth: Analysis of hypothesis testing. The test used is the t-test (partial test) with a confidence level 95% or estimation error (α) of 5%.

Probability Test is conducted to find the effect of each variable towards the dependent variable partially; with the significance level (α) is 0.05 or 5%. The hypotheses are as follows:

- If the probability ≤ α, then H0 is rejected. There is a significant association between independent variable and the dependent variable partially.
- If the probability > α, then it is fail to reject H0. There is no a significant association between independent variable and the dependent variable partially.

Fifth: Conducting Regression Analysis of CSR Disclosure Index (CSDI) towards ROA, ROE, and Stock Return.

4 Findings

Table 1 and 2 below summarized descriptive statistics of variables used in this research.

Table 1. Descriptive Statistics Relating to Variables

	N	Minimum	Maximum	Mean	Std. deviation
ROA	24	-.2223	.4162	.1358	.1632
ROE	24	-1.1817	1.6313	.3003	.4939
SR	24	-.4940	4.0700	.3683	.9236
CSDI	24	.3671	.5063	.425631	.0398232

Variable definitions

ROA	:	Earnings after taxes divided by total assets owned by the company.
ROE	:	Earnings after taxes divided by total equity of the company.
SR	:	Difference between the stock price index of individual companies and the year before divided by individual company stock price index in the previous year.
CSDI	:	Total items disclosed divided by the total score of items disclosed added with the total items that are not disclosed.

Table 2. Cross-Section Random Effects Regressions for ROA

Variable	Exp. Sign	CSDI
C	-	-.260316
		(.2351)
ROA	+	.930653**
		(.0206)
Observations		24
Adjusted R^2		.19
Prob > *F*-test		.02

Based on the results of the regression calculation of ROA in table 2, the value of coefficient of α= -0.260316 and β= 0.930653. Therefore the regression equation now can be formulated as follow:

$$Y1 = -0.260316 + 0.930653CSDI + e \qquad (1)$$

The coefficient of CSDI variable is equal to 0.930653, it means that if the CSR disclosure is increase one unit, then Return on Assets (ROA) will increase by 0.930653 and vice versa. If the CSR disclosure is decrease one unit, then the Return on Assets (ROA) will decrease by 0.930653, with the assumption that the other variable is constant. The value of Constanta is -0.260316, it means that if there is no variable, and then the value of Return on Assets (ROA) is -0.260316.

Table 3. Cross-Section Random Effects Regressions for ROE

Variable	Exp. Sign	CSDI
C	+	.912975
		(.4196)
ROE	-	1.439342
		(.5792)
Observations		24
Adjusted R^2		-.03
Prob > *F*-test		.57

Based on the results of the regression calculation of ROE in table 3, it shows the value of coefficient of α= 0.912975and β= -1.439342. Therefore the regression equation now can be formulated as follow:

$$Y2 = 0.912975 - 1.439342CSDI + e \tag{2}$$

The coefficient of CSDI variable is equal to -1.439342, it means that if the CSR disclosure is increase one unit, then Return on Equity (ROE) will increase by -1.439342 and vice versa. If the CSR disclosure is decrease one unit, then the Return on Equity (ROE) will decrease by -1.439342, with the assumption that the other variable is constant. The value of Constanta is 0.912975, it means that if there is no variable, then the value of Return on Equity (ROE) is 0.912975.

Table 4. Pooled Least Square Regressions for SR

Variable	Exp. Sign	CSDI
C	-	.1797
		(.9329)
SR	+	1.2877
		(.7967)
Observations		24
Adjusted R^2		-.04
Prob > *F*-test		.80

Based on the results of the regression calculation of Return Stock in table 4, it shows the value of coefficient of α= -0.1797 and β= 1.2877. Therefore the regression equation now can be formulated as follow:

$$Y3 = -0.1797 + 1.2877\ CSDI + e \tag{3}$$

The coefficient of CSDI variable is equal to 1.2877; it means that if the CSR disclosure is increase one unit, then Stock Return will increase by 1.2877 and vice versa. If the CSR disclosure is decrease one unit, then the Stock Return will decrease by 1.2877, with the assumption that the other variable is constant. The value of Constanta is -0.1797, it means that if there is no variable, and then the value of Stock Return is -0.1797.

The result proves that the disclosure of CSR only have positive influences toward the companies' ROA, while its not correlate with the companies' ROE and Stock Return. It means that although the results indicate that the improvement of CSR information disclosure makes cigarettes companies has a chance to improve its management efficiency at using its assets to generate earnings (ROA), in a short term it will not affect their efficiency to generate profits from its shareholders investments in the company (ROE) nor affecting an investor's decision to invest on them.

Based on the result of this study, the significance value of ROA is less than 0.05, which is 0.0206. It proves that the disclosure of CSR (CSDI) activities has a positive association on ROA. Therefore the hypothesis (H1) is supported. This result supports the previous research conducted by [14]. The CSR indexes provide support for company to build image, maintaining reputation and legitimacy of investors to increase capacity of company. This was consistent with the statement of [14], that participation in various forms of CSR can provide many benefits for company.

Meanwhile, this study showed an opposite result for ROE. The significance value is greater than of 0.05%, which is 0.5792 to be exact. This indicates that the disclosure of the CSR activities did not have a significant association on the ROE. Thus the hypothesis (H2) is not supported. It is also supported by previous studies conducted by [15].

The disclosure of CSR activities are still less effective to stimulate an increase in ROE because the purpose is to generate profits in the long term; while ROE is an indicator for short term financial performance. The unstable fluctuation of ROE in cigarettes companies is one that can be the cause of this inconsistency results. There is a pretty high standard deviation in each year which means high risk for investment in the tobacco industry because of the uncertainty [16]. Given that, the disclosure of CSR is not the only the reason for investors to invest.

CSR has no significant association toward Stock Return. The significance value is greater than of 0.05%, which is 0.7967. This indicates that the disclosure of the CSR (CSDI) activities did not have a significant association on the company's Stock Return (SR) ratio.

CSR is a long-term investment for the company. By revealing the activity of their social responsibility, the company tried to mingle with the public so that their economy activities can be accepted by society. This supports the stakeholder theory in the sense that the stakeholders have the same rights as the shareholder, who allowed them to put their demands and needs towards the company [17]. In this case the company's relationship with stakeholders is called CSR, and by paying attention to their needs, the demand of stakeholders will assist the management in order to make a better decision. Furthermore, the results can be used as one of the considerations for investors to invest in companies that have a commitment toward CSR activities, and also reduce the assumption which said that the implementation of CSR will reduces the expected return for investors. Similarly for the management division, we hope that in the future they will pay more attention to the effective implementation of CSR, so that it will be able to encourage the companies to have better performances, hoping that in the future the cigarette companies can also compete with other multinational companies to deal with the global competition.

5 Conclusion

This study showed that CSR index can increase company's performance in form of ROA in listed cigarettes companies during 2007-2014. However, the CSR does not have a significant association with the ROE. Moreover, the CSR disclosure does not have a significant association with the Stock Return since this study revealed that CSR index does not increase company's performance in form of Stock Return in listed cigarettes companies during 2007-2014.

References

1. Hamdani. D., *Gatra* (2013)
2. McWilliams. A., D. Siegel, Acad Manag. Rev., **26,** 1 (2001)
3. Khan, R.M. and A. Hassan, IJCRB, **4,** 10 (2013)
4. F. R. Council, *Thinking About Disclosures in A Broader Context. A Road Map for a Disclosure Framework* (2012)
5. Purwanto. A., Akuditi, **8,** 1 (2011)
6. Murtanto, *International Business Management Conference* (2004)
7. Jitaree. W., *School of Accounting, Economics and Finance* (2015)

8. Soemarso. S.R., *Akuntansi: Suatu Pengantar 1* (2004)
9. Sugiono. A., *Manajemen Keuangan Untuk Praktisi Keuangan* (2009)
10. Sudana. I.M., *Manajemen Keuangan Perusahaan; Teori & Praktek* (2009)
11. Samsul. M., *Pasar Modal & Manajemen Portofolio* (2006)
12. Hartono. J., *Analisis dan desain sistem informasi* (2005)
13. Republik, I., *Limited Liability Company (Perseroan Terbatas)* (2007)
14. Setiawan, M., Darmawan, European Journal of Social Sciences, **23,** 2 (2011)
15. Budianto. J.Th., L.N.F., *Majalah Ekonomi*, **18,** 1 (2008)
16. Bains, A. *What Is Standard Deviation?* (2011)
17. Goukasian, L. and L.K. Whitney, *Corporate Socially Responsible Firms Perform Well! Evidence from Financial and Operating Performances* (2008)

DOI: 10.1051/shsconf/20173404005

Understanding Prudence Corporate Governance: An Alternative Approach

Zakaria Abas[1,*]

[1]UUM, College Of Business, 06010 Sintok Kedah, Malaysia

Abstract. In the past decades, the world was stunned by the collapse of few established corporate organization worldwide. These high profile organizations are said to have a weak corporate governance system among others lack of accountability, lack of transparency and poor leadership and decision making that plunge their organization towards an endemic financial and management crisis. Since then and until today, corporate governance received a continuous high attention by all parties concerned, whereby exercising good corporate governance practises is becoming a mandatory within their respective organization. To elaborate further, perhaps maybe there is an abundance of established guidelines and practical perspective of corporate governance within the literature, however, this paper offers an attention-grabbing alternative approach of understanding prudent corporate governance thought that based on Islamic perspective. Perhaps we could revisit the philosophy of corporate governance, review the possibilities in light of adopting a more comprehensive perspective on the concept of development which includes individual, organizational and societal goals.

Keywords: Corporate governance, Islamic corporate governance

1 Introduction

Over the past few decades, much has been discussed on the endemic financial crisis shrouding East Asia with particular focus on the structural weaknesses in the affected countries. Malaysia is no exception. Among these is the governance of the corporate sector. Our economic crisis has shown us how the lack of corporate governance has exacted a heavy toll on the market and as a result depleted investors' confidence. Although there are contending, varied and complex views on the causes of the crisis, many were acquiescent that weak corporate governance was partly to be blamed for poor investment decisions, lower quality public investment and public services, reduced private investment, risky financing practices and the resultant large-scale non-performing loans in the financial sector, growing number of corporate bankruptcies, and poor audits.

* Corresponding author: zakaria@uum.edu.my

Nevertheless, the crisis was somewhat a blessing in disguise, for the excesses of the past and deluge of corporate collapses seem to have led to increasing awareness and emphasis on the issues of corporate governance. Overnight, the term has become a corporate buzzword. Just what is corporate governance? To put it simply, it means that corporations must exercise greater accountability when undertaking their business operations, both to their shareholders and the public at large. It is in fact a systematic framework that provides a blueprint for the companies to manage themselves in the best interest of the shareholders and to manage risks in an increasingly competitive global economy.

Given the background in the preceding paragraphs, this paper promotes a brief and preliminary thought on the basic issues relating to: i) enhanced role of the company directors, ii) whether corporate governance is value-loaded, iii) central element of corporate governance, iv) the dual nature of man, v) the concepts and philosophy of corporate management in Islam, vi) social responsibility and managerial accountability, vii) success and failures in managing business, leadership and decision making process. Finally, this paper provides a conclusion on the related issues relating the alternative approach to corporate governance.

2 Enhanced Roles of the Company Directors

Enhancing corporate governance involves a collective responsibility among regulatory agencies, professional bodies, corporate leaders as well as investors and stakeholders. There is now a greater demand for transparency and accountability from corporate leaders such as company directors. Directors are expected to achieve high performances for their companies while acting in accordance with the prevailing regulatory requirements. A vital element of corporate governance seems to lie with the board of directors and the way management is run, with a sense of stewardship, responsibility and accountability to shareholders, who are the legal owners of corporations. The board of directors is entrusted with the conduct and daily management of business activities and commercial feasibility of the venture by using the assets made available. Directors are, therefore, subjected to higher standards which cover not only the technical efficiency of operations, but also the implementation of an efficient management system through the use of best practices developed from high ethical values.

Moral and ethical values have been much discussed but less seen practiced among the corporate leaders. Company directors are in fact leaders that play a significant role in shaping the colour of their corporations. We have learned from the past that a successful leader is one who builds core values in people he leads. This requires him to have a sound moral education, imbibed with spiritual values, which when combined with leadership, training and delegation, will result in his people being able to clearly and unhesitatingly differentiate between right and wrong, moral and immoral, ethical and unethical practices in their pursuit of achieving success.

3 Is Corporate Governance Value-Loaded?

The aim of this paper is to address the issue of "how do we move towards an alternative approach of corporate governance, say, Islamic Corporate Governance?" Is there any such term as Islamic and non-Islamic corporate governance, that is, is corporate governance value-loaded? Let us revisit at the definition of corporate governance as spelled out by the [3],"Corporate governance is the process and structure used to direct and manage the business and affairs of the company towards enhancing business prosperity and corporate

DOI: 10.1051/shsconf/20173404005

accountability with the ultimate objective of realizing long-term shareholder value, whilst taking into account the interests of other stakeholders".

Clearly, there arise a few issues that need to be addressed; firstly, on the concept of business prosperity and corporate accountability; secondly, on the ultimate objective of corporate management; and thirdly, on the legal structures and administrative process in the decision-making by the company's directors. The Finance Committee in their 275-page reports enveloping 70 recommendations has elaborated each of these. As [4] mentioned, [7] wrote in Corporate Governance (1992) that a good governance system should consist of a system of structure, operating and controlling a company to achieve the objectives of fulfilling long-term strategic goals, building shareholders' value by establishing a dominant market share, or be a leader in a chosen sphere; Consideration and care for the needs of the environment and local community, including the economic and cultural interaction; Maintaining excellent relationship with customers and suppliers in terms of quality of service provided, considerate ordering and account settlement procedures; and Maintaining compliance with all legal and regulatory requirements under which the company operates. In broad sense, an effective corporate governance system should provide a mechanism to regulate duties and responsibilities of directors to restrain them from abuse of powers, ensuring they act in the best interest of the company and to be aware of the impact of the company on the business community and the environment. [2] explains succinctly that if management is about running business, governance is about seeing that it runs properly. He elaborated that governance is concerned with the supervision and control of executive actions, the concern for the effect of the entity on other parties; the acceptance of a duty to be accountable; and the regulation of corporation within a jurisdiction of the state in which it operates.

A careful observation of the description above would easily reveal that the central element of corporate governance is primarily and essentially man, be he the director, the manager, the shareholder, the auditor, the stakeholder, or even the regulator. It is man that is involved in the process of managing the business of the company towards achieving its mainstream goal of maximizing the wealth of its owners (the shareholders). The codes of upholding trust, maintaining integrity, exercising transparency and accountability, prudent management of resources, maximizing returns, caring and concern of the environment are but would remain as mere noble codes if the issues of man, his values, ethics and moral conduct are not tackled in the first instance. In short, corporate governance is basically the moral and ethical dimension of managing a company's business. In this regard, corporate governance is indeed very much value-loaded. Man is essentially an ethical and moral being. For the Muslims, the ethical codes of conduct based on the Tawhidiq worldview and Quranic values are considered more elevating than those, which are detached from religious moorings.

4 The Dual Nature of Man

Islam looks at man as having a dual nature [5], "one is the animal aspect and the other the spiritual aspect, which is sometimes called rational aspect. When is implied and demanded is that this animal aspect of man be subjugated by the rational one and such submission of the former to the latter affirm and reaffirm its covenant with God". Although Islam enjoins the exercise of rationality but such consumption of man's intellectual capacity has its own limits and defined boundary confined by Divine values. The concept of rationality is not confined to the concept of reason towards the understanding of objects in space and time with their relations and events. The early Muslims have always defined this intellect (Aql) as a spiritual substance. Thus, by rational we also mean to include within such context in which rational operates, not only the world of objects but also that of spiritual and truth.

 DOI: 10.1051/shsconf/20173404005

In short, Islam calls for balance equilibrium, harmonizing his rational and spiritual dimensions, integrating values and knowledge and placing things in their appropriate position. The sources of guidance for the Muslims are basically the Quran and the Sunnah, the traditions of the Prophet Muhammad (pbuh). To be a Muslim is to submit and worship Allah but it is important to note that the act of submitting and worshipping in Islam does not, again, confine itself adherence of the rules for religious rituals but instead encompasses the code of rules on all the aforementioned activities. This draws demarcation line between Islam and all other religions of the world and this also underlies the key to comprehend the place and significance of Islamic values in all spheres of human value activities, corporate governance inclusive, within the framework of the Islamic value system. These values are part and parcel of the religion itself and the performances in adherence to them is regarded as the act of ibadah – worshipping and submitting – to Allah, just as the performance of prayer and fasting are [1].

In light of the above, and of Quranic revelations concerning the reality of God, man and the universe, the scope and meaning of corporate governance as far as Muslim managers are concerned assume a broader and more comprehensive import that what is normally conceived by conventional and contemporary schools of thought.

5 The Concept and Philosophy of Corporate Governance in Islam

The philosophy of corporate governance in Islam resides upon three distinct aspects, namely, values, knowledge (intellectual) and competency. The fundamental of all values rested upon the concept of vicegerency (khalifah) or trusteeship. The underlying message in this concept is: man's position in relation to man (enjoining good deeds and forbidding evil doings); secondly, man's position in relation to nature (to work, develop and exploit the natural resources of the earth in accordance with the ordained principles); thirdly and most importantly, man's position in relation to his Creator, Allah. In essence, its chief objective inherent in its philosophy of managing a business is, to organize human society and its main systems and disciplines on the basis of social justice for all, irrespective of color, creed or race and to establish concepts of goodness, virtue and benevolence among people so that they are able to enjoy life of security, peace, prosperity and happiness on an individual as well as at social level.

The concept of managing life, governing a corporation, leading a community or society, rests upon the cardinal principle of submission to Allah. This notion has a deep and broad implications to Muslims for the acceptance of Allah as their Lord implies a binding covenant and what follows is the submission to the rules, desires and the will of their Lord, hence the religion Islam which means submission.

6 Social Responsibilities and Managerial Accountability

Islam looks at accountability and responsibility in a different light. In terms of responsibility, the manager is not merely responsible to his human superiors; he, as a servant and trustee of Allah in all situation, is simultaneously responsible to Allah, the Owner of his very self and the resources he is utilizing and managing. This responsibility to Allah, is in fact a function to the intrinsic quality of the manager's life as an Amanah (trust) from Allah. To forget or to neglect this fundamental aspect of managerial responsibility tantamount to a betrayal (khianat) of the Divine trust, and will result with all the attending consequences in this world and in the next. Here again, we see that the concept of accountability in professional literature and culture is confined only to public accountability or accountability to human superiors. The Islamic concept of accountability extends the scope of human answerability to God as the Judge and Ruler of man, and not merely as a Creator. The concept of Taqwa (mindfulness of the pleasure and displeasure of God for all human actions) has to be understood in the context of the activities of the professional and corporation.

The make up a more wholesome meaning of human existence is defensible. In tandem with this aspiration, the concept of work and dedication to work as a form of ibadah (worship in its wider sense) and 'amal soleh' (virtuous act) together with religious observances are warranted. The concept of Allah as the Real Manager of resources helps to liberate the human mind from the false sense of autonomy and dominion over the resources.

7 Success and Failures in Managing Business

It is without doubt that the ultimate objective of directing and managing the business and affairs of the company is that of enhancing shareholder value. Ultimately, the owners of the business, the shareholders – expect to receive an appropriate return on their investment. Generally, the success and failure of a business organization is usually viewed in terms of profit and loss. Although the objectives of getting higher monetary and material returns are not in themselves objectionable in Islam, provided the means employed are lawful and the gains legitimate, the idea of success being limited solely to materialistic and temporal consideration are alien to the Islamic worldview.

Islamic managers who are imbued with the worldview of Tawhid (ones of God) are not anti-profit or anti-worldly gain within the limits provided by religion. Their vision of success and failure extend beyond worldly existence to the life in the hereafter. In the Quranic definition of success (al-falah), a true gain is a gain that persists to the life eternal and a true loss (khusran) is a loss that is also experienced in the hereafter. Since in the Islamic vision of reality, human life does not begin and end here, but continues unbroken beyond the world of phenomenon, and that life in the hereafter is an abiding eternal life of reward and punishment, it follows that the criteria of success and failure cannot be limited to the here and now, but should be equally applicable to the abiding and eternal life. The safest criterion of success, therefore, must entail the question of Divine pleasure and displeasure. Thus, a gain in any form, however, great it may be, which does not lead to felicity and salvation in the hereafter, is not regarded as true success.

Islamic manager's criteria of success and failure either for themselves or for the organizations are derived from the comprehensive concept of al-falah and khusran. This involves the question of the propriety of ends and means. Both ends and means must be legitimate and justifiable from ethical principles of the Quran. The application of this concept of success implies a more holistic planning, goal setting, and implementation of the plans which keep in mind the omniscience and omnipresence of Allah in His universe. The end result of such a holistic vision and endeavor is greater security for everyone, minimum disruption of societal peace and negligible harm to the environment which has been ravaged by the misuse of technology, human callousness, myopia, greed, rivalry for domination and devastative competition – in short, the consequences of man acting as though he is the Master of universe. However, this holistic vision of success and failures does not in any way impair technical efficiency and managerial excellence.

8 Leadership and Decision Making Process

Much has been said about the concept of vicegerency or trusteeship. In essence, the qualities of an Islamic manager (leader) are the qualities of an Islamic man itself [6]. The Holy Prophet Muhammad (pbuh) once said: "Every one of you is a keeper or a shepherd and will be questioned about the well-being of his fold. So, the Head of the State will be questioned about the well-being of the people of the State. Every man is a shepherd to his family and will be answerable about every member of it. Every woman is a shepherd to the family of her husband and will be accountable for every member of it. And every servant (employee) is a shepherd to his master (employer) and will be questioned about the property of his master".

Islam lays down the followings as among the main attributes that must exist in an Islamic manager or administrator (leader), among others; moderate and consistence in his directions and commandments; exercise process of consultation (shura) in decision making; set himself as an exemplary; humble and helpful; honest, uphold justice, committed and responsible in handling subordinates; strive hard to achieve group's objective; and of prime importance, having strong faith in deciding any action.

9 Conclusions

The paper attempts to propose an alternative approach of corporate governance by specifically discuss on the issues of corporate governance seen from the Islamic perspective. In as much as corporate governance involves the exercise of greater accountability, enhancing transparency, timely disclosure of information, prudent decision making, good management of resources, and employment of the right code of conduct among the company directors and managers, there is nothing that separates these elements from the Islamic moral codes. The boundary that delineates Islamic approach towards corporate governance and its counterpart rests upon the spiritual and philosophical values of the former. Indeed, corporate governance is basically the moral and ethical dimensions of managing business and at the centre of man with its intractable moral and ethical problems plaguing society, which needs to be addressed. While the Muslims may regard the Tawhidiq worldview and Quranic values as more appropriate in justifying the relevance of employing spiritual-based corporate governance, more often than not, most of these moral codes of conduct are in fact universal in values. Hence, one should not be apologetic for its inability of integrating religious faith and values at the workplace, even within a non-Muslim corporation. In fact, this Divine guidance should be seen as morale enhancement, with the Divine commandments as its complementary compelling force.

Nevertheless, corporate governance is but a small slice of the large ethical code of corporate practice. The intention of institutionalizing it is to enhance public and investors' confidence in the market-place. Indeed, the country has enough codes of ethics to cover all the professions but their enforcement and effectiveness leave much to be desired. If laws, by-laws, statutory regulations, general orders and code of ethics are not effective enough to curb unethical conduct, perhaps we need to review and adopt a more comprehensive perspective on the concept of development which includes individual, organizational and societal goals. Rather than striving to develop Malaysia into a mere industrialized nation, robbed of its soul, perhaps we could aim at achieving it as an exemplary developed moral-nation of the East.

The authors would like to thank the anonymous reviewers.

 DOI: 10.1051/shsconf/20173404005

References

1. A. Khurshid, Islamic Foundation. *Its basic Principles and Characteristics.* (1975)
2. B. Tricker, An International Review. Corporate Governance–the subject whose time has come. Corporate Governance: **8(4),** 289-296 (2000)
3. F. Committee, Report on Corporate Governance. *Corporate Governance.* (1999)
4. F.S. Cheah, C. Yeap, MAICSA. Perspectives on Corporate Governance and Management. (1999)
5. S.N. Al-Attas, Berita Publishing: *Faces of Islam.* (1989)
6. S.O Al-Habshi, INMIND. *Islamic Management for Excellence.* (1998)
7. T. Sheridan, N. Kendall, Pitman Pub. *Corporate governance: An action plan for profitability and business success.* (1992)

DOI: 10.1051/shsconf/20173404006

Financial restatement and firm performance in family controlled and CEO duality companies: evidence from post 2007 Malaysian Code of Corporate Governance

Sok Fun Chin[1*], *Kin Boon* Tang[2], and *Ayoib* Che Ahmad[3]

[1]South Australian Matriculation/SACE International, Taylor's College Subang Jaya, Malaysia
[2]Faculty of Social Sciences, Nottingham University Business School Malaysia Campus, Malaysia
[3]Tunku Puteri Intan Safinaz School of Accountancy, College of Business, Universiti Utara Malaysia, Malaysia

Abstract. Do financial restatements dampen firm performance? One argument about this is that restatements hurt investor confidence in the credibility of a firm's disclosure, resulting in a decline in demand for the firm's securities, thereby, leading to a significant drop in asset price. On the other hand, agency theory suggests that family ownership could have potential benefits to a firm's performance. An increase in family ownership will have a greater concern for reputation to the controlling family in producing high quality accounting information and, thereby, reduce the likelihood of financial restatements. We evaluate these arguments by distinguishing the effects of restatements on firm performance under two corporate governance environments; family-controlled and CEO duality companies. Based on a sample of Malaysian listed companies in 2008 after the post MCCG 2007 initiative, our findings suggest that (1) restatements dampen firm performance, (2) the dampening impacts of restatements are completely mitigated in family-controlled companies, and (3) the dampening effects are more pronounced in non-family controlled companies than family-controlled companies in non-CEO duality companies. Using this evidence, we recommend that Malaysian regulators develop policies that are unique to the Malaysian markets so as to curtail accounting irregularities. They should reconsider the relevance of requiring CEO non-duality as a practice of good corporate governance and encourage more investment in family-controlled companies.

1. Introduction

*Corresponding author: sokfun.chin@taylors.edu.my

In Malaysia, the ownership and control of corporations are highly concentrated, and many are in the hands of family-controlled companies [1]. At times, external financial statements prepared as instructed by the controlling family may mislead users and hide opportunistic activities carried out by the dominant family at the expense of minority shareholders [2]. To compel the management team of listed companies to act ethically in the interest of the general public, the Securities Commissions has been issuing the Malaysian Code of Corporate Governances over the years (*viz.,* MCCG 2000, MCCG 2007, MCCG 2012 and MCCG 2016 draft) yet the accounting irregularity-financial restatement still occurs. It is common for the Chief Executive Officer (CEO) in a family-controlled company to hold the position of Chairman (CEO duality) so that this greater voting power makes it easier to appoint more members of the family to the Board of Directors [3]. Since having a separation of roles between the CEO and Chairman is accepted as part of good corporate governance in MCCG in that a person cannot dominate the decisions made by the board. Accordingly, this study is aimed at providing empirical evidence concerning financial restatement and firm performance in family-controlled companies when the CEO holds dual roles, using post MCCG 2007 data.

2. Literature Review

2.1 Financial restatement and firm performance

General Accounting Office (GAO) states that financial restatement "occurs when a company, either voluntarily or prompted by auditors or regulators, revises public financial information that was previously reported" [4,p1footnote].The firm's dismal performance may be the reason the manager restates the account; this typically happens to firms experiencing high growth and improving performance during a period of economic boom. As a result, financial analysts may be guided by unrealistic future results and expectations. When the economy slows down, managers find it hard to achieve the targets set during the boom, and thus, resort to practising accounting irregularities to meet analysts' expectations [5].

Restatement normally indicates that the manager is unable to report the firm's future prospects in a credible way, causing lenders to reassess the credit risk profile of the restating firm, thus, leading to higher interest rates, more covenant restrictions, and a higher likelihood of loans being secured than initiated before the restatement [6-7]. Also, investors, customers and suppliers tend to lose confidence in the firm's future prospects and change their terms of trade with the restating firm [8]. This means restatement will increase the cost of debt, future litigation concerns [9], and lead to meaningful declines in firm value and expected future cash flows [9]. Accordingly, we hypothesise that:

H1. Financial restatement reduces firm performance.

2.2 Financial restatement and firm performance in family-controlled companies

In a highly concentrated ownership structure, the controlling family can seek personal gain through expropriation of minority shareholders, related party transactions [10], managerial entrenchment [11], and have significant effect on the recognition of goodwill impairment[12]. Family-controlled companies often weaken the effectiveness of Anglo-American corporate governance mechanisms [13].

On the other hand, agency theory explains that the proportion of managerial ownership can reduce the agency cost between the managers and shareholders by aligning the interests of managers with those of shareholders [14]. Based on the agency theory, an increase in the family ownership is expected to reduce the opportunistic goodwill impairment recognition" [3, pp.155]. In a multi-generation family-controlled firm, if the family owners are found restating the accounts, they may lose reputation-based benefits, such as high social status and political connections [15]. As such, our testable hypothesis is stated, as below:

H2. The dampening effect of financial restatements is mitigated in family-controlled companies.

2.3 Financial restatement and firm performance in dual CEO companies

Agency theory argues that shareholder interests require protection by instituting separate roles for board chair and CEO [16], as it is commonly believed that CEO (duality) tends to allow for abuse of power for personal gain; this includes increasing compensation or other forms of perquisite consumption [17].On the other hand, Stewardship theory supports the premise that the CEO duality company will enjoy the classic benefits of unity of direction and of strong command and control ([18]. The CEO-Chair appointment is an efficient response to the firm's contracting environment, in view of the increased riskiness of a firm, industry concentration, CEOs' ability and track record, and strong governance[19].

Nevertheless, the impact of CEO duality on the issues of corporate governance and firm performance remains inconclusive. [20] found insignificant association between a CEO holding dual roles as chairman and income-increasing and income-decreasing earnings management among Malaysian firms, supporting the findings of [21]. [22] too yielded insignificant correlation between the CEO-duality role and return on equity. In an earlier study,[23] revealed significant evidence of the relationship between role duality and performance measure in return on assets and return on equity. Accordingly, the following hypothesis is developed.

H3. CEO duality has significant impact on occurrence of financial restatement.

2.4 Financial restatement and firm performance in family controlled and dual CEO companies

In Malaysia, it is common to find the CEO holding dual roles as Chairman and CEO in family-controlled companies [3; 24]. Previous studies suggest that having separate leadership helps to enhance firm performance[24-25], but the CEO duality in family-controlled firms alone does not lead to goodwill impairment recognition [3]. In addition, having a separate CEO and chairman in a family-controlled company is not an effective way to reduce the occurrence of earning management when the CEO is a member of the controlling family (Presley and Abbott 2013[26]).

3. Data and Methodology

3.1 Study period

MCCG 2007 was launched in 2007.Compare to MCCG2000, MCCG 2007 stipulated more stringent compliance in the process of the selection of the CEO candidate. This study has chosen 2008 as the research period to examine the effectiveness of MCCG 2007 in reducing the occurrence of financial restatement in family-controlled companies when qualified CEOs held dual roles.

 DOI: 10.1051/shsconf/20173404006

3.2 Sample selection and operational definitions

The data, sourced from Bursa Malaysia's website. The 2008 annual reports of these firms were examined for the criteria for financial restatement to be in compliance with the General Accounting Office (GAO) definition, and the keywords, "restatement", "restate", and "prior year adjustments", were scrutinised in each annual report for evidence of restatement [27]. Under the GAO definition, restatement does not include changes in accounting policies that lead to restating accounting figures.

3.3 Firm performance (FV) measures

Four performance measures are employed. (a) *Return on assets(ROA)* measuring the percentage of net profit over total asset. (b) *Earning per share(EPS)* calculated as net profit over weighted average of common shares outstanding. (c) *Profit margin(PM)* quantifying the percentage of net profit divided by turnover. (d) *Return on equity(ROE)* measures percentage of the net profit over common shareholders' equity.

3.4 Corporate governance variables

Financial restatement (FR) is financial restatement which coded as 1 if financial restatement occurred, or 0 otherwise.
CEO duality: the CEO has dual roles are coded 1 or 0, if otherwise[18].
Family-controlled companies: The company is deemed a family-controlled company if (1) the founder of the firm is the CEO or its successor is related by blood or marriage, or (2) at least two family members work as managers, or (3) family directors have managerial ownership, either direct or indirect shareholding of a minimum of 20% in the firm[24].
There are two control variables(CV) used; (a) Debt-to-asset ratio and (b) Altman Z-Score.

3.5 Baseline models

We use the following baseline model to examine the effects of financial restatement on firm performance, where FV is firm performance which includes return on assets, earning per share, profit margin, and return on equity.

$$FV_i = c + \beta_1 FR_i + \sum_{k=1}^{n} \delta_k CV_{k,i} + \varepsilon_i \qquad (1)$$

In order to examine the effects of firm performance on restatements, the following logit model is employed:

$$FR(0,1)_i = c + \beta_1 ROA_i + \beta_2 EPS_i + \beta_3 PM_i + \beta_4 ROE_i + \sum_{k=1}^{n} \delta_k CV_{k,i} + \varepsilon_i \qquad (2)$$

The effects of restatement on firm performance will be further examined under two corporate governance environments; family-controlled companies and CEO duality companies.

DOI: 10.1051/shsconf/20173404006

4. Findings and Discussions

4.1 Descriptive statistics

Table 1 shows correlation between financial restatement and both variables of corporate governance (CEO duality and family-controlled companies) is negative but statistically insignificant, indicating there is no direct relationship between the corporate governance variables and financial restatement. Also, the correlation between FR and the three firm performance measures; ROA, PM and ROE is negative and statistically significant, suggesting preliminary evidence in support of H1 that restatement reduces firm performance. Next section, regression and binary analyses will be used to control corporate governance environments that may mitigate or propagate these adverse effects of restatement on firm performance.

Table 1: Means, Standard Deviations, and Correlations

	Mean	Std. Dev	FR	DU	FC	DA	ROA	EPS	PM	ROE
FR	0.0610	0.2408								
DU	0.2317	0.4245	-0.1399							
			(-1.2641)							
FC	0.6220	0.4879	-0.1166	-0.0487						
			(-1.0504)	(-0.4361)						
DA	-1.7688	49.682	0.0117	0.0550	-0.0515					
			(0.1044)	(0.4930)	(-0.4618)					
ROA	-0.0287	0.1802	-0.4915***	-0.0057	0.2151*	0.0033				
			(-5.0476)	(-0.0512)	(1.9702)	(0.0298)				
EPS	0.1667	28.9977	-0.0529	-0.0737	-0.1135	0.0800	0.4615***			
			(-0.4741)	(-0.6615)	(-1.0221)	(0.7184)	(4.6535)			
PM	-0.0854	0.9510	-0.3565***	0.0086	0.2263**	0.0002	0.7848***	0.2286**		
			(-3.4137)	(0.0771)	(2.0776)	(0.0020)	(11.327)	(2.1004)		
ROE	0.0772	0.6738	-0.1937*	0.0040	-0.0070	0.0082	0.3327***	0.1799	0.1679	
			(-1.7664)	(0.0360)	(-0.0628)	(0.0732)	(3.1551)	(1.6357)	(1.5232)	
ZS	0.6212	0.3768	-0.0346	-0.1805	0.0996	-0.0265	0.21299	0.0751	0.0945	-0.0144
			(-0.3100)	(-1.6414)	(0.8956)	(-0.2372)	(1.9485)*	(0.6742)	(0.8487)	(-0.1284)

Notes: *,**, and *** denote significant at 10%, 5%, and 1% respectively. t-statistics are reported in parentheses. FR = financial restatement; DU = CEO duality; FC = family controlled; DA = debt to asset ratio; ROA = return on assets; EPS = earnings per share; PM = profit margin; ROE = return on equity; ZS = z-score.

4.2 Firm performance and financial restatement

Table 2 presents that, with the exception of earning per share, the effects of restatement on firm performance are negative and statistically significant. Restatement will create uncertainty about the firm's future prospect and impede the firm's ability to pursue profitable investment projects and, thereby, lead to a decline in firm performance [8]. We suggest that the adverse impact of restatement on firm performance is a result of negative market reaction towards the firm's credibility, in line with [7,pp44] that argued "restatements may shake investor confidence in the credibility of corporate disclosure,

DOI: 10.1051/shsconf/20173404006

depress demand for a firm's securities, and constrain corporate opportunities thereby leading to a substantial loss in market value".

Table 2: The Effects of Financial Restatement on Firm Performance

	Firm performance measures			
Independent variables	**Return on assets**	**Earnings per share**	**Profit margin**	**Return on equity**
Financial Restatement (FR)	-0.362798 ***	-6.176407	-1.397547 ***	-0.544603 *
Constant	-0.064787 *	-2.984122	-0.129094	0.133777
Control variables				
Debt/Asset	5.15E-05	0.048249	0.000125	0.000134
Z-score	0.093913 **	5.815538	0.207919	-0.037243
$AdjR^2$	0.252462	-0.022871	0.100627	0.001083
F-statistic	10.11856 ***	0.396288	4.020900 **	1.029281

Notes: *,**, and *** denote significant at 10%, 5%, and 1% respectively. The following baseline model is estimated using multiple regressions.

$$FV_i = c + \beta_1 FR_i + \sum_{k=1}^{n} \delta_k CV_{k,i} + \varepsilon_i \quad (3)$$

4.3 Firm performance, financial restatement and corporate governance

Family-controlled

Tables 3a and 3b present the estimates for testing the relationship between firm performance and financial restatement in family-controlled and non-family controlled companies, respectively. It is interesting to note that restatements have no impact on firm performance in family-controlled companies (Table 3a). However, when the companies are not controlled by the family (Table 3b), the results show significant impacts of restatements on performance, suggesting that there is a mitigation effect from family-controlled companies. This evidence is in line with our H2.

CEO duality

Table 3c presents the estimates for the 63 non-CEO duality companies. The remaining 19 CEO duality companies are not examinable as all the companies in this sub-sample are non-financial restating companies. The coefficient for restatements is negative and statistically significant (return on assets, profit margin, return on equity), but much smaller than those in non-family controlled companies, implying that the dampening effects of restatements on firm performance are more pronounced in non-family controlled companies. The results also provide some evidence of the mitigation effects of CEO duality on firm performance in companies. This evidence is also consistent with our H3.

DOI: 10.1051/shsconf/20173404006

Table 3a: The Effects of Financial Restatement on Firm Performance in Family Controlled Companies (n=51)

	Firm performance measures			
Independent variables	**Return on assets**	**Earnings per share**	**Profit margin**	**Return on equity**
Financial Restatement	0.004669	6.082336	-0.094894	-0.462240
Debt/Asset	4.02E-05	0.043907	0.000109	9.94E-05
Z-score	0.015960	0.177551	-0.373424	-0.002101
Constant	-0.009039	-2.573681	0.328348	0.093396
$AdjR^2$	-0.059436	-0.052188	-0.018836	-0.047689
F-statistic	0.064967	0.173341	0.691876	0.241359

Notes: *,**, and *** denote significant at 10%, 5%, and 1% respectively.

Table 3b: The Effects of Financial Restatement on Firm Performance in Non-Family Controlled Companies (n=31)

	Firm performance measures						
Independent variables	**Return on assets**		**Earnings per share**	**Profit margin**		**Return on equity**	
Financial	-	***	-	-	***	-0.614170	*
Restatement	0.604072		18.17152	2.194797			
Debt/Asset	0.000378		-	0.001903		0.002397	
			0.113099				
Z-score	0.196259	**	16.38359	0.952880	*	-0.074659	
Constant	-	**	-	-	*	0.181880	
	0.132700		3.102051	0.696473			
$AdjR^2$	0.538039		-	0.301764		0.011234	
			0.016080				
F-statistic	12.64683	***	0.841745	5.321813	***	1.113619	

Notes: *,**, and *** denote significant at 10%, 5%, and 1% respectively.

Table 3c: The Effects of Financial Restatement on Firm Performance in Non-CEO Duality Companies (n=63)

	Firm performance measures						
Independent variables	**Return on assets**		**Earnings per share**	**Profit margin**		**Return on equity**	
Financial	-	***	-	-	***	-0.568577	*
Restatement	0.367686		7.283374	1.415743			
Debt/Asset	5.13E-05		0.051574	0.000134		0.000139	
Z-score	0.086236		7.297367	0.173534		-0.172812	
Constant	-		-	-		0.235065	
	0.055521		2.723906	0.091292			
$AdjR^2$	0.281421		-	0.101509		0.007973	
			0.029044				
F-statistic	9.093814	***	0.416705	3.334860	**	1.166097	

Notes: *,**, and *** denote significant at 10%, 5%, and 1% respectively.

5. Conclusion and Policy Implications

In this study, four measures of firm performance were used to examine the impact of financial restatement on firm performance based on a sample of Malaysian listed companies in 2008. Three out of the four measurement of firm performance suffer when there is occurrence of financial restatement. Using this evidence, we recommend that investors and analysts may need to lower their expectations on the future prospects of the restating firms after the incidence of restatement. Likewise, since companies with lower performance are likely to be involved in restatement, investors and analysts may need to be cautious when using financial reports of these firms to make decisions.

Our study also finds that the dampening impacts of restatements are completely mitigated in family- controlled companies. The dampening effects are more pronounced in non-family controlled companies than family-controlled companies in non-CEO duality companies. Therefore, our findings support the agency theory which argues that an increase in the family ownership will promote greater concern for reputation, better monitoring of manager and more incentives to produce high-quality accounting information [15].

The implications of this study suggest that the practice of CEO duality and family firms is beneficial in curtailing accounting irregularities and lead to better firm performance. The policy makers for Malaysian listed firms may want to reconsider the relevance of requiring CEO non-duality as a practice of good corporate governance, and to encourage more investment in the family-controlled companies.

References

1. C.A.Mallin. Corp Gov, 4th edn, Oxford University Press, Oxford(2013).
2. A.Prenciple., and S.Bar-Yosef. Corporate governance and earnings management in family-controlled companies. J of Acc Aud and Fin **26**, 199-227. http://dx.doi.org/10.1177/0148558X11401212 (2011).
3. N.M.Salleh., and N.Omar. CEO duality, family-control and goodwill impairment. As J of Bus & Acc, **7**(1), 143-179(2014).
4. General Accounting Office. Financial statement restatements: trends, market impacts, regulatory responses and remaining challenges. Report 03-138,P.1footnote. Government Printing Office, Washington, DC(2002).
5. P.M.Dechow., and C. M. Schrand. Earnings Quality. Research Foundation Publications
2004(3) CFA Institute,1-152 (2004).
6. X.Chen., Q. Cheng., and A.K. Lo. Accounting restatements and external financing choices. Cont Acc Res **30**(2), 750-779(2013).
7. J.R.Graham., S. Li., and J. Qiu. Corporate misreporting and bank loan contracting. J of Fin Econ **89**, 44–61(2008).
8. S.M.Albring., S.X.Huang., R.Pereira., and X.Xu. The Effects of Accounting Restatements on Firm Growth, J of Acc & Pub Pol **32**(5), 357-376(2013).
9. Z.Palmrose., and S.Scholz. The circumstances and legal consequences of non-GAAP reporting: Evidence from restatements. Cont Acc Res **21**(1), 139-180(2004).
10. R.C.Anderson., and D.M.Reeb. Founding-family ownership and business performance: Evidence for the S&P 500. J of Fin **58**, 1301-1328(2003).
11. A.Shleifer., and R.Vishny. A survey of corporate governance. J of Fin **52**, 737-783(1997).
12. N.M.Saleh., and N.Omar. CEO Duality, Family-Control and Goodwill Impairment. As J of Bus and Acc **7**(1), 143-179(2014).

13. R.Rusmin.,G.Tower.,T.Achmad.,and J.Neilson. Concentrated family ownership structures weakening corporate governance: A developing country story. Corp Own & Con **8**(2), 94-105(2011).
14. M.C.Jensen.,and W.H.Meckling. Theory of the firm: managerial behaviour, agency costs and ownership structure. J of Fin Econ **3**(4),305-360(1976).
15. L.Ma., S., and G.Tian. Family control, accounting misstatements, and market reactions to restatements: Evidence from China. E Mar Rev, http://dx.doi.org/10.1016/j.ememar.2016.06.001 (2016)
16. L.Donaldson. and J.H.Davis. Stewardship theory or agency theory: CEO governance and shareholder returns. Aus J of Mgt **16**(1), 49-65(1991).
17. M.C.Jensen. The modem industrial revolution, exit, and the failure of internal control systems. The J of Fin **48**, 831-880(1993).
18. P.L.Rechner., and D.R.Dalton. CEO duality and organizational performance: a longitudinal analysis, Stra Mgt J **12**(2), 155-160(1991).
19. A.Ghosh., and D.Moon. When the CEO is also the chair of the board. W Paper, Zicklin School of Business, Baruch College(2009).
20. H.AHashim., and S. Devi. Board independence, CEO duality and accrual management. As J of Bus & Acc **1**(1), 27-46(2008).
21. R.A.Rahman., and F.H.Mohamed-Ali. Board, audit committee, culture and earnings management: Malaysian evidence. Mgerial Aud J **21** (7), 783-804(2006).
22. S.N.Shukeri., W.S. Ong., and S.M.Shahidan. Does Board of Director's Characteristics Affect Firm Performance? Evidence from Malaysian Public Listed Companies. Int Bus Res ISSN 1913-9004 (Print), ISSN 1913-9012 (Online) (2012).
23. A.Rahman., and R.M.Haniffa. The effect of role duality on corporate performance in Malaysia. Corp Own & Con **2** (2). 40-47(2005).
24. N. Amran., and A.Che-Ahmad. Corporate governance mechanisms and performance: Analysis of Malaysian family and non-family controlled companies. J of Mod Acc & Aud **6**(2), 1-15(2010).
25. A.Che-Ahmad., N.Aziah., A. M., and I.Zuaini. Corporate governance, ownership structure and corporate diversification: Evidence from the Malaysian listed companies. Asi Aca of Mgt J **8**, 67-89(2003).
26. T.J.Presley., and L.J.Abbott. AIA Submission: CEO overconfidence and the incidence of financial restatement. Adv in Acc, incorp Adv in Int Acc **29**, 74-84(2013).
27. S.N. Abdullah., N.Z.M.Yusof., and M.N.Mohamad-Nor. Financial restatements and corporate governance among Malaysian listed companies, Mgerial Aud J **25**(6), 526 - 552 (2010).

SHS Web of Conferences **34**, 04007 (2017) DOI: 10.1051/shsconf/20173404007
FourA 2016

Effect on Value Earnings Management Company with Good Corporate Governance Practices as Moderating Variable

Emylia Yuniarti [1 *], *Mukhtaruddin* [1], and *Nadia* Hanim [1]

[1] Accounting Department Economic Faculty, Sriwijaya University, Palembang Indonesia

Abstract: Firm value reflects the size of the stock market reaction to the company. Increasing the value of the company is an achievement that suitable with the desire of the owner, because as the value of the company increasing as well the welfare of the owner. Earnings management has an influence on the value of the company. Good Corporate Governance to be one way to eliminate the management engineering efforts specifically to make regulations with the requirement for the companies to disclose certain information mandatorily and voluntarily. This research will discuss the influence of earnings management on company value, the effect of Corporate Governance on company value directly, and the effect of Corporate Governance as a moderating variable from the influence of earnings management and firm value. Data analysis technique used in this research is a descriptive data analysis method, analysis model and classical assumption check because this research aims to determine the effect of earnings management, company value and Corporate Governance on Enterprise State-owned Enterprises listed on the Indonesia Stock Exchange. The results of this study are earnings management negatively affect the value of the company, corporate governance itself has no effect on firm value, but as a moderating variable of corporate governance affect earnings management effect on firm value.

1 Background

The financial statements are the source of information for internal and external parties. Internal party here is the management company and its employees. Benefits own financial statements to internal parties is as information for making decisions, and as the evaluation of the financial information to the financial stability and profitability. Then external parties, namely investors, lenders, suppliers, customers and governments. The manager is one of the parties involved in the company's activities. As the parties involved in the activities of the company, managers have an obligation to provide information to external parties.

When the state sector in the management did not achieve the target profit, then management can modify the profit. One of the ways of management in the process of preparing financial statements that may affect the rate of profit is shown Profit Management is expected to enhance the Company's value at any given moment. Earnings

* Corresponding author: emylia_yuniarti@yahoo.com

Management by management companies will increase the company's value (Tobin's Q) and then going down (Morck, Scheifer & Vishny, 1988).

Corporate governance is one key element in improving economic efficiency, which includes a series of relationships between the company's management, board of directors, shareholders and other stakeholders. Corporate governance also provides a structure that facilitates the determination of the objectives of a company, and as a means to determine the performance monitoring techniques (Deni, Khomsiyah and Rika, 2004).

Based on research Herawaty (2008), earnings management has a significant effect on firm value. The asymmetry between the management and the owners provide opportunities for managers to manage earnings (Earnings Management) in order to increase the value of the company at a given time so as to mislead the owners (shareholders) of the company's true value. The increasing value of the company is an achievement, in accordance with the wishes of its owner, due to the increasing value of the company, the welfare of the owners will also increase. Value companies are also shareholder value reflects the size of the stock market reaction to the company. The greater the value of the company that shareholder value also reflects the public has assessed the stock market price over book value (Megawati, 2010).

Maximizing the value of the company is very important for a company, because to maximize value also means maximizing shareholder wealth that is the purpose of the company. The value of the company can deliver maximum shareholder wealth when the company's stock price to rise. The higher the stock price, the higher the wealth of shareholders. To achieve the company's value generally investors hand over its management to the professionals. The professionals are positioned as the manager or the commissioners.

In the process of achieving maximizes the value of the company will appear a conflict of interest between managers and shareholders in this case as the owner of the company (agency problem). Management of the company as the manager has more information about the company and the former more than the shareholders resulting asymmetry of information that enables management accounting practices oriented towards profit to achieve a certain performance

Based on what have described above, some question rise and will be analyse are: 1. Is Profit Management positive effect on company value? 2. Does the practice of corporate governance positively affects firm value? 3. What is the effect of the Value Profit Management companies weakened by the practice of corporate governance proxy for Corporate Governance Perception Index?

2 Literature Review

In understanding the earnings management can be attributed to the agency theory as the theoretical rationale. Agency theory is the relationship between the agent and the principal manager here is that investors (Ujiyantho & Bambang, 2007). Principal delegate responsibility for decision making to the agent, it can be said that the principal give a mandate to the agent to perform certain tasks in accordance with the employment contract has been agreed. Authority and responsibility as well as the principal agent stipulated in the employment contract by mutual consent (Muliati, 2011).

Earnings management is interference in external financial reporting process with the aim to benefit themselves. Earnings management is one of the factors that can reduce the credibility of financial statements, profit management add bias in the financial statements and may interfere with users of financial statements believe earnings figures as a result of engineering the earnings numbers without engineering (Setiawati & Na'im, cited in Muliati 2011). Motivation of earnings management done for reasons of capital markets caused more by the general belief that the figures - accounting numbers, in particular profit is one of the important information that is used by investors to assess the price of shares Mulyono (2011), it is not surprising that there are some managers who seek to make their financial statements look good with the intent to influence the stock prices in the short term.

Corporate governance is a concept based on agency theory, is expected to serve as a tool to provide assurance to investors that they would receive a return on the funds they have invested. Corporate governance relates to how investors are confident that managers will benefit them, confident that the manager will not steal / darken or invest in projects that do not benefit related to the fund / capital already invested by the investor, and is concerned with how the investor control managers (Shleifer and Vishny, in Ujiyantho, 2007). In other words, corporate governance is expected to function to suppress or lower the cost of agency (agency cost). Corporate Governance assessment is currently able to use the data from the Indonesian Institute for Corporate Governance (IICG). Where the assessment CG using several assessment includes four stages: Self-assessment, collection Corporate Documents, Preparation of Papers and Presentations, and observations to the company.

One alternative used in assessing the value of the company is to use Tobin's Q. This ratio was developed by Professor James Tobin (1967). This ratio is a valuable concept because it shows the current estimate of the financial markets on the value of the return on every dollar of incremental investment. Company size is a variable that positively affect earnings management, which means that the larger the company measured by market capitalization, the greater the profit of the company management. This is because managers in large companies will receive incentives when they perform management in order to reduce post positive earnings. In addition, large companies should be able to meet the expectations of investors and many estimates and judgments that must be applied to large enterprises. Thus, the size of the company can be used as a consideration for potential investors to make investments and decisions.

3 Research Framework

This study aims to analyse the Influence of the Value Profit Management Company, the influence of corporate governance practices affect the value of the Company, Corporate practice Governance influence on the relationship between Earnings Management and Corporate Values.

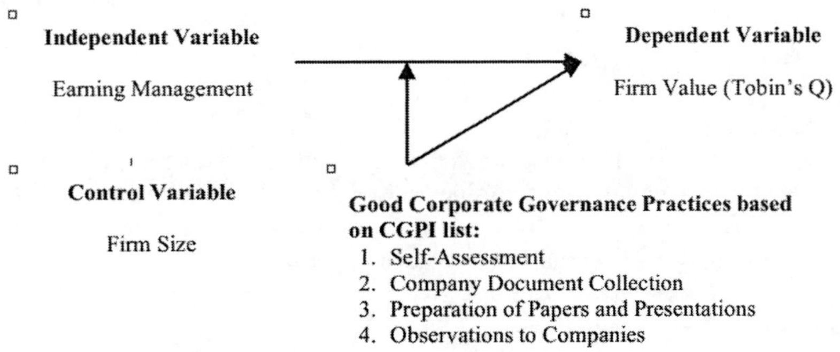

Fig 1. Research Framework

The shareholders or owners of the company can only rely on reports from the company's management to know the state of his company. While managers as managers of the

company to find out more information on the company's internal and prospects for the future, causing information gaps. This condition is often referred to as information asymmetry. Because of the asymmetry of information is the owner of the company cannot know the actual condition of the company so that the company's management has the opportunity to do a profit management. Basically, the manager earning management to enhance shareholder value. This activity can actually increase the value of the company at a certain period but can also reduce the value of the company in the future.
proposed hypothesis is:

H1: Earnings management effect on firm value

In perspective, the agency theory, the risk adverse agent and which tend selfish will allocate resources (investing) which do not increase the value of the company. The agency issues would indicate that the value of the company will rise if the owners of ordinary companies controlling behaviour management in order not to waste resources companies, either in the form of investment that is not feasible, or in the form of shirking.

Corporate governance is a system that regulates and controls the company that is expected to provide and enhance the company's value to its shareholders. Thus, the implementation of good corporate governance is believed to increase the value of the company. Black et al. (2003) argued that first, the company will be better managed so that it can be more profitable higher dividends. Second, caused by outside investors can assess earnings or dividends equal to the higher of companies are implementing better corporate governance. The results showed that there was no evidence that companies with good corporate governance are more or pay higher dividends, but there is evidence that investors assess earnings or dividends equal to the current higher for firms with better corporate governance. the second hypothesis to be tested in this study is

H2: Corporate governance effect on the company value

By reason of increasing the company value, management make a decision by doing earnings management. Therefore, the practice of corporate governance in the company will restrict Earnings Management for their mechanisms of control in the company. Corporate Governance Practices can be proxy by the CGPI.

Earnings management has a close connection with the company's internal conditions, as well as corporate governance. Therefore, in this research Good Corporate Government used as a moderation variable that strengthen the influence between earnings management and corporate value.

H3: Corporate Governance as a moderating variable effect on Earnings Management and Corporate Values

4 Research Methodology

Independent Variables

Independent variables in this study are earnings management. Earnings management here is the action of management to enhance shareholder value. According Rahmawati et al (2006) earnings management can be measured by discretionary accruals which calculated by subtracting the total accruals and accruals non-discreanitory.

Dependent Variables

The company value is the dependent variable, measured by Tobin's Q

$$Q = \frac{\text{MVE} + \text{D}}{\text{BVE} + \text{D}} \tag{1}$$

Control Variables
Company size measured from the natural logarithm of the market value of the company's equity at the end, the number of shares outstanding at year end multiplied by the year-end stock market prices.

Moderating Variable
This variable was measured using an instrument that has been developed by the Indonesian Institute of Corporate Governance (IICG) in the form of Corporate Governance Perception Index (CGPI)

In this study, multiple regression model was used to test hypotheses of the study are:

1. $Q_{it} = \alpha_0 + \alpha_1 EM_{it} + \alpha_2 UP_{it}$ (2)
2. $Q_{it} = \alpha_0 + \alpha_1 CGPI + \alpha_2 UP_{it}$ (3)
3. $Q_{it} = \alpha_0 + \alpha_1 EM_{it} + \alpha_2 CGPI + \alpha_3 EM*CGPI + \alpha_4 UP_{it}$ (4)

EM = Earnings Management proxy by Discretionary Accrual (DA)
CGPI = *Corporate governance*
Q = Tobin's Q proxy by Corporate Value
UP = Company Size proxy by the natural logarithm of the market value

5 Empirical Result

The test results showed that the first model variables Earnings Management, Company Size control variables which were statistically significant. Earnings Management adversely affect the Company Value Management means the use of earnings will decrease the value of the Company. Size may affect the company's earnings management where large enterprises have more complex operational activities allowing for an earnings management, but the company in the sample using the Earnings Management is not a strategy increases the Company Values.

Earnings management is done to raise the company value but in this study the earnings management is proven to reduce the value of the company. Distrust of others and investors against companies doing earnings management can decrease the value of the company itself. From the test results, the first hypothesis said that earnings management has a positive influence on the value of the company is not proved, because the earnings management negatively affect the value of the company. Increasing the possibility of earning management company, the smaller the value of the company.

In which regression model testing, CGPI negatively affect company value means the absence of CGPI effect on firm value indicating no consideration of such information by investors. Perhaps investors still do not trust the results of the survey appear CGPI especially the case of PT. Tin and Bank BNI which entered in the top 10 but the turmoil of performance the following year. Thus, IICG a great struggle to CGPI more credible and can be used as an indicator for investors in making investment decisions (Nuswandari, 2009). The test of the hypothesis 2 which says that corporate governance has a positive effect also not accepted because corporate governance has no effect on the value of the company directly. Corporate Governance corporate governance perception proxied using this index cannot affect the investor directly

The third regression model testing, variable Earning Management Company's positive effect on value. Variable CGPI has a negative effect on the Company Value means CGPI as moderating variables have a relationship. Seen from the value of the variable Management Earning third model to be smaller than the value of the variable Earning Management first model.

Corporate Governance when used as moderating variables have an influence on the value of company this is because investors will consider the company's performance in terms of good corporate governance despite visible indications of earnings management in the company. Good corporate governance can reduce and prevent the occurrence of earnings management in the enterprise

6 Conclusion

Based on the analysis of research that has been done, it can be concluded that the sample of state-owned enterprises listed in the Indonesia Stock Exchange 2009 - 2013 are: Earning Management adversely affect the Company's value due to the effect of earnings management, it will cause the value of the company down. Variable corporate governance proxy for CGPI in this study had no effect on the Company Values. CGPI here considered cannot be considered outsiders to assess a company. Variable corporate governance is moderating variables between Earnings Management and Corporate Values. Indirectly Corporate Governance variables affecting earnings management influence on corporate value, where the value of the company is getting better with the influence of this CGPI. The size of the company affect the company value in each regression models were conducted. This means that the larger the company the greater the Corporate Value.

References

1. G. Boediono, "Kualitas Laba: Studi Pengaruh Mekanisme Corporate Governance dan Dampak Manajemen Laba Dengan Menggunakan Analisis Jalur". SNA VIII (2005)
2. D. DarmawatI, Khomsiyah dan Rika Gelar Rahayu. 2004. "Hubungan Corporate Governance dan Kinerja Perusahaan". SNA VIII (2004)
3. I. Ghazali, *Aplikasi Analisis Multivariate SPSS.* Semarang: BP UnDip. (2005)
4. V. Herawaty, "Peran Praktek Corporate Governance Sebagai Variabel Moderating dari Pengaruh Earnings Management Terhadap Nilai Perusahaan". SNA XI (2008)
5. Megawati. "Pengaruh Corporate Governance, Leverage, dan Manajemen Laba Terhadap Nilai Perusahaan yang Termasuk Kelompok Jakarta Islamic Index tahun 2005-2007". UIN Sunan Kalijaga, Yog (2010) http://digilib.uin-suka.ac.id/3489/1/BAB%20I%2CV.pdf
6. NK. Muliati, "Pengaruh Asimetri Informasi dan Ukuran Perusahaan pada Praktik Manajemen Laba di Perusahaan Perbankan yang Terdaftar di Bursa Efek Indonesia". Universitas Udayana, Denpasar (2011)
7. http://www.pps.unud.ac.id/thesis/pdf_thesis/unud-780-tesisktmuliati.pdf
8. D. Mulyono, Dwi, "Pengaruh Manajemen Laba (*Earning Manajement*) Terhadap Kinerja Keuangan". *Tesis.* Universitas Islam Indonesia Malang. (2011) dari http://lib.uin-malang.ac.id/?mod=th_detail&id=08510022
9. Morck, R. And A. Shleifer, and R.W. Vishny, *Management Ownership and Market Valuation: An Empirical Analysis.* J of Fin Ec, **20**, 293-315, (1988)
10. Sunarto. 2009. "Teori Keagenen dan Manajemen Laba". *Kajian Akuntansi,* Vol. 1 No. 1. ISSN : 1979-4886
11. SWA 27/2009, 21 Desember 2009
12. SWA 26/2010, 9 December 2010
13. SWA 27/2011, 21 December 2011
14. SWA 27/2012, 21 December 2012
15. SWA 27/2013, 23 December 2013

DOI: 10.1051/shsconf/20173404007

16. M. Arief Ujiyantho, dan Bambang Agus P. "Mekanisme Corporate Governance, Manajemen Laba dan Kinerja Keuangan", SNA X Makassar, (2007)
17. T.A. Utama, dan Abdul Rohman. "Pengaruh *Corporate Governance Perception Index, Profitabilitas, Leverage,* dan ukuran Perusahaan Terhadap Nilai Saham". Dip.. J. of Acc.. **2** No. 2, (2013), 1-9

DOI: 10.1051/shsconf/20173404008

The Association between Related Party Transactions and Real Earnings Management: Internal Governance Mechanism as Moderating Variables

Genius Khober Limanto[1], and *Antonius* Herusetya[1,*]

[1]Accounting Department, Business School, Universitas Pelita Harapan, Tangerang, Indonesia

Abstract. We investigate the association between related party transactions (RPT) and real earnings management (REM). We also investigate the role of internal governance mechanism through the effectiveness of board of commissioner and audit committee in mitigating the association between RPT and REM. Our research sample consists of 386 firm-years of manufacturing firms listed in Indonesian Stock Exchange (IDX) from year 2010 - 2014. Using linear regression, we find evidence that RPT has positive association with REM, only when the firm has higher RPT but not in the lower RPT. We find a contradictory result that board of commissioners strengthen the positive association between RPT and REM. Finally, we find evidence that the effectiveness of audit committees weaken the positive association between RPT and REM, both in full sample and in high RPT sample.
Keywords: related party transactions, real earnings management, board of commissioner, audit committee, corporate governance

1 Introduction

One of the consequences of agency problem between controlling and minority shareholders is the practice of earnings management (EM) by controlling shareholders. And related party transactions (RPT) have been used as means of earnings management. Sales to related parties could be used opportunistically to manage earnings upwards and to present more positive operating performance [1]. In addition, certain types of RPT, such as lending to other firms controlled by the same controlling shareholders, have been used by companies to divert earnings from operations to their controlling shareholders and to expropriate their minority shareholders' wealth to maximize the wealth of the controlling shareholders [2]. Firms that engage in RPT which expropriate the minority shareholders are more likely to engage in earnings management than firms without such transactions [3].

*Corresponding author: antonius.herusetya@uph.edu

Weak corporate governance mechanism has a role in allowing the practice of earnings management, while an effective corporate governance functions were found to have significant negative influence on earnings management practice because they play an important role in overseeing the quality of reported earnings [4, 5]. A study also found that companies with weaker corporate governance mechanism have more and higher RPT which imply agency problem between managers and shareholders [6]. Those with better mechanism have less and lower RPT which imply that corporate governance is important in deterring manipulation through RPT.

There are some research gaps in the previous studies. First, only few researches about RPT and EM have been conducted in Indonesia. Second, previous studies had shown mixed results on the association between RPT and EM. Kuan [2] did not found any evidence of the association between aggregated RPT and absolute discretionary accruals. Others found only positive association between purchases and expenses from related party, and abnormal accruals [7]. Third, previous researches applied different methods of measuring RPT and EM, and recognized that those were not good enough to capture the phenomena of RPT [2, 7]. Fourth, Kuan [2] recognized that using only RPT was not enough to explain EM behavior and suggested to include more characteristics, such as Indonesian ownership structure and corporate governance mechanism. Our study contribute to the other measurement of EM i.e., real earnings management (REM), since it is believed that REM is the preferred method to manage earnings [8, 9] We also investigate whether board of commissioners (BCE) and audit committee (ACE) have moderating effect in the association between RPT and REM.

2 Hypothesis development

Prior studies indicates mixed results in the association of RPT and earnings management (EM). Sumiyana and Febrianto [7] for example, discovered that purchase and expenses from transaction with related party had positive influence towards EM. Kuan [2] did not find significant relationship between related party transactions (RPT) and EM. Munir [3] found that RPT that resulted in expropriation activities negatively affect earnings quality of firms with significant family ownership. This low earnings quality was likely resulted from earnings management to mask the firm's poor performance resulted from expropriation. Those prior researches focus only in the accrual earnings management (AEM) while companies prefer using REM because it is more difficult to be detected by auditors [8, 9]. We proposed our hypothesis as follows:

H_1: There is positive association between related party transactions and real earnings management.

Hermawan [5] found that the effectiveness of board of commissioners had positive association with earnings quality. She also found that the effectiveness of audit committee had positive association with earnings quality in a company if board of commissioners were not function effectively. Audit committee characteristic has negative association with earnings management [4]. These studies suggest that if corporate governance functions effectively then the earnings management can be minimize. And because of the corporate strategies in RPT and REM is under the control of internal governance mechanism [5], we expect that the effectiveness of board of commissioners and audit committee could weaken (strengthen) the positive (negative) association between RPTs and REM. So our proposed hypothesis is as follows:

H_2: The effectiveness of board of commissioners weakens (strengthens) the positive (negative) association between related party transactions and real earnings management.

H_3: The effectiveness of audit committee weakens (strengthens) the positive (negative) association between related party transactions and real earnings management.

3 Research design

3.1 Data and sample selection

We use purposive sampling method in our sample selection from manufacturing industries listed in Indonesian Stock Exchange (IDX) from the year 2010 - 2014 with the criteria i.e., the firms have complete disclosure about the nature and extent of their related party transactions (RPT), their boards of commissioners, their audit committee, and their financial performance in their annual reports; and firms used Rupiah as their currency in the financial reports.

3.2 Empirical model

Following Munir [6], we use empirical models to test the first hypothesis (H_1) as follows:

$$REMIDX_{i,t} = \beta_0 + \beta_1 RPT_{i,t} + \beta_2\, ABSDNI_{i,t} + \beta_3 LOSS_{i,t-1} + \beta_4\, MTB_{i,t-1} + \beta_5 LEV_{i,t} + \beta_6 LogASSETS_{i,t} + \beta_7 BIG4_{i,t} + e_{i,t} \quad (1)$$

And using following regression model to test H_2 and H_3:

$$REMIDX_{i,t} = \beta_0 + \beta_1 RPT_{i,t} + \beta_2 RPT_{i,t} {}^*BCE_{i,t} + \beta_3 RPT_{i,t} {}^*ACE_{i,t} + \beta_4 ABSDNI_{i,t} + \beta_5 LOSS_{i,t-1} + \beta_6 MTB_{i,t-1} + \beta_7 LEV_{i,t} + \beta_8 LogASSETS_{i,t} + \beta_9 BIG4_{i,t} + e_{i,t} \quad (2)$$

Based on prior studies, we predict that related party transactions (RPT) have positive association with real earnings management (REMIDX) [3, 7]. The moderating role of board of commissioners and audit committee are represented by RPT*BCE and RPT*ACE, and we predict that these variables have negative associations with REMIDX [4, 5].

3.3 Operational variable

The dependent variable of this research is real earnings management (REMIDX). We use REM in aggregate level (REMIDX) instead of individual level, i.e., the sum of abnormal cash flow from operations (CFO), abnormal production cost (PROD), and abnormal discretionary expense (DISC), because management can engages real transactions by using combination of it [8]. The following is the equation for normal CFO.

$$\frac{CFO_{i,t}}{Assets_{i,t-1}} = k_1 \frac{1}{Assets_{i,t-1}} + k_2 \frac{Sales_{i,t}}{Assets_{i,t-1}} + k_3 \frac{\Delta Sales_{i,t}}{Assets_{i,t-1}} + e_{i,t} \quad (3)$$

Production cost (PROD) consists of cost of goods sold and changes in inventory. The following is the equation for normal PROD.

$$\frac{COGS_{i,t} + \Delta INV_{i,t}}{Assets_{i,t-1}} = k_1 \frac{1}{Assets_{i,t-1}} + k_2 \frac{Sales_{i,t}}{Assets_{i,t-1}} + k_3 \frac{\Delta Sales_{i,t}}{Assets_{i,t-1}} + k_4 \frac{\Delta Sales_{i,t-1}}{Assets_{i,t-1}} + e_{i,t} \quad (4)$$

Discretionary expenses (DISEXP) consist of selling expenses, general and administrative expenses, and research and development expenses. The following is the equation for normal DISCEXP.

$$\frac{DISC_{i,t}}{Assets_{i,t-1}} =$$

$$k_1 \frac{1}{Assets_{i,t-1}} + k_2 \frac{Sales_{i,t-1}}{Assets_{i,t-1}} + e_{i,t} \qquad (5)$$

We run the data through these three equations (3-5) so that we get the residual errors ($e_{i,t}$) of each equation, and the sum of standardized value of each abnormal CFO, PROD, and DISEXP are REMIDX. We follow Chi [10] to determine the aggregated value of REM (REMIDX) that is, – Standardized Abnormal CFO + Standardized Abnormal PROD – Standardized Abnormal DISC. Data for RPT are collected from transactions disclosed in the Related Party Transactions section in Notes to the Financial Statements (i.e., acquisition of assets from related party, assets sales to related party, equity sales to related party, trading relationship with related party, and cash payment or loan to related party). RPT variable is the total RPT that could result in expropriation of minority shareholders divided by total sales in year t. Our sample are grouped into two sub-sample (i) firms with high RPTs (i.e., equal to or more than 1 percent of firm's total sales), and (ii) low RPT (i.e., less than 1% of firm's total sales) [3].

Board of commissioners' effectiveness (BCE) and audit committees' effectiveness (ACE) are moderating variables. The values of these variables are determined by checklists

Table 1. Selected descriptive statistics of key variables

Variables	n	Mean	Min	Max	Std. Dev
REMIDX	386	-0.00	-8.331	11.804	2.231
ABCFO	386	0.005	-1.158	0.960	0.184
ABPROD	386	0.223	-0.653	5.591	0.486
ABDISEXP	386	-0.126	-0.466	1.086	0.164
RPT	386	0.213	0.00	1.488	0.318
BCE	386	0.614	0.00	1.00	0.487
ACE	386	0.850	0.00	1.00	0.358

based on Hermawan [5]. BCE is a dummy variable with the value of 1 if the score of BCE of a company is equal and above 34 and 0 otherwise. ACE is a dummy variable with the value of 1 if the score of ACE of a company equal and above 22 and 0 otherwise. To determine the moderating effect of BCE and ACE, each of BCE and ACE is multiplied by RPT.

4 Results and discussion

4.1 Descriptive statistics

Table 1 presents the mean, minimum, and maximum value of key variables used in the empirical model. The mean of real earnings management in aggregate level (REMIDX) is almost zero, while in individual measures is 0.005, 0.223 and -0.126 each for abnormal operating cash flows (ABCFO), abnormal production (ABPROD), and abnormal discretionary expense (ABDISEXP). The mean of related party transactions (RPT) is 0.213 or 21.33% of firm's total sales.

4.2 Test of association between RPT and REM (H1)

The regression results of hypothesis 1 (H1) are presented in Table 2. Our main variable RPT with all sample (n= 386) has a coefficient of 0.236 (t-stat= 0.78), not significant at 0.10. We perform additional test with our sub-sample of high RPT and low RPT. We find evidence that the coefficient of RPT in sub-sample with high RPT is positive (β_1= 0.922, t-stat= 2.87) and significant at 0.01, but not significant in sample with low RPT (i.e., the coefficient has t-stat of -1.16). Our findings are consistent with our early prediction (H1) that related party transaction has a positive association with earnings management only if in the condition of sub-sample with high RPT, but not with low RPT. This findings indicate that high related party transactions in a firm can be used as earnings management tools, specifically in the aggregate level of real transactions to expropriate minority shareholders.

Table 2. Test of hypothesis 1 (H1)

Dependent variable: REMIDX							
Main variables	Predicted sign	All sample		Sub-sample			
				High RPT		Low RPT	
		Coefficient	t-stat	Coefficient	t-stat	Coefficient	t-stat
RPT	+	0.236	0.78	0.922	2.87***	-7.710	-1.16
Adj. R^2		0.112		0.147		0.014	
Prob (F)		0.000		0.000		0.008	
No. of Obs.		386		256		130	

*** represent statistical significance of t-statistics at 1 percent level with two-tailed test.

4.3 Test of moderating role of internal governance mechanism (H2 and H3)

First, we investigate with all sample on how board of commissioners and audit committees have role in the association between RPT and real activities. Table 3 presents the result of robust regressions of the second and third hypotheses (H_2 & H_3). We found that the moderating variables RPT*BCE has positive coefficient of 1.678 (t-test= 3.08) significant at 0.01, contrary to our hypothesis. On the other hand, the coefficient of RPT*ACE is -1.503 (t-stat= -2.37) significant at 0.05, consistent with our hypothesis. We also perform further tests with sub-sample of high RPT (n= 256) and low RPT (n= 130) to get robust results. With high RPT sample, variable RPT*BCE has positive coefficient of 1.478 (t-stat= 2.42), significant at 0.05, while in the low RPT sample the coefficient is not significant (t-stat= -0.06). Also in Table 3, with high RPT sample variable RPT*ACE has coefficient of -1.400 (t-stat= -2.13), significant at 0.05, while in low RPT sample, RPT*ACE has coefficient of 0.685 (t-stat= 0.28), not significant.

With our full sample and sub-sample tests, we conclude that the board of commissioners (BCE) do not function effectively as we predict to minimize (weaken) the association between RPT and real earnings management, especially with high RPT sample. This result implies that when firm's RPTs are equal to or higher than 1 percent of total sales, board of commissioners strengthens the positive association between RPT and REM. Our results give interpretation that the board of commissioners might get involve with the RPT in the strategic level to increase the performance of the firms by engaging real activities. On the other hand, in general and with high RPT sample, we found consistent results that the effectiveness of audit committees (ACE) weaken the association between RPT and REM, consistent with previous study [5].

DOI: 10.1051/shsconf/20173404008

5 Conclusion

We investigate the association between related party transactions (RPT) and real earnings management (REM). Using linear regressions, we find evidence that there is a positive association between RPT and REM, only when the firm has higher RPT but not in the lower RPT [3, 7]. We also investigate the role of internal governance mechanism through the

Table 3. Test of hypothesis 2 (H2) and hypothesis 3 (H3)

Dependent variable: REMIDX							
Main variables	Predicted sign	All sample		Sub-sample			
				High RPT		Low RPT	
		Coefficient	t-stat	Coefficient	t-stat	Coefficient	t-stat
RPT	+/-	0.190	0.30	0.906	1.49^	-1.758	-0.70
RPT*BCE	-	1.678	3.08***	1.478	2.42**	-0.817	-0.06
RPT*ACE	-	-1.503	-2.37**	-1.400	-2.13**	0.685	0.28
Adj. R^2		0.109		0.166		0.129	
Prob (F)		0.000		0.000		0.000	
No. of Obs.		386		256		130	

***, ** represent statistical significance of t-statistics at 1 percent, and 5 percent levels respectively with two-tailed test, and ^ significant with one-tailed test.

board of commissioners and audit committee in mitigating the association between RPT and REM. We find a contradictory result with our hypothesis i.e., the board of commissioners strengthen the positive association between RPT and REM especially in our high RPT sub-sample. Our alternative explanation to this is that the board of commissioners might involve themselves in the expropriation of minority shareholders' wealth by engaging real activities. Finally we find consistent results with our hypothesis that the effectiveness of audit committees weaken the positive association between RPT and REM [5]. We propose accruals and real transaction to be examined together for further study instead of REM only [9].

References

1. J. J. Chen, P. Cheng, X. Xiao. Related party transactions as a source of earnings management. AFE, **21** (3), 165-181 (2011).
2. L. Kuan, G. Tower, R. Rusmin, J. W. M. Van der Zahn. Related party transactions and earnings management. JAAI, **14** (2), 115-137 (2010).
3. S. Munir, N. M. Saleh, R. Jaffar, P. Yatim. Family ownership, related-party transactions and earnings quality. Asian Academy of Management Journal of Accounting and Finance, **9** (1), 129-153 (2013).
4. A. Klein. Audit committee, board of director characteristics, and earnings management. JAE, **33**, 375-400 (2002).
5. A. A. Hermawan. The influence of effective board of commissioners and audit committee on the informativeness of earnings: evidence from Indonesian listed firms. APJAF, **2** (1), 1-38 (2011).
6. E. A. Gordon, E. Henry, D. Palia. Related party transactions: associations with corporate governance and firm value in EFA 2004 Maastricht Meeting Paper & AFA (2004).

7. Sumiyana, R. Febrianto. Related parties' transaction and earnings management: a case in Indonesia. Jurnal Ekonomi dan Bisnis Indonesia, **27**, 192-209 (2012).
8. D. A. Cohen, A. Dey, T. Z. Lys. Real and accrual-based earnings management in the pre-and post-Sarbanes-Oxley periods. TAR, **83** (3), 757-787 (2008).
9. S. P. Kothari, N. Mizik, S. Roychowdhury. Managing for the moment: The role of earnings management via real activities versus accruals in SEO valuation. TAR, **91** (2), 559-586 (2016).
10. W. Chi, L. L. Lisic, M. Pevzner. Is enhanced audit quality associated with greater real earnings management? AH, **25** (2), 315-335 (2011).

DOI: 10.1051/shsconf/20173405001

Impediments of Environment Management System (EMS) Implementation in Malaysian Local Authorities – A Preliminary Finding

Nero Madi[1*], *Corina* Joseph[1], and *Mariam* Rahmat[1]

[1]Universiti Teknologi MARA Cawangan Sarawak, Kampus Samarahan 1, Sarawak, Malaysia

Abstract. The implementation of environmental management system (EMS) by public sector agencies is in line with the Malaysian Government Transformation Programme (GTP) and the New Public Management doctrine. Local authority is chosen as a sample because of its important role in advancing towards sustainable development. This study aims to examine impediments to the ISO 14001 implementation by Malaysian local authorities using the institutional theory as the underlying framework. The survey instrument was developed and distributed via mail to all 146 local authorities in Malaysia. Finding revealed that the impediments to the EMS implementation were explained by lack of coercive, normative and mimetic isomorphism under the institutional theory. It is expected that the survey findings would fill in the gaps in the literature in better understanding the lack of implementation of IS00 14001 particularly in the emerging economy.

1 Background

Reforms introduced by the Malaysian Government in 2009 and 2010 are set to achieving the nation's bold aspirations of Vision 2020 by joining the high-income nations. The Government has introduced five major initiatives to transform the nation – the New Economy Model (NEM), 1Malaysia, the Government Transformation Programme (GTP), the Economic Transformation Plan (ETP) and the Tenth Malaysia Plan. The NEM lists out 8 Strategic Reform Initiatives (SRI) to drive the Malaysian economy towards a high income, inclusive and sustainable economy. One of the initiatives is strengthening the public sector via numerous public sector reforms agenda, for example, the implementation of environmental management system (EMS), one of the efforts undertaken in advancing sustainable development. The Brundtland Commission defined sustainable development as the "ability to make development sustainable — to ensure that it meets the needs of the present without compromising the ability of future generations to meet their own needs" [1, p.8].

In heading for the developed nation status by 2020, so is the increase in expectation of the more educated societies for efficient service delivery by local government. This is particularly true in the urban local authority jurisdiction areas. [2, p. 8] stated that: "as the

* Corresponding author: neromadi@uitm.sarawak.edu.my

population becomes urbanized, local government grows in direct importance to the welfare and the quality of life experienced by a majority of the citizens ...". In a few years, it is expected that more than two-thirds of Malaysian citizens would reside in the urban areas [3]. The urbanization process would translate to congestion and increased pollution of the water, air, noise and overall environment. It is, therefore, urgent for local authorities to undertake more aggressive and concerted efforts involving the public, in addressing this impending issue.

To forge ahead the several initiatives introduced by the Malaysian government for improved governance and accountability of the administrative system, particularly with regards to environmental responsibilities, the adoption of ISO 14001 Environmental Management System (EMS) is appropriate. However, despite the numerous reports by organisations who have benefitted from implementing the standard (see, for example, the British Standards Institution), only 3 out of 149 Malaysian local authorities are implementing ISO 14001 [4], and the implementation are influenced by merely the internal organizational factors. It is put forward here that, there are possibly several institutional factors that influence the non implementation of EMS by Malaysian local authorities. Hence, this study aims to examine the impediments to the adoption of EMS by Malaysian local authorities

2 Literature review

The implementation of EMS is subject to several challenges. Of concern is the measurement or evaluation of performance [5, 6, 7]. In a survey of ISO 14001 certified companies by [7], the ISO 14001 "does not require companies to reach minimum levels of environmental performances (apart from legal compliance) nor does it supply methods to be used to measure continual improvement". In the previous study conducted by [8], the two most significant impediments expected by organizations during the EMS planning stage include documentation (costs, time for preparation) and training of employees and there was no significant impediments during the implementation stages as the employees were involved in the decision making process. Other impediments include: lack of support and resources; unclear guidelines for EMS implementation for organisations with mobile workforce, such as the construction sector; lack of set guidelines for setting of objectives and targets and extent of involvement of employees, suppliers and other stakeholders and lack of guidelines on how to accomplish continuous improvement elements of the standard Interpretation of terms present within the standard.

In the study on the EMS implementation in Seberang Perai Municpal Council [9] found that, as a whole, there are no obvious challenges faced in the EMS implementation. The main obstacle leading to MPSP slow progress in implementing EMS at its early stage in 2006 was due to psychological factor i.e. lack of confidence to start off the programme. Another barrier in the EMS implementation is resistance from staff during its inception. From the literature review, it is found that there are limited studies conducted on the implementation of EMS in particularly Malaysia.

The reasons for non disclosure of sustainability information on Malaysian local authority websites using the institutional theory were explored by [10]. Lack of support from management towards achieving sustainable development may explain the lack of normative isomorphism. The lack of monitoring by higher authority towards the implementation of the sustainable development program also affects the non–reporting of sustainability information on the websites. This could possibly explain the lack of coercive pressure by higher authority. Hence, drawing from [10], this paper used the isomorphism conception to explain the impediments of EMS implementation.

3 Theoretical framework

To investigate impediments of EMS implementation by the 146 Malaysian local authorities, this study employed the institutional theory, which has been used by many researchers to understand various organizational and individual practices and activities. The process by which organizations tend to adopt the same structures and practices is labelled as 'isomorphism' [11, 12].

Identified by [12], the three forms of institutional pressures causing organisations to conform are coercive, mimetic and normative isomorphism. Coercive isomorphism is the most cited type of institutional pressures. According to [12, p. 149], the pressures can be "exerted by other organizations on which an organization may be dependent, as well as cultural expectations in which the organizations operate". It is advanced in this paper that lack of formal pressure from relevant authority possibly explains the non-existence of coercive isomorphism, in line with [10]. The second type of institutional pressure is mimetic isomorphism. Mimetic isomorphism suggests the 'copying' of best practice of "those organizations perceived to be legitimate and successful" [12, p.150]. The final isomorphism category is normative isomorphism. This element of pressure is normally developed by professional and occupational groups [13]. [12] argued that the more highly professionalized a workforce becomes in terms of academic qualifications and participation in professional and trade associations, the greater the extent to which the organization becomes similar to other organizations in the same field. Further, this paper used the concept of isomorphism to explain the impediments for EMS implementation due to the nature of the local government and because of "organizational competition for political power, social fitness, and institutional legitimacy" [14, p. 566]. The proposed framework for impediments of EMS implementation based on [4, 9] and other literature is presented in Figure 1.

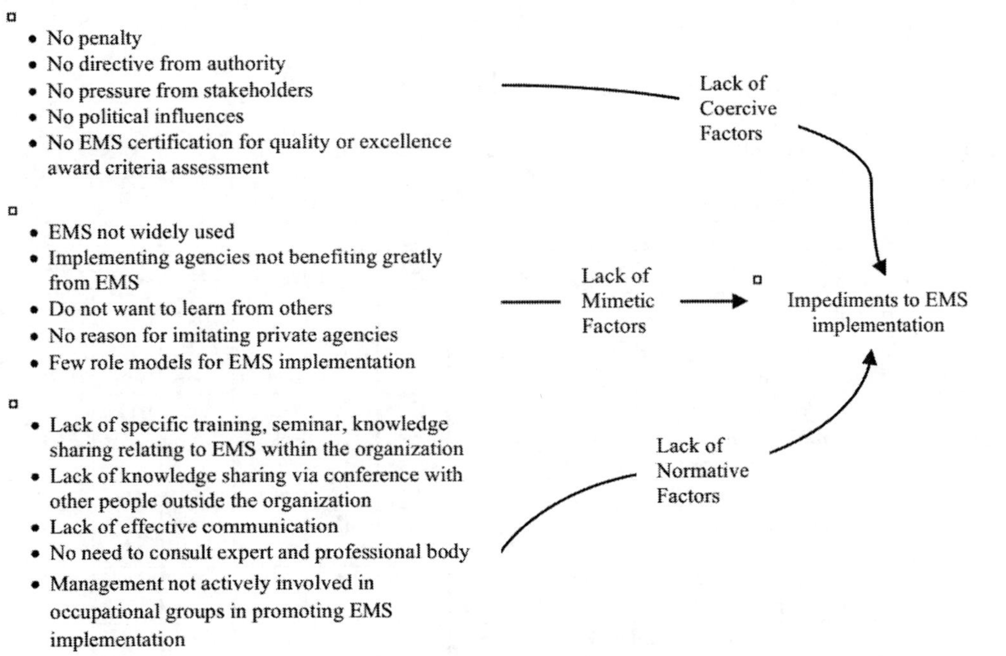

Fig. 1: Framework for impediments of EMS implementation

4 Research methodology

The research instrument was developed based on the findings from [4, 10] that examined the implementation of EMS in selected local authorities in Malaysia, using the institutional theory, particularly the isomorphism mechanisms. The questionnaire consists of three sections: i) Section A (demographic profile of the respondents); ii) Section B (impediments to the adoption of ISO14001 using the three isomorphism mechanisms – coercive, mimetic and normative) and attitudes by using 5-point Likert scale (1=strongly disagree, 2 = disagree, 3 = neither disagree nor agree, 4 = agree, 5 = strongly agree); and Section C which captures the respondent's readiness to implement EMS.

The pilot test was carried out involving councils from Sabah, Sarawak and Selangor. Changes were made to the survey instrument based on the feedbacks to make it more respondent-friendly In the actual data collection, the questionnaire was mailed to the person in charge of the Quality Management Unit, of all 146 local authorities from April to June 2016. The list of local authorities was available from the Ministry of Urban Well Being, Housing and Local Government Malaysia. Data was analysed using the Statictical Package for Social Sciences (SPSS) Version 20.0. The results are explained in the following sections.

5 Result and Discussion

Only 43 (out of 146) questionnaires were completed and returned. This represents 29.5% response rate. The Cronbach Alpha was conducted to examine the internal consistency of all items related to the three isomorphism categories namely: coercive, normative and mimetic. The coefficients for all the three categories were above 0.7 and considered acceptable. Table 1 summarizes the level of agreement of impediments to the EMS implementation.

Table 1. Level of agreement of impediments to the EMS implementation

Isomorphic Pressure	Elements	S D	D	N	A	S A	Total
Coercive	No directive	6	5	11	10	11	43
	No Penalty	2	6	12	8	15	43
	No Pressure	2	6	15	10	10	43
	No Political Influences	1	7	9	11	15	43
	No EMS Certification required for quality competition award criteria assessment	4	7	19	9	4	43
Mimetic	EMS not widely used	2	2	6	21	11	42
	Implementing agencies not benefiting greatly from EMS	6	15	15	5	2	43
	No intention to learn EMS	11	12	15	4	1	43
	No reason to imitate implementing agencies	8	13	13	7	2	43
	Few role models as benchmark for EMS	4	4	19	8	8	43
Normative	Lack of sharing values within organization	5	3	11	16	8	43
	Lack of sharing values & norm with outside organization	5	2	11	17	8	43
	Lack of effective communication	3	4	12	18	6	43

DOI: 10.1051/shsconf/20173405001

	No need consultation with professional bodies	8	12	16	5	2	43
	Inactive management involvement with occupational group	4	6	15	14	4	43

Constructed scale: SD=Strongly Disagree; D=Disagree; N=Neutral; A=Agree; SA=Strongly Agree

Table 1 reveals that more than 46% of the respondents agree that the impediments for the EMS implementation are due to 1) no directive from authority; 2) no penalty; 3) no pressures from shareholders, and 4) no political pressures. Lack of formal pressure from relevant authority possibly explains the non-existence of coercive isomorphism, in line with [10]. As suggested by [10], penalty or sanction could be considered by the higher authority, for example, the Ministry of Urban Well Being, Housing and Local Government for the local authorities that are not implementing any sustainable development initiatives, such as, the EMS. The lack of pressures from stakeholders to explain the impediments for EMS implementation is not consistent with [9] that indicate the community played a major role in pushing the organization to be socially and environmentally responsible. [9] found that educated people surrounding the councils tend to coerce the councils to accept responsibility and be accountable for sustainable development issues, including environment issues. Meanwhile, the results on the agreement that the political pressure does influence the impediments to EMS implementation clearly supported the coercive isomorphism tenet, whereby organizations are competing for political power [14].

Majority of the councils (76%) agreed that EMS is not widely used. However, more than 46.5% disagreed that the impediments for the EMS implementation are due to: 1) Implementing agencies not benefiting greatly from EMS; 2) No intention to learn EMS and 3) No reason to imitate implementing agencies. The findings that there is absence of mimetic isomorphism among the local councils in Malaysia refer to councils known for their advanced sustainability development initiative, which is not consistent with [15]. This indicates the in-sensitivity of the peers' action or activity among councils in the same organizational field [16].

As for the normative isomorphism, more than 46% of the respondents agree that the impediments for the EMS implementation are due to 1) Lack of sharing values within organization; 2) Lack of sharing values & norm with outside organization, and, 3) Lack of effective communication. The above findings support the argument for absence of normative isomorphism which, when applied, indicates that the EMS activity has not become a shared norm of organizational members and therefore no acceptance within organizations. This is not consistent with [17, p. 21)] finding that "there was a process of osmosis and the willingness of key Triple Bottom Line (TBL) reporting instigators to try something new". The findings suggest that the EMS has not been institutionalized within the Councils as it has not become shared values within the organization.

6 Conclusion

The aim of this paper is to examine the impediments to the adoption of EMS by Malaysian local authorities using the institutional theory. The institutional theory offers insightful explanation on the impediments to EMS implementation. The absence of coercive isomorphism is supported by :1) no directive from authority; 2) no penalty; 3) no pressures from shareholders, and 4) no political pressures. The lack of mimetic isomorphism is supported by: 1) EMS is not widely used; 2) Implementing agencies not benefiting greatly from EMS; 3) No intention to learn EMS and , 4) No reason to imitate implementing agencies. Finally, the non-presence of normative isomorphism is due to: 1) Lack of sharing values within organization; 2) Lack of sharing values & norm with outside organization, and, 3) Lack of effective communication.

There are several implications from the findings of this paper. Firstly, to promote coercive isomorphism among local authorities to undertake the EMS implementation, the regulatory authority could be more proactive by including the implementation of EMS or other sustainable development best practices in the award assessment criteria, for example, the PBT Star Rating or Accountability Rating. Secondly, to promote the mimetic isomorphism, the Ministry may encourage the EMS implementing agencies to be the role model for other local authorities. This may promote the learning activity among the councils in the same organizational field. Similarly, benchmarking the role model could also be included in the performance evaluation criteria for local authorities. Finally, normative isomorphism can be promoted by consultation with accreditation body, professional networking via conference presentation and also regular training. This may promote the institutionalization of EMS as the norms and values will be shared among the internal and external participants of the organizations.

It is expected that the survey findings would fill in the gaps in the literature in better understanding the IS0 14001 adoption stage. The findings of this paper are important as no other similar study has been undertaken documenting the impediments to the adoption of ISO 14001 in Malaysian local authorities. Our findings are expected to have practical implications for public sector organizations and government departments implementing sustainability agenda. Our findings also aim to contribute towards the theoretical perspectives that are relevant to encourage the ISO 14001 implementation in developing countries.

This study has thrown up several questions in need of further investigation. A potential avenue for future research is to conduct a longitudinal case study in the council to have more insights into the reasons for not implementing EMS. A longitudinal analysis may encourage research engagement with practitioners and provide opportunity to anticipate and observe the actual EMS implementation, having access to internal documentations, and to attend the council's meeting. Further, the qualitative method such as interview would enable the researcher to answer the research question 'why', thus, providing richer explanation for the non-observable phenomenon while conducting survey.

Acknowledgement

This research was supported by the Excellent Fund grant (600-RMU/DANA 5/3 (7/2015)) received from Universiti Teknologi MARA (UiTM) through the Research Management Unit, UiTM Sarawak.

References

1. The Bruntland Report Our Common Future: Report of the World Commission on Environment and Development, available at http://www.worldinbalance.net/pdf/1987-brundtland.pdf (accessed 1 October 2013) (1987)
2. H. S. Abdullah, M. Kaliannan, ASC, **4 (11),** 87–92 (2008)
3. L. Junus, Masalah yang Dilupakan: Proses Urbanisasi Meningkatkan Jumlah Air Larian Permukaan. *Utusan Online*. Retrieved from http://www.utusan.com.my/sains-teknologi/sains/masalah-yangdilupakan-1.60227#sthash.gLUqBFU6.dpuf. (2015, February 16).

4. C. Joseph. L. Jonathan, E.O. Nichol, *Environment Management System Implementation in Selected Local Authorities in Malaysia*, unpublished research, Research Management Unit, Universiti Teknologi MARA, Sarawak. (2014a),
5. D. Nawrocka, T. Parker, JCP, **17**, 601–607 (2009)
6. E.Perotto, R. Canziani, R.Marchesi, P. Butelli, JCP, **16**, 517–530 (2008)
7. C. Comoglio, S. Botta, JCP, **20**, 92–102 (2012)
8. A. Zutshi, A. Sohal, Technovation, **24**, 335–357 (2004)
9. C. Joseph, L. Jonathan, E.O. Nichol, *Environment Management System Implementation in a Malaysian Local Government: Isomorphism Perspective*, a paper presented at the *International Conference on Governance, Jakarta, 13 – 14 February 2014*. (2014b)
10. C. Joseph, SMRJ, **8(.1),** 1-13 (2011)
11. J. W. Meyer, B.Rowan, TJS, **83(2),** 340–363 (1977)
12. P. J. DiMaggio, W. W Powell, ASR, **48(2),** 147–160 (1983)
13. A. S. Rahaman, S. Lawrence, J. Roper, Crit. Pers. On Acc, **15(1)**, 35–56 (2004)
14. V. L. Carpenter, E. H.Feroz, AOS, **26 (7 and 8),** 565-596 (2001)
15. C. Joseph, R.Taplin, SRJ, **8 (3),** 363 – 372 (2012b)
16. J. Bebbington, C. Higgins, B. Frame, AAAJ, **2(4),** 588-625 (2009)
17. J. Mack. The Influences on Queensland Local Government Authorities to Adopt Triple Bottom Line Reporting, Working Paper, Queensland University of Technology, Australia, Brisbane. (2006)

Female Directors on Corporate Boards: Does Female Leadership Drive Corporate Environmental Transparency?

Michelle Siew Huei Phua [1*], and *Poh-Ling* Ho[1]

[1]Curtin University, Sarawak, Malaysia

Abstract. This paper examines the role of female directors on corporate boardroom on the extent of corporate environmental disclosure (CED) of 260 Malaysian listed companies in year 2013. Resource dependence theory is utilized as the theoretical framework to explain the role of female directors on corporate boards. Content analysis is employed to gauge the extent of CED based on a self-constructed index that is derived from Global Reporting Initiatives (GRI) and prior studies. Multiple regression is conducted and findings revealed that female directors' presence and female holding multiple directorships to be significant predictors of extent of CED. The findings lend support to the resource dependence theory on the contribution of board gender diversity and are useful for both policy makers and regulators.

1 Introduction

Gender equality in corporate boardrooms has been a topic of corporate governance discussions around the world. When Norway first introduced gender quota on its OSE-listed Norwegian companies in year 2005, the move has raised discussions among leaders of the corporate world and academics on female leadership roles [1]. Several countries have since encouraged corporate boardroom gender diversity to diversify the memberships of a board. In a similar move, Malaysia became the first country in the Asia Pacific region to introduce a board gender quota in year 2011.

The board of directors of a firm play important roles in determining a firm's strategic directions and disclosure decisions [2]. Women make up of at least half of the human capital in any nation. By including women into corporate boardrooms, it enables firms to tap into a pool of talent which women can contribute. According to [3], a heterogeneous group of decision makers can enhance competitiveness of a firm in an industry and produce effective decisions for a firm as the inclusion of non-traditional members in a board can tackle the issue of group-think that normally happens in a homogeneous board. Worldwide, there has been constant increasing interest on the issue of environmental sustainability. Pressure from various stakeholders has pushed corporations around the world to embed environmental sustainability practices into their day to day business operations. Investors have shifted from evaluating solely the financial performance of a firm to include non-

*Corresponding author: michelle.phua@curtin.edu.my

financial performance to gauge a firm's long term value creation [4, 5]. The inclusion of women directors can broaden the perspective of a board of directors to facilitate decision making and respond to demands of a diverse group of stakeholders. Prior studies have shown that corporate board gender diversity is positively associated with the extent of sustainability disclosure of firms in developed countries [1, 6]. Hence, this paper focuses on the role of female directors in corporate decision making through the lens of environmental reporting by Malaysian listed companies in year 2013.

The resource dependency theory is used as the framework to explain the inclusion of women into corporate boards. By having diversity in boardrooms, firms can tap into more resources such as having more external networks, different sets of skills, expertise and experiences of female directors which adds value to the firm. The content analysis method is employed to gauge the extent of environmental disclosure by 260 Malaysian firms listed on the Bursa Malaysia stock exchange and multiple regression analysis is carried out to examine the association between corporate board gender diversity and extent of environmental disclosure by the listed firms.

This study finds that the presence of female directors on corporate boardrooms and female directors with multiple directorships are significant predictors of environmental disclosure practices of Malaysian listed firms. The result provides support to the role played by female directors in providing essential resources such as advice, counsel and outlook to firms. These results are of significance to policy makers and regulators to further encourage participation of women in corporate decision making positions.

This study contributes to the body of literature on corporate governance in terms of board gender diversity. First, it provides important insights into the roles played by female directors in Malaysia, a developing country, where board memberships are normally dominated by males. Secondly, a study on the association of board gender diversity and voluntary environmental disclosures provides rich insights into the contribution made by female directors on corporate boards in terms of sustainability issues. Finally, as the Malaysian government has advocated the importance of having board gender diversity and environmental sustainability, the examination of such board compositions and disclosure practices can assist policy makers to deliberate their future policy-making decisions.

The rest of the paper is structured as follows. Section 2 provides the literature review into the development of women in Malaysia and hypotheses developed to examine female roles in corporate decision. Section 3 gives the research methodology and the results of this study are presented in Section 4. Section 5 concludes the paper.

2 Literature review and hypotheses development

Following the initiative taken by Norway in mandating gender quota in Norwegian corporate boards, Malaysia too joined the bandwagon by enacting a policy target through its Securities Commission Corporate Governance Blueprint in 2011. This authoritative document advocates women in decision making positions through a recommendation for public listed companies to achieve 30% female representation in corporate boardrooms by year 2016.

In today corporate world where culture, political and societal views and values are constantly changing and evolving, it is crucial for a firm to be able to manage its business risks and environmental challenges in order to satisfy a wide range of stakeholders. This could be achieved by having an effective board as strategic directions are made at the board level and influences the actions taken by management which in turn affects both shareholders and stakeholders. Hence, an effective board is needed for corporate management to better respond to demands from various stakeholder groups.

DOI: 10.1051/shsconf/20173405002

Resource dependency theory provides an ideal framework to explain the need to include female directors on corporate boardrooms. This theory posits that an organization has to draw crucial and critical resources in order to better meet the demands of the society which will ensure the legitimacy of a company. The board of directors has several functions in an organization which includes monitoring and controlling management's decisions, providing useful information and counsel to the management; ensuring management complies to applicable laws and regulations pertaining to the firm and providing beneficial linkage to their external environment [7]. By having heterogeneous boards, the diversity of memberships in a board enable firms to address specific issues faced by different stakeholder groups and respond to such demands accordingly. Also, by having female directors on corporate boardrooms, a board can better manage and monitor management as female directors normally will bring different sets of skills, experience and expertise which can help to enhance the governance of a corporate board. Board gender diversity are evidenced to provide critical external resources to a firm as the involvement of female directors in decision making positions contribute valuable insights to management, unique perspectives into setting corporate strategy and enhancing corporate reputation in the society [8].

The impact of corporate board gender diversity on environmental sustainability of a firm should not be overlooked. Prior studies found that presence of female directors in corporate board enhanced the level of environmental awareness of a company [9]. According to Park, Choi [10], female directors are more environmentally conscious than their male counterparts and are generally more sensitive towards environmental exploitations issues that happened in the society. Galbreath [11] however documented a different conclusion regarding female directors' contribution towards environmental disclosure in U.S. firms. Hence, in order to examine the association between the presence of female directors on corporate boards on the extent of corporate environmental disclosure, the following hypothesis is developed:

H_1: There is a positive association between presence of female directors on corporate boards and the extent of corporate environmental disclosure by Malaysian listed firms.

The contribution of female directors' influence on corporate decision making can also be evaluated through the number of directorships held by the female director. By holding multiple directorships of different firms, a female director can acquire knowledge on best practices in the industry and can provide this crucial resource to help management better understand the society it operates in. It enables the sharing of timely and relevant information across firms. Previous studies found that interlocked firms practiced similar corporate strategies [12]. In order to formally test the association between the two variables, the following hypothesis is developed:

H_2: There is a positive association between female directors with multiple directorships and the extent of corporate environmental disclosure by Malaysian listed firms.

Board independence is a core to good corporate governance of a board. Independent non-executive directors exist in corporate boards to function as a check and balance mechanism to ensure that decisions made by the board as a whole and management is congruent with the interest of the owners of the firm [13]. Malaysia Code on Corporate Governance 2012 has emphasized on the importance of environmental sustainability in making corporate decisions as it affects both shareholders and stakeholders. In order to test the association between board independence and the extent of corporate environmental disclosure, the following hypothesis is tested:

H_3: There is a positive association between independent non-executive female directors and the extent of corporate environmental disclosures by Malaysian listed firms.

Audit committee is where corporate accounting policies as well as corporate disclosure policies decisions are made. Hence, another way to gauge influence of female directors on

DOI: 10.1051/shsconf/20173405002

corporate reporting is to examine female involvement in corporate board audit committees. Several studies found that female directors are normally seated in board audit committees perhaps due to their sensitive nature towards the needs of stakeholders and community around them [14]. To formally examine the association between female directors on board audit committee and the extent of corporate environmental disclosure, the following hypothesis is developed:

H_4: There is a positive association between female directors seated on board audit committee and the extent of corporate environmental disclosure of Malaysian listed firms.

3 Data and methodology

3.1 Sample source and variables operationalization

The data of this study is taken from the annual reports of 260 non-financial quoted Malaysian companies with financial years ending during 2013. Year 2013 is chosen for the analysis as it is timely to evaluate public companies' respond to the Corporate Governance Blueprint 2011 to include female directors on corporate boardrooms. The dependent variable in this study is the extent of corporate environmental disclosure (CED) by Malaysian listed firms. A researcher-constructed index of 18 items is utilized as a proxy to gauge the extent of environmental disclosure by Malaysian firms. The index is derived from the Global Reporting Initiatives (GRI) to ensure content validity as it is the most authoritative source of corporate social and environmental responsibility reporting globally. Independent variables included in this paper are presence of female directors in corporate boards (FemDir), female directors holding multiple directorships (FemMD), female directors as independent non-executive directors (FemINED) and female directors seated on board audit committee (FemBAC). Variables which are found to be determinants of CED in prior studies are included in this paper as control variables. These variables are firm size (Size), profitability (RoA), leverage (Lev), ownership concentration (OwnerConc), auditors' quality (Big4) and industry membership (Ind).

3.2 Multiple regression model

To test the association between the dependent variable with independent variables and control variables included in the study, the following Ordinary Least Square (OLS) regression with robust standard errors on a basis of cross-sectional analysis is developed:

$$CED_x = \beta_0 + \beta_1 FemDir + \beta_2 FemMD + \beta_3 FemINED + \beta_4 FemBAC + \beta_5 OwnerConc + \beta_6 Big4 + \beta_7 RoA + \beta_8 Lev + \beta_9 Ind + \beta_{10} Size + \varepsilon_x$$

4 Results and discussions

Table 1 summarizes the multiple regression results based on the model developed. The hierarchical multiple regression technique is used in this study in order to control for the confounding effects by control variables identified in the study. When only control variables used in the study is entered, the explanatory power of the model as indicated by the value of adjusted R-square is 21.6%. The second multiple regression is run with both independent and control variables entered and the explanatory power of the model has improved with the adjusted R-square of 28.80%. The model is highly significant ($p<0.01$) suggesting that the model explains a substantial percentage of the variation in the level of environmental disclosure.

Table 1: Summary of Multiple Regression Results

Model	Adjusted R Square	Std. Error of the Estimate	Change Statistics				
			R Square Change	F Change	df1	df2	Sig. F Change
Control variables only	.216	.20024	.234	12.884	6	253	.000
Independent and control variables	.288	.19084	.081	7.390	4	249	.000

H_1 proposes a positive association between female directors in corporate boardroom and the extent of environmental disclosure practices. Table 2 reveals a positive and statistically significant association between CED and female directors on corporate boardroom ($p<0.10$), suggesting that the presence of female directors engender greater environmental disclosure. Hence, H_1 is supported. The result of this study provide evidence that female directors can bring valuable resources to a company by influencing management to be more environmentally sustainable by having environmental-friendly practices and providing more environmental information to the public.

H_2 proposes a positive association between female directors holding multiple directorships and the extent of environmental disclosure practices by Malaysian listed companies. H_2 is supported by the results shown in Table 2 as there is a positive and statistically significant association between a firm's CED and female director holding multiple directorships ($p<0.01$).

Table 2: Multiple Regression Results

Model	Unstandardized Coefficients		Standardized Coefficients	t	Sig.
	B	Std. Error	Beta		
(Constant)	-.691	.183		-3.777	.000***
FemDir	.236	.152	.121	1.554	.060*
FemMD	.498	.191	.194	2.602	.005***
FemINED	.118	.285	.038	.415	.339
FemBAC	-.025	.301	-.006	-.084	.467
OwnCon	.090	.075	.066	1.212	.113
RoA	.004	.063	.006	.066	.474
Leverage	.059	.057	.099	1.034	.151
Industry	.012	.032	.021	.385	.350
Big4	.062	.026	.136	2.364	.009
NLogTotalAsset	.042	.010	.278	4.283	.000

Legend: The independent variables are FemDir= female directors; FemMD= female directors with multiple directorships; FemINED= female directors as independent non-executive directors; and FemBAC= female directors in board audit committee. The control variables are OwnCon=ownership concentration; RoA=return on assets; Lev=leverage; Industry= industry membership; Big4= auditor's quality; LgTotalAsset=log of total assets. Associations *, ** and *** are statistically significant at the 10%, 5% and 1% respectively.

Although the association between the extent of CED and female directors as independent non-executive directors in corporate boardroom is positive, this association does not reach statistical significance. Hence, H_3 is rejected. This may be due to the lack of experience and knowledge of independent non-executive directors on environmental issues.

The influence of independent directors on environmental reporting may be limited as they are not directly involved with the daily operations of the organizations.

H_4 proposes a positive association between female directors as board audit committee members and the extent of environmental disclosure practices by Malaysian listed companies. However, H_4 is, not supported by the results in Table 1 as there is no statistical significance. This could be attributed to the fact that more focus is being directed to the financial risk of a firm, rather than environmental issues facing the firm.

This study also includes control variables to control for any confounding effects they may have on the results of the study. Of the six control variables incorporated, only firm size and auditor's quality are found to have positive and statistically significant impact on CED at $p<0.01$ level. The results suggest that big firms and Big 4 auditors drives environmental transparency.

5 Conclusions

Using a sample of 260 Malaysian listed companies, this paper examines the role of female directors in corporate decision making through the lens of environmental reporting by Malaysian listed companies in year 2013. The multiple regression results indicate that female directors in a boardroom and female directors holding multiple directorships are positively and significantly associated to the extent of CED of Malaysian companies. The findings of this paper highlights the importance of the role played by female directors in providing valuable resources to a company by influencing management to communicate environmental practices of the companies to the public through annual reports. This lends support to the resource dependency theory on the contribution of corporate board gender diversity in a company.

The results of this study provide valuable knowledge and fresh insights into role of female directors and environmental disclosure practices of Malaysian listed companies. Policy makers and regulatory bodies may benefit from the findings of this paper to encourage more active participation of female in corporate decision making positions. Future research can consider examining the impact of the presence of female directors on the overall CSR dimensions and taking into account information disclosed on other medium of reporting such as press releases, media and company's website to have a thorough picture of a company's disclosure practices.

References

1. R. Labelle, C. Francoeur, F. Lakhal, Gend Work Organ, **22,** 4 (2015)
2. O. Bohren, R. Strom, J Bus Finan Account, **37,** 9/10 (2010)
3. W. Amar, M. Chang, P. McIllkenny, J Bus Ethics **129**, 3 (2015)
4. J. Cowton, S. Sandberg, *Encyclopedia of Applied Ethics* 2nd ed (2012)
5. A. Haji, M. Anifowose, SAMPJ, **7,** 2 (2016)
6. M. Larkin, A. Benardi, M. Bosco, Account Publ Interest, **13**, 1 (2013)
7. E. Landry, A. Benardi, M. Bosco, Corp Soc Resp Env Ma, **23**, 1 (2016)
8. D. Seto-Pamies, Corp Soc Resp Env Ma, **22,** 6 (2015)
9. L. Liao, L. Luo, Q. Tang, Br. Account. Rev., **47**, 4 (2014)
10. S. Park, S. Choi, E. Kim, Corp Soc Resp Env Ma, **19,** 6 (2012)
11. J. Galbreath, Bus Strat Environ **19,** 5 (2010)
12. T. Lambooy, Crit Studies Corp Resp Gov Sust, **1** (2010)
13. A. Haji, MAJ, **28,** 4 (2013)
14. T. Hafsi, G. Turgut, J Bus Ethics, **112,** 3 (2013)

SHS Web of Conferences **34**, 05003 (2017) DOI: 10.1051/shsconf/20173405003
FourA 2016

Do Cross-Listed Firms Report Better Social Responsibility Performance?

Guangyou Liu*

Business School and Center for Accounting, Finance and Institutions, Sun Yat-sen University, Guangzhou, 510275, P.R.China

Abstract. This paper investigates the corporate social responsibility (CSR) disclosure of cross-listed companies in China, to bridge the gap between existent CSR literature and the literature on cross-listing. Mann-Whitney-U and Wilcoxon ranking tests document the traditional bonding effect regarding reported corporate financial performance. Furthermore, the test results and findings of this study extend the financial bonding effect to other dimensions of corporate responsibility. In addition to better economic returns to owners, the bonding effect marks better reported CSR performance for cross-listed companies in terms of pay raises for employees, creation of new jobs, desirable stock option plans for management, and increased input in environmental protection and energy-saving activities.

1 Introduction

A rich source of literature in corporate social responsibility (CSR) can be found in the academic dimensions of general management and business ethics. Some international business researchers contribute to the existent literature by extending their investigations into the CSR performance of multinational companies (Egri and Ralston, 2008). Recent research in finance and accounting also shows a trend toward increasing interest in the multinational operations of capital markets, such as cross-listing topics. However, few of the existent studies address the impact of cross-listing on the CSR performance, especially in the context of cross-listing that occurs across different levels of capital market maturity.

In this study, fifty companies listed in both Mainland China and in the Hong Kong Special Administrative Region (HKSAR) that issued CSR reports for the fiscal year 2011 are used to formulate a sample. The aim is to examine whether cross-listing from a developing pre-mature capital market to a developed mature capital market leads to better reported social responsibilities. The bonding effect has traditionally been related to corporate financial performance. The empirical results of this study partially prove the bonding effect hypothesis for the reported CSR performance of companies cross-listed from developing to mature capital markets. Our findings have extended the bonding effect to more extensive indicators of corporate social and environmental performance. More specifically, in addition to better economic returns to owners, the bonding effect indicates

* Corresponding author: mnslgy@mail.sysu.edu.cn

the reported CSR out-performance of cross-listed companies in terms of pay raises for employees, creation of new jobs, desirable stock option plans for management, and increased input in environmental protection and energy-saving activities, as compared to those counterparts that are listed only in the domestic capital markets of Mainland China.

2 Current Studies on Chinese CSR Disclosure

Globalization is the critical reason for many scholars to extend their concerns regarding corporate social responsibility into China, the largest emerging market. Most Chinese scholars in their native language are criticized for writing too many conceptual papers, or using self-developed measurement tools in their studies, and proper research methodologies and internationally adopted theories need to be considered if they want to publish their investigations of Chinese CSR disclosure in English language journals. Language and legislative differences also have been obstacles to including the study of Chinese firms' corporate social responsibility disclosure into the existing literature. As many CSR disclosure texts are narrative, and they are not required by the mandatory annual reports which focus on financial results, it is not easy for interested researchers to acquire the longitudinal data regarding the Chinese corporate social responsibility.

In practice, CSR disclosure was officially noted later in last decade in China. The 2009 CSR Development Index Report marked the establishment of a comprehensive assessment system of CSR disclosure in China, consisting of responsibility management, economic, social, and environmental responsibilities (Blue Book, 2009). This report was issued by the Economics Division of the Chinese Academy of Social Sciences (CASS), aiming to reveal Chinese top 100 firms' overall level of CSR disclosure.

Most studies of CSR disclosure in China are related to Chinese top 100 companies. The content analysis of Chinese firms' social responsibility disclosure in Gao (2009) proposes that the Chinese firms' CSR disclosure can be classified into economic, legal, ethical and philanthropic dimensions. Gao (2009) find that most Chinese top 100 firms are concerned with economic returns and firm value, but few firms give high priorities to ethical and legal issues. Luethge and Han (2012) examine a small sample of social responsibility disclosed in the Chinese firms' 2008 annual reports, and they conclude that there is no relationship between firm's financial performance and reported social responsibility. Both studies have noted the gap in CSR disclosure between Chinese top 100 and the world top 500 firms. Our research intends to make contributions in bridging the gap. China is converging with the rest of the world in social, economic, environmental aspects.

3 Literature Review and Research Hypothesis Development

Earlier literature in capital market research observe cross-listing behaviours of international market participants in order to identify the bonding effect in financial term. The bonding effect of cross-listing was proposed by Coffee (1999) in his studies on the behaviours and consequences of cross-boundary capital market participants. Coffee (1999) explains the motifs of cross-listings in a range from emerging markets to mature and developed capital markets, appreciating the functions of mature and developed capital markets in improving the quality of accounting information, the effectiveness and efficiency of corporate governance, and the mechanisms for protecting investors. In mature and developed capital markets, the cross-listed firms from pre-mature or developing markets are bonded to their well-established counterparts because of the sophisticated systems for supervising securities, and the stringent regulations on information disclosure.

Early CSR literature focuses on the theoretical conceptualization and analytical framework of corporate social responsibility (Carroll, 1979). From that time, dimensions such as economy, ethics, and philanthropy were established as categories of CSR research. Currently, the research makes it clear that a firm can perform CSR in varied ways, either incorporating social features or characteristics in products and operations, or undertaking social or ecological investment. To investigate the bonding effect of cross-listing on CSR disclosure, we adopt the CSR conceptions of Carroll (1979) and classify the CSR indicators derived from the corporate reports in four dimensions, according to the resources that each firm consumes: financial resources, human resources, social resources, and ecological resources. To examine the bonding effect caused by cross-listing in respect to financial, economic, intellectual, social and ecological consequences, we propose the following research hypotheses:

H1: Cross-listed companies tend to report better economic performance than those matched companies not cross-listed.

H2: Cross-listed companies tend to report better human resource management performance than those matched companies not cross-listed.

H3: Cross-listed companies tend to report more significant social contributions than those matched companies not cross-listed.

H4: Cross-listed companies tend to report better ecological performance those matched companies not cross-listed.

4 Research Methodology

4.1 Sample Selection and Data Collection

Our sample is selected out of a pool of A share issuers and A+H share issuers dated from 1991 to 2010, where there 1276 listed firms issuing A shares and 62 listed firms issuing both A and H shares. With registrants in financial and insurance industries excluded, our sample pool is reduced to 838 firms issuing A shares only and 50 firms issuing both A and H shares. Due to the limited number of the A+H share sample, we select the paired sample of listed firms issuing A shares only to explore the impacts of cross-listed firms on the CSR performance and disclosure. Both industry and size are considered in pairing: one listed firm of the same CSSEC Industry Code issuing only A shares with sizes nearest to that of each studied dual listing firm are selected, which results in a paired sample of 50 firms issuing A shares only. In this way, we arrive at the final sample composed of 50 A+H share issuers and 50 firms issuing A shares only.

For empirical testing of four research hypotheses, we collected data corporate annual reports and stand-alone CSR reports, using two Chinese databases, WIND and JUCHAO. Annual reports available through WIND provided financial data such as profit, return on assets (ROA), and dividends, and non-financial data on areas such as job creation, training programs, social donation, and environmental protection contributions are taken from the stand-alone CSR reports of JUCHAO.

4.2 CSR Dimensions and Indicators

As stated in the prior section of research hypotheses development, we aim to test the association between voluntary disclosure of CSR information and cross-listing behavior. To observe the differences of reported CSR performance between cross-listed and non-cross-listed companies, we propose four corporate responsibility dimensions based on the firm's consumption of different types of resources (i.e., financial, human, social and ecological resources). Table 1 summarizes the selected indicators and descriptions of the four dimensions of CSR performance that are disclosed in the annual reports and stand-alone CSR reports under investigation.

Table 1. A summary of CSR indicators and brief descriptions

Indicators/ Variables	**Descriptions**
Panel A Financial Dimension	
lgASSET	Firm size measured by log value of total assets
PROFIT	Log value of dollar value of net income (net loss) from annual report
ROA%	Return on assets calculated based on financial statements
DIV-Y/N(1/0)	Dummy variable indicating whether a firm has declared dividends
DIV(BTAX)	Dollar value of accumulated payment of dividends
A/P turnover	Account payable turnover (including all current payables)
Panel B Human Resource Dimension	
MGTHOLDING	Percentage of management shareholdings
MGTCOMPENSATION	Dollar value of management salaries
PAYROLL	Average dollar value of payroll (scaled by employment size)
PAY-INCREASE	Percentage of payroll increase compared to the prior year
EMP-TRAINING(Y/N)	Dummy variable indicating whether a firm has employment trainings
TRAIN-FEES	Dollar values of corporate employment training investment
TRAIN_TIMES	Frequency of corporate employment trainings
Panel C Social Dimension	
TAXATION	Dollar value of corporate tax payment
NEW_JOBS	Increase of numbers of employment compared to the prior year
R&D	Dollar value of R&D investment
NEG_REPORT(Y/N)	Frequency of negative media coverage for the company
POS_MEDIA(Y/N)	Frequency of positive media coverage for the company
DONATION(Y/N)	Dummy variable indicating whether a firm discloses its donation(s)
DONATION_VALUE	Dollar value of donation(s)
Panel D Ecological Dimension	
ENV_POLICY(Y/N)	Dummy variable indicating whether the environmental policy exists
ENV_INVEST(Y/N)	Dummy variable indicating whether any ecological investment exists
ENV_INVEST	Dollar value of investment in environmental protection
ENERGE_SAVE(Y/N)	Dummy variable indicating whether the firm disclose such a matter
INT_ORG(Y/N)	Dummy, indicating whether international standards are considered

4.3 Test Method

The sample of cross-listed companies included fifty firms, we chose the Mann-Whitney-U and Wilcoxon ranking tests, rather than regression models. our objective was to investigate whether the companies cross-listed in both Mainland China and in Hong Kong outperform their counterparts in the match sample of listed companies that are only listed in Mainland China in terms of the CSR indicators listed in Table 1.

Table 2. Mann-Whitney-U and Wilcoxon ranking test results

Tested Variable	Mann-Whitney U	Wilcoxon W	Z	Asymp. Sig. (2-tailed)
lgASSET	745.000	2,020.000	-3.481	0.000
PROFIT	1,150.000	2,425.000	-2.031	0.042
ROA%	895.000	2,170.000	-2.447	0.014
DIV-Y/N(1/0)	1,125.000	2,400.000	-1.324	0.185
DIV(BTAX)	896.000	2,171.000	-2.844	0.004

AP turnover	1,058.000	2,333.000	-1.324	0.186
MGTHOLDING	730.000	2,005.000	-3.671	0.000
MGTCOMPENSATION	1,093.000	2,368.000	-0.924	0.356
PAYROLL	787.000	2,062.000	-3.192	0.001
PAY-INCREASE	954.000	2,229.000	2.041	0.041
EMP-TRAINING(Y/N)	1,225.00	2,500.000	0.457	0.648
TRAIN-FEES	19.000	40.000	0.645	0.519
TRAIN_TIMES	443.000	821.000	0.597	0.550
TAXATION	236.000	587.000	1.867	0.062
NEW_JOBS	656.000	1,931.000	4.095	0.000
NEG_REPORT(Y/N)	1,225.000	2,500.000	0.583	0.560
POS_MEDIA(Y/N)	1,050.000	2,325.000	1.604	0.109
R&D	1,245.000	2,520.000	0.036	0.971
DONATION(Y/N)	1,100.000	2,375.000	1.330	0.184
DONATION_VALUE	404.000	932.000	1.265	0.206
ENV_POLICY(Y/N)	1,175.000	2,450.000	0.597	0.550
ENV_INVEST(Y/N)	900.000	2,175.000	3.102	0.002
ENV_INVEST	31.500	811.000	3.547	0.000
ENERGE_SAVE(Y/N)	900.000	2,175.000	3.626	0.000
INT_ORG(Y/N)	1,200.000	2,475.000	0.403	0.687

5 Results and Discussion

Mann-Whitney-U and Wilcoxon ranking test methods have been applied to test our four research hypotheses. Table 2 presents the test results under the selected test methods. The first hypothesis is related to the bonding effect of cross-listing in the financial dimension of CSR performance, and the economic consequences reported in the corporate accounting disclosures. According to the test results, both firm value (indicated by Profit/Loss, ROA) and shareholder value (indicated by Dividend Payment) for the cross-listed companies rank significantly higher than those in the matched group. H1 is substantially supported by the sample of Chinese A+H share-issuing companies. Our data confirm the initial premise that the bonding effect of cross-listing creates a positive association between transparency of information disclosure and corporate economic performance.

The hypothesis on the human resource dimension (H2) has been partially evidenced by better managerial incentives and higher payment to employees in the cross-listed companies. Economic indicators such as Management Shareholding, Payroll, and Pay Raises rank higher for the cross-listed sample group than for the matched group. Indicators related to employee training do not show significant differences between the compared groups. The evidence on the hypothesis H2 can be explained by the transmission of the economic bonding effects of cross-listing to the corporate utilization of human resources. With regard to the indicators related to economic rewards, such as management incentives in terms of stock options and employment payment, the cross-listed companies outperform than the matched group. Regretfully, however, performance indicated by employee-training arrangement does not provide evidence for the extension of the bonding effect. This result can be explained by the prevailing situation of many listed companies from Mainland China that were originally state-owned enterprises and are currently seeking privatization. New reforms of business systems in the recent decade have enabled the management team of listed companies to share, in a privatizing incentive approach, the ownership of the equity that was previously wholly owned by the state.

Test results of the third hypothesis, related to corporate responsibility incurred by utilization of social resources, demonstrates the fact that the cross-listed companies have created more job opportunities in the domestic and international labor markets. Meanwhile, they have intent to make best advantage of international capital markets to avoid tax obligations. Active participation in CSR in the social resource dimension tends to be more sophisticated than the other dimensions. With respect to research and development (R & D), both positive and negative media coverage, and philanthropic donations, cross-listed companies do not report performance that is any better than the domestic listed counterparts.

Surprisingly, the hypothesis on the ecological resource dimension of CSR performance (H4) has been reversely supported in terms of environmental protection and energy-saving technology. The test results indicate that domestically listed companies report significantly better figures for their investment in environmental protection and adoption of energy-saving techniques and facilities. One possible reason is that in recent years the Chinese government and the Mainland China stock exchanges have strengthened regulations for and supervision of information disclosure for environmental protection and energy-saving technology, which has led to Mainland China's domestically listed companies reporting significantly better figures for their investment in environmental protection and adoption of energy-saving techniques and facilities. Another possible explanation is related to the economic costs of performing CSR to cross-listed companies, which means both information disclosure and actual ecological CSR performance involve substantive financial expenses, so that the financial results, as reported in the accounting statements, are lower.

6 Conclusions

This paper concludes that the empirical results of this study partially prove the bonding effect hypothesis for the corporate performance of companies cross-listed from under-developed to developed capital markets. On one hand, the traditional bonding effect of cross-listing, in terms of corporate financial performance, has been fully evidenced in the samples of Chinese listed companies. On the other hand, the test results and findings of this study extend the bonding effect to indicators of corporate social and environmental performance. Specifically, in addition to better economic returns to owners, the bonding effect also marks improvement in CSR performance of cross-listed companies in terms of pay raises for employees, desirable stock option plans for management, the creation of new jobs, and increased input in environmental protection and energy-saving activities. Though the test results document no significant evidence for better CSR performance regarding training hours for employees, management cash-compensation, taxation, philanthropic donation, environmental policy, and involvement in international environmental/ecological organizations, the discrepancies may open a range of avenues for future research. Also business CSR practitioners should make marginal efforts to perform better corporate responsibilities for using the human, social and ecological resources. In sum, this study contributes its findings about CSR disclosure of Chinese cross-listed firms to the sustainable developments in Asia and in broader international communities.

References

1. A.B., Carroll, A three-dimensional conceptual model of corporate performance, Academy of Management Review, **4**, 4 (1979).
2. J.C., Coffee Jr., The future as history: the prospects for global convergence in corporate governance and its implications, Northwestern University Law Review, **93**, 3, (1999).

3. Y, Gao, Corporate social performance in China: Evidence from large companies, Journal of Business Ethics, **89**, 1 (2009).
4. C.P., Elgri, D.A., Ralston, Corporate responsibility: a review of international management research from 1998 to 2007, Journal of International Management, **14**, 4 (2008).
5. D., Luthge, H. G., Han, Assessing corporate social and financial performance in China, Social Responsibility Journal, **8**,3 (2012)

DOI: 10.1051/shsconf/20173405004

Could Enterprise Risk Management (ERM) Create, Protect and Enhance Shareholders' Wealth Among Malaysian Listed Issuers

Mohd Ariff Bin Kasim[1*], *and Siti Rosmaini* Binti Mohd Hanafi[1]

[1] Department of Accounting, College of Business Administration, Ajman University, 346, Ajman, United Arab Emirates (UAE).

Abstract. The primary objective of this study is to explore the extent to which ERM implementation is capable of enhancing shareholders' wealth. A sample of 283 companies operating in the Main Market of Bursa Malaysia was selected. This study utilized the definition of ERM as the independent variable and shareholders' wealth as dependent variable. The overall observation revealed that ERM was significant in explaining the variation in shareholders' wealth. Objective setting, event identification, risk response, control activities, information and communication, and monitoring substantiated the hypothesized relationship. The research framework of this study, which utilized assumptions from stewardship and agency theory, could serve as guidance for future research on organizational control and governance. The study fills the gap in literature which mostly concentrated on the USA and western countries. The CRO and the board should focus more advising functions at the earlier stage of the ERM implementation and monitoring functions once ERM matured.

1 Introduction

In light of the rising incidence of corporate financial scandals, regulatory agencies around the world have struggled to focus on strengthening measures to prevent further corporate collapses arising from mismanagement or unexpected financial surprises. Analysts have suggested that ERM could offer a safety net against these types of unexpected surprises [1, 2]. The primary objective of ERM is to create, protect, and enhance shareholders' value by managing the uncertainties surrounding the achievement of the organization's objectives [3]. ERM could be one of the alternative governance tools to be deployed to protect companies from a wide range of exposures to risk [4].

Currently, the majority of studies on ERM have concentrated on the USA and western countries and little is known about ERM in other geographical locations, particularly Asia. This present study aims to bridge the gap in literature by empirically investigating ERM implementation in Malaysia. In its early stage, ERM implementation in Malaysia was mostly driven by consultants; thus, such implementation was lacking in empirical evidence. In this regard, the adoption of ERM in Malaysia could be attributable either to the influence of a general trend or alternatively to legislation imposing on firms an obligation to

[*] Corresponding author: m.ariff@ajman.ac.ae

implement an ERM. Considering the fact that Malaysian listed issuers have started to invest in ERM, it is of paramount importance to shed light on the simple yet critical question that may be contemplated by stakeholders: does ERM implementation improve shareholders' wealth? Therefore, the primary objective of this study was to explore the extent to which ERM implementation is capable of enhancing shareholders' wealth, thus providing them with a reason for investment in ERM beyond mere fulfilment of the mandatory requirements imposed by Bursa Malaysia.

2 Literature review

The COSO suggested eight components of an integrated ERM framework: 1) internal environment, 2) objective setting, 3) event identification, 4) risk assessment, 5) risk response, 6) control activities, 7) information and communication, and 8) monitoring. The main criteria for evaluating the effectiveness of ERM is to measure the extent to which the eight components of the framework are executed. To date, studies providing empirical evidence of the impact of ERM implementation on financial performance are in their infancy [5-9]. The need for more research is critical considering the current tendency of regulatory authorities to shift their attention to ERM. Indeed, advocates of ERM have claimed that ERM is capable of enhancing shareholders' wealth; nevertheless, there have been mixed results surrounding this significant claim. Earlier ERM literature reported favorable impact of the ERM implementation such as those studies by [3, 5, 9-17].

It is interesting to note that ERM practice has evolved over the years and, since 2010, the focus has shifted from mere announcement of implementation of ERM to the appointment of a Chief Risk Officer (CRO) with responsibility for oversight and regulatory compliance. Studies [18, 19] found that the existence of the CRO related to reduction in the volatility of stock prices. The involvement of CRO was also reported to positively influence firm value [5]. It is possible that, over the years, ERM implementation has reached maturity, as recent studies have started to reveal more consistent sets of results on the impact of ERM on organizations. Study [20] suggested that firms demonstrated progressively higher value as the ERM matured. Further, study [21] suggested that the use of a dedicated risk manager is capable of improving the firm's operating performance and emphasized that the existence of a direct reporting line from the risk manager to the board or CEO could lead to superior firm value.

The fact that most of the existing studies concentrated on the USA and western countries obviously created a gap in the literature. Interesting, there are two studies from Malaysia that could shed light on ERM implementation in a different geographical setting. Studies [22, 23] generally discovered that ERM was not the main factor leading to value creation and that ERM implementation was not a significant variable in predicting variations in financial performance. This finding, in fact, contradicted all the previous studies, which reported a connection between ERM and shareholders' value. These empirical findings provide interesting grounds for further investigation as they are contrary both to the objective of ERM as highlighted by the COSO [24] and to various existing studies. These contradictory findings should not be ignored as the world was taken by surprise by the USA credit crisis which occurred in late 2008 and it is possible that the future will yield ever more shocking financial crises.

Remarkably, a recent study on ERM in China substantiated previous ERM studies in the USA and western countries. Study [25], involving 254 Chinese non-financial state-owned enterprises between 2006 and 2011, further confirmed the significance of ERM. They reported that ERM significantly increased firm value. More recent studies in ERM had incorporated corporate governance elements such as the board and senior executive management. Board was reported to be the most significance factor leading to an increase in senior management involvement in risk oversights [26]. Study [27] offers new insights in ERM literature by incorporating multi-theoretical approach utilizing a combination of institutional, agency resource dependency theory. Their study investigates maturity and board involvement in the ERM. It was reported that ERM matured firms tend to have more involvement from the board and senior management in the ERM monitoring task . The latest study [28] also incorporated various corporate governance attributes in an effort to understand ERM implementation among Italian private firms. They had incorporated six control variables, namely, profitability, leverage, firm dimension and complexity, quality of external audit, corporate ownership and industry characteristics. They concluded that ownership structure and size of the firm are capable of enhancing ERM implementation.

3 Hypotheses development

The implementation of ERM among Malaysian listed issuers is mandated by Bursa Malaysia's Practice Note 9 [29] and Listing Requirements, paragraph 15.26 (b) and such requirement therefore validates the control mechanisms highlighted in agency theory. There is apparent justification for the notion that agency theory may be an appropriate tool for theorizing on the impact of ERM implementation on firm value [30].

Study [8] suggested that the use of agency theory is likely to simplify the complexity inherent in real business environments. Indeed, further review of the individual attributes of the COSO's ERM framework revealed not only monitoring attributes but also advisory functions. In fact, seven out of the eight attributes of the ERM framework demanded a higher number of advisory than monitoring functions. Perhaps, the assumptions of stewardship theory provide a more satisfactory explanation of the advisory nature of the functions of the board and CRO in the context of successful implementation of ERM within an organization, particularly in the early stage of implementation. With reference to monitoring, the gap left by stewardship theory could be perfectly covered by the assumptions by agency theory that supports monitoring of overall ERM implementation.

Based on existing literature, this study identified five firms' specific attributes as control variables to overcome any undue influence on the ERM implementation. As the ERM is mandated for all listed companies in Malaysia, for the purpose of developing a comprehensive panoply of ERM tools, large firms may utilize greater financial resources than a small firm [27, 31-33]. Total revenue was used as proxy for firm size. The second and third control variables were the existence of a separate ERM unit and the existence of a CRO [5, 18, 19, 33]. Small firms may not be able to afford a dedicated department focusing on ERM. Instead, these small firms may have just a Chief Audit Executive (CAE) to oversee this function. This is not the ideal concept for a perfect ERM environment. At the other end, firms with a dedicated and well-structured ERM unit are often led by a CRO. The fourth control variable was the duration of the ERM implementation as a proxy for ERM maturity. Despite the mandatory requirement for ERM, not all listed firms had the same maturity or degree of ERM implementation. At the initial stage, firms may allocate most resources to educating staff and creating awareness of the ERM than on the design of an appropriate ERM framework. At the later stage, the focus inevitably shifts to the implementation of the ERM. Studies have revealed that mature firms demonstrate superior ERM implementation then less mature companies [27, 31, 32]. Finally, firms in different industries may be subject to different types and frequencies of risk; the varying level of risk thus served as one of the control variables. Generally, banking and financial institutions would be more exposed to various risks and would be more likely to have a more comprehensive ERM system. Existing literature reported that the degree of the ERM implementation is affected by industry classification [27, 28, 34, 35]. It is hypothesized that, upon controlling for the effect of the above control variables, significant relationships would exist between ERM implementation and shareholders' wealth, as measured by firm value [5, 6, 12], earnings before interest, taxes, depreciation and amortization (EBITDA) [13, 14], and ROE [14]. The equation below depicts the hypothesized relationships. Figure 1 summarizes the expected relationships between ERM implementation and shareholders' wealth.

$$Y = \alpha 0 + \gamma 1Xc1 + \gamma 2Xc2 + \gamma 3Xc3 + \gamma 4Xc4 + \gamma 5Xc5 + \beta 1X1 + \beta 2X2 + \beta 3X3 + \beta 4X4 + \beta 5X5 + \beta 6X6 + \beta 7X7 + \beta 8X8 + \epsilon$$

where Y = Shareholders' Wealth (Proxy: Firm Value, EBITDA and ROE)

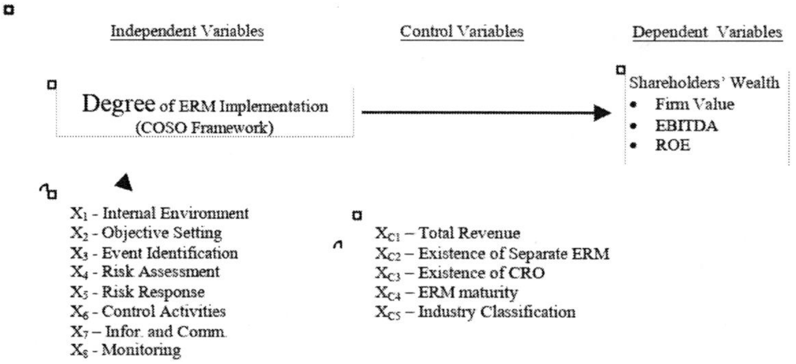

Fig. 1. Expected relationships between ERM implementation and shareholders' wealth

4 Methodology

The study utilized the definition of ERM and the framework proposed by the COSO [24] as the independent variable. A self-reported instrument was used to measure the degree of ERM implementation. The instrument was adopted from [23]. The instrument also used unique measurement scales that enabled ratio data to be collected (as such data measure the degree of implementation or compliance with the COSO ERM framework). There were 823 listed companies on Bursa Malaysia. Ideally a sample size of 263 companies should be sufficient for a satisfactory sample size. Considering the possibility of a lower response rate, a total of 823 copies of the questionnaires were distributed by mail to all the companies in the population. A copy with return paid envelope was mailed to the CRO for companies that had a Risk Management department/ unit; otherwise it was mailed to the CAE heading the Internal Audit department. The data collection procedure took place from December 17, 2014 to June 30, 2015. Significant efforts were devoted to ensuring the highest response rate via multiple emails and text messages. Eventually, 306 copies of the questionnaires were returned of which 23 copies were blank. A total of 269 copies were returned within the data collection period while another 14 surveys were received in July 2015. Finally, the total number of respondent companies was 283. With regard to the dependent variable, firm value, EBITDA and ROE were the proxies used as a basis for measuring shareholders' wealth were obtained from the Osiris database between July 1, 2015 and July 31, 2015. The data were extracted from the latest available financial data, mostly for the year ended December 31, 2014.

Table 1: Descriptive Statistics

Variables	Definition	Frequency	Percent
Industry class	Banking, insurance and financial institution	118	41.7
	Plantation	17	6.0
	Construction	32	11.3
	Trading	60	21.2
	Automotive	56	19.8
ERM maturity (Duration of the ERM)	in the first year of the ERM implementation	24	8.5
	in the 2-3 years of the ERM implementation	167	59.0
	in the 4-5 years of the ERM implementation	40	14.1
	beyond fifth years of ERM implementation	52	18.4
Separation of ERM unit	separate ERM unit/ Department	230	81.3
	no separate ERM unit	53	18.7
Existence of CRO	with CRO	224	79.2
	without CRO	59	20.8

5 Results

Table 1 depicts the respondents' demographic information coupled with descriptive statistics of relevant variables. All variables that were subject to multiple regressions were transformed using Inverse Distribution Function method. Factor analysis was performed, the results of which are presented in Table 2. There were eight components extracted (Varimax with Kaiser Normalization, suppressed at 0.50) and factor scores from each were used to test the hypothesized relationships. Each factor was then named accordingly. Factor 1 was termed Monitoring, Factor 2—Event Identification, Factor 3—Information and Communication, Factor 4—Objective Setting, Factor 5—Risk Response, Factor 6—Risk Assessment, Factor 7—Control Activities and Factor 8—Internal Environment. The factors were then rearranged to reflect the sequence in the COSO ERM framework.

Table 2. Factor Analysis

Panel A: KMO and Bartlett's Test

Kaiser-Meyer-Olkin Measure of Sampling Adequacy.		.935
Bartlett's Test of Sphericity	Approx. Chi-Square	14919.745
	Sig.	.000

Panel B: Rotated Component Matrixa

	Component							
	1	2	3	4	5	6	7	8
ERM25_1)	.716							
ERM30.1_1)	.715							
ERM30.2_1)	.689							
ERM26_1)	.641							
ERM27_1)	.630							
ERM30.3_1)	.625							
ERM28_1)	.594							
ERM6_1)	.565							
ERM1_1)		.761						
ERM11_1)		.706						
ERM14_1)		.643						
ERM2_1)		.624						
ERM9_1)		.583						

ERM20_1)	.559						
ERM12_1)	.513						
ERM23_1)		.865					
ERM22_1)		.826					
ERM24_1)		.787					
ERM21_1)		.725					
ERM29_1)		.544					
ERM7_1)			.661				
ERM8_1)			.650				
ERM5_1)			.601				
ERM15.2_1)			.527				
ERM18.2_1)				.765			
ERM18.3_1)				.744			
ERM18.1_1)				.679			
ERM18.4_1)				.632			
ERM15.1_1)					.759		
ERM15.3_1)					.647		
ERM15.4_1)					.511		
ERM16_1)						.710	
ERM17_1)						.608	
ERM3_1)							.798
ERM4_1)							.510
ERM10_1)							.501

Table 3 depicts the results of the hierarchical multiple regressions on the hypothesized relationships. Multiple analyses were performed to substitute different measures or proxies for shareholders' wealth. There were five control variables in the analyses – 1) Total Revenue, 2) existence of separate ERM unit, 3) Existence of CRO, 4) ERM maturity, and 5) Industry Classification. The overall observation revealed that ERM was significant in explaining the variation in shareholders' wealth. With reference to firm value, X2 – Objective Setting (coeff. 881.576, p-value 0.045), X3– Event Identification (coeff. 95.947, p-value 0.040), X5 – Risk Response (coeff. 1013.621, p-value 0.026), X6 – Control Activities (coeff. 944.610, p-value 0.037) and X7– Information and Communication (coeff. 1128.478, p-value 0.014) were among the ERM variables that reported significant direct/ positive relationships with firm value. It is significant that, after controlling for the effect of five variables, all ERM variables reported positive relationships with firm value, thus supporting the hypothesized relationship that ERM implementation leads to better firm value. The second proxy utilized, EBITDA, revealed an interesting pattern of results. X1 – Internal Environment and X2 – Objective Setting reported negative relationships with EBITDA. All other ERM variables (X3–Event Identification, X4 – Risk Assessment, X5 – Risk Response, X6 – Control Activities, X7 – Information and Communication and X8 – Monitoring) revealed positive relationships with EBITDA, though not at 0.05 levels of significance. The third measure of shareholders' value, ROE, revealed four ERM variables reported significant relationships with ROE. X3 – Event Identification (coeff. 0.409, p-value 0.003), X5 – Risk Response (coeff. 0.093, p-value 0.032), X7 – Information and Communication (coeff. 0.204, p-value 0.047) and X8 – Monitoring (coeff. 0.397, p-value 0.002). All other ERM variables reported mixed (positive and negative) relationships with ROE but not at 0.05 significance levels.

The unique insignificance pattern of results for EBITDA certainly raised some concern. Most probably, this is due to the fact that majority (59 percent) of the respondent were in the 2-3 years period of the ERM implementation which reflect their ERM's maturity. Perhaps the intended impact of the ERM on the primary business operations is yet to be visualized. Interestingly, ERM seems to have a favorable impact on firm value and ROE. Two control variables reported a consistent pattern of results throughout the three regression analyses. XC1 – Total Revenue and XC5 – Industry Classification revealed to be significant variables capable of influencing the overall relationship between ERM variables and shareholders' wealth.

Table 3: Impact of ERM implementation on shareholders' wealth

Type of Observations	Firm Value			EBITDA			ROE		
Variables	Coef.	p-value	VIF	Coef.	p-value	VIF	Coef.	p-value	VIF
Constant	3055.083	.252		141.853	.690		1.961	.010*	
XC1 - Total Revenue	1.397	.000*	1.702	.121	.000*	1.702	.000	.000*	1.737
XC2 - Separate ERM unit	1079.168	.437	1.657	137.428	.457	1.657	-.514	.214	1.826
XC3 - Existence of CRO	1173.936	.401	1.823	162.210	.384	1.823	.160	.696	1.897
XC4–ERM maturity	-623.284	.254	1.206	4.298	.953	1.206	-.090	.590	1.532
XC5 - Industry Class.	-913.455	.010*	1.871	-127.441	.007*	1.871	-.409	.000*	1.861
X1 - Internal Environment	-7.770	.986	1.034	-3.226	.956	1.034	.080	.533	1.090
X2 - Objective Setting	881.576	.045*	1.082	-42.898	.463	1.082	.168	.188	1.112
X3 - Event Identification	95.947	.040*	1.203	22.278	.724	1.203	.409	.003*	1.185
X4 - Risk Assessment	240.428	.601	1.145	70.267	.252	1.145	-.074	.577	1.141
X5 - Risk Response	1013.621	.026*	1.126	38.443	.526	1.126	.093	.032*	1.130
X6 - Control Activities	944.610	.037*	1.133	61.523	.306	1.133	-.140	.273	1.111
X7 – Information and Comm.	1128.478	.014*	1.139	34.133	.576	1.139	.204	.047*	1.304
X8 - Monitoring	369.798	.412	1.102	2.236	.970	1.102	.397	.002*	1.118
R			.700			.528			.754
R^2			.490			.279			.568
Adj. R^2			.465			.244			.547
Std. Error of the estimate			7036.53136			937.77712			1.98414
R^2 change			.038			.010			.043
F change			2.495			.468			3.288
Sig. F change			.013			.878			.001
F value			19.741			7.934			26.532
p-value			.000			.000			.000

* Significance at the 0.05 level

6 Discussion, conclusion and managerial implications

The present results justified the dual-theoretical approach that combines stewardship and agency to theorize the relationships between the advising and control functions exercised by the CRO and board. One may argue that ERM itself is part of a set of external control mechanisms as it is mandated by regulatory agencies. Nevertheless, the implementation of ERM within an organization involves a combination of both advising and control functions. Evidently, overall results supported the hypothesized relationships and further substantiated recent studies on the significance of ERM in enhancing shareholders' wealth [25, 27, 28, 31-33]. Specifically, the following ERM attributes reported significant positive impacts on shareholder's wealth—objective setting, event identification, risk response, control activities, information and communication and monitoring. Though this study was conducted in a similar research setting to two earlier studies on ERM in Malaysia [22, 23], the present results interestingly failed to substantiate the findings of those two studies, particularly the fact that ERM did not have a significant influence on shareholders' wealth.

Up until now, ERM literature concentrated on US and western countries; this study enriches the literature by offering fresh insights in different geographical settings. This study both provided much-needed empirical input concerning the extent and nature of ERM implementation and identified the theoretical foundation underlying ERM as a governance tool. Indeed, the implementation of ERM is capable of enhancing shareholders' wealth.

The present study also substantiated the notion that the primary objective of ERM is to create, protect, and enhance shareholders' wealth. It is hoped that the results of this study will enrich the current ERM literature. The results support the validity of a dual-theoretical construct based on stewardship and agency to theorize the dynamic interrelations between firms' advising and control functions. It is anticipated that this study will shed light on and promote additional empirical studies to further test both its hypothesized relationships and its theoretical justifications. The results rationalized the mandatory requirements imposed by Bursa Malaysia [29] concerning the implementation of ERM. This study is among the very few empirical studies that utilized the COSO's ERM framework to assess the degree of ERM implementation. Nonetheless, it was measured using self-reported questionnaires and hence is affected by various limitations commonly found in studies using self-reported instruments. The study relied solely on self-reported responses and failed to corroborate the results with alternative measures. Future studies may fruitfully explore alternative approaches to self-reported measures— such as those in studies [5, 9, 31], while at the same time retaining a similar theoretical foundation.

This study chose to view ERM implementation as part of the firm's advising and control functions based on stewardship and agency theory; thus, the board and CRO are assumed to be the primary entities responsible for the success of ERM implementation. Further, the CRO could lobby the board and CEO to incorporate elements of ERM into staff performance appraisals, thus indirectly forcing the staff to embrace the ERM philosophy in their daily business operations. A continuous education program across various units in the organization is an indispensable element of an ERM strategy. Ultimately, once the ERM philosophy has pervaded the organization's daily routine, the CRO and the board will be able to visualize not only a significant improvement in business operations, but also an increase in shareholders' wealth.

Note:

The ERM instrument utilized in this study is available upon request. Kindly contact the corresponding author: m.ariff@ajman.ac.ae

References

1 Ye, Y.G., Liu, N., and Cao, R.X.: 'Risk Analysis for Listed Insurance Companies in China: Based on Contingent Claim Approach', in Editor (Ed.)^(Eds.): 'Book Risk Analysis for Listed Insurance Companies in China: Based on Contingent Claim Approach' (2008, edn.), pp. 12-19
2 Culp, C.L., and O'Donnell, K.J.: 'Catastrophe reinsurance and risk capital in the wake of the credit crisis', The Journal of Risk Finance, 2009, 10, (5), pp. 430 - 459
3 Walker, P.L., Shenkir, W.G., and Barton, T.L.: 'Enterprise Risk Management: Pulling it All Together' (The Institute of Internal Auditors Research Foundation (IIARF), 2002. 2002)
4 Burnaby, P., and Hass, S.: 'Ten steps to enterprise-wide risk management', Corporate Governance, 2009, 9, (5), pp. 539-550
5 Hoyt, R.E., and Liebenberg, A.P.: 'The Value of Enterprise Risk Management', Journal of Risk and Insurance, 2011, 78, (4), pp. 795-822
6 McShane, M.K., Nair, A., and Rustambekov, E.: 'Does enterprise risk management increase firm value?', Journal of Accounting, Auditing & Finance, 2011, 26, (4), pp. 641-658

7 Sarens, G.: 'Internal Auditing Research: Where are we going? Editorial', International Journal of Auditing, 2009, 13, (1), pp. 1-7
8 Bromiley, P., McShane, M., Nair, A., and Rustambekov, E.: 'Enterprise risk management: Review, critique, and research directions', Long range planning, 2015, 48, (4), pp. 265-276
9 Gordon, L.A., Loeb, M.P., and Tseng, C.-Y.: 'Enterprise risk management and firm performance: A contingency perspective', Journal of Accounting and Public Policy, 2009, 28, (4), pp. 301-327
10 Cassidy, S.M., Constand, R.L., and Corbett, R.B.: 'The market value of the corporate risk management function', Journal of risk and insurance, 1990, pp. 664-670
11 Smith, C.W.: 'Corporate risk management: theory and practice', Journal of Derivatives, 1995, 2, (4), pp. 4-12
12 Sinclair, S.: 'Essays in multiperiod risk management with information updating'. Ph.D., University of Pennsylvania, 2004
13 Ertugrul, M.: 'Essays on agency problems and corporate governance'. Ph.D., The University of Connecticut, 2005
14 Lin, Wu, S., Penm, J.H.W., and Terrell, R.D.: 'The relationship and causality testing between diversification, risk and financial performance: Empirical examination in Taiwan's banking industry', International Journal of Services, Technology and Management, 2005, 6, (6), pp. 556-574
15 Simkins, B.: 'Enterprise Risk Management: Current Initiatives and Issues Journal of Applied Finance Roundtable', Journal of Applied Finance, 2008, 18, (1), pp. 115-132
16 Andersen, T.J.: 'The performance relationship of effective risk management: Exploring the firm-specific investment rationale', Long Range Planning, 2008, 41, (2), pp. 155-176
17 Andersen, T.J.: 'Effective risk management outcomes: exploring effects of innovation and capital structure', Journal of Strategy and Management, 2009, 2, (4), pp. 353-379
18 Pagach, D., and Warr, R.: 'The characteristics of firms that hire chief risk officers', Journal of Risk and Insurance, 2011, 78, (1), pp. 185-211
19 Pagach, D.P., and Warr, R.S.: 'The effects of enterprise risk management on firm performance', Available at SSRN 1155218, 2010
20 Farrell, M., and Gallagher, R.: 'The Valuation Implications of Enterprise Risk Management Maturity ', Journal of Risk and Insurance, 2015, 82, (3), pp. 625-657
21 Grace, M.F., Leverty, J.T., Phillips, R.D., and Shimpi, P.: 'The Value of Investing in Enterprise Risk Management ', Journal of Risk and Insurance, 2015, 82, (2), pp. 289-316
22 Manab, N.A.: 'Enterprise-wide risk management (EWRM) implementation, compliance and value creation among public listed companies (PLCs) in Malaysia'. Ph.D, Universiti Teknologi MARA, 2009
23 Kasim, M.A., Aziz, A.A., and Kassim, I.: 'Enterprise Risk Management (ERM) in Malaysia: A Study of the Status of Implementation, the Role of Internal Audit and the Impact on Organisational Performance' (LAP LAMBERT Academic Publishing, 2011. 2011)
24 COSO: 'Enterprise Risk Management- Integrated Framework-Executive Summary' (The Committee of Sponsoring Organisations of the Treadway Commission 2004. 2004)
25 Ai, J., and Chen, H.: 'Does ERM Improve Firm Value? Evidence from Listed Chinese Nonfinancial SOEs', Evidence from Listed Chinese Nonfinancial SOEs (December 15, 2014), 2014
26 Beasley, M., Branson, B., and Hancock, B.: 'Current state of enterprise risk oversight: Progress is occurring but opportunities for improvement remain', in Editor (Ed.)^(Eds.):

'Book Current state of enterprise risk oversight: Progress is occurring but opportunities for improvement remain' (2012, edn.), pp.

27 Beasley, M., Branson, B., and Pagach, D.: 'An analysis of the maturity and strategic impact of investments in ERM', Journal of Accounting and Public Policy, 2015, 34, (3), pp. 219-243

28 Mafrolla, E., Matozza, F., and D'Amico, E.: 'Enterprise Risk Management In Private Firms: Does Ownership Structure Matter?', Journal of Applied Business Research, 2016, 32, (2), pp. 671

29 Bursa Malaysia: 'Practice Note 9: Internal Control and Corporate Governance Statement', in Editor (Ed.)^(Eds.): 'Book Practice Note 9: Internal Control and Corporate Governance Statement' (Bursa Malaysia Securities Berhad, 2012, edn.), pp.

30 Jerry, A.M., Hively, K., and Merkley, B.W.: 'Enterprise Risk Management: Trends and Emerging Practices' (IIA Research Foundation, 2001. 2001)

31 McShane, M.K., Nair, A., and Rustambekov, E.: 'Does Enterprise Risk Management Increase Firm Value?', 2011, 26, (4), pp. 641-658

32 Farrell, M., and Gallagher, R.: 'The valuation implications of enterprise risk management maturity', Journal of Risk and Insurance, 2015, 82, (3), pp. 625-657

33 Grace, M.F., Leverty, J.T., Phillips, R.D., and Shimpi, P.: 'The value of investing in enterprise risk management', Journal of Risk and Insurance, 2015, 82, (2), pp. 289-316

34 Beasley, M.S., Clune, R., and Hermanson, D.R.: 'Enterprise risk management: An empirical analysis of factors associated with the extent of implementation', Journal of Accounting and Public Policy, 2005, 24, (6), pp. 521-531

35 Beasley, M., Pagach, D., and Warr, R.: 'Information conveyed in hiring announcements of senior executives overseeing enterprise-wide risk management processes', Journal of Accounting, Auditing & Finance, 2008, 23, (3), pp. 311-332

SHS Web of Conferences **34**, 06001 (2017) DOI: 10.1051/shsconf/20173406001
FourA 2016

Employability Skills Required of Accountants

Lin Mei Tan[1] and *Laswad* Fawzi[1]

[1]School of Accountancy, Massey University, Private Bag 11222, Palmerston North 4442, New Zealand

Abstract. Employability skills have become increasingly important in a competitive job market in a globalised world of advanced technology. Our study examines employability skills of accountants as indicated in job advertisements in Australia and New Zealand, as they provide a window into the skills that are most valued by employers. The results show that of the 31 identified skills, 13 were most cited. Importantly, the most sought after skills included the ability to collaborate with colleagues, present, discuss and defend views, and having a positive attitude. Overall, a team player with a positive attitude and good communication skills appeared to be the most valued behavioural skill as perceived by employers. This probably reflects a change in the work of accountants, from being 'backroom number crunchers' to engaged business professionals.

1 Introduction

The environment in which accountants work is rapidly changing particularly with advances in information technology and globalisation of economies. To meet the challenges brought about by these changes, it is essential that University accounting programmes provide graduates with strong technical knowledge and the essential skills to gain employment and make an immediate contribution to a business [1]. The increasing importance of skills development is reflected in the International Accounting Education Standards Board (IAESB)'s 2015 release of *International Education Standard (IES) 3 Initial Professional Development – Professional Skills*. This educational standard indicates the level of professional skills needed to perform the role of a professional accountant and classifies the skills under four competency areas: intellectual, interpersonal and communication, personal, and organisational.

There is an extensive literature on the generic employability skills expected of accounting graduates. The majority of these studies were carried out in the last three decades and focused on the importance of various skills as viewed by academics, employers, students or graduates and professional accounting bodies. These studies suggest that accounting programmes are failing to meet the expectations and needs of employers or are not in harmony with the requirements of the real world [3]. An important implication is that a number of vital skills need to be developed within the accounting curriculum [12].

Although there is consensus among accounting academics that employability skills need to be developed in the accounting curriculum, it is a challenge to decipher what and how to incorporate the development and assessment of a wide range of employability skills. One of the main constraints as identified in the literature is that there is insufficient time to help students develop a wide range of employability skills as the accounting curriculum is already crowded with required technical content [11]. Apart from the time factor, other possible constraints include the lack of expertise and resources in assisting students with skills development. Further, there are suggestions that some of the generic skills are best developed and applied in the workforce rather than taught in a university. As it is not practical to cover all the required skills expected of employers, accounting educators perhaps need to prioritise certain skills that are deemed most important in the workplace. However, as the global economy evolves, it is essential to continuously ascertain and articulate employers' opinions on what makes a graduate employable so that accounting educators are kept abreast of the skills and attributes that are most valued in the workplace. This study contributes to the literature on the employability skills required of accounting graduates by examining the essential skills listed in job advertisements for accountants in Australasia.

2 Literature Review

There are various definitions of employability skills. They are commonly termed as generic skills, non-technical skills, capabilities, key competencies, personal transferable skills, soft skills and attributes, and are considered relevant to both entry-level and established employees [12]. Since the mid-1980s, the literature on employability skills has grown in Australia, NZ and other countries indicating that the issue is not unique to a particular country but is an area of growing concern worldwide and in all disciplines including accounting. The accounting curriculum has long been criticised for being too narrow and outdated, and failing to prepare students adequately for employment in the modern business environment [1]. The 20th century witnessed the accounting profession in numerous countries calling for further development of generic skills in the accounting curriculum to enhance a capacity for lifelong learning. In the US, The American Education Change Commission (AECC, 1990) set out a number of desired capabilities which accounting courses should develop in students, including: intellectual skills, interpersonal skills, communication skills, personal capacities and attitudes. The Australian and NZ accounting bodies have supported similar views regarding the importance of these various skills. To gain a better understanding of what skills are required of accounting graduates, numerous studies have been conducted in various countries, raising the question: "What are the important employability skills required of accounting graduates

Studies on graduate employability skills in general have identified a number of skills as important; however, personal, interpersonal, and intellectual capabilities usually rank highly [4]. Studies conducted in the accounting disciplines showed that graduates need more than technical skills to be employable. Communication skills and interpersonal skills are generally rated by employers as the most important characteristics [7]. Knowledge of software packages such as spreadsheet software, Microsoft Excel, Windows and word processing software are also in high demand [7]. However, the importance of individual skills varied for different groups of graduates/students, and educators and employers [8]. For example, working independently is rated highly by employers whereas academics rated teamwork and communication skills as more important. Stoner and Milner [8] show a gap exists between the skill expectations of academics and graduates/students. The study on graduates by de Lange et al. [4] found significant expectation gaps particularly in interpersonal skills, oral expression and computing/information technology skills. Regarding technological skills, Rackliffe and Ragland [10] pointed out that there is a possible disconnect between Excel skills needed in practice and those developed in accounting education. Bui and Porter's [3] study shows that there are differences in educators and employers' perceptions of the knowledge, skills and competencies required of graduates. Employers in particular expressed their disappointment that the new graduates they have employed lack numerous essential skills or do not match their expectations [5]. Wells et al.'s [13] study indicates a gap exists between expectations of graduates and what university course programmes are providing in NZ universities. The greatest gap identified by graduates, employers and academics was the development of team skills [4].

The issue of employability skills has been debated in the literature for over three decades. Although most universities have adopted a variety of ways to develop various skills, there is little evidence that accounting education has changed much to meet the demands of key stakeholders like employers and professional accounting bodies [4]. There is also little evidence from employers of accounting graduates that the level of generic skills development has improved over the past two decades. Although there is consensus that generic skills are needed to be developed in the accounting curriculum, the extent to which each skill should be developed at tertiary institutions remains debatable. Therefore, it is not surprising that the skills expectation gap continues to be an area of concern in the 21st century.

3 Objectives of Our Study

Employers are key stakeholders in accounting education and they play an important role identifying the skills that are most relevant for accounting graduates [12]. To examine the current employer demands for workplace skills, we developed the following research questions:

- What are the current skill requirements for accountants as indicated in job advertisements?
- Are there any differences in the skills demanded by employers in jobs advertised for more experienced versus less experienced accounting positions?

In the early 1990s, Birkett developed a set of competency based standards for professional accountants in Australia and NZ, which were ultimately adopted by the major accounting bodies in Australia and NZ (CPA Australia, ICAA and ICANZ). Birkett's [2] taxonomy of skills is classified into two main groups: cognitive and behavioural. He identifies three dimensions of cognitive skills: technical, analytic/constructive and appreciative, and three dimensions of behavioural skills: personal, interpersonal and organisational. According to Birkett [2], "professional work requires action, supported by decision processes; it involves the exercise of instrumental skills, supported by intellectual skills." He suggests that both cognitive and behavioural skills are likely to parallel typical patterns of career progression – with technical and personal skills drawn on initially, then analytic/constructive and interpersonal skills, and appreciative and organisational skills coming to the fore at more advanced career stages.

The IAESB of IFAC recently revised its standard on skills to meet the challenges of the changing environment in which professional accountants work. The IES 3 *Initial Professional Skills* came into effect on 1 July 2015. It identifies four skills: the intellectual, interpersonal and communication, personal and organizational skills which a professional accountant integrates with technical competence, and professional values, ethics and attitudes to demonstrate professional competence. In summary, most skills framework have some common elements that concern basic/fundamental skills, people related skills, conceptual/thinking skills, personal skills and attributes and skills related to the business world and community. For this study, we draw on the framework developed by Birkett [2] and IES 3 as both have identified a comprehensive list of skills required of accounting graduates. Figure 1 shows the employability skills framework used in this study. Based on

DOI: 10.1051/shsconf/20173406001

Birkett's framework [2], skills are considered under two broad categories: cognitive and behavioural.

4 Research Method

We examined job advertisements in the accounting area in Australia and NZ as there is a high degree of labour mobility between the two countries. There are no restrictions for New Zealanders and Australians to live and work in the other country. We used a major online job site, www.seek.co.nz (jobs in NZ and Australia), which is one of the most popular job sites used in Australasia. From this website, we selected accounting as the main job classification and screened the job subcategories shown. Then we selected 16 out of 24 job sub classifications we considered were more suitable for applicants with at least an undergraduate accounting qualification. We captured and examined the job advertisements for the six-month period, from July 2015 to December 2015, and only focused the content analysis on employability skills required for the job. Over this period there were 18,439 and 3,282 accounting advertisements in Australia and NZ, respectively, after removing any repeated advertisements. We used a content analysis software package, Wordstat, 2004) to generate the frequencies of specific categories of skills included in the advertisements. Drawing from the literature, we initially built a categorisation dictionary of employability skills that fall under each of the five categories. We started with a list of well-known words or phrases commonly used in the literature to denote skills for each of the five main categories and their respective subcategories. The initial categorisation dictionary was then applied to the database of job advertisements to identify the frequency with which specific categories were listed in the advertisements. We then examined the most frequently stated words and phrases in the advertisements which were not in our initial dictionary. Words or phrases that we considered relevant were subsequently included in our subcategories after having considered their context in the entire advertisement's text.

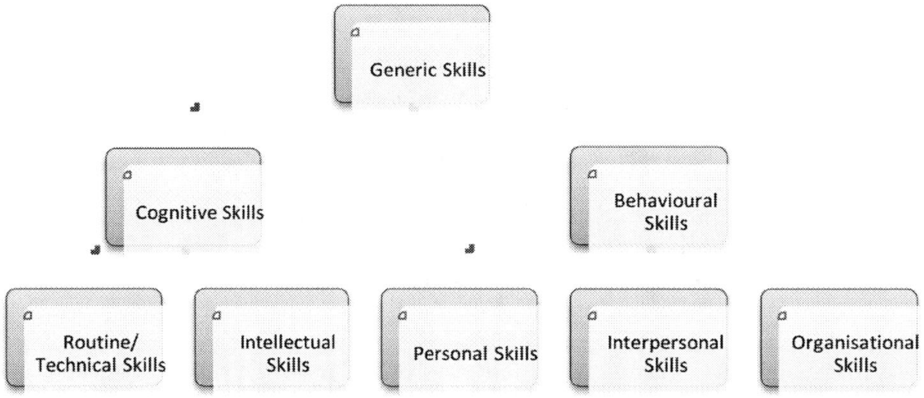

Fig. 1. Employability skills framework

5 Findings

The most frequently sought after job categories were in business services and corporate advisors, financial accounting and reporting, taxation, management accounting and budgeting, analysis and reporting, financial managers and controllers, and assistant accountants. However, there was a much higher proportion (13%) of taxation jobs in Australia as compared to NZ (2.9%). This is likely to reflect the much simpler tax system in NZ as compared to the more complex tax system in Australia. Behavioural skills dominated the top requirement as the number of words or phrases cited that fall into this category represented about 77% of all word/phrases used in the advertisements for jobs in Australia and New Zealand. Table 1 shows the frequency of each of the skills in the five broad categories that were mentioned in the advertisements. Interpersonal skills were the most frequently mentioned set of skills in the job advertisements. The second most frequently cited set of skills in the advertisements were personal skills. These results suggest that interpersonal and personal skills were in high demand by employers in Australia and NZ. The frequencies of citations of intellectual, routine/ functional, and organisational categories in both countries were very similar. These findings are generally consistent with the broad categories of professional skills identified in IES 3.

Table 1. Broad Categories of Skills and the Extent of Citation in Advertisements

	Australia		New Zealand	
	No	%	No	%
Interpersonal	16,750	91%	2,888	88%
Personal	11,516	62%	2,243	68%
Intellectual	8,180	44%	1,424	43%
Routine/functional	7,964	43%	1,409	43%
Organisational	7,740	42%	1,303	40%

To gain further insights into the occurrence of specific skills cited in the advertisements, we examined the percentage of advertisements that have words/phrases shown in the dictionary. The results show that 13 out of the 31 specific skills in the dictionary were cited in more than 10% of the total advertisements. Table 2 shows the skills ranked by the percentage of citations in job advertisements. The top 13 skills mentioned in the advertisements for jobs in Australia and NZ were found to be similar except for some slight differences in the ranking of some of skills within the top most cited 13 skills. The top four skills were ranked in the same order for Australia and NZ. Collaborating with colleagues was the most cited interpersonal skill and was mentioned in more than 70% of the job advertisements. This finding suggests that employers value team work and prefer a candidate that is a team player and can collaborate with colleagues. This aspect of interpersonal skills has become more and more important as it is critical to an organisation's productivity and profitability. The second ranked skill which is also an interpersonal skill, i.e., the ability to present, discuss and defend views, encompasses communication, presentation and speaking skills. One aspect of personal skill that is positive attitude was ranked third with employers expressing a desire for employees who are committed, passionate, dynamic, energetic, vibrant, and self-driven. The fourth ranked skill is the use of information technology. Not surprisingly, although use of information technology is a routine skill, it is an important skill in today's global, technology-demanding economy. Further, it indicates that information technology is a core competency for accountants.

The next three frequently cited skills were: leadership skills, understand group dynamics, and meeting deadlines. Understanding group dynamics is about people skills and relationships and is an important aspect of interpersonal skill which is highly valued in the accounting work environment. The ability to lead and meet datelines indicates the importance of good organisational skills which is vital for accountants in managing their work. The three skills ranked 8th to 10th were observant and awareness, act strategically, and think and act independently. Acting strategically refers to the employers' preference for a person who is proactive, has the quality of acumen and good work ethics, and one who is commercially and intellectually astute. Employers also desire a recruit who has the skill of being able to think and act independently, i.e., one who is able to apply professional scepticism, is self-motivated and inquisitive, and is a practical person. Both of these skills are important personal skills. Another highly desired intellectual skill by employers is being observant or aware, i.e., a person who pays attention to details and is accurate. This is an essential skill as accountants are expected to be detail oriented so as to be able to detect errors, discrepancies or inconsistencies.

Table 2. Top 13 Skills Ranked by Percentage of Job Advertisements

		Australia	New Zealand
Interpersonal:	Collaborate with colleagues	1 (77.3%)	1 (72.5%)
Interpersonal:	Present discuss defend views	2 (58.1%)	2 (46.7%)
Personal:	Positive attitude values	3 (39.6%)	3 (45.4%)
Routine:	Use of information technology	4 (31.8%)	4 (35%)
Organisational:	Apply leadership skills	5 (23%)	7 (19.1%)
Organisational:	Meet deadlines	6 (21.7%)	5 (22.9%)
Interpersonal:	Understand group dynamics	7 (20.6%)	6 (22.5%)
Intellectual:	Observant and aware	8 (17.5%)	10 (16.1%)
Personal:	Act strategically	9 (16.2%)	9 (17.4%)
Personal:	Think and act independently	10 (16.4%)	8 (18.9%)
Intellectual:	Analyse, reason and conceptualise issue	11 (15.1)	13 (13.3%)
Personal:	Be flexible	12 (13.8)	11 (15.7%)
Intellectual:	Solve problems and construct arguments	13 (12.7)	12 (13.8%)

The last three skills are: analyse, reason and conceptualise issues, be flexible and solve problems, and construct arguments. Analyse, reason and conceptualise issues refers to analytical skills and logical reasoning skills. These are intellectual skills expected of accountants who frequently have to deal with issues concerning costs, revenues, investments and profit. Solve problems and argument is another highly cited intellectual skill. Employers seek to employ people who are forward thinking, lateral thinking, and who can think "outside the box"; accountants are expected to contribute to add value to their work. Being flexible is a personal attribute which refers to someone who is able to adapt to circumstances. With the accounting industry being highly dynamic and the rapid advancement of technology, the ability to adapt quickly to changes is an important skill cited in many advertisements.

We also examined whether the skill requirements cited in the job advertisements differ by experience. We used those advertisements that cited the requirement of Chartered Accountant (CA) or Certified Public Accountant Australia (CPA) to proxy for more experienced recruits. The mean number of times skills are cited in the advertisements for each one of the five broad categories. A t-test was used to compare differences between means of the two groups (CA/CPA and non-CA/CPA).The results show that there were significant differences in the frequency of interpersonal and intellectual skills cited in advertisements that required a qualified CA/CPA and those that did not require such qualifications in Australia and NZ. Interpersonal skills were cited more frequently for jobs requiring a CA/CPA than those that did not require such professional qualifications. This result suggests that interpersonal skills are considered as very important for jobs that require more experienced accountants. It is very likely that those advertisements that specified the CA/CPA requirement are for top level positions. On the other hand, intellectual skills were cited more frequently for jobs that did not require a CA/CPA qualification, indicating that employers emphasised intellectual skills as very important for less experienced recruits which includes entry level accountants. This may also suggest that employers assume that CA/CPAs have those skills and therefore there would be no need to emphasise such skills in the advertisements. Two other differences were found in the Australian advertisements only. Personal skill and routine skill appeared more times in the advertisements for less experienced accountants than for experienced accountants. Again, this result perhaps suggests that employers are more particular that a less experienced recruit is skilled in routine task and possesses good personal attributes.

6 Conclusions

In a globalised business and employment world and with advances in technology, the issue of employability skills has become increasingly important over the years. To identify the most in demand skills required of accountants, our study examines the employability skills indicated in job advertisements for accountants in Australia and NZ as they provide an objective indication of the important skills required in the work place.

Our analysis of the job advertisements shows that a mix of skills is required of accountants in Australia and New Zealand. The top 13 skills most cited out of the 31 we identified in the dictionary were: collaborate with colleagues, present, discuss, and defend views, positive attitude values, using information technology, meet datelines, understand group dynamics, apply leadership skills, think and act independently, act strategically, observant and aware, analyse, reason and conceptualize issues, be flexible, and solve problems and construct arguments, suggesting they were the most in demand by employers. Although some of the rankings of skills were slightly different between the Australian and New Zealand advertisements, the top 13 skills were the same. Employers, therefore, seek a multitude of skills but placed more emphasis on interpersonal and personal skills. Most importantly, they valued a recruit who can collaborate with colleagues, present, discuss, defend views and have a positive attitude. Overall, a team player with a positive attitude and good communication skills appeared as most desired by employers. This probably reflects a change in the work of accountants from being the "backroom number cruncher" to an engaged business professional.

The results further show that in both the New Zealand and Australian advertisements, good interpersonal skills were more often cited for an experienced accountant than a less experienced accountant. In contrast, intellectual skills were cited much more frequently for less experienced or entry level accountants than jobs requiring CA/CPAs. Furthermore, the occurrence of two other skills only in the job advertisements in Australia suggests that employers emphasised personal and routine skills for less experienced Australian accountants than for CA/CPAs.

What implications do our results have for accounting academics and accounting education? Some academics believe that universities should not be fully responsible for helping students develop all the employability skills expected by employers. Employers generally tend to expect a work-ready graduate whereas academics tend to have a stronger focus on the development of intellectual capability and ability [3]. Unequivocally, it is not possible for tertiary institutions to teach all the skills that are demanded by employers. Academics play a key role in delivering the initial professional development whereas an employer's role is to help in nurturing the professional development of accounting graduates. Accounting qualifications are important in providing a broad introduction to a

DOI: 10.1051/shsconf/20173406001

potential career in business or management. IES-3 expects that IFAC member bodies design their professional accounting education programmes so that graduates achieve an intermediate level of proficiency in four professional skills: intellectual, interpersonal and communication, personal and organization.

This study shows that interpersonal skills are very important for accountants. The literature indicates that this is the skill with the greatest gap as identified by graduates, employers and academics. Academics certainly need to focus on the development of this skill. Employability skills are also interconnected in some way. So by helping students develop teamwork skills, it will also help them develop people and communication skills. Employers also want people with good personal attributes such as an energetic, dynamic, passionate and adaptable personality. This is an important skill as it is very often cited for less experienced accountants which would include entry level accountants. Some of these attributes are harder to help develop as they are more personality traits influenced by genetics or culture, but perhaps could be indirectly influenced during the development of other employability skills. In contrast, attributes like good work ethics, professional scepticism are easier to integrate into the curriculum. Routine skills like the use of information technology must also not be underestimated as they are particularly important for entry level accountants and academics should continuously support students' development of these skills. Information technology competence is an important criteria and employers desire those who are proficient in using various accounting tools such as Excel, MYOB, Xero and other database systems. The next group of skills which accounting academics need to help students develop is organisational skills which include leadership and meeting deadlines. Intellectual skills are also required by employers and most academics should have already integrated the development of analytical and problem solving skills, as well as the need to pay attention to detail, into the accounting programme. Many universities' accounting curricula have already incorporated a number of employability skills in their programmes although they may vary in the emphasis of certain skills over others and the way that these skills are integrated into academic programmes. This study provides some insights to academics into the important employability skills that should be integrated into accounting programmes. Depending on how they are developed and delivered, the study's findings may assist in narrowing the gap between the competencies and skills developed at university and the skills required of accounting graduates by prospective employers.

7 Limitations

A limitation of the study is that the job advertisement is taken as a valid representation of the requirements of employers. It is possible that some skills that are required of a potential employee might not be specifically mentioned in job advertisements. Their non-occurrence does not mean that they are not required. However, job advertisements certainly tell us those skills that are most desired or valued by employers as generally employers seek to minimise recruitment costs by appealing only to potential applicants who have the required skills. Perhaps what is more intriguing is the level of competence actually required by employers which is hardly mentioned in the job advertisements. Furthermore, Hancock et al. [5] interviewed employers and found that they had modest expectations for new graduates' technical skills but viewed non-technical skills as very important. Future research could explore job advertisements and interview the respective employers to tease out their perspectives on the level of competence in employability skills they seek from new accounting recruits. After all, employers also play a crucial role in developing and enhancing the employability skills of their staff.

DOI: 10.1051/shsconf/20173406001

References

1. W. Albrecht, R.Sack. *Accounting education: Charting the course through a perilous future. Accounting Education Series, 16.* American Accounting Association, Florida, (2000).
2. W. Birkett. *Competency based standards for professional accountants in Australia and New Zealand* (Sydney: Australian Society of Certified Practising Accountants, the Institute of Chartered Accountants in Australia and the New Zealand Society of Accountants., (1993).
3. B. Bui, B. Porter. The expectation-performance gap in accounting education: An exploratory study. Accounting Education: An International Journal, **19**,23–50 (2010).
4. P. De Lange, B. Jackling, A. Gut, Accounting graduates' perceptions of skills emphasis in undergraduate courses: an investigation from two Victorian universities. Accounting and Finance, **46**, 365-386 (2006).
5. P. Hancock, B.A. Howieson, M. Kavanagh, J. Kent, I. Tempone, N. Segal, *Accounting for the future: more than numbers: volume 1, Final Report.* Canberra, Australian Learning and Teaching Council (2009).
6. B. Jackling, P. De Lange, P. Do accounting graduates' skills meet the expectations of employers? A matter of convergence or divergence. Accounting Education: An International Journal, **18**, 369–385 (2009).
7. L.M. Johnson, V. E. Johnson, Help wanted – accountant: what the classifieds say about employers' expectations, Journal of Education for Business **70**, 130–134 (1995).
8. M. Kavanagh, L. Drennan. What skills and attributes does an accounting graduate need? Evidence from student perceptions and employer expectations. Accounting and Finance **48**, 279-300 (2008).
9. M. Keneley, B. Jackling, The acquisition of generic skills of culturally-diverse student cohort. Accounting Education: An International Journal, **20,** 605-623 (2011).
10. U.R. Rackliffe, L. Ragland, Excel in the accounting curriculum: perceptions from accounting professors. Accounting Education 25, 139-166 (2016).
11. G. Stoner, *M. Milner, Embedding* generic employability skills in an accounting degree: development and impediments. In K. Watty, B. Jackling, & R. M. S. Wilson (Eds.), *Personal Transferable Skills in Accounting Education.* London: Routledge, (2012).
12. K. Watty, B. Jackling, R.M.S.Wilson, (eds,). *Personal transferable skills in Accounting Education*, London: Routledge (2012).
13. P. Wells, P. Gerbic, P, I. Kranenburg, J.Bygrave, Professional skills and capabilities of accounting graduates: the New Zealand expectation gap? Accounting Education: An International Journal, **18**(4) (2009).

DOI: 10.1051/shsconf/20173406002

Job Design, Training Effect and Job Satisfaction: Evidence from Work Placement at Audit Firms

Guangyou Liu*

Business School and Center for Accounting, Finance and Institutions, Sun Yat-sen University, Guangzhou, 510275, P.R.China

Abstract. This paper aims to test the relationships between job design aspects and job satisfaction, as well as the mediating role of training effect in these relationships. Regression and mediation analyses were performed based on the data collected from a questionnaire-based survey on the senior accounting students' audit work placement at audit firms. I conclude that repeated tedious non-professional job aspect is negatively related to job satisfaction, whereas judgmental professional job aspect is positively related to job satisfaction. I also conclude that training effect of work placement is playing a partial mediating role in the identified positive relationship while having no mediation in the negative one.

1 Introduction

Work placement for accounting graduates is a critical stage for them to perceive as new experience to start their career development in the very near future. The work-integrated learning experience through work placement can help them shape the cognition of future public accounting profession. Particularly, job satisfaction during work placement will have critical impacts on the career decision facing these accounting students shortly before their graduation. In business education, work placement is becoming an important and effective learning tool for business students to bridge the gap between classroom learning and real world practices [1]. The term 'work placement' is equivalent to 'internship' and 'on-the-job training' used by other scholars. Maertz Jr. et al. suggest that successful internships can benefit the interns, schools and employers, and especially the need of students for experiential and vocational learning can be fulfilled [2].

This paper aims to test the relationships between job design in work placement and job satisfaction, as well as the mediating role of training effect in these relationships, based on the data collected from a questionnaire-based survey was completed among a group of senior accounting students who had just finished their work placement at Audit firms. This study makes its unique contribution to the existing accounting education literature by applying the behavioural and cognitive perspectives such as job design, training effect and

* Corresponding author: mnslgy@mail.sysu.edu.cn

DOI: 10.1051/shsconf/20173406002

job satisfaction in accounting research. My findings are implicative to accounting schools and audit firms in improving job training for their students and future employees.

2 Some Institutional Features of Chinese Accounting Education

At least three institutional features of the Chinese accounting university education offer a significant opportunity of studying Asian accounting education. First, China is the largest economy in Asia, fully representative of the emerging and transitory markets in the world. Second, there is an emerging and increasing demand for quality audit professionals in the course of rapid social and economic developments in China. Lastly, Chinese educational systems are similar to those in other Asian countries by sharing cultural values. The Chinese rapid economic growth in most recent four decades calls for the powerful support and safeguard from a well-educated auditing profession. The behavioural and cognitive aspects in accounting education and audit skills training are crucial to the formation of a well-functioning auditing profession in China. Liu points out that nowadays in the key Chinese universities that focus on the training effects of placing their accounting students in the field practices at audit firms, and that the work placement through internships is a compulsory part of the accounting schools' curriculum [3]. Pedagogically speaking, audit work placement via the university-and-audit-firm cooperative education programs not only helps students to enhance their understanding of the textbooks taught in the classroom, but also lays solid foundations for successful future professional careers [3]. This study focuses on job design in audit work placement and job satisfaction for the senior accounting students in the Chinese auditing professional context. Table 1 lists the common characteristics of audit work placement for the Chinese accounting students as summarized in comparisons with the job aspects described in the work of Maertz Jr. et al. [2].

Table 1. Characteristics of Audit Work Placement for Accounting Students in China

Dimension	Description of Work Placement Features
Payment	Paid at the minimum wage level
Time schedule	Full time work
Participant	Undergraduate and professional internship
Credit	Academic course credit
Academic requirement	High formal academic requirements and on-the-job experiences
Organizer	Arranged officially between intern-employer and school
Internship duties	Well defined and well planned, not "do whatever is needed or asked"
Work format	Project-based format, not job-based format
Mentorship	Work sponsor and mentor (i.e. experienced CPAs in the audit firm)
Future job opportunity	Implication regarding future bilateral job/recruitment opportunity

3 Literature Review and Development of Research Hypotheses

As it is widely accepted by both business researchers and practitioners, job design aspects and characteristics are significantly related to job satisfaction [4]. Although there are different ways to categorize job aspects, job design has been generally defined as designing task and role content of jobs to fit in the social and psychological needs of job takers [5]. Theoretically there shall be some associations between job design as work inputs and job satisfaction as social and cognitive outcomes.

Some scholars conclude that job aspects such as autonomy, variety and significance can increase positive impacts of job design on job performance and satisfaction [6] [7]. Others find that higher educated workers like university graduates are more satisfied with their jobs because they have a job of better quality, however, the association between the level of

 DOI: 10.1051/shsconf/20173406002

education and job satisfaction is negative for the first job [8]. Humphrey et al. criticize that most of the existing job satisfaction studies only focus on work aspects, ignoring social and contextual aspects of work [9]. To follow this criticism, I include in the regression model two team aspects as social contexts in addition to those job aspects. My research hypotheses on the relationships between job aspects and job satisfaction are presented in Figure 1.

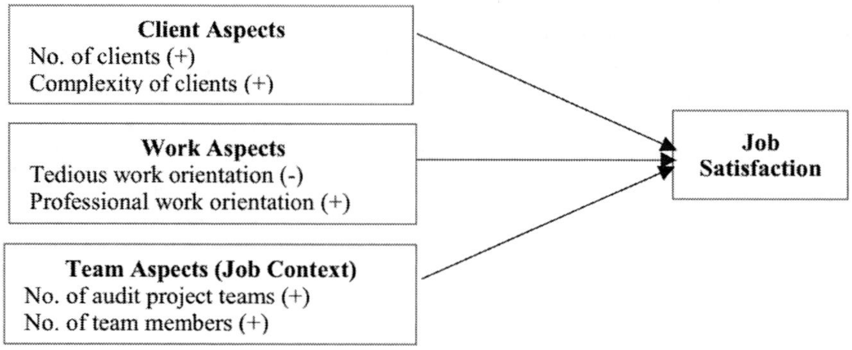

Figure 1 Hypotheses on Relationships between Job Design Features and Job Satisfaction

Further investigations concern how people are satisfied with their jobs. Some scholars prove that jobs with high problem-solving requirements increase job satisfaction because these job provide a chance of demonstrating and reinforcing the sense of professional and judgmental competence on the job [10] [11]. Some studies emphasize the training effect. For example, Howard et al. conclude that the on-the-job training effect is especially important to motivate and satisfy the new entry-level professionals, and that a strongly motivated learner at workplace evaluates highly the change to learn new things and to extend the range of professional competencies [11]. To grasp these implications in prior literature regarding the impacts of on-the-job training effect, I conjecture that training effect may play the mediating role in the relationship between job aspects and job satisfaction. My related hypotheses are expressed by the SEM in Figure 2.

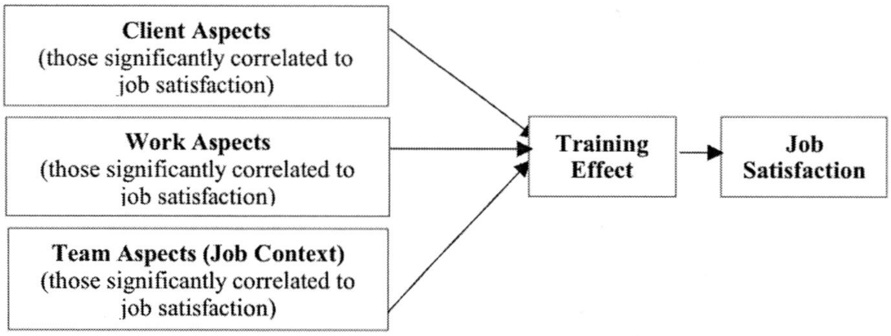

Figure 2 Hypotheses on Mediating Role of Training Effect

4 Research Methodology

4.1 Data Collection

The research data for this study have been collected from a two-year questionnaire-based survey on the cooperative accounting education arrangement in a top-tier Chinese university. The 30-minute surveys were administered anonymously in mid-May 2013 and 2014 respectively, among 350 interns in total who returned from the four-month work placement at Audit firms officially assigned by the university and 15 contracted Audit firms. 208 effective responses were collected. All the questionnaire items were either adapted from the existing literature or designed based on the interviews with 20 representatives from the target population who participated in a pilot pre-test before the formal surveys.

4.2 Variables

This study investigates four groups of variables, including dependent, explanatory, potential mediating, and control variables. Job satisfaction is the only dependent variable; six job design features of three dimensions are used as explanatory variables; gender and firm size are included in the regression model as control variables, and training effect is expected to be a mediating variable. Table 2 presents a summary of all the variables in this study.

Table 2. Variables and Descriptions

Variable	Description
SATISFACTION	Dependent variable indicating job satisfaction with work placement
TRE	Expected mediating variable indicating training effect of work placement
CL_NO	Number of clients served during work placement
CL_CMPLX	Degree of client complexity experienced in work placement
WK_TEDI	Extent of repeating tedious non-professional jobs during work placement
WK_PROF	Extend of performing judgmental professional tasks during work placement
TM_NO	Number of teams attended during work placment
TM_PEERS	Average number o peers on audit engagement teams
GENDER	Control variable, indicating gender feature of questionnaire respondents
FIRMSIZE	Control variable, indicating small, large and international Audit firms

4.3 Testing Methods

The testing methods used in this study include two parts. The following regression model is used to test the hypothesized relationships between job design aspects and job satisfaction, and a structural equation model (SEM) based on Figure 2 is used to test whether training effect mediate the proven significant relationship between job design and job satisfaction.

$$SATISFACTION = \alpha + \beta_1 CL\_NO + \beta_2 CL\_CMPLX + \beta_3 WK\_TEDI + \beta_4 WK\_PROF + \beta_5 TM\_NO + \beta_1 TM\_PEERS + \beta_2 GENDER + \beta_3 FIRMSIZE + \varepsilon$$

5 Empirical Results and Analyses

5.1 Descriptive Statistics and Bivariate Correlation Analysis

Tables 3 reports high job satisfaction (mean=4.99) and training effect (mean=5.26). It also shows the statistical characteristics of other variables under investigation.

Table 3. Descriptive Statistics

Variable	N	Minimum	Maximum	Mean	St.Deviation

SATISFACTION	208	1.86	7.00	4.9918	1.00297
TRE	208	2.44	7.00	5.2569	.85275
CL_NO	206	.00	25.00	5.8641	4.70609
CL_CMPLX	201	.00	1.00	.4174	.34691
WK_TEDI	208	1.30	7.00	4.8644	1.07253
WK_PROF	208	1.00	7.00	3.8344	1.18299
TM_NO	205	1	25	5.80	4.416
TM_PEERS	204	2.0	48.0	7.030	5.0756
GENDER	208	0	1	.76	.425
FIRMSIZE	208	1.00	3.00	1.9087	.64928
Valid N (*Listwise*)	198				

Table 4 reports the Pearson bivariate correlations between the studied variables. *SATISFACTION* is significantly associated with all the other variables except *GENDER*, while *TRE* is very significantly ($p<0.01$) correlated to *WK_PROF* and *FIRMSIZE*. It can be preliminarily observed that job satisfaction is related to most job design aspects, and that only one job aspect *WK_PROF* is significantly correlated to job satisfaction.

Table 4. Pearson Bivariate Correlation Analysis

Variable	**1**	**2**	**3**	**4**	**5**	**6**	**7**	**8**	**9**	**10**
1SATISFACTION	1									
2TRE	.472***	1								
3CL_NO	-.238***	-.063	1							
4CL_CMPLX	.184***	.133*	-.331***	1						
5WK_TEDI	-.207***	.071	.237***	-.011	1					
6WK_PROF	.385***	.468***	.014	.014	.214***	1				
7TM_NO	-.244***	-.095	.805***	-.303***	.167**	-.034	1			
8TM_PEERS	.196***	.115	-.218***	.279***	-.218***	-.051	-.250***	1		
9GENDER	-.063	.097	.114	.017	.122*	.032	.109	-.051	1	
10FIRMSIZE	.497***	.263***	-.395***	.419***	-.164*	.240***	-.397***	.222***	-.043	1

Note: *** $p<0.01$ (2-tailed); ** $p<0.05$; * $p<0.10$.

5.2 Tests on the Correlations between Job Design and Job Satisfaction

Table 5 reports the regression results. The empirical results prove that tedious non-professional tasks are negatively correlated to job satisfaction ($p<0.01$), whereas correlation between judgmental professional tasks and job satisfaction is positive ($p<0.01$). No other variables significantly contribute to job satisfaction except for *FIRMSIZE*.

Table 5. Results of Regression on Job Satisfaction

Explanatory Variables	**Standardized Coefficient**		
	Beta	**t**	**Sig.**
(constant)		9.370	.000
CL_NO	-.020	-.189	.850
CL_CMPLX	-.011	-.170	.865
WK_TEDI	-.207	-3.309***	.001
WK_PROF	.318	5.127***	.000
TM_NO	-.003	-.025	.980
TM_PEERS	.084	1.358	.176
GENDER	-.018	-.305	.760
FIRMSIZE	.375	5.240***	.000

DOI: 10.1051/shsconf/20173406002

Model Summary &	**R**	**R^2**	**Adj.R^2**	**S.E.**	**F**	**Sig.**
	.616	.380	.353	.81306	14.455	.000

Note: *** p<0.01 (2-tailed); ** p<0.05; * p<0.10.

5.3 Tests on the Mediating Role of Training Effect at Work Placement

As it is indicated in Table 5 that among those explanatory variables, only *WK_TEDI* and *WK_PROF* have significant correlation with *SATISFACTION*, structural equation modeling is run on these two variables. Figure 3 describes the SEM results of the conceptual modelling on the potential mediating role of training effect. A significant (p<0.01) partial mediating function of *TRE* exists in the relationship between *WK_PROF* and *SATISFACTION*, not in that between *WK_TEDI* and *SATISFACTION*. These results lead to the conclusion that training effect plays a significant mediating role only in the relationship between judgmental professional tasks and job satisfaction, not in that between tedious non-professional tasks and job satisfaction.

Note further that the mediating function of *TRE* is partial, which means that the judgmental professional job aspect significantly increases job satisfaction and good on-the-job training effect in work placement is enhancing this increase.

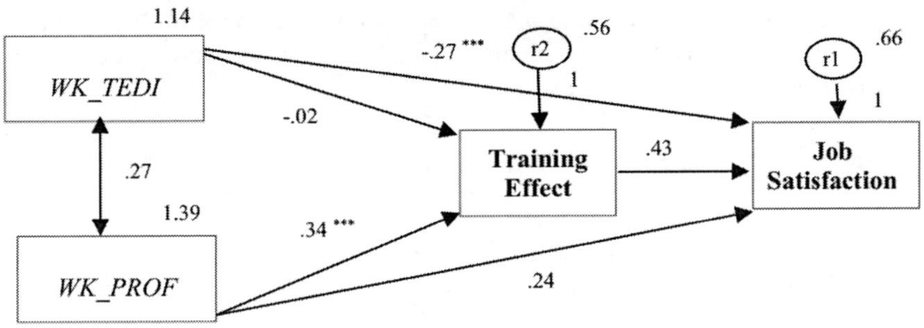

Figure 3 SEM Results regarding Mediating Role of Training Effect

6 Conclusions and Implications

The conclusions of this study include two parts. First, tedious non-professional tasks in work placement decrease job satisfaction, whereas judgmental professional tasks increase job satisfaction. Second, the partial mediating role of training effect significantly enhances the relationship between judgmental professional tasks and job satisfaction, in other words, if more judgment professional tasks were given, they could improve the training effect of work placemen, and eventually better job satisfaction.

My conclusions are quite implicative to accounting schools and audit firms, suggesting that students trained in audit work placement be given more judgmental professional tasks, which would lead to improvement in training effect and job satisfaction.

References

1. C.P., D'Abate, M.A., Youndt, K.E.Wenzel, Making the most of an internship: An empirical study of internship satisfaction. Academy of Management: Learning and Education, **8**, 4 (2009).

2. C.P., Maertz Jr., P.A., Stoeberl, J., Marks, Building successful internships: Lessons from the research for interns, schools, and employers. Career Development International, **19,** 1 (2014).
3. G., Liu, A survey on student satisfaction with cooperative accounting education based on Audit firm internships. Asian Review of Accounting, **20,** 3 (2012).
4. S., Mills, Job design. Personal Review, **2,** 2 (1973).
5. J.R., Hackman, G.R., Oldham, Motivation through the design of work: Test of a theory. Organizational Behavior and Human Performance, **16,** (1976).
6. Y., Fried, G. R., Ferris, The validity of the job characteristics model: A review and meta-analysis. Personnel Psychology, **40,** (1987).
7. J., Sutherland, Employment status and job satisfaction. Evidence-based HRM: a Global Forum for Empirical Scholarship, **1,** 2 (2013).
8. E., Verhofstad, H., De Witte, E. Omey, Higher educated workers: Better job but less satisfied? International Journal of Manpower, **28,** 2 (2007).
9. S.E., Humphrey, J.D., Nahrgang, F.P. Morgeson, Integrating motivational, social, and contextual work design features: A meta-analytic summary and theoretical extension of the work design literature. Journal of Applied Psychology, **92,** 5 (2007).
10. E. L., Deci, R. M., Ryan, The "what" and "why" of goal pursuits: Human needs and the self-determination of behavior. Psychological Inquiry, **11,** (2000).
11. J.K., Howard, A.N., Raymond, C. Wang, Motivation to learn and course outcomes: The impact of delivery mode, learning goal orientation, and perceived barriers and enablers, Personnel Psychology, **59,** 3 (2006).

Diploma students' perspective: Pursue or not to pursue a degree in accountancy?

Nor'aini Awab[1,*], and *A.H.* Fatima[2]

[1]MARA Professional College
[2]International Islamic University Malaysia

Abstract. This is an exploratory study which investigates the reasons influencing Diploma in Accountancy (DIA) students' intention to pursue or not pursue a Bachelor of Accountancy (BAcc) degree programme. The Theory of Reasoned Action (TRA) is used as a basis for the study. The research involves collecting primary data through survey questionnaires. 250 questionnaires were distributed to the final year DIA students in MARA Professional College. Based on the findings, the reasons that influence students' intention to pursue are: the hope to earn a good salary, higher future earnings potential and greater employability in the future. In contrast, the reasons for non-pursuance of a BAcc programme are because students think that they can develop their potential for personal growth and career development with other degree programmes or they want to get jobs and start working after obtaining the DIA. These findings would be useful to the Ministry of Education, the universities as well as the professional bodies to ensure that the students continue their studies in the accounting discipline to achieve their full potential.

1 Introduction

In Malaysia, a diploma programme is one of the many entry routes to a degree programme. This route is an additional opportunity for, mainly Bumiputera, students who could not enter into a degree programme due to academic or financial reasons. Similarly, students who complete their Diploma in Accountancy (DIA) programmes can opt to pursue the Bachelor of Accountancy (BAcc) programme. MARA Professional College or Kolej Professional MARA (KPM) is one of the Malaysian institutions that offer the DIA programme. KPM is managed by the *Majlis Amanah Rakyat* or Public Trust Council (MARA). One of the main objectives of KPM is to provide technical and skill based competence to Bumiputera students, who fail to meet entry requirements into the matriculation programme at any university [1]. These students, who are normally from rural

* Corresponding author: norainiawab08@gmail.com

and financially disadvantaged backgrounds are given an alternative route to better employment opportunities or higher education through KPM.

Unfortunately, the report from KPM to the Ministry of Education (MOE) on 2005 – 2012 statistics revealed that only about 19% of the DIA graduates pursued their studies to a bachelor degree. Out of this percentage, some of the DIA graduates chose to continue their studies in other fields, such as Bachelor in Management, Business Studies, Marketing and Finance. Thus, a very small percentage pursued the BAcc programme.

Since a degree in BAcc is normally a stepping stone towards attaining a professional accounting qualification, particularly for Bumiputera students, the concern is that these students will fail to proceed towards a career as a professional accountant. This problem is considered dire, as only 8 percent of professional accountants registered by the Malaysian Institute of Accountants (MIA) are Bumiputera. Therefore, the Malaysian government and the MIA are encouraging more Bumiputera accountants. However, their plans may be hampered if a significant percentage of these Bumiputera students from the DIA programme do not follow through by pursuing a BAcc degree nor a professional qualification later.

Therefore, it is vital for this study to investigate and explore the possible reasons that could lead these students to either pursue or hinder them from pursuing the BAcc programme. This study is essential as it provides needed baseline information to institutions that offer the DIA programme, particularly KPM. The lecturers from the DIA programmes could highlight the facts that are in line with reasons for DIA students to pursue the BAcc programme. At the same time, DIA programme lecturers could pacify concerns that the students may have to enable them to pursue a BAcc programme.

Moreover, this study is timely because accounting departments in higher learning institutions can use these findings in their strategies to increase students' enrolment. Based on the reasons provided by the students, the MOE and universities could then promote the factors that would enable these students to pursue a BAcc degree and alleviate the factors that would deter them from doing so. Furthermore, DIA students from KPM tend to come from less financially advantaged family backgrounds, thus this may also affect their pursuance of a BAcc programme. Hence, this study extends the literature, which mainly focuses on factors that influence the decisions of secondary school students to further their studies in accounting at the tertiary level, whereas others focus on factors that influence accounting graduates from pursuing a professional qualification.

The next section reviews the relevant literature, followed by a discussion on the theory and research method. Subsequently, the findings are discussed prior to concluding the study.

2 Literature review

Several factors influence decision making [2], thus understanding these factors is important to comprehend the outcome. Prior studies have investigated intrinsic and extrinsic factors in various contexts [3], however the review of the literature will focus on Malaysian students' decisions to major in accounting.

Factors influencing students' intention to enrol in BAcc degree were investigated on 173 students from public and private higher learning institutions in Penang and Kedah [4]. The results logically showed that students intend to enrol in the BAcc when they have a positive attitude towards it. On the other hand, a survey was conducted on the first batch of 50 accounting students in Universiti Malaysia Sarawak, to identify the factors influencing the decision of these students to accept the offer of accounting as a major [5]. The study revealed that job prospect has a significant influence over the students' decision to choose accounting. Similarly, a survey was carried out on 400 students from the International

Islamic University Malaysia and Tunku Abdul Rahman College [6]. The study was conducted on the influential factors of students' intention to enrol in the accounting programme. The factors were divided into three categories: attitude towards accounting (ATT), subjective norm (SN) and amount of information on accounting (AIA). Their findings showed that ATT, SN and AIA significantly influenced the male students' intention to enrol in the accounting programme; while the female students were mostly influenced by ATT and SN.

In examining the Bumiputera DIA students in Malaysia, 65 questionnaires were distributed to selected respondents in final year DIA programme [7]. It was found that the decline in the number of students enrolling in the BAcc programme was due to the decline in the number of students achieving the required CGPA to be eligible to enrol in the programme. The findings also reveal that the university, lecturers and students all have significant roles to play in helping the students to obtain the desired result. Another study conducted a survey on 66 final year DIA students in UiTM Kedah [8]. However, this study focused on the factors influencing the students' decision to pursue professional accounting qualification. The two main influencing variables were CGPA and the availability of a study grant.

The review of the literature indicates that in the Malaysian context, the studies tend to investigate accounting students' decision to select accounting as a major subsequent to their joining the BAcc programme. Thus, the sample may be positively biased. However, DIA students come from a different route, hence may have other motivations to pursue or not to pursue a BAcc programme. The literature review has shown that there are limited studies on DIA students, particularly in relation to their pursuing a BAcc programme.

3 Theory and research method

This study refers to the Theory of Reasoned Action (TRA), which was theorised by Martien Fishbein and Icek Ajzen [9, 10]. TRA suggests that a person's behaviour is determined by his or her intention to perform the behaviour and that this intention is, in turn, a function of his or her attitude towards the behaviour and subjective norms. In this study, a DIA student's intention to pursue a BAcc programme is predicted by his or her attitude towards the programme, and the student's perception towards it may be influenced by several factors. However, since DIA students are a largely unstudied population, in this context this study is deemed exploratory in nature. Since the aim of the study is to explore and discover the factors that influence DIA students' decision, the TRA is only referred to as a basis of study, thus the research has not specifically categorised the influencing factors into attitude nor subjective norms. Furthermore, being exploratory, the study was unable to draw a hypothesis.

The questionnaire survey method is used to collect this study's data. The questionnaire was adapted from prior literature [8, 11, 12, 13]. Subsequent to the demographic information in Section A of the questionnaire, the study enquires on the respondent's intention to pursue a BAcc programme in Section B. If the respondent's answer is a `Yes' in this section, the respondent will proceed to answer Section C. In contrast, if it is a 'No', the respondent will answer Section D. There are 18 items each in Sections C and D to measure the reasons to pursue or not to pursue a BAcc programme, respectively. Both these sections use a five-point Likert scale ranging from 1 = 'No Influence' to 5 = 'Very Strong Influence' and 1 = 'Strongly disagree' to 5 = 'Strongly agree', respectively.

After conducting the pilot test, the questionnaires were distributed to a total of 250 final year DIA students in four KPMs: KPM Indera Mahkota, Kuantan, KPM Beranang, Selangor, KPM Ayer Molek, Melaka and KPM Bandar Penawar, Kota Tinggi, Johor. Final year students were the target respondents as they were contemplating their future following

 DOI: 10.1051/shsconf/20173406003

graduation. The response rate was 82%. Then, the data was analysed primarily using 'means' to determine the extent of influence of each factor.

4 Results and discussion of findings

Reflecting the general demographics of the population of DIA students, 73% of the respondents were female students. Out of the 205 students who responded to the questionnaire, 153 students (75%) intend to pursue a BAcc programme. The remainder (25%) do not intend to pursue a BAcc programme. From the 52 students who do not want to pursue a BAcc programme, 19.2% and 55.8% intends to pursue a professional qualification or other degree programmes, respectively. Means have been used to measure the reasons to pursue and not to pursue a BAcc degree. The standard deviations for these reasons are minimal and ranges between 0.363 and 0.639. Having acceptable standard deviation is logical as this data is primary data gathered through questionnaires. Furthermore, the Likert scale ranges from between 1 and 5, hence there are no extreme values.

The means in Table 1 have been ranked from the highest to lowest to indicate the most to least influential reasons. Analysing Table 1, the main factor to motivate students to pursue a BAcc programme is a good salary as shown by the first 3 items. The potential for personal and career development is the next reason, followed by genuine interest in figures. It is interesting that the students consider their interests and self development first before employability opportunities, which is ranked next. Then, the students consider contribution to the society and prestige as a BAcc degree holder. The importance of income potential and job prospects is consistent with prior literature [7, 14].

Table 1. Reasons to pursue BAcc programme

Statement	Mean
I hope to get a good salary after obtaining a BAcc degree	4.177
The salary offered to BAcc graduates is higher compared to other graduates	4.078
There is future high earnings potential with a BAcc degree	3.961
There is potential for personal growth and career development with a BAcc	3.935
I genuinely am interested in numbers and calculations	3.895
There are greater employability choices with a BAcc degree	3.83
I will be able to positively contribute to the society if I have a BAcc degree	3.771
It is prestigious to have a BAcc degree (status highly valued by the public)	3.732
I am quite confident that I will be able to attain financial assistance or study grant to pursue my studies	3.719
I have to satisfy my family's hopes and dreams	3.68
Obtaining a BAcc ensures good physical working environment and condition	3.66
Obtaining a BAcc enhances decision making ability	3.654
I am interested in course or subjects taught in the programme	3.536
There is job security for a BAcc holder	3.569
I am influenced by my teachers / lecturers / instructors / counsellors	3.484
Obtaining a BAcc enables the attainments of power and authority in an organization	3.458
I am influenced by my friends	3.007
I may not be able to get a job after the Diploma	2.928

DOI: 10.1051/shsconf/20173406003

Other factors that are taken into consideration in deciding to pursue a BAcc programme is the availability of scholarships. According to the results, family plays a more influential role than lecturers and friends. The students are also relatively confident about getting a job upon graduating from DIA.

In contrast, Table 2 lists the reasons not to pursue a BAcc programme. The top five items that influences students not to pursue is basically related to two factors: (1) They can still attain desired goals (i.e. personal and career development, enhance decision making ability and contribute to society) with other degree programmes. (2) They need to get a job, which they are quite certain they will get, after their DIA to help their families financially. The two items (items 2 and 3), which make up the second factor have identical means. The second factor is logical as these DIA students generally come from rural areas and relatively financially disadvantaged families.

Table 2. Reasons not to pursue BAcc progamme

Statement	Mean
I can develop my potential for personal growth and career development with other degree programmes	3.365
I think that I will be able to get a job after the Diploma	3.346
I have to start working with a Diploma to help and satisfy family's financial needs	3.346
I will be able to positively contribute to the society if I have other degrees.	3.25
I can still enhance my decision making ability even without a BAcc degree	3.115
Obtaining other degree programmes enables the attainments of power and authority in an organization	3.039
I can obtain future high earnings potential even without a BAcc degree	3.039
There is greater employability with other degree programmes.	2.942
The salary for a Diploma holder adequate and I am satisfied with it.	2.846
I am not interested in course or subjects taught in the programme	2.808
Obtaining a BAcc degree doesn't ensure good working conditions.	2.673
My friends told me not to pursue BAcc programme as it is difficult to pass	2.596
There is no difference in salary of BAcc graduates compared to other graduates	2.558
I don't think I will be able to get financial assistance or study grant to pursue my studies	2.558
There is no job security even with a BAcc degree in the current competitive job market	2.442
There is nothing prestigious about a BAcc degree holder.	2.423
My teachers/lecturers/instructors/counsellors advised me NOT to pursue a BAcc degree	2.346
I dislike numbers and calculations	2.154

The means close to '3' indicate that the students were 'neutral' in relation to other items. As for the last four items, although these students do not intend to pursue a BAcc degree, they still consider that a BAcc degree holder has job security and prestige. This is because means close to '2' is a 'disagree' to the statements. These students also disagree that their lecturers advised them not to pursue a BAcc and that they disliked numbers.

5 Conclusion

In analysing the main reasons for both pursuing and not pursuing a BAcc programme, the students seem to be fixated on earning a salary. This is reasonable if they need to help their families' financial circumstances. Having to earn a salary is also a deterrent to pursuing a BAcc programme as these students need to start working immediately after graduating from DIA and help their families financially. Therefore, in addition to the financial assistance and scholarships given by the government and various institutions, the universities that offer the BAcc programme may need to become more innovative and offer part-time BAcc programmes that cater to full-time working students.

The limitation of this study is that it only focused on DIA students from KPM, thus future studies could expand the sample to other diploma programmes and higher learning institutions. Furthermore, the findings of this study have to be confirmed through focus group discussion or interviews with the students. Nevertheless, the findings are useful to the MOE, KPM and universities to ensure that more Bumiputera students pursue a BAcc degree in their career path to becoming a professional accountant.

References

1. Majlis Amanah Rakyat. *Higher Education Division*. http://www.mara.gov.my/en/bah.-pendidikan-tinggi (2015)
2. C. Dietrich. *Student Pulse* **2**, 1 (2010).
3. A. Djatej, Y. Chen, S. Eriksen, D. Zhou. Global Perspectives Acc. Ed. **12**, 53 (2015)
4. B. Zainin. *University Science Malaysia Doctoral Thesis* (2001)
5. S. Ali, M. IUP J Accounting Research & Audit Practices **12**, 25 (2013).
6. G. Zandi, M. Sulaiman, B. Naysary. irep.iium.edu.my/30401/1/The_Academic_gender.pdf (2013)
7. S. Mohamad, F. Bujang. J of Fin Reporting & Accounting **5**, 41 (2007)
8. N. Omar. *Master Dissertation Universiti Utara Malaysia* (2009)
9. M. Fishbein, I. Ajzen. *Belief, attitude, intention & behaviour: An introduction to theory & research* (1975)
10. I. Ajzen, M. Fishbein, M. *Understanding attitudes & predicting social behaviour* (1980)
11. S. Felton, N. Buhr, M. Northey. Issues in Acc. Ed. **9**, 13 (1994)
12. F. Abdul Halim, B.W. Yap, M.A. Lazim. *Conference on Scientific & Social Research 2005 Proceedings* (2005)
13. M.R. Ramlan, R. Ishak, M.'A. Md Yusof. *Universiti Utara Malaysia Project Report* (2003)
14. F.A. Gul, B.H. Andrew, B.H., S.C. Leong, Z. Ismail. Acc & Fin **29**, 93 (1989)

Perception of Diploma in Accountancy students on ethics education

Nor Sharizad Zainal Abidin[1,*], and *A.H. Fatima*[2]

[1]Kolej Profesional Mara Indera Mahkota
[2]International Islamic University Malaysia

Abstract. The aim of this study is to determine the perception of Diploma in Accountancy (DIA) students on whether ethics should be integrated into the DIA curriculum. This study also investigates the reasons for their opinion. In order to meet these objectives, 100 questionnaires were distributed to final year DIA students of Kolej Professional MARA (KPM). Based on the findings from 96 respondents, two-thirds majority were in support of ethics education in the DIA programme. These students, who agree to ethics integration, were primarily of the opinion that ethics was crucial to the accounting profession, and if ethics were not taught, ethical scandals involving the profession would persist. On the other hand, the one-third minority, who disagree that ethics should be integrated, were mainly of the opinion that there were insufficient materials to teach ethics, accounting lecturers were unwilling or lacked qualifications to teach ethics, and ethical training could not take place in a classroom setting. These concerns of the DIA students should be taken into consideration by the Ministry of Education (MOE) and KPM, should they want to be in line with *Halatuju* 3.

1 Introduction

The importance of professional ethics has been spurred by global business scandals. Ethics in the accounting profession is vital, hence the International Federation of Accountants (IFAC) has stressed the integration of ethics into the accounting curricula [1]. The Reassessment Report on Accounting Programmes (*Halatuju* 3), which was issued by the Ministry of Higher Education (now Ministry of Education, MOE) and the Malaysian Institute of Accountants (MIA), in collaboration with the universities in Malaysia, reemphasises the requirement to integrate ethics into the accounting curriculum [2]. In Malaysia, accredited accounting programmes are required to be in line with *Halatuju* 3. This reassessment report is only for accounting programmes, whereas some other professional programmes have their own reassessment reports. Since *Halatuju* 3 affirms the importance of ethics, it is necessary for all accounting programmes in Malaysia to integrate ethics into their curriculum.

Kolej Profesional Mara (KPM) offers Diploma in Accountancy (DIA) programme. The students that graduate from the DIA programme can pursue a degree in Bachelor of Accountancy or sit for professional examinations en route to becoming a professional accountant. Graduates from the DIA programme can also start working in accounting related jobs. Thus, it is equally important to integrate ethics into the DIA programme, in line with the aspirations of IFAC and *Halatuju* 3.

Currently, the DIA programme has a standalone ethics course, i.e. Business ethics, to impart ethics education to DIA students. This course mainly provides the fundamentals and principles of ethics in general and does not necessarily put ethics into context of practical

* Corresponding author: author@email.org

accounting. Previously, lessons on ethics were also integrated into the Accounting Theory and Practice (ATAP) course. The integration of ethics into courses is essential [3], particularly a course on accounting practice, to ensure that the students do not compartmentalise ethics separately from accounting.

Regardless of ATAP being the only course in the DIA curriculum that formally integrates ethics into accounting subject matter, in July 2015 the ATAP course was replaced by Taxation 2. As a result of this change to the curriculum, the only course that formally integrates ethics into the programme is Business ethics, as mentioned above. Consequently, the integration of ethics into the standard DIA curriculum seems to be diminishing. Moreover, by only having a standalone course, without formal ethics integration into accounting subject matter, there is a concern that students may consider ethics to be idealistic and fail to establish the connection between ethics and accounting. Failing to integrate ethics into accounting would jeopardise ethics education in the DIA programme, which would be contrary to the aims of *Halatuju* 3.

As the DIA students are affected by this change in curriculum, it is necessary to get their viewpoint on ethics education. Therefore, the aim of this study is to determine whether in the opinion of DIA students, there is a need for ethics education in the DIA programme and the reasons for their opinion. Although studies have been conducted on ethics education [4], none have considered the viewpoint of students at the diploma level. This study would thus be beneficial to KPM to attain the opinions of their key stakeholders, i.e. the students. Moreover, the findings may also be useful to KPM and MOE for them to consider in their evaluation of the new curriculum.

The next section reviews the literature. Then, the discussion proceeds with the theory generally referred to in ethics studies as well as the research method. The findings are analysed prior to concluding the study.

2 Literature Review

There are two main areas in ethics education literature: (1) ethics education on moral reasoning, ethical awareness or ethical sensitivity [5, 6, 7], and (2) perception on ethics education. Generally, the first group of studies find that ethical reasoning, awareness and sensitivity are enhanced subsequent to certain forms of ethics education. In the context of Malaysia, the ethical attitude of 210 final year accounting students from a public university was tested [8]. The study investigated whether the students were willing to take a bribe or defraud shareholders (under two circumstances: will not get caught and 10% chance of getting caught) and their attitude towards whistle blowing. The study concluded that it is crucial for universities to inculcate ethics, in line with *Halatuju* 3.

In relation to studies on the perception of ethics education, these studies tend to focus on the educators' perspective [3, 4, 9, 10]. These studies are carried out in the United States of America (US) [3], Japan [10] and Malaysia [4, 9]. The educators generally perceive that ethics should be integrated into the accounting curricula [3, 4, 9, 10]. Ethics is embedded in relevant courses [9] or both, taught as a standalone course as well as integrated into various courses [4, 10]. The accounting educators deem that they are most appropriate to teach ethics to accounting students [3].

The review of the literature above has highlighted that the perception of DIA students on ethics education is lacking and this gap needs to be filled.

DOI: 10.1051/shsconf/20173406004

3 Theory and Research Method

Kohlberg theorised the "Cognitive Moral Development" of an individual, which was later developed into a four-component moral process model by Rest [11]. Kohlberg's and Rest's moral development processes were further extended into Thorne's Integrated Model (TIM). The model prescribes gradual development of ethical behaviour through ethics education. Thus, even though TIM is not a theory, it is referred to by several prior studies [12, 13, 14] in relation to ethics, as well as in support of ethics education.

Based on the above, although TIM provides support for ethics education in general, it does not explain the possible perception of students on ethics education. Therefore, the current study is exploratory in nature as it aims to determine the perception of DIA students on ethics education. Considering that it would be difficult to predict the perception of these students, hence a hypothesis cannot be framed in this study.

This study collected primary data from DIA students using survey approach. The questionnaire was adopted from prior Malaysian literature [4] that adapted the instrument from an earlier US study [3]. The questionnaire comprise four sections: Section A gathers the demographic details of the respondents. Section B consists of a filtering question. The respondents were required to answer "Yes" or "No" to this question. If the respondents perceived that ethics should indeed be integrated into the DIA programme, they were required to answer Section C, if not, they were requested to proceed to Section D.

Section C solicited DIA students' reasons for ethics integration. On the other hand, Section D enquired on students' reasons for disagreeing to ethics being integrated into the programme. Both Sections C and D used a five-point Likert scale ranging from (1) for 'strongly disagree' to (5) for 'strongly agree' to elicit the responses of students.

Subsequent to performing a pilot test, 100 copies of the questionnaire were distributed to final year DIA students from four KPMs: KPM Beranang, Selangor; KPM Air Molek, Melaka; KPM Bandar Penawar, Johor Bharu and KPM Indera Mahkota, Kuantan. The final year students were selected as the sample of study as they had undergone the DIA programme and were better able to provide an opinion on whether ethics education would be warranted in the programme. Since the questionnaires were distributed through the Heads of department, it enabled a response rate of 96%.

In addition to the descriptive statistics to analyse the demographics of the respondents, Cronbach's alpha was used to test the reliability of the questionnaire instrument. Then, means were used to measure the responses on reasons to have or not to have ethics education in the DIA programme.

4 Results and Discussion of Findings

Based on the Cronbach's alpha, the questions on reasons for having ethics education and not having ethics education were 0.809 and 0.804, respectively. Cronbach's alpha of above 0.8 is considered to be high in reliability [15]. Out of the 96 respondents, 59.4% were female students. Then, in answering Section B, 66.7% of the respondents were of the opinion that ethics should indeed be integrated into the DIA programme. Surprisingly, a third of the respondents were of the opinion that ethics should not be integrated into the DIA programme. The percentage of dissenting respondents is high and contrary to prior literature [4]. Their opinion also contradicts the recommendation by IFAC and *Halatuju* 3.

Table 1 below lists the reasons, based on DIA students' perception, for integrating ethics into the DIA programme. The list is no longer in the order as it is in the questionnaire, as it has been ranked in accordance to the highest to lowest mean.

Table 1. Reasons in support of integrating ethics in the DIA programme

Statements	Means
Failure to teach ethics in the accounting curriculum contributed to recent business scandals	4.25
It is important for accountants to be ethical	3.74
Ethical development is crucial to the accounting profession since it is required by MIA	3.74
Students' ethical awareness has declined over the years	3.66
MIA may lose its authority to regulate the profession if ethical lapse continues	3.59
There seems to be a decline of ethical awareness amongst accounting students in Malaysia	3.57
To prepare students to make ethically sensitive decisions instead of just focusing on figures	3.34
Students do not seem to be getting adequate ethical training	3.33
Public trust needs to be re-established due to recent accounting scandals	3.13

The means above indicate that the DIA students are concerned that the failure to teach ethics would contribute to business scandals. These students perceive, and rightly so, that being ethical and ethical development is crucial to the accounting profession. The students generally consent that students' ethical awareness has declined over the years, but their agreement is marginally less on accounting students' ethical awareness has also declined. Finally, the DIA students are concerned that MIA may lose their authority if ethical scandals involving the accounting profession persist.

The DIA students tend to be more 'neutral' about the last three statements as their means are close to '3'. These findings could be interpreted as the students being relatively indifferent about enabling students to make ethically sensitive decisions. In general, they also seem undecided on whether there is adequate ethics training and on regaining public trust subsequent to ethical scandals.

In contrast, Table 2 below lists the reasons of the DIA students who are of the opinion that ethics education should not be included in the DIA programme.

Table 2. Reasons in support of not integrating ethics into the DIA programme

Statements	Means
There is inadequate materials for teaching ethics	4.50
Normally, accounting lecturers are not willing to teach ethics	4.40
Ethics cannot be taught	4.30
Accounting lecturers may not be qualified to teach ethics	4.00
Students cannot adequately be trained in classrooms to deal with "real world" ethical dilemmas	3.80
Responsibility of encouraging ethical values should be left to parents and family members	3.80
Teaching ethics in the Diploma Accounting curriculum is only a strategy to please the public as a result of recent accounting scandals	3.79
Diploma Accounting curriculum is not an appropriate place to teach	3.70

ethics	
There is little or no time to teach ethics in the current Diploma Accounting curriculum	3.61

The students that perceive that ethics should not be integrated into the DIA accounting programme generally were of the opinion that there were inadequate materials for teaching ethics. They also perceived that accounting lecturers are not willing to teach ethics or were not qualified to do so. This is the perspective of the DIA students, thus future research should also consider seeking the perspective of the DIA lecturers to determine if they have the same opinion.

These students also perceived that ethics cannot be taught, or at least facing 'real world' ethical dilemmas cannot be taught in classrooms. These DIA students were also of the opinion that some responsibility of imparting ethical values to the students should be borne by the parents and family members.

These DIA students deem that teaching ethics in the DIA programme is merely a ploy to appease the public's reaction to the business scandals. They consider that the DIA curriculum is not the appropriate place to integrate ethics, possibly because there is little time to add lessons on ethics in the current DIA curriculum.

It is difficult to compare the findings of the perception of DIA students on ethics integration as prior literature tend to be on the perception of educators at the undergraduate level of study. Also, the findings from prior literature tend to be in support of ethics education [4].

5 Conclusion

In analysing the results above, the DIA students, who support the integration of ethics education, realise the importance of ethics to the accountancy profession. Their perception is in line with those of IFAC and *Halatuju* 3. However, the findings on the dissenting students' reasons are more imperative. This is because, although a minority, a third of the respondents is still a significant proportion of the sample, and considerably higher to prior studies [4]. Therefore, should the MOE and KPM want the DIA curriculum to be in line with IFAC and *Halatuju* 3 guidelines, then the concerns of these students should be addressed.

Firstly, the MOE could consider giving research grants to academics and researchers to develop ethical cases in the context of Malaysia to alleviate the concern on lack of teaching materials in relation to ethics. Moreover, KPM should collaborate with MOE to train all teachers and lecturers at school and tertiary institutions on approaches to teaching ethics, imparting ethical values and becoming role models for students. Furthermore, the perception that accounting lecturers are unwilling to teach ethics should be studied further, as it could be a misconception of students. This is because prior literature suggests that the accounting educators perceive that they are the most appropriate to teach ethics [3].

Then, the issue that ethics cannot be taught has to be alleviated. The concept and models developed by Kohlberg, Rest and Thorne propose that ethical development is in stages, hence should take place as early in an individual's life as possible. Therefore, it may not be appropriate to leave ethics education to be integrated only at the tertiary level in the Bachelor of Accountancy (BAcc) programme. Thus, the KPM may have to consider integrating ethics into accounting courses as the ATAP course has been removed from the DIA curriculum. A standalone course on Business Ethics may be insufficient.

Certain limitations of the study must be noted. Firstly, the study was only conducted on a sample of DIA students. Hence, future studies can be extended to also consider other diploma students as well as the educators' perspective on ethics education. Future studies could also consider having focus group discussions and interviews with students as well as educators to confirm the findings of this study.

References

1. C. O'Leary, S. Mohamad. Mal. Acc. Rev. **7**, 1 (2008)
2. *Halatuju 3*, Ministry of Higher Education (2015)
3. C. Blanthorne, S.E. Kovar, D.G. Fisher. Iss. Acc. Ed. **22**, 355 (2007)
4. Y. Y. Win, S. Ismail, A.H. Fatima. Mal. Acc. Rev. **13**, 1 (2014)
5. M. J. Abdolmohammadi. R. Prof. Res. & Ethics Acc. **10**, 37 (2005)
6. M. P. Coyne, D.W. Massey, J.C. Thibodeau. Adv. Acc. Ed. **7**, 171 (2005)
7. M. J. Billiot, D. Daniel, S. Glandon, T. Glandon. Am. J. Bus. Ed. **5**, 277 (2012)
8. N. Abu Bakar, S. Ismail, S. Mamat. J. Fin. Rep. & Acc. **6**, 21 (2008)
9. J. Singh, M. Poduval. *Proceedings 2nd Conf. Teaching & Learning* (2009)
10. S. Sugahara, G. Boland. R. Prof. Res. & Ethics Acc. **15**, 193 (2011)
11. J.R. Rest. *A psychologist look at the teaching of ethics* (Hastings Center Report 1982)
12. M. B. Armstrong, J.E. Ketz, D. Owsen. (2003). J. Acc. Ed. **21**, 1 (2003)
13. C. Y. Loh, J.B. Wong. Aus. Acc. Bus. & Fin. J. **3**, 3 (2009)
14. N. L. Ahmad. Mal. J. Society & Space, **11** (2015)
15. U. Sekaran, R. Bougie. *Research methods for business* (John Wiley & Sons, UK 2013)

DOI: 10.1051/shsconf/20173406005

Positive and Negative Perceptions of Bumiputra And Non-Bumiputra Students on Professional Qualification

Noor Asidah Abd. Rashid[1,*], and *A.H. Fatima*[2]

[1]Sekolah Menengah Kebangsaan Jengka Pusat
[2]International Islamic University Malaysia

Abstract. Bumiputra and non-Bumiputra students may come from various economic backgrounds and culture. This may influence their perception on their career choice of pursuing a professional accounting qualification. Thus, this study investigates the difference in positive and negative perceptions of Bumiputra and non-Bumiputra students on pursuing a professional qualification upon graduation. A questionnaire survey method was used to collect the data from final year accounting students from five public and three private universities in Malaysia. Means and independent sample t-tests results were analysed. Results indicated that there are only a few significant differences between Bumiputra and non-Bumiputra students on positive and negative perceptions on becoming professional accountants. As perception frames action, these findings would be useful to the Malaysian Institute of Accountants as well as professional bodies to attract both Bumiputra and non-Bumiputra graduates to become professional accountants.

1 Introduction

Globalisation, advances in information technology, the multidisciplinary economic environment and new business relationships have changed the accounting profession. Due to these changes, the accounting profession requires a new breed of professional accountants with multi-skills, competence and integrity. In Malaysia, the title "professional accountant" is awarded to a member of the Malaysian Institute of Accountants' (MIA), having acquired education from a recognized programme and having attained the required training [1].

The need for professional accountants has been recognized and declared by the International Federation of Accountants [2]. This need is echoed by Malaysia; Malaysia has announced that it would require about 60,000 accountants by the year 2020 [3]. In Malaysia, graduates from accredited bachelor of accounting programmes are the primary source, contributing to 59 per cent of newly registered accountants in 2013 [4]. However, this approximates to only 18 percent of accounting graduates each year. Thus, there is a dire need in Malaysia to attract accounting graduates to get professionally qualified.

* Corresponding author: puan_noora@yahoo.com

However, 'one-size-fits-all' approach to attracting graduates may not be conducive in the Malaysian environment as Malaysia is a multi-racial country.

Since the intention to pursue a professional qualification would be framed by positive and negative perceptions towards it, this study aims to investigate these perceptions from both Bumiputra and non-Bumiputra students. This is because based on Malaysia's census in 2010, about 60 percent of the population are Bumiputras (Malays and aboriginal tribes), thus the remaining 40 percent are non-Bumiputras (Chinese, Indians and others). This study also compares the difference in perceptions between the Bumiputra and non-Bumiputra groups on pursuing a professional qualification. The findings of this study would be useful to professional bodies to attract graduates from both Bumiputra and non-Bumiputra groups by promoting graduates' positive perceptions and appeasing their negative perceptions on professional qualifications. The MIA would also benefit by having more members.

Subsequent to the introduction, the next section briefly reviews the relevant literature to identify the gap. Then, theory, hypotheses and research method are discussed prior to the discussion of results and findings. The study concludes by contemplating the implications, limitations and suggestions for future research.

2 Literature Review

Students' perception can be expressed as students' attitude [5] or beliefs [6]. This study accepts the latter definition. Understanding students' perceptions of accountancy profession is an important step in identifying the students' intention in pursuing the professional qualification as students' career choice is highly influenced by students' belief that was formed through perception [7, 8]. Literature on perception in relation to accounting may be classified into perception on accounting as a profession [9, 10], which would affect perception of students in taking accounting as a major [11, 12] as well as perception of students to become professionally qualified. The review of the literature will focus on relevant studies related to perception of students on becoming professional accountants, particularly in the Malaysian context.

In Malaysia, a prior study found that job security and stability, opportunities for advancement and students' perception towards the profession were the main variables that influenced students' decision to sit for professional examinations [13]. Nevertheless, the learning environment plays a role of moderating the attitude of students, and it significantly moderates the attitude-intention relationship [14]. However, non-commitment by accounting students could still exist although undergraduate accounting programmes are designed to prepare students for a career in the accounting profession [15]. Hence, in addition to the learning environment, environmental factor (such as expected income), people related factor (such as the accountants work that apply what they have learnt during college) and behaviour capability (such as type of academic institution and self-motivation) have significant effects on accounting undergraduates' preference to practice as Chartered Accountants [16].

Literature also shows that there are limited studies that compare various ethnicities or racial groups in Malaysia, although there are several in other countries [17, 18, 19]. Nevertheless, there are studies that compare the differences between the perception of Malay and Chinese accounting students [20], Malaysian students and international students [21], and Malaysian Chinese accounting students and English students [22]. In general, the findings revealed that there were differences in perception by students of different racial and ethnic background.

The literature review clearly indicates that there is a gap in the study on the perceptions of Bumiputra and non-Bumiputra students. This study is warranted as the above literature shows that there are differences in perception between different ethnic or cultural groups.

3 Theory, Hypothesis and Research Method

The Theory of Reasoned Action (TRA) model emphasizes on the specific process by which an individual makes choices, and is applied in understanding students' career decisions [23]. The TRA model was developed in order to establish links between attitudes, intentions, norms, beliefs and behaviours of individuals [24]. Generally, attitudes and social norms form beliefs, which would lead to intention and later may result in performance of behaviour. However, in this study, only the initial stage of the TRA model is focused on, i.e. perceptions (positive and negative). This is because positive perception is likely to influence positive intention to pursue a professional accounting qualification, whereas negative perception is likely to deter the student from pursuing the qualification.

Although the theory cannot be used to specifically explain a difference in perception between Bumiputra and non-Bumiputra students, there is however evidence from prior literature [17, 20], which supports a difference between perceptions of different racial or ethnic groups. Thus, based on the above, the following hypothesis is framed:

H_1: There is a significant difference between perception of Bumiputra and non-Bumiputra students on pursuing professional qualification.

The study used a survey instrument adapted from prior literature [25, 26, 27] to gather the positive and negative perception of students on pursuing a professional qualification. The responses were measured using a five-point likert scale from 1 (strongly disagree) to 5 (strongly agree).

Prior to distributing the questionnaires, a pilot test was performed to enhance the validity and understandability of the questionnaire. Then, a total of 500 questionnaires were distributed to 5 public and 3 private universities in Malaysia. 352 usable responses (70.4%) were received. Of these, 66.5 percent were from Bumiputra and 32 percent from non-Bumiputra, which approximates the population of Malaysia.

Test of reliability was performed using Cronbach's alpha. Then, in order to test H_1, means were used to calculate the responses and independent sample t-tests were used to measure the differences.

4 Results and Discussion of Findings

Out of the questionnaires returned, 344 were useable, resulting in 68.8 percent of the initially distributed questionnaires. The Cronbach's alpha for positive and negative perceptions were 0.749 and 0.935, respectively. Therefore, the questionnaire items are considered acceptable in terms of reliability as Cronbach's alpha in excess of 0.7 is acceptable [28]. From the 344 useable questionnaires, 67 percent and 23 percent of the respondents were Bumiputra and non-Bumiputra, respectively. The majority of the respondents were female students (75 percent). As for Cumulative Grade Point Average (CGPA), 25 percent, 49 percent and 26 percent of the respondents had a CGPA of 3.5 and above, between 3 and 3.49, and below 3, respectively.

Table 1 provides the results on the differences in perception between Bumiputra and non-Bumiputra students. Based on the p-value of the Independent sample t-test, the results generally indicate that the perception of Bumiputra and non-Bumiputra are not significantly different. However, there are some exceptions on "gives you authority/power", "positively contribute to society" and "allows for independent autonomy". In all three cases, the average perception of the Bumiputra students was significantly higher, but only to a weak extent (significant at 10 percent significance level) than their non-Bumiputra counterpart. The results of the Independent sample t-test were confirmed by the Mann-Whitney test.

In analyzing the means, 'good salary', 'greater choice for employability' and 'prospects for promotion' are the top three benefits of becoming professionally qualified as perceived

DOI: 10.1051/shsconf/20173406005

by the latter group. Contrarily, 'allows independence', and 'having job security' is perceived to be the top benefits by the Bumiputra students. A 'good salary', 'gives authority, 'interesting and exciting profession' and 'enhances creativity' ties for the third positive perception for the Bumiputra group. The findings as presented in Table 1 suggest that there are indeed some differences in the perception of Bumiputra and non-Bumiputra students, but they are minimal. However, the differences in perception seem to increase on the potential negative aspects of pursuing a professional qualification.

Table 1. The difference in positive perception between Bumiputra and non-Bumiputra students on becoming a professional accountant.

STATEMENTS	Bumiputra	non-Bumiputra	p-value
Gives a good salary	4.039	4.093	0.507
Prestigious lifestyle and social status	3.885	4.009	0.168
Future high earnings potential	4.000	3.932	0.386
Promotion prospects or opportunities	4.017	4.042	0.765
To have greater employability choices	3.944	4.051	0.220
Potential for personal growth and development	3.992	3.873	0.154
Gives you authority / power	4.039	3.881	0.058
Positively contribute to society	4.021	3.873	0.071
Opportunity to work for a large corporation	3.996	3.864	0.118
Challenging, interesting, satisfying and exciting profession	4.039	3.983	0.504
Opportunity to choose career specialization	3.996	3.873	0.141
Opportunity to apply skills and abilities	4.000	3.873	0.126
Allows you to enjoy a quality family life	4.004	3.958	0.571
Assure good physical working conditions	4.009	3.975	0.686
Enhance your creativity	4.039	4.009	0.724
Has job security	4.043	3.958	0.307
Allows for independence	4.060	3.898	0.057
Gives you a chance to interact with various people	4.004	3.983	0.801

The results in Table 2 indicate that there is significant difference (significant at 5 percent significance level) in perception on 'there are too many hurdles to qualify', 'there is no time to relax in the first few years' and 'it costs a lot of money to become a Professional Accountant'. The means on 'skills and maths background is required to be a Chartered Accountant' are only weakly significantly different (significant at 10% significant level).

Overall, the non-Bumiputra students seem to perceive these negative aspects of pursuing a professional qualification more negatively, based on the means. Furthermore, 'limited free time' and professional accounting qualifications being expensive seem to be the primary concern for Bumiputra and non-Bumiputra students, respectively. Meanwhile, 'recent negative image of the profession', and 'A Chartered Accountant is not as exciting as other careers' are the other main concerns of the Bumiputra students. On the other hand, for the non-Bumiputra students, their other main concerns are 'there are too many hurdles to qualify', 'low earnings in the initial employment years', 'personal liabilities for malpractice', 'market for a professional accountant is saturated', 'there is no time to relax in the first few years' and 'skills and math background is required to be a Chartered Accountant'. Based on the overall findings of perceived benefits and negative aspects on becoming a professional accountant, Hypothesis 1 is partially supported as there are only some significant differences between the perception of Bumiputra and non-Bumiputra students. In referring to TRA, these perceived benefits (positive perception) are likely to influence an intention to pursue a professional qualification, whereas negative perceptions are likely to deter it. However, as this paper's objective is to gather these positive and negative perceptions, thus a future study should explore whether these perceptions do indeed have an effect on intention of students to pursue a professional qualification.

Table 2. The difference in negative perceptions between Bumiputra and non-Bumiputra students on becoming a professional accountant.

STATEMENT	Bumiputra	non-Bumiputra	p-value
It costs a lot of money to become a Professional Accountant	3.85	4.364	0.037
It takes too much time to qualify as a Professional Accountant	3.80	4.227	0.100
There are too many hurdles to qualify	3.75	4.318	0.029
Limited free time	4.15	4.046	0.685
Low earnings in the initial employment years	3.90	4.318	0.092
Recent negative image of the profession	4.05	4.227	0.487
Personal liabilities for malpractice	3.90	4.318	0.092
Market for a professional accountant is saturated	3.90	4.318	0.092
Chartered Accountant firms demand long and inflexible hours	3.90	4.273	0.144
A Chartered Accountant is not as exciting as other careers	4.00	4.227	0.366
There is no time to relax in the first few years	3.80	4.318	0.036
Skills and math background is required to be a Chartered Accountant	3.85	4.318	0.054

5 Conclusion

The findings of the study have certain implications. Firstly, the positive perception of Bumiputra and non-Bumiputra students on pursuing an accounting qualification is minimal. Thus, professional accounting bodies can use similar marketing strategies to attract both Bumiputra and non-Bumiputra students. However, the professional accounting bodies as well as MIA have to realise that the negative perceptions perceived by these students are equally important. Hence, accounting bodies should also attempt to address these concerns by the students. Moreover, more of the negative perceptions are significantly different between Bumiputra and non-Bumiputra students. Therefore, these concerns have to be addressed based on ethnic groups and cultural background, instead of assuming that all students have the same concerns.

The study's limitations can be overcome by future studies by extending the investigation to students from all universities in Malaysia. Also, a larger sample size would enable future studies to attain the perception of the main racial groups in Malaysia, instead of just grouping them into Bumiputra and non-Bumiputra. Moreover, since this study only investigates the initial stage of TRA, which is perception, future studies could extend this study by determining whether positive perception does indeed lead to the intended performance, i.e. pursuing professional qualification. Nevertheless, the study has produced findings that would be useful to accounting professional bodies.

References

1. Malaysian Institute of Accountants. *By-Laws (On Professional Ethics, Conduct and Practice)* www.mia.org.my/ (2010)
2. IFAC. *Definition of Professional Accountant.* https://www.ifac.org/system/files/meetings/ files/Definition Professional Accountant Presentation.pptx (2011)
3. M. Abdullah. Utusan Malaysia. http://www.utusan.com.my/utusan (2001)
4. Malaysian Institute of Accountants. *Annual report 2013.* (2013)
5. T. McDowall, B. Jackling. Asian Rev. Accounting, **18,** 30 (2010)
6. B.P. Green, P. Graybeal. J. Education for Business, **86,** 100 (2011)
7. A. Ferreira, A. Santoso. Accounting & Finance **48,** 209 (2008)
8. H. Dalci, M. Arasli, S. Tümer, J. Baradarani. Acc. Emerging Economies, **3,** 145 (2013)
9. W.S. Albrecht, R.J. Sack. Accounting Education **16,** 5 (2000)
10. J. Mbawuni. Int. Education Studies **8,** 9 (2015)
11. P. Marriott, N. Marriott, N. Accounting Education: Int. J. **12,** 113 (2003)
12. B. Jackling, P.D. Lange, J. Philips, Sewell. Accounting Research J. **25,** 113 (2012)
13. M. Mustapha, M.H.A. Hassan. Int. J. Education **4,** 1 (2012)
14. S.Y. Foong, C.H. Khoo. J. Acc. Emerging Economies **5,** 202 (2015)
15. Z. Ahmad, H. Ismail, R.N. Anantharaman. Education & Training **57,** 360 (2015)
16. N. Jaffar, N. Ismail, S.M. Zahid. Accountancy Bus. & the Public Interest 43 (2015)
17. P. Auyeung, J. Sands, J. Accounting Education: Int. J. **6,** 13 (1997)
18. J.E. Myburgh. Meditari Accountancy Research **13,** 35 (2005)
19. P.L. Wessels, L.P. Steenkamp. Meditari Accountancy Research **17,** 117 (2009)
20. E.K. Ghani, J. Said. Canadian Social Science **5** (2009).
21. S. Abidin, N.A. Zaluki, N. Salmi. Malaysian Accountancy **1,** 4 (2011)
22. E. Germanou, T. Hassall, Y. Tournas. Asian Rev Accounting **17,** 136 (2009)
23. M. Strader, B. Katz. J Social Psy **130,** 141 (1990)
24. M. Fishbein, I. Ajzen. *Belief, attitude, intention & behavior: Introduction to theory & research* (1975)
25. J. Paolillo, R. Estes. Accounting Review **57,** 785 (1982)
26. K. Ahmed, K. Alam, M. Alam. Accounting Education **6,** 325 (1997)
27. S.M. Jackman, A. Hollingworth. http://files.iaaer.org/publications/COSMOSvol15no2.pdf?1403722838 (2004)
28. U. Sekaran, R. Bougie. *Research methods for business: A skill building approach (5th ed.)* (2010)

Fostering values: four stages towards developing professional ethics for future accountants

Zaleha Othman[1], *Fathilatul Zakimi Abdul Hamid*[2] *and Rob Melville*[3]

[1]Othman Yeop Abdullah Graduate School of Business, Sintok, University Utara Malaysia, Sintok, Malaysia
[2]Tunku Intan Shafinaz School of Accountancy (TISSA), College of Business, University Utara Malaysia, Sintok, Malaysia
[3]Cass Business School, City University London, United Kingdom

Abstract. The many accounting scandals occurred in the last three decades have change the perspective of accountant globally. As such, the higher institutions have to play their role in nurturing professional ethics in order to change the misconception towards the profession. Our observation of the literature indicates that incorporating professional ethics in higher institutions is a way forward towards developing future accountants with values. Henceforth, we conducted a generic inquiry study to explore how higher institutions could inculcate accounting graduates with professional ethics. Our findings show a conceptual framework which depicted four stages towards incorporating professional ethics at tertiary level education there are: 1) value development, 2) ethics maturation, 3) professionalism development and 4) ownership through effective implementation and enforcement. Consequently, the findings contribute to expanding the current knowledge in our conceptualisation of the professional ethics concept. In addition, the findings support the development of ethics education for accounting graduates in higher institutions in Malaysia. We consider that this study provides evidence to educators and policy makers that teaching methods and pedagogical policies should ensure professional ethics education in business schools in Malaysia is treated as a pervasive element of curricula rather than an optional choice.

KEY WORDS: Accounting students, ethics educational strategy, professional ethics, professional ethics development, professional ethics model, qualitative research

1. Introduction

In recent years, the accounting profession has experienced a damaging threat to perceptions of its conscientiousness in dealing with ethical issues. This threat has developed as a result of both collective and individual characteristics of accountants. Collectively, there is a view that accountants lack professional ethics [1]. At an individual level, the role of accountants and their firms in the global accounting fraud of the last two decades and their failure to recognise and report the management weaknesses suggest significant shortcomings in the professional ethics of the accounting profession. Some asserts that the many accounting fraud showed an ethical lapses among accountants [2].

Several scholars have identified dysfunctional or ineffective ethics education as one of the causes of a perceived decline in ethical behaviour [3]. Other scholars have called for higher education institutions to revisit the position of ethics education and the development of graduates' knowledge of and competency in ethics [4].

Given the magnitude of the perception of the poor professional ethics standards among the accounting profession, we propose that professional ethics need to be more effectively inculcated and that this can be undertaken through formal education at higher institution. Henceforth, we attempt a study that suggests strategy for the development of professional ethics. Our direct experience of teaching in accounting and our familiarity with various accounting curricula also underpin our view that there is a need to change the philosophy of education in ethics for accounting graduates. Similar voice of concerns has been pointed by Dellaportas [5], who claimed that giving attention to ethics awareness and sensitivity is not enough for today's accounting graduates. He found that focusing on how ethics should be taught is not enough: what matters is whether students are able to make moral judgments after they have attended discrete and dedicated ethics courses. Thus, the intention of this paper is to provide support for educators in higher education institutions in Malaysia particularly to enable accounting programmes to embrace the nurturing approach for professional ethics in order to develop graduates with a more effective ability to address ethical issues and uphold ethical values. Hence, we construct a research question, that is: how to foster professional ethics among accounting students in Malaysian higher institution. Using the generic inquiry method we explore the phenomenon with the aimed of providing conceptual framework for fostering professional ethics for future accountants prior to joining the work force. Based on our endeavour of seeking knowledge related to the research inquiry, our findings show a conceptual framework which depicted four stages towards incorporating professional ethics at tertiary level education they are: 1) value development, 2) ethics maturation, 3) professionalism development and 4) ownership through effective implementation and enforcement.

2. Literature review

Scholars have claimed that violation of ethical norms in business in general and among accountants specifically has motivated higher institutions to improve their ethics teaching methods [4,5].

Some scholars suggest that teaching ethics is an effective mechanism for developing ethical principles in accounting students and O'Leary and Mohamad [6] argue that teaching ethics plays an important part in achieving the goal of having ethical practitioners. Their study of a group of final year accounting students focused on the impact of ethics instruction on future accountants and which instructional methods – lectures, tutorials or group assignments – were effective in developing ethical practitioners. They found that teaching ethics has an impact on student ethical decision making. Others, such as Klimex and Wenell provide evidence concerning the mode of teaching ethics. Their study assessed the ethical reasoning ability of 60 senior undergraduate accounting students in a higher education institution in the United State [7]. They found that the group that followed the stand-alone ethics course showed higher ethical reasoning ability than the group that followed the integrated ethics course. This suggests that in accounting education stand-alone ethics courses might be a more effective teaching mode than an integrated ethics approach. Similar findings were noted in other study. Others also found that exposure to ethics is necessary for accounting graduates prior to their becoming accountants [8].

Slocum et al. conducted a study of integrated micro-insertion for the teaching of business ethics and explored the researchers' experiences in teaching ethics in an undergraduate business class [9]. They developed a conceptual model of the micro-insertion method, which they believed led to greater efficiency in teaching efforts aimed at inculcating moral reasoning. Jun Gu and Neesham also found similar outcome which support the assertion that exposing students to identify-based and rule-based teaching improved student ethical decision making [10].

In the Malaysian context, studies have predominantly been conducted in relation to the perception of ethics in general, showing different levels of perceptions of ethics. One such example is Rashid and Alsagoff's study of perceptions of ethics among managers in Malaysia. Their most significant finding was that Malaysian managers had high ethical values at the time, although there were differences in the level depending on the manager's position, the type and size of the business and the job specification [11].

The closest links in studies between ethics and the accounting profession in a Malaysian environment can be found in the research of Yunanda and Abdul Majid [12] and O'Leary and Mohamad [6]. The former focused on religion as an element of ethics in the accounting profession and the latter examined the influence of ethics teaching in accounting courses. In their investigation, Yunanda and Abdul Majid emphasized the importance of religion as an element that guides ethical reasoning [12]. Interestingly, they argued that a code of ethics in itself is not enough to provide a framework for ethical and moral reasoning for accountants. According to Yunanda and Abdul Majid, religion promotes ethics in business through guidance towards principles of personal and professional ethics. The principles of personal ethics were identified as having the characteristics of concern for the well-being of others, respect for the autonomy of others, trustworthiness and honesty, willingness to comply with the law, justice and fairness, refusing to take any unfair advantage, benevolence and preventing harm [12]. For professional ethics the characteristics were identified as objectivity, impartiality, full disclosure, openness, confidentiality, due diligence, fidelity to professional responsibilities and the avoidance of potential or apparent conflict of interest. Yekta et al. also found similar results in that codes of ethics do not have a significant impact on reporting wrongdoing in organizations [13]. They suggest that codes of ethics can work effectively and will prove positive in the reporting of wrongdoing if an organization invests time and effort in building good employee relationships. Although there is great concern on developing professional ethics, there has as yet been little academic focus on professional ethics, particularly in Malaysia. It is thus timely to explore this issue in order to provide a better understanding of professional ethics for the accounting profession.

3. Methodology

Considering the nature of the research questions, we examine the phenomenon using qualitative research approach. Specifically, we employed the Generic Inquiry Method (GIM). GIM is an approach that does not confined to any philosophic assumption as other qualitative method such as phenomenology, grounded theory and ethnography [14]. Generic qualitative approach merely seek to understand how people interpret, construct or make meaning from their world and their experience [15] which fit into our aim of study, seeking information from the people based on their experiences in academic and practiced on how to foster professional ethics to our future accountants at higher institution prior to joining the work force.

The sample of respondents was selected based on the purposive sampling method. Purposive sampling, as explained by Ritchie et al. [16], allows the researcher to choose respondents based on particular features or characteristics which enable detailed exploration of the research topic. Hence, based on the suggestion we interviewed seven respondents, they were: three academics, one practitioner, two members of professional accounting bodies and one team member from the My3S program.

The interviews conducted sought information on the participants' views of the process of nurturing professional ethics. Following completion of the interviews, the recorded data were transcribed verbatim. These were then analysed using Spencer et al.'s method of analysis, incorporating the three processes of explanatory accounts, data description and data management [17]. Based on our analysis, we developed a model that represents the process of nurturing professional ethics, which is presented and discussed in the results section of this paper (see Figure 1).

4. Results

Based on our interpretation of the data, we developed a model mapping the voices of the respondents to the process of nurturing professional ethics. This model comprises four stages, there are: value development, ethics maturation, professionalism development and ownership. These four elements need to be implemented and enforced in order to foster professional ethics. Figure 1 shows how these four elements form the environment in which professional ethics develop.

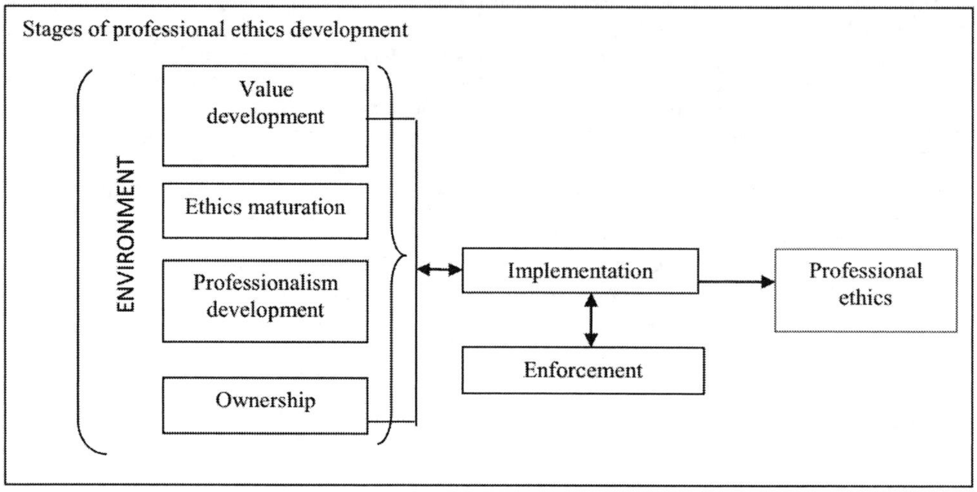

Fig.1. Stages of professional ethics development

The goal of our study was to gain understanding of the process of nurturing professional ethics among undergraduate accounting students. In so doing, we generates a conceptual model as shown in Figure 1, of nurturing professional ethics at tertiary level. Our study aimed to demonstrate that fostering professional ethics is a new strategy with potential in the education of accounting students. We assert that value development, ethics maturation, professional development and ownership are the four key strategic elements of fostering professional ethics and that these have the potential to support the development of future accountants who will uphold ethical values.

Our study has found that the ethical canons and standards of professional codes of conduct provide fundamental roadmaps for the individual, although the principles of the individual are equally significant. Individual principles are groomed and nurtured by significant figures and forces in the individual's background. Here, there is a notable linkage with value development during childhood. The data indicated that the respondents perceived that values within the context of home, school and referent others (peers and role models) influenced the fostering and acceptance of ethical principles.

The ethics maturation process is a process of inculcating an ethical culture with the objective of nurturing ethical reasoning. A fundamental part of the ethics maturation process, as represented in our data, is learning from peers and others prior to starting work in the profession. This is a notion closely allied to what Merton and Rossi [17,18] term 'anticipatory socialisation', defined as "the process of adopting attitudes and beliefs of a reference group before obtaining membership" (p. 85). We posit that this is a process of learning that can take place in the university environment.

From another perspective, Garavan and Morley defined anticipatory socialisation as "all learning that took place prior to a graduate's first day on the job" [19] and which guides the professional socialisation of students. Our respondents emphasised that professional ethics can be nurtured through an integrated mode of formal and informal teaching and learning. Thus, the respondents highly recommended that university authorities not only employ the classroom approach as a formal mode of learning that inculcates professional ethics, but also support informal modes of learning, such as self-policing, as a means of anticipatory socialisation.

Another point raised was the role of self-development programmes, such as internships and training. Internships provide opportunities for students to develop moral reasoning and moral decision making. Most of the respondents concurred that internships develop interactive skills and inject professionalism. These skills are identified as components of the ethics maturation process. Internships would provide chances for students to gain familiarity with the profession before enrolling as members.

Our study also supports the strategy of inculcating ethics by introducing ethics-based learning, integrating the traditional rules-based teaching approach and a contextual ethics-based approach. Whereas the traditional rules-based teaching approach relates to formal classroom learning, embedding ethics in the curriculum with the aim of developing intellectual reasoning, the context-based approach involves learning outside the classroom, for example in extra-curricular activities, with the aim of building moral development. By integrating the two modes, it might be possible to overcome the shortcomings of a single approach. Dellaportas [5] and Sisaye have argued that the traditional modes of integrating ethics in the accounting curriculum need revision [20]. Sisaye for example, opines that the traditional mode of learning strategies is not effective as it is a single loop approach [20]. He proposes the use of sociological organisational learning strategies, i.e. functional-institutional and consequential-conflictual strategies, contending that these are effective as they exhibit positive relationships with moral and ethical reasoning. Both learning strategies are useful, as they may change the prevailing pedagogy and contribute towards positive results.

Furthermore, this study highlights the value of fostering a sense of 'ownership' in the profession as a means of enabling professional ethics development. Ownership relates to sense of pride in the profession and of stewardship. As part of this, we would point to the salience of adhering to the code of conduct of the profession and upholding its values (i.e. integrity and being morally upright) as elements in our data related to honouring the profession.

5. Conclusions

We have undertaken an exploratory study that support towards developing strategy for fostering professional ethics at higher institutions. We have developed a contemporary framework, which identifies the key factors to be considered in developing professional ethics and the various stages in the process of nurturing ethical principles that could be useful guide for policy setting in the accounting profession. Given the limitation of the literature in this aspect, we concur that the findings contribute to expanding the current knowledge where it provides conceptualisation of the professional ethics concept. In addition, the findings support the development of ethics education for accounting graduates in higher institutions in Malaysia. We consider that this study provides evidence to educators and policy makers that teaching methods and pedagogical policies should ensure professional ethics education. Our findings are especially significant for accounting educators and business schools in terms of the implications for education aimed at developing professional ethics. Our findings provide a guide for educators in higher institutions, proposing a new strategy for ethics education encompassing four processes: value development, ethics maturation, professional development and fostering ownership. Most importantly, higher education is identified as the ideal environment in which to foster professional ethics (notwithstanding the role of early development). In sum, this study serves to highlight the conceptual and practical benefits of integrating strategies for the development of ethical principles in higher education institutions generally and accounting faculties specifically in Malaysia. We suggest that future research should consider exploring the phenomenon using a case study.

References

1. M. Metzger, Accounting ethics: Imagine that, J Gov.Fin.Mgt **53**, 1 (2004).
2. N. Koumbiadis, G. Conway, J. Angel, An empirical study in ethical perception: will the mandated AICPA curriculum change differ between those who have taken an ethics course or not? W J Mgt **2,** 3 (2010).
3. S. Ghosal, Bad management theories are destroying good management practices, Aca. Mgt L Edu **4** (2005).
4. E.L. Felton, R.R. Sims, Teaching business ethics: Targeted outputs, JBE **60** (2005).
5. S.Dellaportas, Making a difference with a discrete course on accounting ethics, JBE **65** (2006).
6. C. O'Leary, C, S. Mohamad, The successful influence of teaching ethics on Malaysian accounting students, Mal.Acctg Rev 7, 2 (2008).
7. J. Klimex, J., K. Wenell, Ethics in accounting: An indispensable course? Aca.Edu.L J **15** (2011).
8. N.J. Mastracchio, Teaching CPAs about serving the public interest, CPA **75**, 6 (2005).

a. Slocum, S. Rohlfer, S. C. Gonzalez-Canton, Teaching business ethics through strategically integrated micro-insertions, JBE **125** (2014).

9. G. Jun, C. Neesham, Moral Identity as leverage point in teaching business ethics, JBE **124** (2014).
10. M.Z.A Rashid, S.K. Alsagoff, Perceived ethical values of Malaysian managers, JBE **12** (1993).
11. A.R. Yunanda, N. Abdul Majid, The contribution of Islamic ethics towards ethical accounting practices, I Social Env. Acctg **5** (2011).
12. A.Z. Yekta, A.Z. Ahmad, R. Kaur, Corporate code of ethics and reporting wrongdoing in private sector organizations in Malaysia, Int. J B Mgt **5**, 7 (2010).
13. K. Caelli, L. Ray, J.Mill, Clear as mud: Toward greater clarity in generic qualitative research, Int. J Qual. M **2,** 2 (2003).
14. S.B. Merriam, *Basic interpretive qualitative research.*(1998). S. B. Merriam (Ed.), *Qualitative research in practice* (Jossey-Bass, San Francisco, CA, 2002).
15. J. Ritchie, J. Lewis, G. Elam, G, Designing and selecting samples. J. Ritchie, J. Lewis (eds.), *Qualitative research: A guide for social science students and researchers* (Sage Publication Ltd, London, 2003).
16. L. Spencer, J. Ritchie, J, W. O'Connor, *Analysis: Practices, principles and processes*. J. Ritchie, J. Lewis (eds.), *Qualitative research: A guide for social science students and researchers* (Sage Publication Ltd, London 2003).
17. R.K. Merton, A.K. Rossi, *Contributions to the theory of reference group behaviour, Reading in Reference Group Theory and Research* (New York, 1968).
18. Z.R.Elias, The impact of professional commitment and anticipatory socialization on accounting students' ethical orientation, JBE **68** (2006).
19. T.N. Garavan, M. Morley, The socialization of high-potential graduates into the organization: Initial expectations, experiences and outcomes, J Man. Psy **12,** 2 (1997).

20. S. Sisaye, The functional-institutional and consequential-conflictual sociological approaches to accounting ethics education: Integrations from sustainability and ecological resources management literature, Man. A J **26**, 3 (2010).

SHS Web of Conferences **34**, 06007 (2017) DOI: 10.1051/shsconf/20173406007
FourA 2016

Malaysian accounting: an inquiry into the struggle, the crisis, and the future

Hasri Mustafa[1,*]

[1]Department of Accounting and Finance, 43400, Universiti Putra Malaysia,

Abstract. Are we satisfied providing of what accounting has provided? How many books about accounting and accountability have we read and forgotten the instant we finished reading them? What difference they have made then? Will readers understand the deepest emotion of the texts? So many years have passed and many witness what happen to Malaysian history to accounting field was left unnoticed. Attempt to answer the questions will subsequently implicate the conditions of theorizing accounting based upon an inquiry into Malaysians own struggle, crisis and future prospect. The aim of the study is to encourage accounting researchers, especially Malaysians, to view accounting with responsibility, answerability, blameworthiness, liability, and other terms associated with the expectation of account-giving.

1 Introduction

Does it really matter who were native and who were not to a particular territory (or State) to the field of accounting? The school systems and universities leave no comprehensive discussion in regard to accounting knowledge. Pre-independent Malaysia literature in respect to social conceptions and groupings of races that define essential types of individuals has received rather reluctant to the attention of Malaysian accounting scholars. Prior to the making of the State, Malaysia was ruled by a number of kingdoms and colonial periods. Essentially, Chinese and Indians comes to Malaysia from a set of unsystematic organisation where legislations to regulate them were from colonial vision of 'divide and rule'. This is particularly problematic from the context of Malaysia historical setting which was simply irrelevant to many young accounting students and lecturers. Most research in accounting see accounting from the 'perfect' market condition that is based on utilitarian in function.

Interestingly, scholars who studied accounting found the imperfect market condition is actually a good point of departure for a good research [1]. Because history is the study of the past, particularly how it relates to the memory, discovery and collection, crisis, fighting and the other struggling are invaluable for the very basis of inquiry. In accounting, a good illustration of exposing the faulty of ideal market may be found in the Enron case of 'cooked the books' [2, 3]. Our failure to notice early of Enron's predicament was because

* Corresponding author: hasyie@upm.edu.my

accounting researchers correspond to Enron within the consciousness of the limited generalised observations, such as in the words 'success', 'measure', and 'earnings per share' [4]. A research on these identified independent variables may popular to some scholars, but they are limited in providing a better understanding of accounting knowledge. If this world accounting research community is to survive, the current disinterest of researching accounting chronicles and old historical account of events would be enlarged and many will still see accounting in the orderly form.

Accounting is primarily a conception about civilization and organized society, even though accounting was primarily theoretical. The elaboration about the diversity of the society is critical so that efforts in discussing, arguing, thinking and contemplating of accounting knowledge are obtained at best. Regarding the three main groups of Malays, Chinese and Indians, they have influenced continuous debates amongst historians from many spheres: politics, economics, ideology and religion [5, 6, 7, 8, 9]. Malaysia's plural society has provoked exploration of topics such as the State, the regime, the capitalists, the nationalists, the enactment and the self-accountability as designated contexts within which the dominant theorising of motivation, contract, production, agency and organisation of Malaysians are discussed. The failure to include them, bring us no sufficient condition to examine matters of power and conflict hence avoiding critical discussion in accounting literature.

2 The spirit

In July 2011, I started with my first plan of producing such study. I wanted to produce a study that would contribute to the understanding of accounting in Malaysia but based on a collective portrait of national quarrels, riots, communism, political movement and the establishment of social justice and democracy societies from the very first to the newest Malaysian historical settings. I hope the historiography would contain a sound reflection on theoretical conception on accounting in a clearer and weightier presentation than my doctoral thesis. In so doing, although some of the data was from my PhD thesis, most contents are new. I had a feeling the significance of the study lies, besides events and new settings, with the effort to bring the historiography to which a social mix of suggestions would be invited. As such, I required more open and rational dialogues or exchange of views.

Fortunately, on 1 September 2011, with a strong recommendation from my Head of Department and Dean, I was granted to carry my sabbatical leave for that works for six months in Christchurch and three months in London, the places where I was not a student but would meet me with a number of accounting scholars. If I were to retrospect, I am glad that I had opportunities to discuss with friends, colleagues and scholars about how to engage Malaysia in a more informative form within its own proper frame of reference. Beginning in Australia, a country with a second-highest human development index globally[2], I was introduced in my PhD to 'new', 'alternative' and 'critical' views of accounting at a broader social level. Then, in Britain, where the rise of nationalism and movement is remarkably dramatic, I came to realise that methodology alone is too situation-specific to research so to be bound by researcher's own avowed aspirations. Different approaches may necessary to attain a full description of the world. Later in New Zealand, I took an unpaid leave from 1 August 2012 to 31 May 2015. I felt that a passionate desire for entirely accepting liberalisation revolution as the validity claim for the so-called critical accounting is not really helpful. I believe Malaysia's growth, inasmuch as New

Zealand's agricultural territorial-in-chief, is powered by its own characteristics and therefore somehow individualistic.

I should add here that I find myself constantly asking: to which part of the world I could know better about Malaysia? Although there were a number of historical details I gained from their libraries, history is not only the *dis*covery but also the *re*covery. Malaysian history is not out there, it is in the common sense[3] of Malaysians. Admittedly, in Malaysia not many references in accounting I could follow. Just look at the other ancient kingdoms such as ancient Greek [10, 11], Roman [12], India [13, 14] Mesopotamia [15, 16, 17, 18], and in the early Muslims accounting especially the Middle and Near East Countries [19, 20, 21]. With their publication number increases I afraid the younger generation of accounting would have developed tastes in non-Malaysian history, leaving the Malaysian history into separate discipline. A new generation like me has to go out to get the idea of Malaysian accounting from the philosophy of internationalism in the foreign material and spiritual culture. There was bread and maybe sugar in the West, or any other places around the world, but then the flour was gone. There is no need here to recapitulate in any detail the arguments that I have received with my non-Malaysians discussants but what in Malaysia do I have out there?

It is from this frame of references I develop the contents of this study. The next section includes the conditions and characteristics of what Malaysia has come. It contains a base of support in what I call the societal program of Malaysia from the period of Malay States, Malaya, Malaysia and 1-Malaysia. It recognises 'struggle for survival', 'crisis of growth' and 'future of ethical institutions' views from the actual history and nature of Malaysia that have potential contribution to the conception of accounting and the development of accounting theory. Malaysia had wealthy resources, but there were the reasons for conflicts and political differences. To escape these characteristics is to ignore the common production of Malaysia. There seems little reason to deny accounting conception a glimpse of this aspect. The final section discusses the theoretical implications and the limitations of the research as a whole.

3 An inquiry into Malaysian accounting

Malaysian Dilemma shows that although Malaysia's plural society is in principle universal, they involve the individual struggle and an optimisation directed towards the future of each ethnic group. In 1957, Rupert Emerson, a Harvard political scientist, made these remarks about Malaysia: '[t]he country is a 'perfect example of the plural society'. The different peoples have in common only the fact that they live in the same country. In race, religion, language, culture, economic interests…their outstanding characteristic is not unity but profound diversity' [22, p. 5].

The key points; 'plural society' and 'race, religion, language, culture, economic interests', which Emerson underscored, are the landmarks of Malaysia's societal program. In principle, almost any discussion on Malaysia's societal program must be structured within the three main ethnic groups: Malays, Chinese and Indians, who all tend to remain more dominant than other ethnic groups. These three main groups are the ones to which Emerson refers 'their outstanding characteristic is not unity but profound diversity'. The definition of Malays, Chinese and Indians suggests that the three dominant groups are heterogeneous in the same sense that Malaysians at large are. According to Malaysian historian Khoo Kay Kim, 'in the case of the Chinese and Indians, the heterogeneous character of each group [is] derived from the land of its origin. This was reinforced in the land of its adoption' [23, p. 251]. To Khoo, 'no simple explanation can be given for this. But one plausible explanation is that their influence was thwarted by the fact that they had to operate in a plural society which was also under colonial rule' [24].

Now, how then am I to speak about accounting in response to the above accounts? Over its short history, Malaysia has experienced dilemmas that were initially casual but were soon to be cumulative across social capital, ideology and basic material needs. All these factors have placed accounting into the much wider context that I had ever anticipated. For I see them as genuine and sincere. They are the real battlefields for Malaysians. They offer me to see accounting on societal program from the imperfect market condition as point of departure.

3.1 A State and the struggle for survival

Table 1 summarises my observations on the factors cited above but includes the other more critical contents about the first production (construction) of today's Malaysia. The table I use to restrict my accounting journey from the references of monarchy, wars, colonial powers, and labour capital for international trade and local production. I would call these references *the struggle for survival period* as the events involve Malays, colonies and merchant-migrants into the state of continuing to live, typically in spite of an accident, ordeal, or difficult circumstances.

Table 1. Societal program of the production (construction) of Malaysia from the 'struggle for survival' view

Years	Struggle for survival
1351 – 1767	A tributary gold and silver tree from Malay State Rulers of Patani, Kedah, Kelantan, Trengganu and Perak to Siam Kingdom
1402 – 1511	Malacca Sultanate empire
1402 –	Malacca 'Empat pembesar Lipatan' system
1414 – 1946	Malacca traders from Europe, Persia, Arabia, Tamil and other Native Indians and Chinese
1511 – 1641	Portuguese Malacca colonisation
1511 –	Indian Chetty (moneylender) System
1606 – 1641	The Dutch East India Company began the campaign to destroy Portuguese power with the first serious attempt being the siege of Malacca
1641 – 1842	Dutch Malacca colonisation
1819 –	Chinese Kongsi (clan) System
1826 – 1946	Malacca was governed by the British East India Company that formed part of the Straits Settlements, together with Singapore and Penang
1874 – 1946	British Malaya colonisation
1850 – 1880	Tin mining industry
1870 – 1890	Chinese financial mining enterprise truck system
1860 – 1910	Indian labourer broker Kangany System
1890 – 1940	Rubber plantation industry
1929 – 1942	Rubber coupon system in Malaya
1941 – 1945	Japanese invasion of Malaya
1942 – 1945	Japanese Banana Money
1944 – 1945	Severe food production crisis
18 Dec 1941	Malayan Peoples' Anti-Japanese Army (MPAJA)
23 Feb 1950	Communist and Bukit Kepong blood incident

3.2 Individualism and the crisis of growth

Accounting is a product of the professionalism of a social class, elite group and specific agent [To name a few, 25, 26, 27]. There are examples that many of the relationships between accounting and society are accounted around these thought but, in major part, they inevitable come with disagreement to reach an understanding when necessary. One of the examples is the role played by political systems which were often accused of intervention and exaggerating contexts of reality. In the reconstruction of Malaya, these systems included the UMNO, MCA, DAP, MIC, and other political voices. These organisations reasonably pointed out the possibility that someone belonging to Malaya may participate in society but feel alienated due to their values and principles. As common for political ideologies, these movements focused upon questions of class, power, conflict and the reproduction of hierarchy and hegemony, and the concepts remained relevant until they were disrupted by notions of corruption and nepotism. Table 2 lists the significant events that deployed these central concepts from which comes the taint of political movement, ideology, and unionism. These events which were configured following the struggle for survival I refer as *the crisis of growth*.

Table 2. Societal program of the production (construction) of Malaysia from the 'crisis of growth' view

Years	Crisis of Growth
1930	MCP (Malayan Communist Party)
1946 – 1948	Malayan Union
31 Jan 1948	The Federation of Malaya
1 March 1946	UMNO (United Malays National Organization)
August 1946	MIC (Malaysian Indian Congress)
27 Feb 1949	MCA (Malaysian Chinese Association)
1949 –	Student politics in University Malaya
1950	Chinese People's Action Party (PAP)
24 Nov 1951	Pan-Malaysian Islamic Party (PAS)
1954	UMNO, MCA and MIC jointly formed the National Alliance (Perikatan Kebangsaan)
31 August 1957	The Federation of Malaya achieved independence
16 Sept 1963	The Federation of Malaya was reconstituted as "Malaysia" to include Singapore, Sabah, and Sarawak
18 March 1966	Democratic Action Party (DAP)
1963 – 1966	Indonesia–Malaysia confrontation
9 August 1965	Singapore out of the union
24 March 1968	Chinese Gerakan Party (Chinese Liberal Party)
13 May 1969	Chinese – Malays racial riot
31 August 1970	Rukunegara ("the Principles of Nationhood")
1971	The New Economic Policy (NEP) and the development of Bumiputera (son of the soil) entrepreneurship
1983	The 1983 Constitutional Crisis
1988	Bahasa Baku (standard pronunciation of Malay Language)
2 Dec 1989	The Malaysian Communist Party formally ends 41 years of armed struggle against the government
1993	The loss of the Malay Rulers' immunity
2 Sept 1998 –	Deputy Prime Minister and United Malays National Organization (UMNO) Cases
4 April 1999	Parti Keadilan Rakyat (People's Justice Party) (PKR)
Nov 2007	Hindu Rights Action Force (Hindraf)
2011 - 2014	Bersih II and III Demonstration, Blackout 505 Demonstration, Kajang Move

3.3 Institutions and the future of ethical world

Accounting is a field of ethical practices. One can recognises throughout accounting in the literature the concept of accountability that lies at the heart of accounting. There is no doubt that what has been achieved so far in accounting treatises began with this concept – about principles and guidelines that explain the account for one's actions. It is precisely embedded in the agricultural stock counting, bookkeeping, budgeting, auditing, etc. Corruption, fraud and scams are not part of accounting as it should be but a kind of deceptions of gaining an advantage through wrong accounting practice by certain individuals. For Malaysia, anticipated societal advancement must absorb the crisis of growth to incorporate an increasing accountability world order which requires transparency, answerability and other expectations of account-giving. As a developing country, growth in Malaysia should not only focus on the situation of those who are by definition positioned within the social hierarchy (which is in theory a matter of individualism such as Mahathir, Anwar, Lim Kt Siang, Karpal Singh, etc). Because Malaysians do not occupy identical or even similar positions in social contexts (i.e. 'Malays', Chinese' and 'Indians'), there is a greater chance of finding contradictory and to be antipathetic. Apparently, there is evidence when situating accounting practice within the politics and conflicts of individualisms, individualism has done little to encourage explain critically the unintended consequences of human actions in institutional contexts. There is evidence in accounting literature that the quality of institution offer sufficient conditions to examine matters of power and conflict hence including critical discussion in making sense of the past and to explain why there is coherence to such practices and proceedings[4]. Table 3 contains significant events that envisage different signs of ethical institutions that came into being as a result of modernisation. I would call these events *the future of ethical institutions* for they produce a variety of compounds for accountability elements.

Table 3. Societal program of the production of Malaysia from the 'future of ethical institutions' view

Years	Future of Ethical Institutions
4 March 1993	Skim Perbankan Islam (SPI) (Islamic banking windows)
1 May 1997	The National *Shariah* Advisory Council (SAC) on Islamic Banking and Takaful
1997 – 1998	Economic crisis
Sept 1998	Ringgit was pegged at 3.80 to the US dollar value
1998	The establishment of Corporate Debt Restructuring Committee (CDRC)
June 1998	The incorporation of Pengurusan Danaharta Nasional Berhad (Danaharta)
Sept 1998	Malaysia's exchange rate regime pegs the Ringgit at 3.8 to the US dollar
17 April 1999	The KLSE Syariah Index
12 Dec 2001	The Malaysian Accounting Standards Board (MASB) on the Presentation of Financial Statements of Islamic Financial Institutions?
April 2000 – March 2003	The merger of banks and financial institutions
Dec 2004	The Guidelines on the Governance of *Shariah* Committee for the Islamic Financial Institutions
2005	The propagation of Islam Hadhari

[4] Some accounting scholars view that there are flaws in the mode of individualism for the critical analysis of accounting practice. For example, Carruthers in his review of Walter Powell and Paul DiMaggio's 1991 edited volume: *The New Institutionalism in Organizational Analysis* concludes that 'one must reject the assumptions of methodological individualism and individual rationality' [28].

 DOI: 10.1051/shsconf/20173406007

Jan 2009	1Malaysia Development Berhad
Sept 2010	The propagation of 1-Malaysia
7 Jan 2010	1Malaysia Clinic
Jan 2010	1Malaysia Poverty Eradication (1AZAM or Akhiri Zaman Miskin 1Malaysia)
1 June 2011	1Malaysia Training Scheme (SL1M or *Skim Latihan* 1Malaysia)
Dec 2011	1Malaysia People's Aid (BR1M or *Bantuan Rakyat 1Malaysia*)
2 July 2012	KADS1M or *Kad Diskaun Siswa* 1Malaysia
March 2015	SR1M or *Sim Kad Rakyat*
July 2015	Accountability issues on 1Malaysia Development Berhad

4 Theoretical implications and the three-transformative event of accounting

So what are the ways to research? Would the above kind of ontological view of viewing accounting work for me? For the moment, let us simply emphasise the following questions: How much can *the struggle for survival*, *the crisis of growth* and *the future of ethical institutions* contribute to accounting theory? What kinds of accounting can I introduce from this three-transformative event? Below, via Table 4, I respond the questions concerning on how to analyse Malaysia's societal development as I have defined above. Table 4 incorporates the three-transformative event. It consists of three main columns which are to be denoted in respect of the societal program of concern; 'struggle for survival', 'crisis of growth', and' future of ethical institutions'. For each of them, the societal program flows following a genealogical analysis from the period of Malay States to Malaya, Malaya to Malaysia and Malaysia to 1-Malaysia. On this arrangement, a moment between the hierarchical power relation between those who used and employed the sources of raw materials and human forces, the organisers and regulators, and those who are thus concerned the moral rights embedded in the principles of economic, can be seen. In the middle row, themes such as 'the State", 'the regime', 'the capitalists', 'the nationalists', 'the enactment' and 'the self-accountability' are put not in specific to certain perspective of the societal program as by no means they can be isolated. The next row contains the said perspectives of the societal program. Note that the lines of them are not parallel with those of the organisations, but on the whole order in which the societal program become unified, runs from episode to episode, from climax to climax.

Table 4. Theoretical interpretation of accounting in accordance to the societal program in the production of Malaysia

<table>
<tr><th colspan="7">Readings: The State, the regime, capitalism, nationaliss, the enactment of self-accountability</th></tr>
<tr><th colspan="3">Struggle for survival</th><th>Crisis of growth</th><th colspan="3">Future of ethical institutions</th></tr>
<tr><td>MALAY STATES</td><td>Accounting is read as a field in which its operation is essentially characterised by the notion of an utopian State ideology to promote stability and security</td><td>MALAYA</td><td>Accounting is read as a field in which its operation is essentially characterised by political survival in which its parent and the ruling party impose a promote hegemony</td><td>MALAYSIA</td><td>Accounting is read as a field in which its operation is essentially characterised by the accountability concepts derived from Shariah law and commitment to Fardhu Kifayah</td><td>1-MALAYSIA</td></tr>
<tr><td>Types</td><td colspan="2">Wealth</td><td>Momentum accounting</td><td colspan="3">Force accounting</td></tr>
</table>

DOI: 10.1051/shsconf/20173406007

	accounting		
Societal Program	Struggle for survival	Crisis of growth	Future of ethical institutions
Conception	Trust, security	Political, hegemonic	Religious: Shariah law, Fardhu Kifayah
Dimension	Wealth magnitude – 'what' amount	Wealth change – 'how' to deal with	Change of wealth change – 'why' dialectical
Scope	Writing and managing money	Rigid accounting control and human capital	Specialist accountant: innovation capital, intellectual property, intangible asset
Form of accounting	Bookkeeping	Income generator	Auditor and governor
Centre	Responsible centre	Cost centre	Accountability centre
Problem	Different units have own self-definition	The unhealthy company will be absorbed by the healthy one	No previlaged accounting framework. Not accounting is also accounting
Motive	Action at a distance, translate wealth into observable and measurable categories	Maintaining hegemony of the preferred monarchy	Corporate accountability

4.1 'Struggle for survival' – Wealth accounting

Ijiri refers 'wealth accounting' as the first dimension for the administration of the magnitude of wealth and wealth composition, with accounting variables that are set apart as net wealth (equity) and liabilities (sources of capital), or as assets (uses of capital) [29]. Accounts of this dimension are meant to emphasise the anticipation of the wealth magnitude and composition of 'what' amount. Cause and effect rationalization in this accounting treatment is very important as it explains the objective and the struggle of a State. Accounting is aimed at achieving a new equilibrium for the State. It started with an exploration of a sense of "unsatisfactory" which is linked with beliefs reflecting negative feelings about people and their wealth, in particular between the poor and the rich, between the sons of the soil (Bumiputera) and the Immigrants. Exploring this institutional setting requires that we think about trust and security.

Writing and managing of money are theorised with bookkeeping to represent 'accounting at a distance', allowing the materials to congregate mobility, stability, collectability, examinability, and combinability [30, 31, 32, 33]. They should become centers of integrity and trust. Centers of integrity and trust, as opposed to "cost centers," develop in an organisational village where there are fences and differences between households but plenty of transactions across them [34, p. 102].

The accountants' functions are plainly constrained to oversee the characteristic of combinable inscriptions and "emerge [when] if the report is seen as illogical, the figures too big, printing problem, and errors" . Accountants have limited freedom and are exposed to the observable and measurable categories.

DOI: 10.1051/shsconf/20173406007

4.2 'Crisis of growth' – Momentum accounting

A discussion of the Federation of Malaya now turns to the vexed issue of political language in a more contemporary setting. It presents the interpretation of accounting in intellectual outlook in a vocabulary of political survival, in the virtue of 'crisis of growth' societal program. Within the political crisis, I view accounting as a locus of power, in a revolutionary or momentum form, and in the concept that may oppose public rationality, in the conception of political and hegemonic. It is from this perspective that the context of accounting is not similar to that in *the struggle for survival*. Accounting is bound to have priority which is based on maintaining the rules of the game of the established community. The result of this process is a condition of the second dimension for the administration of the change in magnitude of wealth (the first dimension), with accounting variables that are set apart as cost (outflows) and income (inflows) of what Ijiri terms 'momentum accounting' [35].

Accounts of this dimension are not meant to emphasise the anticipation of the wealth magnitude and composition of 'what' amount but the wealth change which is explained by its 'how' to deal with. This consideration produces the rigid accounting control by individualists, especially cash and extra remuneration, in order to make sense of the preference of the preferred monarchy. The conception of accounting is acting essentially in the momentum of the political scenario from the particular intellectual context peculiar to movement, ideology, and union since Malaya. The problem of 'momentum accounting', in a nutshell, is that when it comes to the question 'who is the subject of scarification?', the unhealthy company will be absorbed by the healthy one.

4.3 Future of ethical institutions – Force accounting

When examining the production of Malaysia, the discussion should not end to various accounts of the relation between the concepts of struggle for a survival and the political right only. There were clearly a number of early law codes, embodied in particular inherited principles of colonists, which will be seeing inaccurate, too communitarian and lack justice. The concept of 'divide and rule' by colonists since the duration of the old Malay States, then Malaya, until the current Malaysia State are found fundamentally to contain racial characteristic. Political and capitalist theory carried a number of 'social justice questions' to human being.

In the modern presence, fortunately there are concerns to capture the many formulations of the more ethical perspective in Malaysian laws and enactments. More educated Malaysian people see ethics should become the essence which cannot be sacrificed. However, none of this is likely to happen in sincere if communitarian and personalist socialism are the condition for the ethical consideration. Religion should be the source for the idea. Ethics must be institutionalized through religion to include all practices and structure of dominations. This includes accountants, auditors, and council members. The interpretation of them and their system are therefore should be a reference to accounting. Religious auditor or in Islamic Bank is the correct example. Their jobs are to promote the ethical institution and actualization of morale that make a bank worthy of esteem and respect. The taking of *riba* (usury), *gharar* (speculation) and the investing and financing of certain activities, such as, among others, the production of liquor, pork, tobacco; the entertainment activities of brothels and night clubs; and the distillery and gambling houses is neither conditional nor optional. Religious auditor is responsible for the well being of the Malaysian society and the money invested is counted to the productive business venture not to meet personal accumulation wealth. From this standpoint, accounting is set apart to administer internal and external forces. At this stage, accounting is in the third dimension for the administration of the change of capacity to acquire new wealth (the second dimension). Theoretically, this Ijiri calls 'force accounting' as accounting is on force statements that include impulse and action [36]. Innovation capital, intellectual property, intangible asset such as *Mudharabah*, *Musharakah*, *Wadiah*, *Bai Bithamin Ajil*, *Ijarah* and *Wakalah* are the product from this kind of accounting. There could be more specialist accountants.

In the process of transforming Malaysia to 1-Malaysia, the key is then equality and non-discrimination. As Islam prohibits injustice, exploitation, discrimination, and all sort of unfairness, accounting should therefore be applicable to all and should be seen to fundamentally consistent to seek for the best for the society. The problem of 'force accounting' is that 'not accounting' could also a form of accounting too. As such, although specialists are being promoted, there is no previlaged accounting framework however for accountant in the name of accountability. Religious auditor, for example, is accountable not only with the task but with the 'no task' they do since the implication of doing and not doing is very big – to customers, Islamic banking industry, and society at large. Because accounting responds to accountability at issue, matters are divided into necessary, recommended, permissible, dislike or forbidden in the name of corporate accountability.

References

1. O. Marnet, History repeats itself: The failure of rational choice models in corporate governance. Critical Perspectives on Accounting **18**, 2 (2007)
2. C. R. Baker, R. Hayes, Reflecting form over substance: the case of Enron Corp. Critical Perspectives on Accounting **15**, 6–7 (2004).
3. N. Chabrak, N. Daidj, Enron: Widespread myopia. Critical Perspectives on Accounting **18**, 5 (2007)
4. R. J. Craig, J. H. Amernic, Enron discourse: the rhetoric of a resilient capitalism. Critical Perspectives on Accounting **15**, 6-7 (2004)
5. A. Gurowitz, Migrant rights and activism in Malaysia: Opportunities and constraints. The Journal of Asian Studies **59**, 4 (2000)
6. K. S. Jomo, Whither Malaysia's new economic policy? Pacific Affairs **63**, 4 (1991)
7. Ye Lin-Sheng, *The Chinese dilemma* (New South Wales, East West Publishing Pty Ltd., 2003)
8. R. S. Milne, Malaysia-beyond the new economic policy. Asian Survey **26**, 12 (1986)
9. Shukor Omar, *Malay business: Revivalism through entrepreneurship* (Kuala Lumpur, Pelanduk Publications, 2006).
10. G. de Ste Croix, *Greek and Roman accounting*, in A. C, Littleton, B. S. Yamey (eds.), Studies in the history of accounting (London, Sweet and Maxwell Limited, 1956)
11. G. de Ste Croix, *The class struggle in the Ancient Greek World from the Archaic age to the Arab conquests* (Gerald Duckworth and Co. Ltd. London, 1981)
12. R. H. Macve, *Some glosses on Greek and Roman accounting*, in B. S. Yamey, R. H. Parker (eds.), Accounting history: Some british contributions (Oxford, University Press, 1994)
13. B. L. Nigam, Bahi-Khata, The pre-Pacioli Indian double-entry system of bookkeeping. Abacus **22**, 2 (1986)
14. M. E. Scorgie, Indian imitation or invention of cash-book and algebraic double-entry. Abacus **26**, 1 (1990)
15. G. Vollmers, *Accounting historiography using ancient sources: Problems and rewards*, in R. Fleiscliman, V. Radclifle, P. Shoemaker (eds.). Doing accounting

history: Contributions to the development of accounting thought (Durham, Duke University Press, 2003)
16. R. Mattessich, Prehistoric accounting and the problem of representation: On recent archaeological evidence of the Middle East from 8000 B.C. to 3000 B.C. Accounting Historians Journal **14**, 2 (1987)
17. R. Mattessich, Recent insights into Mesopotamian accounting of the 3rd millennium B.C - Successor to token accounting. Accounting Historians Journal **25**, 1 (1998)
18. R, Mattessich, *The beginnings of accounting and accounting thought: Accounting practice in the Middle East (8000 B.C. to 2000 B.C.) and Accounting thought in India (300 B.C. and the Middle Ages)* (New York, Garland Publishers, 2000)
19. O. A. Zaid, *The historical and theoretical framework of financial accounting in the muslim society* (Amman - Jordan, Darul- Bashier for Publishing and Distribution), Second Edition, 1997)
20. O. A. Zaid, Were Islamic records precursors to accounting books based on the Italian method? Accounting Historians Journal **27**, 1 (2000)
21. O.A. Zaid, The appointment qualification of Muslim accountants in the Middle Ages. Accounting Education **9**, 4 (2000)
22. R. Emerson, Foreword. In Frank H. H. King. *The new Malayan nation: a study of communalism and nationalism* (New York, Institute of Pacific Relations, 1957)
23. Khoo Kay Kim, Recent Malaysian historiography. Journal of Southeast Asian Studies **10**, 2 (1979)
24. Khoo Kay Kim, Recent Malaysian historiography. Journal of Southeast Asian Studies **10**, 2 (1979)
25. A. Bhimani, Accounting and the emergence of "economic man". Accounting, Organizations and Society **19**, 8 (1994)
26. A. J. Richardson, Corporatism and intraprofessional hegemony: a study of regulation and internal social order. Accounting, Organizations and Society **14**, 5 (1989)
27. T. Tinker, A. Koutsoumadi, A mind is a wonderful thing to waste: "think like a commodity", become a CPA. Accounting, Auditing & Accountability Journal **10**, 3 (1997)
28. B. G. Carruthers, Accounting, ambiguity, and the new institutionalism. Accounting, Organizations and Society **20**, 4 (1995)
29. Y. Ijiri, A framework for triple-entry bookkeeping. The Accounting Review **61**, 4 (1986)
30. W. F. Chua, Teaching and learning only the language of numbers - monolingualism in a multilingual world. Critical Perspectives on Accounting **7**, 1 (1996)
31. T. M. Porter, Quantification and the accounting ideal in science. Social studies of Science **22**, 4 (1992)
32. K. Robson, Accounting numbers as "inscription": action at a distance and the development of Accounting. Accounting, Organizations and Society **17**, 7 (1992)
33. H. Vollmer, How to do more with numbers: Elementary stakes, framing, keying, and the three-dimensional character of numerical signs. Accounting, Organizations and Society **32**, 6 (2007)
34. S. Jönsson, Accounting for improvement: Action research on local management support. Accting., Mgmt. & Info. Tech. **2**, 2 (1992)
35. Y. Ijiri, A framework for triple-entry bookkeeping. The Accounting Review **61**, 4 (1986)
36. Y. Ijiri, A framework for triple-entry bookkeeping. The Accounting Review **61**, 4 (1986)

SHS Web of Conferences **34**, 06008 (2017) DOI: 10.1051/shsconf/20173406008
FourA 2016

Why do Accounting Students at Higher Learning Institutions Conduct an Academic Dishonesty?

Arizona Mustikarini[1*], *Rijadh Djatu* Winardi [1], and *Maria* Azalea [1]

[1]Department of Accounting, Faculty of Economics & Business, Universitas Gadjah Mada, Indonesia.

Abstract. Academic dishonesty is a serious educational problem. Moreover, the tendency to cheat at workplace appertain with the frequency of cheating in college. This study aims to empirically examine the influence of individual factors and situational factors to the intention of accounting students to conduct an academic dishonesty. This study uses survey as data collection technique by employing a set of a questionnaire. This result of this research finds that attitudes, subjective norms, and perceived behavioural control have positive and significant effects to the intention of the accounting students to commit an academic dishonesty. However, among three situational factors only pressure and definitional ambiguity that have a positive and significant effect to the intention of accounting students.

1 Background

Crimes in the financial sector have become the world's attention in recent years. After a series of corporate crimes that began to surface since the end of 2001-2003, the financial industry experienced some high-level financial scandals, mismanagement, fraud, embezzlement, as well as many cases of fraudulent in financial reporting and audit failures on large multinational corporations around the world [1].

Major financial disasters caused by high-level scandals such as Enron and WorldCom have raised many questions on how they were happened and who were the key actors. In the last five years, fraud perpetrators consistently are mostly from those who work in the accounting department of the company. Fraud committed by executives and upper-level management cause the most amount of losses, of which 66% of them have a college degree or postgraduate [2]. This is interesting because the educated person should demonstrate their educated personality. Educated personality is a power, a mindset, a mental attitude, and a certain wisdom that belongs to those who studied in a college [3].

Many research was conducted to find a relationship between unethical behaviours in the workplace with dishonesty attitude by college students. Several studies have found a strong relationship between academic dishonesty by students with unethical behaviour conduct on the job. [4] find that the tendency to cheat in the workplace highly correlates with the frequency of dishonesty in college, as well as the results of research by the [5, 6, 7] who

*Corresponding author: arizona.mustikarini@ugm.ac.id

DOI: 10.1051/shsconf/20173406008

find significant relationships between the unfair practices in the workplace with dishonest students.

Several studies also conducted to determine the factors that influence students to do academic dishonesty. [8] categorize two groups of factors related to ethical decision making which are individual and situational factors. Academic dishonesty in accounting education describes the violation of academic ethics by accounting students, where many of them will eventually become professional accountants and business leaders in the future [9]. Therefore, it is important to know the causal factors of academic dishonesty so the accounting education can determine the appropriate attempts and solutions to prevent it.

Studies on academic dishonesty among students, especially students of accounting, are still being conducted even though many have been carried out. Research on academic dishonesty mostly undertaken in western educational setting e.g. [6, 10, 11]. Few studies, however, have been undertaken in emerging economies e.g. [12, 13], particularly in Indonesia. The purpose of this study is to provide empirical evidence of individual and situational factors that affect the intention of accounting students in Indonesia conducting academic dishonesty. The key contributing factors might be different from those found in western countries because of different context, cultural background, religion, and attitudes of Indonesian educational context.

2 Hypotheses development

2.1. Attitude

According to [19], attitudes towards the behaviour is the positive or negative evaluation of the individual to such behaviour. Students who believe that the conduct of academic dishonesty will produce a good result, will have a positive attitude to academic dishonesty. Previous studies find that attitudes towards behaviour is one of the most influential factors of intention to perform the behaviour. A study from [20] find that attitude is the strongest predictor towards intention in twenty-nine of thirty research. [11] find a positive and significant relationship between attitudes towards academic dishonesty and the intention to commit academic dishonesty. These results are consistent with the previous studies [14, 15, 16]. Based on these arguments, there is a positive relationship between attitudes towards academic dishonesty and the intention to commit academic dishonesty. Therefore, the first hypothesis is:
H1: Students with more positive attitude toward academic dishonesty will show a higher intention to commit academic dishonesty.

2.2 Subjective norms

[19] posit that the subjective norm is the perceived social pressure or not perceived to perform or not perform a behaviour. Subjective norm is defined as a student's perception that most people who become a reference for him to think that he or she should or should not perform certain behaviours [21]. Previous study from [11] find that subjective norm is a significant predictor of intention to commit academic dishonesty. Similar results were found by [14, 15, 16]. [13] find that the view of the behaviour of peer learning provides normative support to academic dishonesty. Based on these explanation, this research propose that subjective norm can predict the intention to commit academic dishonesty. Therefore, the second hypothesis is:
H2: Students with higher subjective norm on academic dishonesty will show a higher intention to commit academic dishonesty.

2.3 Perceived behaviour control

According to [19] perceived behavioural control refers to the ease perceived by students to do academic dishonesty. A study from [11] find that behavioural control is perceived in a positive and significant effect on the intention to commit academic dishonesty. Several studies also find similar results that behavioural control affects the intention to commit academic dishonesty [14, 15, 16]. Based on above arguments, the third hypothesis proposed is:
H3: Students with higher perceived behavioural control will show a higher intention to commit academic dishonesty.

2.4 Academic integrity culture

Academic integrity culture refers to the values of the institution in increasing the academic honesty at the same time the prevention and punishment of violations of academic [10]. Academic integrity culture of an institution is very important and is a key driver of the perception on the intention to commit academic dishonesty [22]. Previous research conducted by [22, 23] show that the culture of high academic integrity will lead to fewer academic violations. [22, 23] conclude that academic integrity culture was the best predictor on academic dishonesty. Moreover, [10] explain that academic integrity culture affects the intention to commit academic dishonesty in students with low levels adjustments. Therefore, the fourth hypothesis in this study is:
H4: Higher academic integrity culture will lead to the lower intentions to commit academic dishonesty.

2.5 Definitional ambiguity

Definitional ambiguity effect on academic dishonesty committed by students. [12] reveal that when students do not fully understand what creates plagiarism (a form of academic dishonesty), and what penalties for plagiarism, students will not see it as a problem. A study from [13] find similar evidence with [12] whereas definitional ambiguity occurs when a student does not have a complete awareness of the code of conduct, or when they do not receive guidance from faculty about academic dishonesty. [13] find that definitional ambiguity strongly predicts rationalization against dishonest academic behaviour. Therefore, the fifth hypothesis in this study is:
H5: Higher definitional ambiguity will cause a higher intention to conduct academic dishonesty.

2.6 Pressure

Pressure is the motivation for dishonesty that may come from within the students themselves or from another person [17]. The pressure examined in this research is the pressure from another person (outside the student). [24] elaborate these pressures as student rate pressure, time pressure, and the pressure of the task given to them. [13] argue that stress encourages the tendency of students to commit academic dishonesty. [18] find evidence that pressure, in the form of deadline pressure, is a motivator for plagiarism (a form of academic dishonesty). Nevertheless, [9] and [12] find different results that no direct relationship between pressure and academic dishonesty among accounting students in the United Kingdom [9] and Malaysia [12]. Despite of the mixed results, this study attempts to provide empirical evidence whereas the higher the pressure faced by accounting students in Indonesia, the higher the intention to commit academic dishonesty. Therefore, the sixth hypothesis in this study is:

H6: Higher pressure will cause a higher intention to commit academic dishonesty.

3 Research methods

This research is conducted using a survey method. The survey within this study is conducted using questionnaires. Questionnaires are distributed directly by the researcher to the respondents. This method is widely used in previous studies e.g. [13]. The population of the respondent is all undergraduate accounting students at a state university in Yogyakarta, Indonesia year 2013 to 2015. Students with the enrolment year in 2013 to 2015 have been selected because they are categorized as active students.

4 Data analysis

4.1 Characteristics of respondents

The number of respondents in this study were 352 undergraduate accounting students of a state university in Yogyakarta, Indonesia. From the total questionnaires returned, only 342 questionnaires that can be used in research because 10 students do not fill the full research instruments. Total population in this research were 420 students. The total number of respondents are 342 students constitute of 81.43% of the total population. The type of academic dishonesty that mostly conducted is cheating with a percentage of 38.9%. A total of 70.2% of respondents conduct academic dishonesty since the first year of study. A total of 67.9% respondent mostly conduct academic dishonesty for Financial Accounting course. The most impactful effect perceived by respondents in conducting academic dishonesty are reducing the effort required to perform tasks with the percentage of 66%.

4.2 Validity, reliability, and classical assumption test

Testing the validity of the instrument was performed using Pearson bivariate correlation. To test the 342 samples, the terms R tables that must be met is 0.106. In testing the instrument of reliability, Cronbach's Alpha is used. The variables will be reliable if the value of Cronbach's Alpha is greater than 0.7. The normality test method used in this study is One-Sample Kolmogorov-Smirnov Test. The test results show the number 0.057, it is concluded that unstandardized residual of this study is normally distributed. Based on the multicollinearity test results, all independent variables have a VIF (variance inflation factor) less than 10 and a tolerance value greater than 0.1. It shows that there is no multicollinearity on all independent variables. Heteroscedasticity test is also conducted, showing that the significance value of all independent variables is greater than 5%. Thus, it can be concluded that there is no heteroscedasticity.

4.3 Hypothesis testing and interpretation

The hypothesis test used in this research is multiple linear regression. The multiple linear regression test with SPSS 23 is presented in the Table 1.

Table 1. Results of Multiple Linear Regression Analysis

Model	Unstandardized Coefficients		t	Sig.
	B	Std. Error		
(Constant)	-0,101	0,157	-0,646	0,519
Attitude (A)	0,365	0,057	6,350	0,000
Subjective Norms (SN)	0,284	0,056	5,064	0,000
Perceived Behavior Control (PBC)	0,199	0,039	5,042	0,000

Academic Integrity Culture (AIC)	0,118	0,071	1,652	0,100
Definitional Ambiguity (DA)	0,077	0,042	1,833	0,068*
Pressure (P)	0,129	0,050	2,603	0,010

*Supported on a significance level of 10%.

Based on regression analysis on Table 1, it can be formulated using a multiple linear regression equation, the beta coefficient that has been generated. The beta coefficient describes the influence of independent variables on the dependent variable as follows:

$$\mathbf{I = -0{,}101 + 0{,}365\ A + 0{,}284\ SN + 0{,}199\ PBC + 0{,}118\ AIC + 0{,}077\ DA + 0{,}129\ P + e}$$

Multiple linear regression model has the R value of 0.580, it means the level of the relationship between variables is 58%. In addition, the adjusted R-square of 0.325 indicates that the independent variables in the model can explain 32.5% change in the dependent variable. The remaining 67.5% is explained by other variables not included in this study.

From the calculation using SPSS 23, the F value is 28.321 and significance of 0.000. From table F with a 5% significance, the F value is 2.12337. From both the F values, F statistic is larger than F table. Moreover, the significance value of 0.000 is smaller than 5%. Thus, it is concluded that the independent variables which are attitude, subjective norm, perceived behavioral control, academic integrity culture, definitional ambiguity, and the pressure is jointly affect intentions significantly.

5 Conclusion

The results of this study indicate that individual factors which are attitudes, subjective norms, and perceived behavioural control significantly and positively affect the intention of accounting students to commit academic dishonesty. Furthermore, this study gains the results that academic integrity culture has no effect to the intention of accounting students to do academic dishonesty. In addition, this study also obtains the results that definitional ambiguity has effect to the intention of accounting students to do academic dishonesty significantly. Next, this study also obtained the results that pressure positively and significantly affect the intention of accounting students to commit academic dishonesty.

References

1. B. Soltani. The Anatomy of Corporate Fraud: A Comparative Analysis of High Profile American and European Corporate Scandals. *Journal of Business Ethics*, 120(2), 251-274. (2013)
2. Association of Certified Fraud Examiners. *Report to The Nation on Occupational Fraud and Abuse* [online]. Retrieved on January 19, 2016 from http://www.acfe.com/rttn/docs/2014-report-to-nations.pdf. (2014)
3. Suwardjono. *Belajar-Mengajar di Perguruan Tinggi: Redefinisi Makna Kuliah* [online]. Retrieved on February 28, 2016 from http://inparametric.com/download/Artikel1.pdf. (2005)
4. S. Nonis and C. Smith. Personal Value Profiles and Ethical Business Decision. *Journal of Education for Business Ethics*, 76(5), 251–256. (2001)
5. D. F. Crown and M.S. Spiller. Learning from the Literature on Collegiate Cheating: A Review of Empirical Research. *Journal of Business Ethics*, 17(6), 683–700. (1998)

6. R. A. Lawson. Is Classroom Cheating related to Business Students' Propensity to Cheat in the "Real World"? *Journal of Business Ethics*, 49, 189–199. (2004)
7. Z. Ma. Business Students' Cheating in Classroom and Their Propensity to Cheat in the Real World: a Study of Ethicality and Practicality in China. *Asian Journal of Business Ethics,* 2, 65-78. (2013)
8. R. C. Ford and W. D. Richardson. Ethical Decision Making: A Review of the Empirical Literature. *Journal of Business Ethics,* 13, 205-221. (1994)
9. X. Guo. Understanding Student Plagiarism: An Empirical Study in Accounting Education. *Accounting Education,* 20, 17-37. (1994)
10. J. L. Kisamore, T. H. Stone, and I. M. Jawahar. Academic Integrity: The Relationship between Individual and Situational Factors on Misconduct Contemplations. *Journal of Business Ethics,* 75(4), 381-394. (2007)
11. T. H. Stone, I. M. Jawahar, and J. L. Kisamore. Predicting Academic Misconduct Intentions and Behavior Using the Theory of Planned Behavior and Personality. *Basic and Applied Social Psychology,* 32, 35-45. (2010)
12. M. Smith, N. Ghazali, and S. Minhad. Attitudes Towards Plagiarism among Undergraduate Accounting Students: Malaysian Evidence. *Asian Review of Accounting*, 15(2), 122-146. (2007)
13. A. Ellahi, R. Mushtaq, and M. B. Khan. Multi campus investigation of academic dishonesty in higher education of Pakistan. *International Journal of Educational Management,* 27(6), 647-666. (2013)
14. I. Ajzen and L. Beck. Predicting Dishonest Actions Using the Theory of Planned Behavior. *Journal of Research In Personality,* 25, 285-301. (1991)
15. M. J. Mayhew. S. M. Hubbard, C. J. Finelli, and T. S. Harding. Using Structural Equation Modeling to Validate the Theory of Planned Behavior as a Model for Predicting Student Cheating. *The Review of Higher Education,* 32(4), 441-468. (2009)
16. T. P. Cronan, J. K. Mullins, and D. E. Douglas. Further Understanding Factors that Explain Freshman Business Students' Academic Integrity Intention and Behavior: Plagiarism and Sharing Homework. *Journal of Business Ethics,* 1-24. (2015)
17. D. Becker, J. Connolly, P. Lentz, and J. Morrison. Using the Business Fraud Triangle to Predict Academic Dishonesty Among Business Students. *Academy of Educational Leadership Journal*, 10(1), 37-54. (2006)
18. H. P. Koh, G. Scullyb, and D. R. Woodliffa. The Impact of Cumulative Pressure on Accounting Students' Propensity to Commit Plagiarism: an Experimental Approach. *Accounting and Finance,* 51, 985-1005. (2011)
19. I. Ajzen. *Attitudes, Personality and Behaviour*. Maidenhead: Open University Press. (2005)
20. D. Trafimow. The Importance of Attitudes in the Prediction of College Students' Intention to Drink. *Journal of Applied Social Psychology*, 26(24), 2167–2188. (1996)
21. I. Ajzen. The Theory of Planned Behavior. *Organizational Behavior and Human Decision Processes*, 50(1), 179–211. (1991)
22. D. L. McCabe, L. K. Trevino, and K. D. Butterfield. Honor Codes and Other Contextual Influences on Academic Integrity: A Replication and Extension of Modified Honor Code Settings. *Research in Higher Education*, 43(3), 357-378. (2002)
23. D. L. McCabe, L. K. Trevino, and K. D. Butterfield. Academic Integrity in Honor Code and Non honor Code Environments: A Qualitative Investigation. *The Journal of Higher Education*, 70(2), 211-234. (1999)
24. P. G. Love and J. Simmons. Factors Influencing Cheating and Plagiarism Among Graduate Students in a College of Education. *College Student Journal*, 32(4), 1-8. (1998)

Intellectual capital disclosure determinants and its effects on the market capitalization: evidence from Indonesian listed companies

Angga Arifiawan Sudibyo[1,*], and *B.* Basuki[2]

Faculty of Economics and Business, Universitas Airlangga, Surabaya, Indonesia

Abstract. Currently, intellectual capital (IC) is the main source of company's competitive advantage. The role of IC has become one of the main factors of company's value creation process. Since companies rely more on their intangibles (including IC), management must provide an appropriate annual report in which discloses information regarding IC in order to give better information to the related parties. The aim of this study is to examine variables that expected to have a significant influence to the level of intellectual capital disclosure (ICD) practices in Indonesian listed companies' annual reports. A content analysis method is used to examine the ICD level, based on the framework developed by [1]. The IC information was collected from 135 Indonesian LQ45 listed companies, which derived from their annual reports from the year 2012 to 2014. Sample companies are classified into two groups of industries: high- and low profile industries in order to analyze differences in IC reporting practices between the two groups. The empirical results proved that company size, industry type and market capitalization are significantly associated with the ICD, while the influence of company profitability on the level of ICD is not significant. Furthermore, it is revealed that there is no significant difference in the intellectual capital disclosure practices between companies in high- and low-profile industries.

Keywords: Intellectual capital disclosure; Content analysis; Annual report; Industry type; LQ45; Indonesia Stock Exchange; Independent t-test; Market Capitalization, ROA

1 Background

Recently, new economies are shifting towards a knowledge-based economy or knowledge economy (KE) in which companies' competitiveness and sustainability are increasingly dependent of knowledge based resource. The dramatic shift from material sources to knowledge, from hardware to software is actually experiencing by companies across the world. Their key economic resources are no longer natural resource or physical capital, but the knowledge itself [2]. The transition from manufacturing-based economies towards knowledge-based economy makes a continuous enhancement of IC importance in company's value creation process.

[3] found that the relevance of traditional financial accounting model is unable to represents the key factors of company's long term value (i.e., intangible assets). Companies with traditional financial accounting valuation method are based on balance sheet, income statement or statement of cash flow. Stakeholders have been dissatisfied with this traditional financial reporting made by companies and its capability to present users with sufficient information about company's ability to create wealth. [4] as one of the pioneers in intellectual capital studies are argued that reporting intellectual capital of companies' intangible assets is a substantial way of connecting the information gap that may exist between owners and managers.

There are numerous prior study related to IC and even they have taken similar variables, but the empirical results were varying and several studies have contradictory results. This condition is likely happened because until now there is still no

*Corresponding author: anggasudibyo@live.com

regulation or standards of reporting of IC information on the annual reports. In case of Indonesia, the regulation regarding to components of Intellectual Capital is stated in [5]. However, the scope is very limited only for intellectual property that can be included in traditional financial statement. The list of IC items that should be reported in the company's annual report until now is still not regulated yet.

This study concentrates on empirical study on the firms' characteristics (company size, industry and profitability) that becomes the determinants of ICD. It is aimed to also find empirical evidence on the influences of Intellectual Capital of company that represented in their annual report to its market capitalization during the year 2012 until 2014. Companies listed in LQ45 index selected as the sample in this study. LQ45 companies are known for their high market capitalization and transaction value, and also have a good prospect of growth. Thus, those companies are expected to be more active in disclosing their IC. These companies come from various industries so then it is expected to be able in representing two types of industries (high- and low-profile) that become one of the variables in this study. This study also hopes to be able in comparing the level of intellectual capital disclosure between companies in high-profile and low-profile industries.

2 Literature review and hypotheses development

Under the stakeholder theory, a company should consider that stakeholders have a powerful competency to influence management in managing the resources owned by the company. The greater power of stakeholders, the greater company's effort in creating value will be. By appropriately managing the resources, the company will be able to create more value and enhance the financial performance. A company's strategic information including IC information should be communicated to the stakeholders in order to meet their needs. Therefore, as the company disclose their IC information, it is expected to raise the confidence of stakeholders regarding to the company and reduce the risk and uncertainty faced by investors.

A company's resources consist of all assets, both tangible and intangible, capabilities, organizational processes, human and nonhuman identified and controlled by the company, and that permit it to consider and apply value-enhancing strategies [6]. Based on Resource-based View (RBV) Theory, a company will have a competitive advantage when it is able in implementing a value creating strategy, not by simultaneously being implemented by any current or potential competitors. [7] explained about difference strategies (defender and prospector) can be implemented by company to reach its competitive advantage. Knowledge based resource is mainly derived from the RBV ideas. Knowledge resource within a company will allow company to compete either by being a defender or prospector. Intellectual capital, according to RBV point of view, is considered as an important source of competitive advantage because IC items are mostly able to fulfil the attributes of good resource. Thus, it is expected to make a competitive advantage for the company.

Moreover, the basic ideas of signalling theory stated in [8] signalling is a reaction to information asymmetry, it was also used in many accounting studies to explain voluntary disclosure practices of accounting information [9]. [8] proposed that between two parties could get asymmetric information problem if one party send a signal that would reveal some relevant information to the other party. Thus, that informed party will interpret that given signal and adjust the purchasing behaviour. Information of IC in the annual report can be interpreted as a signal. Companies should signal their IC to their stakeholder or to the capital market as "good news" in order to increase the share price or attract new investors.

2.1 Intellectual Capital Disclosure

Intellectual Capital is often divided into different components: human capital (HC), relational capital (RC), and structural capital (SC). HC includes knowledge, skills, and experience owned by the employees of the company. Structural capital (SC) refers to knowledge that attached to the structure and organization process, including corporate culture, share of knowledge, technology, and availability of information systems. Relational or Customer capital is a good relation (association network) owned by company to its stakeholders. The relation here refers to the supplier relations, customer loyalty, goodwill and a good relation with government and society in general. Petty and [10] argues that external capital defined as company's ability to identify the need of the market, so that a good relation may be build between company and external parties. It is argued that between these two IC components is dependent-independent each other. [10] stated that the creation of patent highly depends on the HC, but after it is completed, it considered as company's SC. Information on intellectual capital is important to stakeholders in their decision-making. The greater disclosure provide by a company might reduces the uncertainty be faced by investors and thus it will reduce company's cost of capital. Even though disclosing intellectual capital still voluntarily done by the company, but the number of companies that realized the importance of reporting non-financial information is increasing over years.

2.2 Company Size

Company sixe is reflected by the value of total assets stated in company's balance sheet. The greater amount of total assets, the larger the size of a company, and vice versa. The size of company may affect the company's operating capabilities that will affect the share returns of companies. Studies by [11], [12] and [13] found that size is a significant determinant for intellectual capital disclosure. [12] notes that large companies are basically have more value-driven activities and also larger companies being more visible and obtain more attention by the stakeholders, such as public and government.

H1: Company size has influences on the intellectual capital disclosure

2.3 Industry Type

Information disclosed by companies in particular industry which they consider as relevant, is might not always considered as relevant information for companies in another industry [14]. This study propose that type of industry is a key factor for intellectual capital disclosure recall that intellectual capital among companies in some industries are more important than companies in other industries. The sample is into two groups of industries which are high-profile and low-profile industry. High-profile industries as companies with high consumer visibility, political risk and strong competition among its competitors and whose economics activities more likely to modify their environment, more often discloses the environment impacts rather than other industries, while low-profile industries are the otherwise of high-profile industries. According to [15] companies classified into high-profile industries are mining, chemical, wood, automotive, aviation, agriculture, tobacco and cigarettes, food products and beverages, media and communications, energy (electricity), engineering, health and transport and tourism. Meanwhile, companies in low-profile industries include real estate industry, finance and banking, healthcare, consumer goods, retail and household products. The general expectation is that companies in high-profile industries will disclose more IC information than companies in low-profile industries. High profile companies are more likely to be reliant on intellectual capital and therefore high profile companies are likely to disclose more intellectual capital information under the annual reports.

H2.1 Industry types has influences on the intellectual capital disclosure

H2.2 There is a significant difference of the level of intellectual capital disclosure between high-profile and low-profile industries

2.4 Profitability

Companies with higher profitability are expected to disclose more information to the public in comparison to the companies with lower profitability. According to [6] resource-based view (RBV) theory, companies are able to generate value and will indicate a profitable growth if they follow those growth opportunities that will fit to their resources base. Higher profit companies are mainly attributed to their competitive advantage. Companies that have competitive advantages assumed that they have more financial resources so then they are able to disclose more.

H3 Company's profitability has influences on the intellectual capital disclosure

2.5 Market Capitalization

Market capitalization can be defined as the total worth of companies' securities in the capital market [16]. IC is a part of companies' capital and resources, in which contributes to the wealth creation process. Information regarding IC in annual reports helps to make capital markets more efficient by reducing the information asymmetry problems between management of company and its investors. Additionally, IC disclosure helps the capital market to provide a more accurate market capitalization of firms [17]. When stakeholders have better understanding about the company, they will belief in company's future capabilities in creating more profit, so then it might raise the company stock price (market capitalization increase).

H4 Company's Market Capitalization is influenced by the disclosure of Intellectual Capital

2.6 The differences of ICD between High Profile and Low Profile Industry

This study divided the sample companies into two groups of industries which are high-profile and low-profile industry. Thus, from the study it is expected to be able in explaining any differences of the level of ICD between those two types of industries. The general expectation is that companies in high-profile industries will disclose more IC information than companies in low-profile industries. High profile companies are more likely to be reliant on intellectual capital and therefore high profile companies are likely to disclose more intellectual capital information under the annual reports. However, a study by [11] has been found that companies belonging to high-profile and low-profile industries disclose the same level of ICD under the annual reports.

H5 There is a significant difference of the level of intellectual capital disclosure between high-profile and low-profile industries

3 Research methods

3.1 Sample Selection

DOI: 10.1051/shsconf/20173407001

Data collected are obtained by using documentation technique which comprising the company's annual reports. A sample of Indonesian LQ45 companies was selected for content analysis of the annual reports from the three-year periods from 2012 until 2014. The reason behind the selection of LQ45 is due to it is the 45 most liquid and most heavily traded stocks in the Indonesia Stock Exchange (IDX). Additionally, it represents various industries in the stock market.

3.2 Content analysis of annual report

This study uses content analysis to collect empirical data about ICDs [18]. Content analysis has been argued as the most reasonable methodology for data collection. The ICD index used in this study replicates the index constructed to measure both the extent and quality of IC disclosures (Li *et al.*, 2008). This index contains of 61 items that covered the three IC subcomponents of structural, relational, and human capital. 22 items are HC elements, 18 are SC, and the remaining 21 represent RC elements (see Table 3.1). IC related items appears on the annual report will be scored 1, if there is no disclosure will be scored 0. Thus, the sum of all obtained scores represents company's ICD score. The level of disclosure for each company's annual report is then calculated by dividing the sum or total score of disclosure with the total number of items scored.

$$ICD\ Score = \frac{\Sigma di}{M}\ x\ 100\% \text{ .... [Eq X]}$$

There are three independent variables: company size, industry type, profitability, and intellectual capital disclosure, while the dependent variables are intellectual capital disclosure and market capitalization.

Table 3.1: IC Framework

Classification of intellectual capital, Li et al (2008)	
Human Capital	**Relational Capital**
Number of Employee	Customers
Employee age	Market presence
Employee Diversity	Customer relationships
Employee Equality	Customer acquisition
Employee Relationship	Customer retention
Employee Education	CTE
Skills/ know-how	Customer involvement
Employee Competences	Company image
Employee Knowledge	Company awards
Employee Attitudes	Public relation
Employee Commitments	Diffusion & networking
Employee Motivation	Brands
Employee Productivity	Distribution channels
Employee Training	Relationship with suppliers
Vocational Qualifications	Business Collaboration
Employee Developments	Business agreements
Employee Flexibility	Favourite contract
Entrepreneurial Spirit	Research collaboration
Employee Capabilities	Marketing
Employee Teamwork	Relationship with stakeholders
Employee Community Involvement	Market leadership
Other Employee Features	
Structural Capital	
Intellectual Property	Technology
Process	Financial dealings
Management Philosophy	Customer support function
Corporate Culture	Quality management & improvement
Organization Flexibility	Knowledge-based infrastructure
Structure	Accreditations (certificate)
Learning	Overall infrastructure/ capability
Research & Development	Networking
Innovation	Distribution Network

Results and discussions

The results of multiple linear regression analysis show that variable SIZE and TYPE has significant influence towards the level of ICD, while company's profitability, denoted by ROA has no significant influences towards the level of ICD. The second model examines the level of IC owned by company presented by ICD on annual reports towards company's Market Capitalization (MCAP) proved that the relationship between these two variables is significant. Table below provides the result of independent sample t-test in regard to compare the level of ICD between Indonesian companies in high- and low-profile industries.

Simple Regression Results (Model 2)		
Independent Variables	**MCAP**	
	t-statistic	Sig.
ICD	5.082	0.001
Multiple Regression Results (Model 1)		
Independent Variables	**ICD**	
	t-statistic	Sig.
SIZE	3.639	0.001
TYPE	1.868	0.064
ROA	0.263	0.793
Independent Sample t-test	0.217	1.541

4 Research Discussions

4.1 The influence of Company Size towards Intellectual Capital Disclosure

The result indicates that SIZE has a significant positive influence towards ICD. The result is consistent with some previous studies conducted by [11], [19], [12] and [13], which shows the company size has significant positive effects on the intellectual capital disclosure. But this result is in contrast to study conducted by [20]. The result of this study proves that the larger a company, the higher level of reporting conducted (including IC items). In other words, the larger LQ45 companies disclose more intellectual capital information than the smaller companies. Furthermore, larger companies are generally becoming a subject to special attention from financial analysts, so that in order to maintain their credibility, they have to be transparent to the public.

4.2 The influence of Industry Type toward Intellectual Capital Disclosure

The result also showed that the industry type having a significant impact on intellectual capital disclosure at the level of significance 10%. In regard with the signaling theory, company that hopes to send a positive signal to its stakeholders will tend to disclose more information. This study also supports a study by Bozzolan et al. (2003) which argues that companies within a same industry have the same motive in the reporting activities, so when a company show a lower level disclosure compared to others within the same industry, it is a signal that company hiding some information.

4.3 The differences of ICD between companies in High-Profile and Low-Profile Industries

This study used independent-samples t-test to see the difference of level of intellectual capital disclosure between companies in high-profile and low-profile industries. Table above shows that the value is greater than the level of significance (0.217 > 0.10). Even tough according to the results of the prior hypothesis (H2) proved that high-profile companies exhibit greater disclosure than low-profile companies, however the result of independent t-test indicated that there is no statistically significant difference between high-profile and low-profile industries. This result is similar to the study by [21]. In this case, the insignificant differences of ICD between this two types of industries is probably occurs

 DOI: 10.1051/shsconf/20173407001

because each company whether in high-profile or low-profile industries are tend to provide a complete information to the stakeholders. Therefore, companies across the industries are equally eager to disclose information as complete as possible, including information related to IC items as the company's business strategy.

4.4 The influence Profitability towards Intellectual Capital Disclosure

The result of t-test shows that level of significance of ROA is above the significance value 0.1; it means that profitability has no significant influence to ICD. The result is consistent with the study by [13] and [19] which also failed to support the influence of profitability towards ICD.

The signaling theory by [8] explained that high profitability companies use their financial information to send signals to the market. Furthermore, high profitability companies may have advantage to signal their stakeholder that they are a good company by giving more information about the IC items in its annual reports. However, this study indicates that profitability does not significantly influence the level of ICD. It related with the signaling theory, companies with high profitability tend to not disclose more information related to intellectual capital because they already send a positive signal to the market regarding the financial information. Thus, any additional voluntary disclosure made by company will no longer necessary.

4.5 The influence of Intellectual Capital toward Market Capitalization

The result of the t-test showed that there is a high significant relationship between these two variables. The findings are similar to study by [22] and [23]. The more IC items owned by company, the higher market capitalization would be. Under the stakeholder theory, when a company has higher level of IC, the managers will let the public (stakeholders) know about how is the management performance. The stakeholders would value more regarding to the information disclosure, and thus it might decrease the misevaluation of the company's share price. Moreover, eventually company's share price would increase. As a result of the higher share prices, company's market capitalization would be also affected.

5 Conclusion, limitation and suggestion

The results proved that company size, industry type and market capitalization are significantly associated with the ICD, while the influence of company profitability on the level of ICD is not significant. Furthermore, it is revealed that there is no significant difference with regard to the intellectual capital disclosure practices between companies in high- and low-profile industries.

There are several limitations of the study that influenced the interpretation of the results from the study. Firstly, only three independent variables are used in the study (model 1) in order to examine the level of ICD. R^2 result shows that those variables are only able to explain the model at 8.1 %, and 91.8% are explained by other variables which have better capabilities in explaining the ICD. The second limitation is the use of ROA. The study reveals that the ROA is not significantly influence into the level of ICD.

Based on those limitations, this study proposed some suggestion to mitigate the limitations. First, future studies could also analyses controlling factors to provide additional insights into the relationships between the IC's determinants and ICD. Even tough the role of control variables is not a primary concern in the results, but by adding some control variables hopes that the value of coefficient of determinants will increased. Second, further research could use another measurement of profitability such as ROE or NRM. Ideally, the different measurements of company profitability might offer a different statistical result.

References

1. Li, J., Pike, R., & Haniffa, R. Intellectual capital disclosure and corporate governance structure in UK firms. *Accounting and Business Research,* **38**(2), 137-159 (2008)
2. Drucker, P. The Future That Has Happened Already. *Harvard Business Review*, 7**5**(5), 20-22 (1997)
3. Lev, B. and Zarowin, P. The boundaries of financial reporting and how to extend them. *Journal of Accounting Research*, **37**(2), 353-83 (1999)
4. Eccles, R. and S. Mavrinac. Improving the Corporate Disclosure Process. *Sloan Management Review,* Summer, **36**(4), 11-25 (1995)
5. IAI. The Indonesian Financial Accounting Standards (PSAK) 19
6. Barney, J. Firm Resources and Sustained Competitive Advantage. *Journal of Management,* **17(**1), 99-120 (1991)
7. Miles, R. E., & Snow, C. C. Organizational Strategy, Structure, and Process. New York: McGraw-Hill (1978)
8. Spence, Michael. Job Market Signaling. *The Quarterly Journal of Economics*, **87** (3), 355-374 (1973)

9. Yi, A. & Davey, H. Intellectual capital disclosure in Chinese (mainland) companies. *Journal of Intellectual Capital,* **11**(3), 326-347 (2010)
10. Basuki, B & Kusumawardhani, T. Intellectual Capital, Financial Profitability, and Productivity: An Exploratory Study of the Indonesian Pharmaceutical Industry. *Asian Journal of Business and Accounting* **5**(2), 41-68 (2012)
11. Bozzolan, S., Favotto, F., & Ricceri, F. Italian annual intellectual capital disclosure. *Journal of Intellectual Capital,* **4**(4), 543-558 (2003)
12. Ousama, A. A., Fatima, A. and Hafiz-Majdi, A. R. Determinants of intellectual capital reporting Evidence from annual reports of Malaysian listed companies. *Journal of Accounting in Emerging Economies*, **2**(2), 119-139 (2012)
13. Ferreira, A. L., Branco. M. C. and Moreira, J. A. Factors influencing intellectual capital disclosure by Portuguese companies. *International Journal of Accounting and Financial Reporting*, **2**(2), 278-298 (2012)
14. Ousama, A. A. and Fatima, A. H. Factors influencing voluntary disclosure: empirical evidence from Shari'ah approved companies. *Malaysian Accounting Review*, **9**(1), 85-103 (2010)
15. Utomo, Muhammad Muslim. Praktek Pengungkapan Sosial Pada Laporan Tahunan Perusahaan di Indonesia (Studi Perbandingan Antara Perusahaan High Profile dan Low Profile). *Simposium Nasional Akuntansi* IV, IAI (2000)
16. Ang, Robbert. Buku Pintar Pasar Modal Indonesia (The Intelligent Guide to Indonesian Capital Market, Mediasoft Indonesia
17. Guthrie, James and Petty, R. Intellectual Capital: Australian Annual Reporting Practices. *Journal of Intellectual Capital*, **1** (3), 241-251 (2000)
18. Guthrie, James and R. Petty, K. Yongvanich, F. Ricceri. Using content analysis as a research method to inquire into intellectual capital reporting. *Journal of Intellectual Capital*, Vol. **5** Iss: 2, pp.282 – 293 (2004)
19. Oliveira, L., L. L. Rodrigues, and R. Craig. Firm-specific determinants of intangibles reporting: Evidence from the Portuguese stock market. *Journal of Human Resource Costing and Accounting*, **10**(1), 11-33 (2006)
20. Bukh, P. N., C. Nielsen, P. Gormsen, & J. Mouritsen. Disclosure of information on intellectual capital in Danish IPO prospectuses. *Accounting, Auditing and Accountability Journal,* **18**(6), 713-732 (2005)
21. Bronzetti, G, Mazzotta, R, Sicoli, G, & Baldini, M.A. Intellectual Capital Disclosure in Sustainability Reports. Intellectual Capital Strategy Management for Knowledge-Based Organization. Hershey, PA, USA: IGI Global. 2013. pp. 195-214 (2013)
22. Abdolmohammadi, M. J. Intellectual capital disclosure and market capitalization. *Journal of Intellectual Capital,* **6**(3), 397-416 (2005)
23. Abeysekera, I., & Guthrie, J. An empirical investigation of annual reporting trends of intellectual capital in Sri Lanka. *Critical Perspectives on Accounting,* **16**(3), 151-163 (2002)
24. Donaldson, T., & Preston, L. E. The Stakeholder Theory of the Corporation: Concepts, Evidence, and Implications. *The Academy of Management Review,* **20**(1), 65 (1995)

 DOI: 10.1051/shsconf/20173407002

Investors' response on the deviation between quarterly and annual earnings

Saidatunur Fauzi Saidin[1,*], *Mazrah Malek*[2] , *Daing Nasir Ibrahim*[3] and *Phua Lian Kee*[4]

[1]Universiti Putra Malaysia, Department of Accounting and Finance, Faculty of Economics and Management, 43400 Serdang, Selangor, Malaysia
[2]Universiti Utara Malaysia, Tunku Intan Shafinaz School of Accountancy, Faculty of Business, 06010 Sintok, Kedah, Malaysia
[3]Universiti Malaysia Pahang, Canseleri Tun Abdul Razak, 26600 Pekan, Pahang, Malaysia
[4]Universiti Sains Malaysia, School of Management, 11800 USM, Penang, Malaysia

Abstract. Despite the benefits of timely information, concerns have been raised on the reliability of the quarterly accounts. The occurrence of deviation between audited annual accounts and cumulative quarterly accounts indicates the misstatements in the latter. This study examines investors' response towards the occurrence of the deviation. Data is based on 792 listed companies of Bursa Malaysia in 2012. The results of the OLS regression show that the companies without earnings deviation have significantly higher earnings response coefficient than companies experiencing earnings deviation. It is also found that understated quarterly earnings companies have higher earnings response coefficient than overstated companies. Results imply that investors place higher reliability on the quarterly accounts produced by companies without earnings deviation. Higher reliability is placed on the understated quarterly earnings companies as compared to overstated companies. Findings suggest that investors do value the occurrence and types of earnings deviation. Therefore, steps should be taken to overcome the occurrence of deviation.

1 Introduction

The quarterly accounts which cover an entity's activities for a three-month period rather than twelve-months as reported by the annual accounts allow for a more frequent and timely dissemination on the progress of an entity. Frequent information helps to reduce uncertainties [1]. Timely information increases the relevance of the information [2]. It can be observed that many countries have mandatorily required their listed companies to produce quarterly reports. In Malaysia, the listed companies on the Bursa Malaysia are mandatorily required in March 1999 to produce quarterly accounts for quarters ending on or after 31st July 1999 as a response to the Asian Financial Crisis in 1997/98 and in replace of the half-yearly reporting which was made mandatory since 1987 [1,3]. Currently, the requirement is regulated under Chapter 9.22 of the Listing Requirements of Bursa Malaysia, where listed companies are mandated to produce their quarterly accounts not later than two months after end of each quarter. Part A of Appendix 9B of the Listing

* Corresponding author: saidatunur@upm.edu.my

Requirements further elaborates the information that needs to be disclosed in the accounts. However, like in many other countries, the accounts are not required to be audited.

Despite the benefits of frequent and timely information, many have questioned the reliability of quarterly accounts [2,4,5]. Earlier studies by [5] and [6] assess the reliability of quarterly accounts based on the occurrence of earnings deviation between audited annual accounts and cumulative quarterly accounts. While audited accounts are considered as high quality, the studies argued that the occurrence of earnings deviation represents the misstatements in quarterly accounts. The concern by the Bursa Malaysia has led to mandatory disclosure announcement, whereby companies with 10-percent or more in deviation are required to make immediate announcements concerning the deviation together with complete explanations. While theoretically the occurrence of earnings deviation is an evidence of low quality quarterly accounts, users' perspective on the deviation is still unclear. As the main users of quarterly accounts, investors' perspective towards the occurrence of earnings deviation is important to be examined. An earlier study in Malaysia by [5] examines the effect of audit committee characteristics on the occurrence of earnings deviation. Meanwhile, the study by [6] only examines the occurrences of earnings deviation among the U.S listed companies. This study extends both studies by focusing on the effect of earnings deviation on investors' reliance to the subsequently produced quarterly earnings.

2 Literature review

Financial reporting is a main mechanism in dissemination of asymmetry information between the managers and the stakeholders [7]. Even though, financial information can also be relieved by other means, financial accounts provide the most comprehensive and reliable source of information. Many have shown that the accounts help in the economic decision-making process [8, 9]. Therefore, it is very crucial for the users to have quality financial accounts to avoid misleading judgments. Traditionally, financial accounts are prepared for external users on an annual basis and this has been a common practice across countries. However, annual accounts suffer from timely problem, whereby some of the information reported by the accounts may become irrelevant at the time of production [2]. To overcome this timely problem, interim reporting is required. Countries such as Malaysia, Singapore and the U.S are among those which require their listed companies to produce interim reporting on a quarterly basis, while in the U.K, only half-yearly reporting is required. Interim reporting helps to release frequent information to stakeholders which then reduces uncertainties and enhances confidence over the company's state of affairs [1].

Despite the benefits of frequent and timely information, many are sceptical about the reliability of information provided by the quarterly accounts. As noted by [2]: "while interim reports increase the relevance of the financial statements through more timely communication of position and results, their usefulness to users is also a function of their reliability" (p.268). The concern on the reliability of quarterly accounts arises mainly due to nature that quarterly accounts are not required to be audited by external auditors in most jurisdictions. Unaudited accounts expose them to the risk of errors and manipulations by the managers [3]. Empirical evidence by prior studies has also indicated for the need of auditors' involvement in the quarterly accounts. For example [10] and [11] found that the quarterly earnings which are audited have a higher earnings response coefficient than those without auditor involvement. A study by [12] of financial analysts' perceptions had found that the mean reliability increases with increasing auditor association. Attempts have been made by regulators in some countries such as in the U.S and Thailand by requiring limited audit review of quarterly accounts. Many have highlighted the fourth quarter settling-up phenomenon in the preparation of quarterly accounts where it is argued that the fourth quarter accounts are used by companies to reverse out the misstatements in the earlier three quarters [1, 3]. Conceptually, researchers have argued that since only annual accounts are audited, a company may intentionally misstate their earlier three quarterly accounts and then, make adjustments or corrections in the fourth quarter accounts to avoid the deviation between the quarterly (cumulative) and annual accounts. As claimed by [4], companies generally do not publish the fourth quarter accounts per se, but simply the difference between the annual and the cumulative of the first three quarters' accounts.

Theoretically, the occurrence of earnings deviation between quarterly (cumulatively) and audited annual accounts represents low reliability of quarterly accounts [5,6]. It denotes the occurrence of misstatements in quarterly earnings where higher (lower) earnings reported in quarterly accounts than in audited accounts represents overstatements (understatements). Even though, the occurrence of deviation indicates that necessary adjustments have been made in audited annual accounts to correct the accounts, negative perceptions towards the reliability of quarterly accounts by investors may still exist. They may still use the subsequent quarterly earnings produced by these companies in the equity valuation but the reliance is lower than the quarterly earnings produced by companies without deviation. Therefore, it is postulated that companies without earnings deviation will have higher earnings response coefficient than companies with earnings deviation.

3 Methodologies

Data is based on listed companies of Bursa Malaysia in year 2012. As at 31 December 2011, 956 companies are listed but 109 companies are excluded due to inability to determine the occurrence of earnings deviation. Another 5 companies which have changed their financial year, 50 companies which have been delisted during or after year 2012 are also excluded and thus, bringing the sample companies to only 792 companies. The sample consists of 317 (40 percent) without earnings deviation companies, 275 (35 percent) overstated quarterly earnings companies and 201 (25 percent) understated quarterly earnings companies. Data on quarterly earnings, audited annual earnings and asset size are collected through the Bursa Malaysia's website, while data on share prices and composite index are collected from Bursa Station. The earnings response model as proposed by [13] is used in testing the hypothesis. The model is originated from the semi-strong efficient market model by [14]. [13] postulates that by holding the prior uncertainty about cash flows constant, the earnings response coefficient will increase with the perceived quality of the earnings' signal. Thus, in this study, the occurrence of earnings deviation acts as a moderating variable to the relationship between earnings performance and abnormal return. Therefore, the Ordinary Least Square regression is used and takes the following form:

$$CAR = EP + PROFIT + SIZE + EP*DEVIATE \tag{1}$$

CAR covers for two days of event window; on announcement date and a day after announcement date ($t0$, $t1$) measured by [15] market model. EP is measured by dividing the difference between the earnings per share of current quarter and prior quarter with the market value of share two days prior to the earnings' announcement. PROFIT is measured by dichotomous measurement, where the value of 1, if the observed company has profit and 0, if the observed company incurred a loss in the quarter. SIZE is measured by natural logarithm of total assets. DEVIATE is measured by the difference between profit attributable to owners of the company reported in cumulative quarterly accounts and audited annual accounts. Dichotomous measurement is used where the value of 1, if the observed company does not have earnings deviation and 0, if the observed company has earnings deviation in prior year.

DOI: 10.1051/shsconf/20173407002

4 Results

Univariate analyses are conducted by separating the sample into two groups; companies with earnings deviation and companies without earnings deviation, to examine the mean and frequency differences between the two groups. As can be observed from Table 1, both types of companies are found to have identical mean for CAR. However, earnings deviation companies have better performance (EP), but have smaller assets than without earnings deviation companies. However, only the difference in SIZE is significant between the two groups.

Table 1. Descriptive and univariate analyses (continuous measurements)

Variable	Deviation Companies Mean (std. deviation)	Without Deviation Companies Mean (std. deviation)	T-test
CAR	.001 (.058)	.001 (.048)	.047
EP	.245 (35.815)	-.045 (13.035)	.276
SIZE	8.451 (.628)	8.565 (.765)	-4.576*

* significant at 1 percent level

Meanwhile, from Table 2, it can be observed that the percentage of without deviation companies with positive CAR, negative EP and has PROFIT are larger than deviation companies. However, only the distribution based on profit is significantly different between the two groups.

Table 2. Descriptive and univariate analyses (dichotomous measurements)

Variable		Deviation Companies Number (Percentage)	Without Deviation Companies Number (Percentage)	Chi-square
CAR	0	987 (51.947)	631 (49.763)	1.452
	1	913 (48.053)	637 (50.237)	
EP	0	973 (51.210)	664 (52.366)	.407
	1	927 (48.789)	604 (47.634)	
PROFIT	0	529 (27.842)	292 (23.028)	9.178*
	1	1,371 (72.158)	976 (76.972)	

*significant at 1 percent level

Based on the univariate analyses, it can be concluded that companies without earnings deviation have bigger size of assets and are more likely to have profit than companies with earnings deviation. On the other hand, both types of companies have identical cumulative abnormal return and earnings performance.

Table 3 presents correlation results and as can be observed, CAR are all positive and significantly correlated with EP, PROFIT, and EP*DEVIATION. The correlation between CAR and SIZE is negative. Meanwhile, the correlations among the independent variables are considerably low. While the threshold of 0.8 is usually used for possible multicollinearity problem, this indicates that multicollinearity issue is not critical [16].

Table 3. Correlation analyses

Variable	CAR	EP	SIZE	PROFIT
EP	.126*	-	-	-
SIZE	-.020	-.004	-	-
PROFIT	.152*	.109*	.250*	-
EP*DEVIATION	.115*	.285*	.018	.099*

*significant at 1 percent level

Meanwhile, Table 4 presents the results of OLS regression. The adjusted R-squared is 0.043 and is significant at one percent. It implies that the independent variables used in the model explain about 4 percent of the variation in the CAR. The low adjusted R-squared is consistent with the earlier studies in Malaysia by [17] of 0.10 and in other countries such as by [18] and [19] of around 0.03. The coefficients of EP and PROFIT are both positive and significant at a one percent level. The positive coefficient of PROFIT implies that profiting companies have positive abnormal returns and losses companies have negative abnormal returns of shares. It is consistent with the earlier findings by [19], [20] and [21]. Meanwhile, the positive coefficient of EP implies that the higher the earnings performance, the higher the abnormal returns, which is consistent with the findings by earlier studies [17, 22, 23]. Both of the results are consistent with the argument by [14] that the positive information received by capital market will result in a positive abnormal return, while negative information will result in a negative abnormal return of shares. Meanwhile, the coefficient SIZE is negative and significant at a one percent level. The negative coefficient implies that the bigger the size of assets, the lower the abnormal return of shares which is consistent with the findings of earlier studies by [20] and [25].

As hypothesized, the coefficient of the variable representing the interaction of earnings performance and earnings deviation, EP*DEVIATION is positive and significant at a one percent level. This implies that companies without deviation have higher earnings response coefficients than companies experiencing deviation. The investors of the capital market place higher reliability on quarterly earnings produced by companies without deviation than companies experiencing deviation. The result is consistent with the findings by earlier studies such as [18], [19] and [24] on earnings response coefficients and argument by [13] that investors' response to the earnings will increase with the perceived quality of earnings performance. The investors may still rely on the quarterly accounts produced by companies experiencing deviation in their equity valuation decision, but the reliance is lower than the quarterly accounts produced by the companies without deviation. While quarterly earnings provide value relevance information to investors, the occurrence of earnings deviation decreases investors' reliability to the earnings' numbers.

Table 4. Ordinary Least Square Regression statistics

Variable	Expected sign	t	Coefficient
EP	+	4.85	.0001*
SIZE	-	-3.24	-.0046*
PROFIT	+	8.23	.0185*
EP*DEVIATION	+	4.17	.0005*
Constant		2.22	.0263**
Adjusted R-squared			.0425*

*significant at 1 percent level

An additional analysis is conducted to examine the difference in response of different types of deviation by using the sample of deviation companies only. In this analysis, DEVIATION is measured by 1, if the companies experience understated deviation and 0, if the companies experience overstated deviation. As presented in Table 5, the result is consistent with the earlier regression. The coefficients for EP and PROFIT are significant and positive, while SIZE is significant and negative. Meanwhile, EP*DEVIATION is positive and significant, which implies that companies experiencing understated deviation have higher earnings response coefficient than companies experiencing overstated deviation. Investors place higher reliability on earnings produced by understated companies than those from overstated companies. The types of earnings deviation is valued by the investors in placing their reliability to the quarterly earnings.

Table 5. Ordinary Least Square Regression statistics (deviation companies)

Variable	Expected sign	t	Coefficient
EP	+	2.87	.0002*
SIZE	-	-2.42	-.0051*
PROFIT	+	6.44	.0194*
EP*DEVIATION	+	2.42	.0002*
Constant		1.71	.0002**
Adjusted R-squared			.0369*

*,** significant at 1, 5 percent level respectively

5 Conclusions

The reliability of quarterly accounts has been criticised and prior studies have indicates questionable reliability of the accounts. Earlier studies have argued that the occurrence of earnings deviation between cumulative quarterly accounts and audited annual accounts as an evidence of the misstated quarterly accounts. However, investors' views on the deviation are still unclear. This study argues that the investors may still rely on the quarterly accounts produced by companies experiencing earnings deviation, but the reliance is lower than the quarterly accounts produced by companies without earnings deviation. Using data of listed companies on Bursa Malaysia for year 2012, OLS regression results show that companies without earnings deviation have higher earnings response coefficient than companies experiencing earnings deviation. Furthermore, it is also found that companies experiencing understated quarterly earnings have higher earnings response coefficient than companies experiencing overstated quarterly earnings. The results suggest for the need of companies and regulators to enhance the quality of quarterly accounts. Perhaps by requiring external auditors' involvement in the production of quarterly accounts may resolve the occurrence of earnings deviation.

This paper is funded under the Fundamental Research Grant Scheme (FRGS) by the Malaysia Ministry of Higher Education.

References

1. K. N. I. K. Ismail, S. N. Abdullah, Asian J. of Fin. and Acctg., **1** (2009)
2. K. Lightstone, N. M. Young, T. McFadden, Acctg. Pers., **11** (2012)
3. K. N. I. K. Ismail, R. Chandler, IIUM J. of Econ. and Mgmt., **13** (2005)
4. W. A. Collins, W. S. Hopwood, J. C. McKeown, J. of Acctg. Res., **22** (1984)
5. D. N. Ibrahim, K. J. Raman, S. F. Saidin, Sing. Mgmt. Rev., **31** (2009)
6. M. Al-Darayseh, J. F. Brown Jr., J. of App. Bus. Res., **8** (1992)
7. G. Whittington, Acctg. Bus. Res., **23** (1993)
8. B. Lev, J. A. Ohlson, J. of Acctg. Res., **20** (1982)
9. M. Penno, J. of Acctg. Res., **23** (1985)
10. D. Manry, S. L. Tiras, C. M. Wheatley, Acctg. Rev., **78** (2003)
11. M. Malek, S. F. Saidin, M. N. H. Osman, DLSU Bus. and Econ. Rev., **25** (2016)
12. K. Pany, C. H. Smith, J. of Acctg. Res., **20** (1982)
13. R. Holthausen, R. Verrecchia, J. of Acctg. Res., **26** (1988)
14. E. F. Fama, J. of Fin., **25** (1970)
15. W. F. Sharpe, J. of Fin., **19** (1964)

16. D.N. Gujarati, McGraw Hill, 2003.
17. K. Ismail, R. A. Rahman, J. of Bus. Mgmt., **6** (2012)
18. J. Krishnan, H. Sami, Y. Zhang, Auditing: A J. of Prac. and Theo., **24** (2005)
19. J. R. Francis, B. Ke, Rev. of Acctg. Stud., **11** (2006)
20. F. A. Gul, J. Tsui, D. S. Dhaliwal, Acctg. and Fin., **46** (2006)
21. K. E. Hackenbrack, C. E. Hogan, Cont. Acctg. Res., **19** (2002)
22. J. Lee, C. W. Park, J. of Bus. Fin. and Acctg., **27** (2000)
23. W. R. Landsman, E. L. Maydew, J. of Acctg. Res., **40** (2002)
24. I. Haw, D. Qi ,W. Wu, J. of Acctg. and Fin., **23** (2008)
25. S. W. Teoh, T. J. Wong, Acctg. Rev., **68** (1993)

CEO origin, CEO tenure, and earnings quality: empirical evidence from Indonesia

Poppy Nurmayanti M[1,*] and *Fu'ad* Rakhman[2]

[1] Doctoral Student in Accounting Department, Faculty of Economics and Business, Universitas Gadjah Mada, , Jl. Sosio Humaniora No.1 Bulaksumur,Yogyakarta 55281, Indonesia
[2] Accounting Department, Faculty of Economics and Business, Universitas Gadjah Mada, Jl. Sosio Humaniora No.1 Bulaksumur, Yogyakarta 55281, Indonesia

Abstract. This study examines whether CEO origin and tenure affect earnings quality. Using firms listed in the Indonesia Stock Exchange between 2012 and 2014, we provide evidence that firms with insider CEOs report earnings of higher quality relative to those with outsider CEOs. We further show that as CEO tenure increases and CEOs obtain more experience, the effect of CEO origin on earnings quality becomes less significant.

1 Introduction

This study examines the effects of Chief Executive Officer (CEO) origin and tenure on earnings quality. CEO origin is associated with whether from the CEO was hired from inside or outside the company. An insider CEO is one that has been working for the company prior to the appointment to the CEO position. An outsider CEO, on the contrary, is hired from outside the company. CEO origin is one of the demographic characteristics that potentially affect the crucial decision-making by CEOs, such as accounting decisions [1]. The upper echelons theory states that managers' demographic characteristics are associated with their cognitive styles and values that affect the manager managerial decision-making [2]. Further, Shen and Cannella found that outsider CEOs has a stronger association with corporate performance than insider CEO [3]. The study also finds that outsider CEOs tend to have a shorter tenure than insider CEO [3, 4, 5].

Outsiders CEOs face stronger pressure to demonstrate their capabilities, and thus have more incentives to manage earnings, especially in their early years. Kuang et al. find that outsider CEOs have stronger incentives to manage earnings at the beginning of the tenure years following the appointment as CEO, when compared to insider CEO [5]. This study extends Kuang et al. to test whether CEO origin affects earnings quality by incorporating CEO tenure as a moderating variable [4].

This study uses discretionary accruals as a proxy for earnings quality by using accrual estimation error from the Dechow and Dichev model. We also use the modified Jones model to measure accruals quality as an additional analysis.

We find that in general, insider CEOs report earnings of higher quality than outsider CEOs. Further tests indicate that the result is prevalent only when CEO tenure is less than the median value (less than five years). When CEO tenure reaches five year or longer, the superior earnings quality associated with insiders CEOs is no longer present. This is

*Corresponding authors: poppynurmayanti@yahoo.com, frakhman@ugm.ac.id

consistent with the view that since outsider CEOs face greater pressure to show results, they tend to manage earnings early in their tenure years, lowering earnings quality. This results in a lower earnings quality for firms with outsider CEOs relative to those with insider CEOs. However, when they survived the early years, they no longer have as great incentives to manage earnings. Consequently, after five years, there is no longer any significant difference in earnings quality between firms with insider and those with outsider CEOs.

This study has several important contributions. This study contributes to the accounting literature by providing evidence on the effect CEO origin on earnings quality and how CEO tenure moderates the association between the two variables. This study provides insight to investors and board of directors that hiring CEOs from within provides an advantage in the form of higher earnings quality, at least in the early years, relative to hiring outsiders.

2 Literature Review and Hypothesis Development

2.1 CEO Origins

High earnings quality reflects the company's operational performance more accurately [6]. Dechow et al. argues that higher quality earnings provide more information about the features of a firm's financial performance that are relevant to a specific decision made by a specific decision maker [7]. Previous studies have examined the determinants of earnings quality with a focus on the characteristics or background of top managers. The results of these studies provide mixed results [8].

One aspect stakeholders need to consider when appointing a new CEO is CEO origin. Shen and Cannella argue that outsider CEOs are usually hired when companies desire that their leadership take a new direction [3]. Further, CEOs are hired from outside when the company has a poor performance. This situation requires an outsider CEO to be hired to immediately address the problems and take actions quickly. Consequently, outsider CEOs from are likely to face greater pressure from board of directors and the market to show his managerial ability to achieve results immediately.

Davidson et al. define that a CEO is an insider when the CEO has worked for the company for more than six years. CEOs working for the company for less than six years are classified as outsiders [9]. Kuang et al. [5] describe that CEOs recruited from outside the companies often take advantage of his opportunity as CEO to demonstrate their capability to improve corporate performance [5]. This is useful for his career, so that the labor market can assess and evaluate the CEO's capabilities by comparing it with the CEO recruited from inside the company. Wieserma and Shen and Cannella argued that CEOs from outside the company have stronger interests to report better performance in the long term however, the outsider CEOs tend to have expectations for survival in the company lower than the insider CEOs [4, 3].

There are several advantages to recruit outsider CEO; (1) CEO from outside company more demonstrated his/her ability to improve his/her career [5, 10], and (2) CEO from outside company has the advantage that a company's reputation is maintained especially when companies facing financial distressed or poor performance [11]. Appointment of insider CEO has two advantages, are: (1) inside CEO is to have a better understanding about the internal condition of the company, product, market and competition, and a closer relationship with the client, and (2) inside CEO is to have better social networks, so that they can obtain specific information [12].

Kuang et al. [5] found that CEOs recruited from outside the company have stronger incentives to manage earnings at the beginning of the tenure years when compared to

DOI: 10.1051/shsconf/20173407003

insider CEOs [5]. This implies that outsider CEOs manage earnings more than insider CEOs. Outsider CEOs feel more pressure to to demonstrate good performance early in their tenur years. Therefore, it is expected that outside CEOs will manage earnings more intensively than insider CEOs, especially in the beginning of their tenure years.

2.2 CEO Tenure

Existing empirical evidence find that the market cannot accurately judge the ability of a CEO at the beginning his tenure [12], especially when the CEO does not have previous experience as a CEO. Newly hired CEOs are aware that the market evaluates their ability based on the past and current performance [13]. Thus, to build a reputation in the market and to avoid being labeleld as "low skill" managers, CEOs in early tenure years are likely to have a strong incentive to report good performance. Thus, if the actual performance of certain periods is lower than expected, CEOs have strong incentives to puff up the company's income through aggressive reporting.

Zhang concludes that CEOs with longer tenure are less likely to be aggressive in financial reporting compared CEOs with shorter tenure [13]. CEOs who have longer tenure will try to maintain their reputation, so the CEOs are less aggressive. Further, Ali and Zhang found that CEOs at the beginning of their tenure overstate reported accounting earnings instead of at the end of tenure [14]. The new executives need to prove to the stakeholders of their performance at the beginning of their tenure to quickly obtain recognition and job security.

Kim and Yang provides empirical evidence that a board of directors with longer tenure has a positive impact on financial reporting quality [15]. Specifically, these findings concluded that the boards of directors with longer tenure are associated with lower discretionary accruals. On the contrary, Santoso provides different evidence that CEOs who have a longer tenure actually reported higher discretionary accruals than CEOs with a shorter tenure [16].

2.3 Hypotheses Statements

Based on the arguments in the previous sections, we propose the following the hypothesis:
H1: Insider CEOs report earnings of higher quality than outsider CEOs.
H2: As CEO tenure increases, the effect of CEO origin on earnings quality become less significant.

3 Research Methodology

3.1 Population and Sampling

The study uses firms listed in the Indonesian Stock Exchange (IDX) from 2012 to 2014 as the sample. Following other studies, firms in the financial sector are excluded from the sample. Table 1 Panel A describes the sample selection procedure. The initial sample size is 1,448 firm-years. After eliminating financial firms, public firms younger than two years old, those with missing financial data or with insufficient information about CEO details, and outliers, we obtain a total sample of 632 firm-years for the study. Table 1 Panel B shows that the majority of the CEOs are insiders (88.9%), while the longest tenure is 48 years and the shortest tenure is 1 year. Further, the table shows that the average age of CEO is 52.87 years old, with the oldest is 85 years old and the youngest is 29 years old.

DOI: 10.1051/shsconf/20173407003

Table 1
Sample Selection and Descriptive Statistics

Panel A: Sample Selection

Procedure	2012	2013	2014	Total
Total public firms listed in IDX	459	483	506	1448
Less:				
Financial sectors	(75)	(81)	(87)	(243)
Delisting public firms in IDX	(4)	(7)	(1)	(12)
Public firms listed less 2 years	(23)	(31)	(24)	(78)
Missing financial data or CEO demographic data	(207)	(109)	(115)	(424)
Outlier	(18)	(15)	(19)	(38)
Final sample	132	240	260	632

Panel B: Descriptive Statistics (n = 632)

Variable	Min	Max	Mean	Median	SD
ABS_DD	6.21E-05	0.551	0.053	0.038	0.059
ORIGIN	0	1	0.889	1	0.314
TENURE	1	48	8.677	5	9.515
AGE	29	85	52.866	52	8.448
FirmSize	5.674	19.122	13.857	13.997	1.924
LNSDOCF	4.071	15.909	10.737	10.848	1.980
ROA	-0.297	0.613	0.062	0.049	0.096
LEV	0.007	0.986	0.459	0.461	0.210
LOSS	0	1	0.141	0	0.348
YEAR	0	1	0.791	1	0.407
INDUSTRY	0	1	0.704	1	0.457

Variable definition:

ABS_DD	: discretionary accruals as earnings quality proxy using Dechow and Dichev model,
ORIGIN	: CEO origin determined by a dummy variable equal to 1, if insider CEO; 0 if otherwise,
AGE	: CEO age determined by a dummy variable to 1, if age more than the median years; 0 if otherwise,
TENURE	: CEO tenure determined by a dummy variable to 1, if tenure more than median year; 0 if otherwise,
FIRMSIZE	: natural logarithm of sales,
LNSDOCF	: operating cash flow volatility estimated by standard deviation of operating cash flow in year t-1, t, and t+1 divided by total asset in year t-1,
ROA	: return on assets measured by net income divided by total assets in year t,
LEV	: leverage measured by total liabilities divided by total assets in year t,
LOSS	: financial condition measured by a dummy variable equal to 1, if net income is negative; 0 if otherwise,
YEAR	: years observation as measured by a dummy variable equal to 1, if observation in 2014; 0 if otherwise,
INDUSTRY	: industrial sector, as measured by a dummy variable equal to 1, if industry sector is excluded from trade, services, and investment sectors, services, and investment sectors; 0 if otherwise.

3.2 Research Model and Variable Measurement

We follow Dechow and Dichev (2002) to calculate the discretionary accruals [17]. The following equation (1) presents the research model, with absolute value of accrual estimation error or *ABS_DD* obtained from Dechow and Dichev (2002) as the dependent variable. A higher value of *ABS_DD* indicates lower earnings quality, and vice versa [18, 13].

$$ABS\_DD_{it} = \propto_0 + \propto_1 ORIGIN + \propto_2 AGE + \propto_3 TENURE + \propto_4 Firmsize + \propto_5 LNSDOCF + \propto_6 LEV + \propto_7 ROA + \propto_8 LOSS + \propto_9 Industry + \propto_{10} Year + \varepsilon \quad (1)$$

To estimate ABS_DD, we calculate total accruals (TA_{it}) with the following equation (2) below.

$$\frac{TA_{it}}{A_{it-1}} = \alpha_0 + \alpha_1 \frac{OCF_{it-1}}{A_{it-1}} + \alpha_2 \frac{OCF_{it}}{A_{it-1}} + \alpha_3 \frac{OCF_{it+1}}{A_{it-1}} + \varepsilon_{it} \quad (2)$$

TA_{it} is the total accruals calculated as net income minus operating cash flow; A_{it-1} is the total assets of the firm i from year $t-1$, OCF_{it-1} is the operating cash flow in the year $t-1$; OCF_{it} is the operating cash flows of year t; OCF_{it+1} is the operating cash flow in the year $t+1$; and ε_{it} is error term. CEO origin is a dummy variable set to 1 if the CEO is an insider and 0 otherwise. Insider CEO is defined as in Davidson et al. [12]. We also define CEO tenure as a dummy variable set to 1 if CEO tenure is less than the median tenure of CEO (shorter tenure), and if 0 otherwise. Following previous studies (e.g. Wang 2006), this study includes some control variables such as CEO's age, firm's size, the volatility of cash flow operating (*LNSDOCF*), *ROA*, leverage (*LEV*) and negative earnings (*LOSS*) [19]. _

4 Result

Table 2 shows that the variable *ORIGIN* for the regression with the full sample has a negative impact on the size of absolute discretionary accruals. The result implies that the insider CEOs report a lower discretionary accruals than the outsider CEO. In other words, insider CEOs report higher earnings quality than outsider CEOs. The result supports H1.

Table 2
Regression Results for the Impact of CEO Origin on Earnings Quality

Independent Variable	**Dependent Variables (*ABS_DD*)** Short-term (< 5 years) Coefficient	t-statistic	Long-term (≥ 5 years) Coefficient	*t-statistic*	Full Sample Coefficient	*t-statistic*
ORIGIN	-0.021 **	-2.414	0.024	1.421	-0.016 **	-2.205
TENURE					-0.008 *	-1.757
Control Variables						
AGE	0.006	0.803	-0.007	-1.361	-0.001	-0.156
FIRMSIZE	-0.009 ***	-3.728	-0.004 **	-2.022	-0.006 ***	-4.204
LNSDOCF	0.002	1.050	0.000	0.152	0.001	1.017
LEV	0.038 **	2.319	0.016	1.132	0.029 ***	2.582
ROA	0.209 ***	5.774	0.162 ***	4.896	0.194 ***	7.621
LOSS	0.044 ***	4.664	0.023 **	2.518	0.039 ***	5.790
YEAR	0.025 ***	2.862	0.009	1.509	0.017 ***	3.182
INDUSTRY	-0.010	-1.366	-0.014 ***	-2.567	-0.013 ***	-2.772
n	348		284		632	
Adj R2	16.91%		10.27%		15%	
F-statistic	8.846		4.597		12.494	

***, **, and * indicate significance at the 1 percent, 5 percent, and 10 percent levels, respectively.
See Table 1 for the variable definitions.

To test hypothesis 2, we split the sample into two groups based on the CEO tenure. The first group consists of observations whose CEO tenure is shorter than five years (N=348), while the second group consists of those with five years or longer (N=284). The result suggests that insider CEOs are associated with higher earnings quality relative to outsider CEOs only in the early years of their tenure. Table 2 shows that for CEOs whose tenure is less than five year, insider CEOs generate higher earnings quality. However, for CEOs whose tenure is five years or longer, we do not find any association between CEO origin and earnings quallity. This could be interpreted as that earnings quality gap between insider and outsider CEOs are more prevalent in the first years. As outsider CEOs have more experience after holding their position for five years or longer, the earnings quality gap between insider CEOs and outsider CEOs become less significant. The result supports H2.

We also conduct additional analysis using accrual quality measure from the model developed by Dechow et al [20]. Generally, the results of the additional analysis (not tabulated) are consistent with the results reported in this study.

DOI: 10.1051/shsconf/20173407003

5. Conclusion

This study investigates the effect of CEO origin and CEO tenure on earnings quality. We find that compared with outsider CEOs, insider CEOs tend to report earnings with higher quality. However, we find that the effect of CEO origin is prevalent only in the early years of CEO tenure. After five years, we do not find any significant difference in earnings quality between firms with insider CEOs and those with outsider CEOs.

This study contributes to the accounting literature by providing evidence on the effect CEO origin on earnings quality and how CEO tenure moderates the association between the two variables. This study provides insight to investors that hiring CEOs from within provides an advantage in the form of higher earnings quality, at least in the early years, relative to hiring outsiders.

Further studies could investigate more deeply on the relationship of the origin of the CEO and the company's financial reporting behavior in the context of horizon problems. Researcher might also consider factors such as CEO educational background in accounting, accounting certification, whether CEO turnover was routine or nonroutine, and their impacts on earnings quality.

References

1. Q. Cheng and K. Lo. Journal of Accounting Research. **44**, 5, 815-848. (2006)
2. D. Hambrick, and P. Mason. Academy of Management Review. **9**: 193 – 206. (1984)
3. W. Shen and A. A. Cannella. The Academy of Management Journal. **45**, 4, 717 – 733. (2002)
4. M. F. Wieserman. Human Resource Management. **34,** 1, 185 – 202. (1995)
5. Y. F. Kuang, B. Qin, dan J.L. Wielhouwer. Accounting Horizons. **28**, 3, 605 – 626. (2014)
6. P. R. Demerjian, B. Lev, M. F. Lewis, S. E. McVay. The Accounting Review. **88**, 2, 463 – 498. (2013)
7. P. Dechow, W. Ge., and C. Schrand. Journal of Accounting and Economics. **50**, 344 – 401. (2010)
8. H. Li. International Journal of Financial Research. **5**, 2, 79 – 86. (2014)
9. W. N. III Davidson, C. Nemec, and D. L. Worrell. Journal of Management and Governance. **10**, 35 – 57. (2006)
10. D. Setiawan.International Business and Economics Research Journal. **4**, 2, 119 – 128. (2008)
11. Isabelle Dherment-Ferere, and Luc Renneboog. Working papers from www.researchgate.net, issue 2000-70. (2000)
12. R. Gibbons and K. J. Murphy. Working papers from www.researchgate.net. (1991)
13. W. Zhang. Working papers from www.psu.edu. (2009)
14. A. Ali, and W. Zhang. Journal of Accounting and Economics. **59**,1,1 – 118. (2015)
15. K. Kim and J. S. Yang. Review of Integrative Business and Economics Research. **3**,1, 237 – 256. (2014)
16. R. D. Santoso. Skripsi FEB UGM. (2013)
17. P. M Dechow, and I. Dichev. The Accounting Review. **77**, Supplement, 35 – 59. (2002)
18. J. Francis, A. Huang, S. Rajgopal, and A. Zang. Contemporary Accounting Research. **25**, 1, 109 -147. (2008)
19. D. Wang. Journal of Accounting Research. 44, 3, 619 – 656. (2006)
20. P. M. Dechow, G. Sloan, and A. P. Sweeney. The Accounting Review. **70**, 2, 193 – 225. (1995)

A closer look at accounting for Islamic financial institutions

Nurazleena Ismail[1] *and Zulkarnain* Muhamad Sori[2]

[1]Senior Lecturer, Universiti Teknologi MARA, Kota Bharu, Kelantan
[2]Professor, INCEIF, The Global University of Islamic Finance, Kuala Lumpur

Abstract. This paper provides a discussion of a study that examines accounting in Islamic perspectives. The discussion includes an overview of the development of Islamic finance, accounting in Islamic perspectives, and four key accounting assumptions, namely 'substance over form', 'time value of money', 'fair value' measurement and 'recognition based on probability', which serve as important guides to preparers of financial information.

Keyword: Islamic Accounting, Substance over Form, Financial Information, Fair Value, Time Value of Money

1 AN OVERVIEW OF ISLAMIC FINANCE DEVELOPMENT

Islamic finance has been growing quickly and its importance as an element of the global economy has increased in recent years. According to Ernst and Young's World Islamic Banking Competitiveness Report 2016, Islamic banking assets marked at US$990 billion in 2015. The market share of Islamic banking assets in QISMUT+3 countries (QISMUT+3 is an acronym for Qatar, Indonesia, Saudi Arabia, Malaysia, United Arab Emirates, Turkey, and Kuwait, Bahrain and Pakistan) is exceeding US$920 billion (93% share). Currently, there are over 600 Islamic financial institutions (IFIs) globally and Islamic banking assets are expected to reach USD3 trillion in 2020.

This growth is being achieved due to the involvement of Islamic finance in the provision of financial products and services by IFIs for Shariah (Islamic Law) approved underlying transactions and economic activities. Shariah is a basis for finance that meets the religious requirements of Muslims in line with their 'aqidah as a factor that differentiates Islamic finance from conventional finance (Tajudin, 2010). The current Shariah-compliant products have covered the full range of banking, capital markets, asset management and insurance (takaful) businesses (Azmi, 2011). Shariah complaints' products refer to product with contracts that comply with Shariah and with the prohibition of riba (which literally means 'increase' or 'addition'), maisir (speculation or gambling), and gharar (uncertainty or ambiguity), and permit risk sharing. The basic sources of Shariah are the Qur'an and the Sunnah (Thajudeen, 2012). It follows the consensus of the jurists (Qiyas) and interpreters of Islamic law ('ijmak). The principles that are held within the kingdom of Shariah are absolute, not only in the details of its transactions but also in the extent of its role in implementing the Maqasid al-Shariah. Maqasid al-Shariah or the objectives of Islamic law is one of the most important concepts in strengthening modern Islamic finance and its examination will lead to Maslahah (public interest). Therefore, the Shariah principles will ensure that Islamic finance transactions are current and transparent, and forbid involvement in illegal activities that are harmful to society.

The alignment of accounting for Islamic finance with conventional business without compromising Shariah principles is a challenge for global groups with varied international stakeholders and extensive financial reporting. Even the Qur'an considers this issue as a serious problem, as revealed in this verse:

> *"O ye who believe! When ye deal with each other, in transactions involving future obligations in a fixed period of time, reduce them to writing let a scribe write down faithfully as between the parties; let not the scribe refuse to write: as Allah has thought him, so let him write. Let him who incurs the liability dictate, but let him fear his Lord Allah, and not diminish aught of what he owes. If the party liable is mentally deficient, or weak or unable himself to dictate, let his guardian dictate faithfully. And get two witnesses, out of your own men, and if there are not two men, then a man and two women, such as ye choose, for witnesses, so that if one of them errs, the other can remind her. The witnesses should not refuse when they are called on (for evidence)…." (2: 282)*

This verse has led to the development of a branch of accounting literature to highlight the differences between Western and Muslim business environments (Mirza and Baydoun, 1999). As a result, the regulators of Islamic Financial Institutions (IFIs) have endorsed a major research undertaken by the International Shariah Research Academy (ISRA) to examine four key assumptions that serve as important guides to preparers of accounts and have significant impact on Islamic financial transactions: these are i) substance over form; ii) the time value of money; iii) fair value measurement; and iv) recognition based on probability (Shafii et al., 2013). In addition, this paper has reviewed the literature in relation to the topic of financial reporting in Islamic finance. Accounting theory in conventional and Islamic perspectives has also been examined.

The paper is divided into six sections. The second section illustrates accounting from an Islamic perspective. The third section provides guidelines on financial reporting for Islamic financial institutions. The fourth section provides data for the study. The fifth section discusses on analysis of the four accounting assumptions. Finally, section six concludes the paper.

2 ACCOUNTING IN ISLAMIC PERSPECTIVES

This section discusses accounting based on the Islamic perspective. It considers decision usefulness, stewardship, accountability, financial measurement and disclosure and presentations, and confers about them in the light of the Quran and Sunnah.

Accounting is a book-keeping system, known as "double entry". According to the known history of accounting, there are many books on "Accounting Theory", which first appeared in Italy in the 13th century, born from the hands of an Italian Priest named Luca Pacioli. He wrote the book "Summa de Arithmatica Geometria et Propotionalita" with a chapter on the "Double Entry Accounting System". The development of accounting theory subsequently improved towards the barter economy issues pertaining to understanding the meaning and scope of application of theories from the humanities and management sciences to develop accounting theory. According to Hendriksen (1977) and McDonald (1972), cited in Unegbu (2014), the development of an accounting theory should be possible. McDonald claims that such a theory must have three elements: i) encoding of phenomena to symbolic representation; ii) manipulation or combination according to rules; and iii) translation back to real-world phenomena. In addition, Unegbu (2014) stated that accounting theory is a material field in accounting. In history, accounting predates the monetary economy from the era of barter economy. Through the barter trade (exchange of goods for goods), transactions were pre-determined not only by measurement but also by exchange values. This economy system contributed inherent problems in terms of measurement inequality, was unwieldy in the variety of production and was limited by the problem of coincidence of wants.

Conventional accounting is a method generally used by the managers within a company. There are five principles underlying conventional accounting (Baydoun and Willet, 2000). These are firstly, clear separation from economic intercourse with religious considerations; secondly, the paramount importance of individual satisfaction; thirdly, profit maximization as the gauge of successful performance; fourthly, acceptance of the rule of the survival of the fittest as the best strategy, and finally, a focus on the result of the process without regard for its wider impact on the environment. It also focuses on aspects such as decision-making, the future of the company, and timeliness. This method gives the managers a better idea of the current company situation and how it will affect finances in the future. As mentioned by Remi (2006), accounting has had a significant impact on the decision-making process towards socio-economic and political development, especially in terms of recording, preparing, interpretation, auditing and management and investment. It also has an impact on merger and acquisition, planning, controlling and storage of business operations. Therefore, in most cases, the accounting system has allowed the equity owners to have confidence in the works and reports of the stewards. They are also entrusted with the owners' capital assets (Unegbu, 2014).

Even though conventional accounting is a well-established field, it has been criticized among scholars (Parker, 2014; Lev, 2001) in relation to accounting in Islamic views. The criticisms of conventional accounting include the following: i) externalities are not included in financial reports; ii) intangible assets cannot be measured adequately; iii) human resources and employee reporting interests are not included; iv) social interests are overlooked; v) they promote the exploitation of capitalism over labor and society; and vi) they promote the concentration of wealth and power in the hands of the rich. In light of these criticisms, Ahmed (2012) highlighted that accounting from an Islamic perspective is all about practicing positive norms and bringing self-transcendent values into everyday life, while seeking the will of God and following his orders in the form of the Prophet's (PBUH) Sunnah.

The debate between best practices in terms of accounting from the Islamic perspective or conventional accounting in the same accounting line still continues. This is due to the development of IFIs and a growing belief that the assumptions underpinning Western financial accounting systems are not compatible with Islamic beliefs and values (Haniffa and Hudaib, 2007). Ahmed (2012) added that there have been debates and discussions in literary circles around the world, particularly in Islamic nations, about how modern accounting practices can be transformed to synchronize with the Islamic philosophy of life. Conventional accounting may differ from accounting from the Islamic perspective in terms of its objectives. The objective of conventional accounting is focused on identifying economic events and transactions, whereas accounting from the Islamic perspective must identify socio-economic and religious events and transactions. It has to fulfill the ultimate accountability to Allah and to ensure fair and just financial transactions between human beings. This is because the goals of accounting and management from the Islamic perspective are not merely worldly and money-oriented. Therefore, the Accounting and Auditing Organization for Islamic Financial Institutions (AAOIFI) was established in 1990 to meet the requirements of accounting standards from the Islamic perspective. It was registered in Bahrain in 1991. The objectives of the AAOIFI are i) to develop accounting and auditing thoughts relevant to IFIs; ii) to disseminate accounting and auditing thoughts relevant to IFIs and its applications through training, seminars, publication of periodical newsletters, carrying out and commissioning of research and other means; iii) to prepare, promulgate and interpret accounting and auditing standards for IFIs; and iv) to review and amend accounting and auditing standards for IFIs.

3 FINANCIAL REPORTING FOR ISLAMIC FINANCIAL INSTITUTIONS

The emergence and growth of Shariah business in the modern economy has led to the rise of accounting from the Islamic perspective. This is because accounting from the Islamic perspective is based upon ethical laws originating in the Qur'an and Sunnah, requiring full disclosure of information in accordance with Shariah law and public accountability that focuses on the community that participates in exploiting resources. In order to recognize the Shariah transactions in Shariah business and IFIs, there is a need for accounting treatment and standards devoted to Shariah. This is because conventional accounting is based upon modern commercial law, which is permissive rather than ethical, with limited disclosure of information (public interest) and personal accountability (focus on individuals who control resources). Alim (2014) revealed that accounting from the Islamic perspective has to transact based on Shariah, including a profit-sharing account, a margin revenue account and an additional accounting report such as a statement of the sources and uses of Zakat funds early in its development. It also implies a Shariah-based accounting theory for the concept of Shariah transactions, called the epistemology of Shariah on accounting. There are several epistemology approaches in Islamic accounting, such as empiricism, rationalism, phenomenalism, positivism and post-modernism.

In relation to Islamic financial transactions, acceptance among the relevant players in the global Islamic finance industry is a key challenge, especially in accounting and reporting (Shafii et. al, 2013). Recently, there have been two main challengers for the role: i) the IFRS, issued by the International Accounting Standards Board (IASB); ii) the FAS issued by the AAOIFI. In September 2009, the MASB issued Statement of Principles i-1 (SOP i-1) entitled 'Financial Reporting from an Islamic Perspective'. SOP i-1 served to inform Malaysian constituents that IFRS shall apply to Islamic financial transactions in the absence of any Shariah prohibition to doing so. In addition, the MASB issued a new approved accounting framework – the Malaysian Financial Reporting Standards (MFRS) framework – in November 2011 (Shafii and Zakaria, 2013). The purpose of this issuance is in conjunction with the MASB's plan to converge with the IFRS in 2012. The MFRS framework comprises standards issued by the IASB which are effective as from January 2012. At the same time, it also comprises new or revised standards recently issued by the IASB, such as standards on financial instruments, consolidation, joint arrangement, fair value measurement and employee benefits, among others.

In fact, the AOSSG as a working group on financial reporting relating to Islamic finance raises issues regarding the application of IFRS to Islamic financial transactions. There are two contrasting views on how to account for Islamic financial transactions, which are i) that a separate set of Islamic accounting standards is required; or ii) that IFRS can be applied to Islamic financial transactions. The literature on accounting for Islamic financial transactions represents a spectrum of views wherein transactions can generally be accounted for using IFRS. In 2014, a study conducted by the AOSSG Islamic Finance Working Group found that many standard-setters, regulators and IFIs do not share the AAOIFI's views on accounting or what constitutes compliance with Shariah. Therefore, AAOIFI FAS are not universally accepted according to this study, which was conducted in more than 600 IFIs from 75 countries. The study indicates that IFRS are the more commonly used standards for financial reporting by IFIs. However, 24 IFIs from six countries, namely Bahrain, Jordan, Lebanon, Oman, Qatar and Sudan, asserted compliance with AAOIFI FAS. Furthermore, a study of compliance with local GAAP also found that 26 of 45 IFIs complied with differential requirements for Islamic transactions; these IFIs were from five countries, namely Bangladesh, Egypt, Indonesia, Pakistan and Yemen. The remaining IFIs, from Brunei, India, Thailand, the Philippines, Iran, Sri Lanka, Turkey, the USA and Bangladesh, did not indicate compliance with differential requirements for Islamic transactions.

In particular, analysis of various sources indicates that accounting principles generally do not contradict the Shariah. This is why most of the abovementioned countries do not accept the AAOIFI FAS but rely more on the IFRS to report their financial transactions in IFIs. Thus, four IFRS principles are used by accountants as vital guides that influence transactions for IFIs; these are i) substance over form; ii) the time value of money; iii) fair value measurement; and iv) recognition based on probability (Shafii et al., 2013).

4 DATA

Data for the discussion on key accounting assumptions presented in the following sections is sourced from financial statements of two IFIs that operate under two different financial reporting regimes namely the MASB (IASB based) and the AAOIFI. It is believed that IFIs that operate in a specific country like Malaysia and Bahrain would be bind by the rules and regulations promulgated by their respective regulators in the country. Thus, observing a single IFI in Malaysia and Bahrain would represent the financial reporting practices of the other Islamic banks. The financial statements of Bank Islam Malaysia Berhad (BIMB) and Bahrain Islamic Bank (BiSB) that represent the application of accounting standards issued by the MASB and the AAOIFI respectively, were downloaded from the respective IFIs website.

5 ANALYSIS OF FOUR ACCOUNTING ASSUMPTIONS

5.1 Substance over form

In order to illustrate this issue, the Institute of Chartered Accountants of Pakistan (ICAP) is resistant towards reporting the financing effect arising from trade-based transactions and has even suggested that a financing effect does not even arise (AOSSG, 2010). They mentioned that:

> *"In Islamic finance, (one) cannot have a transaction whose substance is different from its legal forms. In other words, if a trade transaction is not a genuine trade transaction and is just a financing transaction then it is not acceptable in Islamic finance".*

This means that the form and substance of the transaction must be consistent and shall not contradict one another. In the event of inconsistency between substance and form, the Shariah places greater importance on substance (Shafii et al., 2013). This is in line with the majority of scholars from the Hanafi, Maliki and Hanbali schools, which considered the intention and meanings of contracts, unless there is a legal impediment that opposes their consideration. The Shariah Advisory Council (SAC) of Bank Negara Malaysia (BNM) also has adopted the majority view of recognizing substance over form. Others believe that it is acceptable and would benefit users more to show the economic substance of an Islamic financial transaction. Information about legal form may be relegated to the notes to the financial statements. However, Shafi'i scholars favored the principles of form over substance. The evidence shows that there are those who believe that the recognition and measurement of an Islamic financial transaction should give prominence to its legal form to differentiate it from a perceived conventional equivalent. It claims that substance over form is 'an unconcealed violation of Shariah'.

The diversity of reporting practices is evident from our analysis on the IFIs financial reporting practices. Analysis of financial reporting on leasing transactions reveals that IFIs that adopted the MASB (IASB based) accounting standards reported leased asset consistent with the 'substance over form' principle as shown in Figure 1 below. Leased assets (Al-Ijarah) are not recognized as the bank's assets; instead, they are disclosed as financial assets under the Al-Ijarah contract. This indicates that economic substance of the lease transaction is recognized as more important than the legal form. On the other hand, IFIs that adopted the AAOIFI accounting standards used the 'form over substance' principle as indicated in Figure 2 below.

Financing, advances and others

(a) By type and Shariah contract

Group and Bank 31 December 2015	Bai' Bithaman Ajil RM'000	Murabahah RM'000	Bai Al-Inah RM'000	At-Tawarruq RM'000	Ijarah Muntahiah Bit-Tamleek RM'000	Ijarah Thumma Al-Bai RM'000	Istisna' RM'000	Ar-Rahnu RM'000	Total RM'000
At amortised cost									
Cash line	-	-	62,300	997,097	-	-	-	-	**1,059,397**
Term financing									
House financing ^	4,851,790	-	-	6,628,865	-	-	62,580	-	**11,543,235**
Syndicated financing	8,603	-	164,301	1,021,805	-	127,399	-	-	**1,322,108**
Leasing financing	-	-	-	-	81,223	1,769	-	-	**82,992**
Bridging financing	-	-	-	-	-	-	87,630	-	**87,630**
Personal financing	-	-	82,054	10,247,851	-	-	-	-	**10,329,905**
Other term financing	2,583,462	685,973	18,444	5,385,013	-	-	1,762	-	**8,674,654**
Staff financing	96,919	1,515	-	68,007	-	-	16,401	-	**182,842**
Credit cards	-	-	12,695	430,848	-	-	-	-	**443,543**
Trade bills discounted	-	1,139,827	-	-	-	-	-	-	**1,139,827**
Trust receipts	-	20,210	-	-	-	-	-	-	**20,210**
Pawn broking	-	-	-	-	-	-	-	73,883	**73,883**
	7,540,774	**1,847,525**	**339,794**	**24,779,486**	**81,223**	**129,168**	**168,373**	**73,883**	**34,960,226**

Figure 1: Extract BIMB's Financial Statements

As shown in Figure 2, the BiSB recognized leased assets as the bank's assets and subjected to normal provision for depreciation and assets impairment. This indicates that the bank claims ownership over the leased assets, which explains the importance of legal form over economics substance of the transaction[1].

8. IJARAH MUNTAHIA BITTAMLEEK

	2015 Jointly financed					2014 Jointly financed				
	Land BD'000	Buildings BD'000	Aviation related assets BD'000	Others BD'000	Total BD'000	Land BD'000	Buildings BD'000	Aviation related assets BD'000	Others BD'000	Total BD'000
Cost:										
At 1 January	44,305	65,112	7,876	10,762	128,055	40,588	56,058	7,287	9,101	113,034
Additions	12,094	56,553	8,033	4,960	81,640	6,824	37,766	589	2,548	47,727
Settlements	(33,299)	(16,287)	(8,200)	(4,966)	(62,752)	(3,107)	(28,712)	-	(887)	(32,706)
At 31 December	23,100	105,378	7,709	10,756	146,943	44,305	65,112	7,876	10,762	128,055
Depreciation:										
At 1 January	-	10,044	1,792	2,229	14,065	-	12,406	1,245	1,273	14,924
Charge for the year	-	7,901	275	272	8,448	-	3,431	547	959	4,937
Relating to settled assets	-	(3,127)	(1,791)	(1,903)	(6,821)	-	(5,793)	-	(3)	(5,796)
At 31 December	-	14,818	276	598	15,692	-	10,044	1,792	2,229	14,065
Provision for impairment	(7,642)	(2,213)	-	(3,335)	(13,190)	(9,175)	(2,538)	-	-	(11,713)
Net book value:										
As at 31 December	15,458	88,347	7,433	6,823	118,061	35,130	52,530	6,084	8,533	102,277

Figure 2: Extract BiSB's Financial Statements

5.2 Time value of money

With regard to the time value of money concept, where a transaction involves time deferment, the asset (liability) is carried at the present discounted value of the future net cash inflow (outflow) that the transaction is expected to generate in the normal course of business. The application of the time value

[1] the AAOIFI only recognized operating lease and rejected the financial lease in their accounting standards.

DOI: 10.1051/shsconf/20173407004

of money is permissible only for exchange contracts that involve deferred payment and is strictly prohibited in loan transactions (Qard). This is in line with Shafii et al. (2013), who stated that the concept of time value of money is acceptable in contracts of exchange whereby the deferred price will be higher than the spot price in order to uphold the element of justice ('adl) between contracting parties. However, it is not acceptable to charge an extra sum for the deferred payment of a qard (loan). Therefore, this principle is only applicable in exchange contracts, not in loan contracts. This is because interest-based transactions are prohibited in Islamic finance.

In fact, Islamic law recognizes the selling price differences between a cash payment sale and a deferred payment sale such as Murabahah (ISRA, 2011). The majority of Muslim jurists agree that a higher price can be charged in a deferred payment sale due to the time factor involved in determining the value of the contracted item. Also, it can be said that Shariah does not completely rule out the concept of time value of money. This is evidenced by the fact that Islam does not prohibit any increment in a sale contract to be paid at a future date. This is because a repayment of a financing involves paying more than the principal to compensate for the time involved. Hence, the time value of money is accepted for the purpose of pricing of goods/usufruct only and not for pricing of money or debts (Ayub, 2007). So, what is not permissible is its use in compounding the interest for the lending of money.

Analysis of financial statements of Bank Islam Malaysia Berhad reveals the application of the time value of money concept as shown in Figure 3 below. However, there is no disclosure on the concept by the BiSB.

2.7 Leased assets - Finance lease

Leases in terms of which the Group or the Bank assumes substantially all the risks and rewards of ownership are classified as finance leases. Upon initial recognition, the leased asset is measured at an amount equal to the lower of its fair value and the present value of the minimum lease payments. Subsequent to initial recognition, the asset is accounted for in accordance with the accounting policy applicable to that asset.

Figure 3: BIMB's Disclosure on the use of Time Value of Money

5.3 Fair value measurement

Fair value refers to the amount at which an asset could be bought or sold in the current market (Shafii et al., 2013). Islam emphasizes fair dealing and just commercial transactions. However, there are various ways of manipulating market conditions to affect the market price. This indicates that Islamic commercial transactions must be carried out based on fair value.

There are two prohibited sales namely Talaqqi al-rukban (meeting caravans before they reach the market) and najsh (false bidding). The Prophet (PBUH) prohibited such sales in order to prevent city-dwellers from meeting caravans in order to buy from them before they reach the market and find out the prevailing price for their goods. He also said "Do not intercept goods before they are offloaded in the market". If this does occur, the importer of the goods has the option to annul the sale once he reaches the market and finds out the going price (ISRA, 2010). However, jurists have different views on the wisdom behind this prohibition. Some of them say that it is to prevent harm to the importer; some say it is for the benefit of the market and residents of the city; and some say that it is for the benefit of both the importer and the people. Others say that it is in order to prevent deception of the importer through false information about the market prices.

A similar consideration is operative in the prohibition of najsh, which is where a person bids in an auction without intending to actually buy, merely to drive up the price (ISRA, 2010). This is prohibited by consensus. It can occur without the seller's knowledge, in which case the sin is restricted to the bidder who drove up the price. However, if both of them knew, then they share in the sin. In fact, there are clear examples of prohibition of talaqqi al-rukhban and najsh incorporation of the fair value principle in Islamic financial transactions. A review of the classical fiqh literature reveals many instances where this principle was given consideration by jurists in deducing rulings. For example, the fair value benchmarked with the market price was used to determine the subsistence (nafaqah) of the agent-manager while travelling in a mudarabah contract. Scholars also agree on the role of experts and the use of estimation techniques in determining the fair value when there is no clear market price for a commodity. Nevertheless, controversy apparently prevails on the use of discount rates within the estimation techniques.

As shown in Figure 4 below, the BIMB used market information as well as the valuation techniques to estimate market value of assets or liabilities. On the other hand, the BiSB used market information and in the event there is no market information, they will use market information of similar item. If both of the above information are not available, then, historical cost would be utilized as illustrated in Figure 5 below.

2.22 Fair value measurements

'Fair value' is the price that would be received to sell an asset or paid to transfer a liability in an orderly transaction between market participants at the measurement date in the principal or in its absence, the most advantageous market to which the Group has access at that date. The fair value of a liability reflects its non-performance risk.

When available, the Group measures the fair value of an instrument using the quoted price in an active market for that instrument. A market is regarded as active if transactions for the asset or liability take place with sufficient frequency and volume to provide pricing information on an ongoing basis.

If there is no quoted price in an active market, then the Group uses valuation techniques that maximise the use of relevant observable inputs and minimise the use of unobservable inputs. The chosen valuation technique incorporates all of the factors that market participants would take into account in pricing a transaction.

Figure 4: BIMB's Disclosure on Fair Value

k. Determination of fair value

For investments traded in organised financial markets, fair value is determined by reference to quoted market bid prices at the close of business on the consolidated statement of financial position date.

For investments where there is no quoted market price, a reasonable estimate of the fair value is determined by reference to the current market value of another instrument, which is substantially the same or is based on the assessment of future cash flows. The cash equivalent values are determined by the Group at current profit rates for contracts with similar terms and risk characteristics.

Investments at fair value through equity where the Bank is unable to determine a reliable measure of fair value on a continuing basis, such as investments that do not have a quoted market price or other appropriate methods from which to derive reliable fair values, are stated at cost less impairment allowances.

Figure 5: BisB's Disclosure on Fair Value

5.4 Recognition based on probability

Probability is where the degree of uncertainty that the future economic benefits associated with the transaction will flow to or from the licensed person is considered in reference to the recognition criteria. The principle of probability in accounting and financial reporting from the Shari'ah perspective refers to the practice of recording a transaction even though the contract's effects have not yet been entirely concluded (Shafii et al., 2013). In this regard, assets will be recorded once there is a probability of inflow of economic resources, while liability will be recorded once there is a probability of outflow of economic resources due to current obligations and once its amount can be estimated with reasonable certainty. The examples of recognition based on probability in an IFI are: i) recognizing a liability for expected payments to mudarabah account holders; ii) recognizing impairment of financial assets based on an impairment model; iii) recognizing mudarabah receivables using the expected rate of profit; and iv) recognizing income and expenses based on expected future cash flows.

It is recognized that certainty is the highest level of confidence in Islam. It should be noted that jurists have established certain parameters in adopting the concept of probability in fiqh. The SAC of BNM also resolved that the application of the probability principle in Islamic financial reporting is permissible, as it does not contradict the general fiqh principles. Other fatwa-issuing bodies have not yet issued any specific fatwa on the principle.

Figure 6 illustrates the approach taken by the BIMB in applying the principles of probability, while, Figure 7 shows the BiSB disclosure on principle of probability. It is evident that both of the regimes i.e. MASB and AAOIFI accepted and implemented the concept.

(d) Use of estimates and judgement
The preparation of the financial statements requires management to make judgements, estimates and assumptions that affect the application of accounting policies and the reported amounts of assets, liabilities, income and expenses. Actual results may differ from these estimates. Estimates and underlying assumptions are reviewed on an ongoing basis. Revisions to accounting estimates are recognised in the financial statements in the period in which the estimates are revised and in any future periods affected.

Figure 6: BIMB's Accounting Policy on Estimates and Judgement

aa. Revenue recognition Murabaha receivables

Income is recognised by proportionately allocating the attributable profits over the deferred period whereby each financial period carries its portion of profits irrespective of when cash is received. Income related to accounts that are 90 days overdue is excluded from the consolidated statement of income.

Figure 7: BiSB's Accounting Policy on Revenue Recognition

This section has discussed the Financial Reporting assumptions for Islamic Financial Institutions. It has considered four assumptions: substance over form, the time value of money, fair value measurement and recognition based on probability. The next section will conclude the paper.

6 CONCLUSION

This section will provide a summary and the key findings of the paper based on an in-depth review and discussion. Financial reporting in Islamic society should be more detailed as compared to the current form of reporting that is prevalent in Western societies, and should consider moral values. There should be an emphasis on transparency, while manipulation of the values of assets should be avoided. As well as an emphasis on the profit and loss statement, the balance sheet and the cash flow statement, a considerable amount of further information should be provided. This could include a value-added statement and disclosure about the firm's social performance activities. The reports are also likely to include information about zakat, qard and charitable distributions. Items that involve the use of interest will be excluded.

Differences of opinion on how to account for Islamic financial transactions have led to divergent treatments of various transactions in various jurisdictions. Thus, it is necessary to enhance cross-border comparability by standardizing the reporting of Islamic transactions. While this could easily be achieved by the use of the IFRS, it would result in the reporting of the economic substance of transactions and of their financing effect. This may not appeal to certain stakeholders. Therefore, the challenge to standard-setters and stakeholders is to enhance the cross-border comparability of Islamic financial transactions, while being mindful of religious sensitivities. Although the IFRS are internationally accepted, there is resistance by those who believe that some IFRS principles are irreconcilable with their interpretation of Shariah. As a result, a separate financial reporting framework for Islamic financial transactions is warranted. However, the Malaysian regulators have issued directive to the IFIs to comply with the MASB standards due to the key accounting principles are complied with the Shariah requirements.

REFERENCES

1. AAOIFI. (1996). *Accounting and auditing standards for Islamic financial institutions*. Manama, Bahrain: Accounting and Auditing Organization for Islamic Financial Institutions.
2. ACCA & KPMG,. (2012, November). Global alignment:bringing consistency to reporting of Islamic finance through IFRS. *The Association of Chartered Certified Accountants* .
3. Ahmed, A. A. (2012). Accounting in Islamic perspective: A timely opportunity a timely challenge. *ASA University Review* , 11-31.
4. Alim, M. N. (2014). The development of Shari'ah accounting: epistemology perspective. *European Scientific Journal* , 110-117.
5. AOSSG Islamic Finance Working Group. (2015). *Financial reporting by Islamic financial institutions: a study of financial statements of Islamic financial institutions.* Asian-Oceanian Standard-Setters Group.
6. AOSSG Working Group. (2010). *Financial reporting issues relating to Islamic finance.* Asian-Oceanian Standard-Setters Group.
7. Ayub, M. (2007). *Understanding Islamic Finance.* Asia: John Wily and Sons (Asia) Pte Ltd.
8. Azmi, M. F. (2011). Shari'ah Challenge. *Global Islamic Finance Team* .
9. Bank Negara Malaysia. (2013, June 28). Financial reporting for Islamic banking institutions. *Islamic Banking and Takaful Department* , pp. 1-29.
10. Baydoun, N., & Willett R. (2000), Islamic and accounting: ethical issues in the presentation of financial information. *Accounting, Commerce & Finance: The Islamic Perspective,* 1-25.
11. Haniffa, R., & Hudaib, M. (2007). Exploring the ethical identity of Islamic banks via communication in annual reports. *Journal of Business Ethics*, *76*(1), 97-116.
12. ISRA. (2011). *Islamic Financial System: Principles and Operations* . Kuala Lumpur: International Shari'ah Research academy of Islamic Finance.
13. ISRA. (2010). *ISRA Compendium for Islamic financial terms.* Kuala Lumpur: International Shari'ah Research Academy of Islamic Finance.
14. Lev, B. (2001). *Intangibles: Management, measurement, and reporting.* Washington, D.C. Brookings Institution Press
15. Malaysia Accounting Standards Board. (2009). Financial reporting from Islamic perspective. *International Accounting Standards Committee Foundation* .
16. MASB. (2011). *The conceptual framework for financial reporting.* IFRS Foundation.
17. Mirza, M., & Baydoun, N. (1999). *Accounting policy choice in an interest-free environment.* Working Paper No. 1999-014.
18. Parker, L. (2014). Constructing a research field: a reflection on the history of social and environmental accounting. *Social and Environmental Accounting Journal,* 34(2), 87-92.
19. Remi, A., (2006), *Advanced financial accounting*, Second Edition, Lagos Master Stroke Consulting.
20. Shafii, Z., & Zakaria, N. (2013). Adoption of international financial reporting standards and international accounting standards in Islamic financial institutions from practitioners' viewpoint. *Middle-East Journal of Scientific Research* , 42-49.
21. Shafii, Z., Zakaria, N., Shaharuddin, A., Sairally, B. S., Khir, M. F., Hussain, L., et al. (2013). An apparaisal of the principles underlying international financial reporting standards: A Shar'ah Perspective. *ISRA Research Paper* 54/2013, 1-36.
22. Tajudin, A. A. (2010). *Overview of Islamic finance.* Kuala Lumpur, Malaysia: Bank Islam.
23. Thajudeen, K. S. (2012). Maqasid al-Shariah is one of the very important Shariah aspects in Islamic Finance. *INCEIF* .
24. Unegbu, A. O. (2014). Theories of accounting: evolution & developments, income-determination and diversities in use. *Research Journal of Finance and Accounting* , 1-15.

SHS Web of Conferences **34**, 07005 (2017) DOI: 10.1051/shsconf/20173407005
FourA 2016

Fair Value Accounting and the Cost of Equity Capital: The Moderating Effect of Risk Disclosure

Ashwag Dignah[1], *Radziah* Abdul Latiff[1,*] *Zulkefly* Abdul Karim[2], and *Aisyah* Abdul Rahman[3]

[1]School of Accounting, Faculty of Economics and Management, Universiti Kebangsaan Malaysia, Malaysia
[2]School of Economics, Faculty of Economics and Management, Universiti Kebangsaan Malaysia, Malaysia
[3]School of Management, Faculty of Economics and Management, Universiti Kebangsaan Malaysia, Malaysia

Abstract. Evidence thus far suggests fair value accounting poses risk and affects firms' returns in some ways. This research, on a sample of Asian banks, improves the understanding of the information risk effect of fair value accounting by examining the moderating role of risk disclosure in the relationship between fair value accounting and the cost of equity capital. The results from a generalised method of moments on dynamic panel data analysis, show that risk disclosure mitigates the asymmetric information problem. Thus the findings contribute towards the standard setters' effort in improving the practice of fair value accounting, and suggest that there are benefits in mandating disclosure especially for banks.

1 Introduction

The cost of equity capital (COE) is one of the main concerns of firms (1). Information asymmetry and risk could increase the COE, the cost of raising funds, as suppliers of capital reduce the stock price. In this regard, fair value (FV) accounting is an effort to provide relevant information to investors for decision making and reduce their risk of making wrong decisions based on outdated information, such as that based on historical cost. However, Barth (2), Lins et al. (3) and Christensen and Nikolaev (4), among others, criticized FV measurement for introducing more volatility and generating the information asymmetry problem and risk.

Past research finds proportion of fair valued assets (FVA) is positively related to the COE. It is a conjecture at this point that the effect of FVA on the COE could be moderated by practising risk disclosure. Risk disclosure will offer the stockholders justification about the risk and provide them with more relevant information; this action reduces the information asymmetry and the risk as reported by Amran et al. (5). Linsley and Shrives [(6),(7)], Aljifri & Hussainey (8), Abraham and Cox (9) discuss the importance of disclosing information on risk when the level of risk is expected to be high, such as when FV accounting is adopted. Botosan (10) examine the influence of disclosure on the COE. Consequently, practising risk disclosure is expected to reduce the effect of asymmetric information and information risk with FV accounting on the COE. However previous studies report that as the level of risk increases risk disclosure decreases because bank managers are less likely to disclose information about risk that shows a weakness [(11);(12);(13)]. Bank managers prefer to reveal a favourable situation that shows their ability to users and helps them to differentiate their bank from other banks.

Thus, according to the previous discussion, a considerable debate has arisen between academic papers regarding the direction of the influence of risk disclosure in the relationship between FVA and the COE. These conflicts between academic researchers make the impact and benefit from risk disclosure unclear. Thus it is important to examine whether risk disclosure moderates the relationship between FVA and the COE, which is the objective of this paper.

Risk disclosure is expected to play a role differently for different types of FVA, quoted and unquoted FVA. The demand for risk disclosure increases differently with different types of FVA, because each type has different kinds and degrees of risk. Consequently, these different levels of FVA emphasize the importance of investigating whether risk disclosure moderates the relationship between different levels of FVA (quoted and unquoted FVA) and the COE.

Furthermore, the intention of previous studies and regulators to decrease the COE by practising disclosure is the primary motivation for this study. Diamond and Verrecchia (14), Clarkson et al. (15), Botosan (10), Richardson and Welker (16), Botosan and Plumlee (17), Hail (18), Chen et al. (19), Cheng et al. (20), Francis et al. (21), Armitage and Marston (22), Zhaohui (23); Lopes and De Alencar (24), Lopez (25) and Shehata et al. (26) are the previous researchers who focused on the effect on the COE of the quality of disclosure. These previous studies related the COE to disclosure, but their results were inconsistent and they did not use disclosure as a moderating variable.

* Corresponding author: radziah@ukm.edu.my

Moreover, such research sustains the standard setters in developing and initiating other solutions that improve the practice of FV accounting and the quality of information and protect the interests of investors. The lack of literature on the topic relating to risk disclosure, the COE and FV accounting is the second motivation of this research; the aim is to extend the understanding of the effect of moderating risk disclosure on the relationship between FV accounting and the COE and to clarify the issues pertaining to FV accounting.

Findings from the analysis using generalised method of moments (GMM) support our hypotheses and the discussion in previous studies stating the benefit of practising risk disclosure. To sum up, this study utilized risk disclosure as a tool that could minimize the information asymmetry problem, mitigate risks under FVA and improve risk management, thus reducing the COE.

2 The Role of Risk Disclosure

Risk disclosure could contribute to the reduction of future risks arising from asymmetric information. Moreover, the disclosure of information risk is very important, especially after the requirement for FV accounting is mandated and it is especially so for Level 3 type of FVA (27; 28). Furthermore, the IFRSs highlight the importance of risk management and disclosure in firms (29) to decrease the asymmetric information problem and thus the COE (30).

The FV measurement involves estimation and thus is expected to result in volatile earnings, errors both intentional and unintentional [(31); (32); (3); (4)]. These issues concerning FV accounting could mislead investors and regulators in making the right investment decision (33) and make raising fund capital difficult, thus increasing the COE (22). So, bank managers who engage in a high level of risk disclosure are expected to disclose more details and information that justify and explain their risks and regain investors' confidence and improve transparency as discussed in Linsley and Shrives (6). Consequently, practising risk disclosure could moderate the relationship between FVA and the COE since it could reduce the information asymmetry in FVA and the probability of the risk premium. Then, the expected rate of return and the COE will also be influenced.

Moreover, theoretical papers by Merton (34), Diamond and Verrecchia (14), Clarkson et al. (15), Healy and Palepu (35), Lambert et al. (36), Bertomeu et al. (37) and Shehata et al (26) indicate the importance of disclosure in determining the COE. Furthermore, Botosan (10), Richardson and Welker (16), Hail (18), Chen et al. (19), Ashbaugh et al. (38), Francis et al. (21; 38), Armitage and Marston (22), Zhaohui (23), Lopes and De Alencar (24) and Lopez (25) empirically reported that disclosure needs to provide more information about the risk and asymmetric information and therefore it can decrease the COE.

However, there are different degrees of information asymmetry and risk associated with each level of FV accounting. The FV accounting estimation models, such as unquoted FVA, have a high level of errors and the estimation of potential unreliability is expected to be high. This is because the management must estimate unquoted assets at FV, which can be subject to discretion or manipulation and many judgements are needed when the model is estimated [(30);(12); (39)]. The error and asymmetric information problem for FV accounting obtained from market prices (quoted FVA) are smaller than those for FV accounting obtained from estimation models (unquoted FVA) (2). Therefore, the cost of raising capital is expected to be high for banks that engage in a high proportion of unquoted FVA; consequently, the need for risk disclosure is greater with unquoted assets at FV than with quoted assets at FV. In this regard, Liao et al. (27) also documented that more disclosure for Level 3 FVA is required by standard setters. The above discussion suggests that the quality and level of risk disclosure differ with the level of FVA; thus, it is expected that the risk disclosure affects the relationship between FVA and the COE differently.

3. Research Methodology

3.1 The research model:

$$COE_{it} = \alpha_0 + \beta_1 COE_{i,t-1} + \beta_2 FVA_{i,t} + \beta_3 DIS_{i,t} + \beta_4 FVA * DIS_{i,t} + \beta_5 BETA_{i,t} + \beta_6 BTM_{i,t} + \beta_7 SIZE_{i,t} + \eta_i + v_{i,t} \quad (1)$$

Variable description:

$DIS_{i,t}$ - disclosure index representing disclosure level of bank i at time t

The selection included in the index risk disclosure is based on the Basle Committe (1999), IFRS 7 and13, KPMG(2012) and the previous literature [(10); (6); (40)]. For each item of information disclosed, the symbol '√' is given and it will subsequently be awarded 1 mark. However, if the item is not disclosed in the annual report, the symbol 'X' is assigned and 0 marks will subsequently be awarded. At the end of the marking process, the total information available in each organization is divided by the total information shoul be reported: index of disclosure which calculated by score earned by bank / maximum score.

COE_{it} - dependent variable cost of equity of bank i at tine t

This is calculated as $\sqrt{(EPS_{t+2} - EPS_{t+1})/Pit}$

$FVA_{i,t}$ - the proportion of the total of the asset at FV to total asset of bank i at time t

$BETA_{i,t}$ - systematic risk,

$BTM_{i,t}$ - book to market value
$SIZE_{i,t}$ - normal logarithm of sum assets.

The error expression ($\mathcal{E}_i$ is function of the company particular outcome (η_i) and time specific outcome. However, present research based on error term $\mathcal{E}_i = \eta_i + v_{i,t}$ and $\eta_i$symbolize error term that is supposed to be founded on unobserved firm particular outcome that lead to hetrogeneity within $COE_{i,t}$ and $v_{i,t}$ symbolizes reminder stochastic allocations term.

This research employs a dynamic form of the relationship between COE and other explanatory variables as it is expected that the current COE is influenced by the previous year COE. The coefficients of FVA, BETA and BTM are expected to be positive and significant as these variables are different forms of risk factors. The term FAV*DIS is the testb for the moderating effect.

3.2 Data and Sampling

This research extends an earlier research using the same population of listed banks in Asia. Asian market is an ideal setting as the relationship between asymmetric information and COE is more likely in imperfect markets (30) such as those in Asia. Armstrong et al (30) refers to perfect market as when the number of traders is large or infinite and action by any trader is not likely to affect price. This study draws a sample consisting of the largest five Asian (top five in each country) banks for the period 2007 – 2013 because these large banks are the ones holding large assets and resources (especially FVA). the sample starts from 2007 because the FV accounting standards became effective for fiscal year 2008 and early adoption is permitted in 2007 (41). And also in 2007, companies are allowed to practice FV measurement for all financial assets and liabilities as recorded by Fahnestock & Bostwick (42). The relevant data is extracted from Thomson DataStream and annual reports. Initial data collection yielded an initial sample of 995 bank-year observations. The sample size was subsequently reduced to 509 for a number of reasons. Some financial reports could not be obtained for different reasons such as financial reports were not in the English language and financial reports were unavailable. Some banks are deleted due to missing data required to estimate COE and FVA. The final sample involved 112 banks from 28 countries in Asia.

4. Research Findings and Discussion

4.1 Multivariate Analysis

The relationship between the variables is based on the result of the dynamic model specifically, GMM, because the GMM model developed by Arellano and Bond (43) was designed for short panel data (44), which is the nature of our sample. Nickell (45) stated that the fixed-effect model estimator is inconsistent for short panel data. The Sargan test shows the model meets the requirement under the hypothesis related to FVA with a p-value of 0.35 in the table below. This indicates that the instruments are valid. Lastly, the research tests the autocorrelation, AR (1) and AR (2). The resulting *p*-values of 0.00 and 0.28 for the tests of AR (1) and AR (2) indicate that the estimates of the dynamic panel with the hypothesis related to FVA do not suffer from first- and second-order serial correlation. As a result, all the assumptions for testing the validity of the dynamic GMM model are met.

4.2 Discussion

The table below presents the findings. The second to fourth columns show the results without the disclosure variable. As expected and previously reported FVA and lagged COE are positively and significantly related to COE. More importantly the results of the dynamic model show a negative and significant relationship between the moderating variable FVA*DIS and the COE at the 0.05 level (Table- the last three columns). Thus, the GMM models show that a higher FVA will lower the estimated cost of equity when the level of information about risk disclosure is high. The relationship between the COE and FVA supports the moderating risk disclosure effect because it is positive before adding risk disclosure.

Moreover, the positive and significant relationship between FVA and the COE supports the moderating risk disclosure effect since, it changes to negative when risk disclosure is added. It emphasizes the significance of involving a tool such as risk disclosure that provides information about FVA and thus reduces the COE. This result also indicates that high FVA ratios in banks cause higher risk, which, therefore, lead to a higher COE, since this FV accounting includes a higher level of error and the asymmetric information problem according to the papers by Armitage and Marston (22) and Barth (2).

Consequently, disclosing more information about risk and the way in which banks control it could help to decrease the COE. Further, disclosure is an important factor for reducing the information asymmetry among parties and investors, which is one of the issues involved in FV accounting. This finding is consistent with Linsley and Shrives (6,7); Abraham and Cox (9), Aljifri and Hussainy (8), Amran et al. (5) and Vandemaele et al. (46), who reported that the demand for a high quality of risk disclosure increases when the percentage of risk increases, such as that produced by FVA.

Because this disclosure makes the project more visible to investors, producing more accurate pictures of firms' economic position and future cash flows, it enhances the stock market liquidity. Consequently, it reduces the rate of return demanded by investors and the COE. The dynamic result indicates that risk disclosure has a strong ability to reduce the COE, which is influenced by FVA.

Table 1: COE, FVA AND RISK DISCLOSURE

	COE with FVA			**COE with Risk Disclosure**			**COE, FVA and moderating effect of risk disclosure**		
	COEF.	**STD ERR.**	**P-VALUE**	**COEF.**	**STD ERR.**	**P-VALUE**	**COEF.**	**STD ERR.**	**P-VALUE**
$COE_{i,t-1}$	0.10	0.14	0.04	0.20	0.08	0.01	0.23	0.11	0.04
FVA	0.10	0.04	0.02	-	-	-	0.63	0.28	0.02
BETA	0.09	0.10	0.36	0.00	0.05	0.92	-0.09	0.07	0.20
BTM	0.03	0.01	0.02	0.05	0.01	0.00	0.04	0.01	0.00
SIZE	-0.14	0.05	0.00	0.03	0.05	0.53	-0.17	0.09	0.06
DIS	-	-	-	-0.12	0.10	0.25	0.29	0.21	0.16
FVA*DIS	-	-	-	-	-	-	-0.85	0.44	0.05
Number of Observations	509			509			509		
Observations per Group	115			115			115		
AR (1)	0.00			0.00			0.00		
AR (2)	0.21			0.20			0.28		
Sargan Test	0.10			0.05			0.35		

Thus, this result is also consistent with previous research, such as Botosan (10), Richardson and Welker (16), Hail (18), Chen et al. (19), Ashbaugh et al. (37), Francis et al (21,38), Armitage and Marston (22), Zhaohui (23), Lopes and De Alencar (40) Lopez (41). These studies show that high levels of information and disclosure reduce the information asymmetry and improve the coordination between firms and investors, thus decreasing the COE.

Further, this significant and negative result for moderating risk disclosure is consistent with our primary correlation, which shows the importance of risk disclosure in reducing the COE. Moreover, this dynamic model, which includes the moderating risk, is also consistent with hypothesis and the expectations. Therefore, the moderating risk disclosure acts as a buffer and reduces the impact of FVA on the COE.

Moreover, most of the control variables are consistent with the expectations when moderating disclosure is integrated into the model. The BTM is positively and significantly related to the COE at the 0.00 level. Size is another control variable that is significantly and negatively related to the COE at 0.06 level. However, beta is insignificant and not consistent with the expectations. Gebhardt et al. (47) and Hail (18) documented some issues relating to beta that could make it poorly suited to testing the influence of factors on the COE. The lack of significance of the beta is not surprising as it was found by Allee (48).

The table presents an insignificant relationship between disclosure and the cost of equity. This result could also refer to issues of variation between minimum and maximum index disclosure. Thus, the results show a negative and significant relationship between moderating risk disclosure and the COE and indicate that a high level of moderating disclosure affects the relationship between FVA and the COE. And this GMM result is consistent with hypothesis.

5 Conclusion

Risk disclosure is believed to be useful for investors' decision making and decreasing the COE. Risk disclosure reports complement the role performed by accounting numbers in FV accounting and produce more accurate pictures of firms' economic position. Accordingly, this research is concerned with the risk disclosure effect on the association between FVA and the COE. Based on the findings of this research, a significant and negative relationship was found between FVA and the COE when risk disclosure moderates. This result suggests generally that a bank with a higher level of FVA will hold a lower level of risk and asymmetric information issues, leading to lowering of the estimated COE when the level of information about risk disclosure is high. The results apply to when FVA is calculated in total or separately as quoted FVA or unquoted FVA.

References

1. Daske,H. Journal of Business Finance and Accounting, **33**(4):329-373.(2006)
2. Barth, M.E. The Accounting Review, 69(1): 1-25 (1994)
3. Lins, K.V., Servaes, H. and Tamayo, A. Financial Management, 40(3):525-551.(2011)

4. Christensen, H.B. and Nikolaev, V. Review of Accounting Studies, 18(3): 734-775. (2013)
5. Amran, A., Bin, A.M.R. and Hassan, C.H.M. Managerial Auditing Journal, 24(1):39-57 (2009)
6. Linsley, P.M. and Shrives, P.J. Journal of Financial Regulation and Compliance, 13(3):205-214. (2005)
7. Linsley, P.M. and Shrives, P.J. The British Accounting Review, 38(4):387-404. (2006)
8. Aljifri, K. and Hussainey, K., 2007.Managerial Auditing Journal, 22(9):881 -894.
9. Abraham, S. and Cox, P. The British Accounting Review, 39(3):227-248 (2007)
10. Botosan, C. The Accounting Review, 72(3): 323-349. (1997)
11. Ball, R., Robin, A. and Wu, J.S. Journal of Accounting and Economics, 36(1):235-270 (2003)
12. Verdi, R. Financial Reporting Quality and Investment Efficiency. SSRN. (2006)
13. Beyer, A. and Guttman, I. Journal of Accounting Research (2012) 50 (5):1141-1177
14. Diamond, D.W. and Verrecchia, R.E. The Journal of Finance, 46(4):1325-1359. (1991)
15. Clarkson, P., Guedes, J. and Thompson, R. Journal of Financial and Quantitative Analysis, 31(1):69-84. (1996)
16. Richardson, A.J. and Welker, M. Accounting, Organizations and Society, 26(7):597-616. (2001)
17. Botosan, C. and Plumlee, M. Journal of Accounting Research, 40(1):21-40 (2002)
18. Hail, L. European Accounting Review, 11(4):741-773. (2002)
19. Chen, K.CW, Wei, K.C., and Chen, Z. Disclosure, Corporate Governance, and the Cost of Equity Capital: Evidence from Asia's Emerging Markets. SSRN (2003)
20. Cheng, CS A., Denton., and Huang,H. Review of Quantitative Finance and Accounting, 27(2): 175-204. (2006)
21. Francis, J., Khurana, I.K. and Pereira, R. The Accounting Review, 80(4):1125-1162. (2005)
22. Armitage, S. and Marston, C. The British Accounting Review, 40(4):314-336 (2008)
23. Zhaohui, R.X. Journal of Business Research, 3(1):15-26. (2009)
24. Lopes, A.B. and De Alencar, R.C. The International Journal of Accounting, 45(4):443-464. (2010)
25. Lopez Voluntary Disclosure and the Cost of Equity Capital in the Netherlands, Erasmus University, hesis.eur.nl/pub/10248/M818-Lopez_323267.doc. (2011)
26. Shehata, N. F. Accounting and Finance Research, 3(1):18-26. (2014)
27. Liao, L., Kang, H.H.J., Morris, R. and Tang, Q. Information Asymmetry of Fair Value Accounting and Loan Loss Provisions During the Global Financial Crisis. SSRN(2010)
28. Ryan, S.G. Fair Value Accounting: Understanding the Issues Raised by the Credit Crunch. White paper for the Council of Institutional Investors, New York University Stern School of Business. (2008)
29. Taylor, G., Tower, G. and Neilson, J. Accounting and Finance, 50 (2):417-446. (2010)
30. Armstrong, C.S., Core, J.E., Taylor, D.J. and Verrecchia, R.E. Journal of Accounting Research, 49(1):1-40 (2011)
31. Landsman, W.R. Fair Value Accounting for Financial Instruments: Some Implications for Bank Regulation. Workshop on Accounting Risk Management and Prudential Regulation. Basel, Switzerland: Bank of International Settlement. (2006)
32. Laux, C. and Leuz, C. Journal of Economic Perspectives. (2010)
33. Barth, M.E.,Wayne,R., Landsman,B. and James, M.W. Journal of Banking and Finance 19 (3):577-605. 1995
34. Merton, R.C. The Journal of Finance, 42(3):483-510 (1987)
35. Healy, P.M. and Palepu, K.G. Journal of Accounting and Economics, 31(1):405-440. (2001)
36. Lambert, R., Leuz, C. and Verrecchia, R.E.Journal of Accounting Research, 45(2):385–420. (2007)
37. Bertomeu, J., Beyer, A. and Dye, R.A., The Accounting Review, 86(3):857-886 (2011)
38. Ashbaugh, H., Collins, D.W. and LaFond, R. Corporate Governance and the Cost of Equity Capital. Working paper, University of Wisconsin, and University of Low (2004)
39. Ashbaugh, H., Collins, D.W. and LaFond, R. Corporate Governance and the Cost of Equity Capital. Working paper, University of Wisconsin, and University of Low (2004)
40. Sundmacher, M. Consistency of Risk Reporting in Financial Services Firms. SSRN. (2006)
41. Li, F. And Shroff, N. Financial Reporting Quality and Economic Growth. SSRN.(2010)
42. Fahnestock, Robert T. and Bostwick, Eric D. Journal of Finance & Accountancy, 8, p1. (2011)
43. Arellano, M. and Bond, S. Review of Economic Studies, **58**: 277–297 (1991)
44. Roodman, D. How to do xtabond2: An Introduction to Difference and System GMM in Stata. Center for Global Development working paper 103. (2006)
45. Nickell,S. Journal of Econometric Soceity. 1981
46. Vandemaele, Sigrid, Vergauwen, Philip and Michiels, Anneleen, 2009. *Management Risk Reporting Practices and their Determinants.* University Hasselt.
47. Gebhardt, W.R., Lee, C. and Swaminathan, B., 2001. Toward an Implied Cost of Capital. Journal of Accounting Research, 39(1):135-176.
48. Allee, K.D., 2008. *Estimating Cost of Equity Capital with Time-Series Forecasts of Earnings*. Working paper, Michigan State University.

SHS Web of Conferences **34**, 07006 (2017) DOI: 10.1051/shsconf/20173407006
FourA 2016

Do Aggregate Earnings Components Predict GDP Growth Better? An International Setting Comparison

Sari Atmini[1,*], and *Sumiyana* Sumiyana[2]

[1]Faculty of Economics and Business, Universitas Brawijaya
[2]France Faculty of Economics and Business, Universitas Gadjah Mada

Abstract. This study investigates the usefulness of aggregate accounting earnings and their components to predict GDP growth. This study shows evidence that net income, net operating income, other comprehensive income, and net income change at aggregate level could predict GDP growth. This study splits sample into developed and developing countries. Almost all of aggregate earnings components could predict GDP growth in developed countries. On the contrary, only other comprehensive income and net income growth could predict it in developing countries. Moreover, sensitivity analysis shows evidence that the predictive power of accounting numbers are more consistently found in developed countries. It implies that all economic activities of firms listed in developed countries' stock exchanges fully reflect to that of macro economy. This reflection is better than those of in developing countries. We infer that the value relevance of accounting earnings and their disaggregation have reached not only in the capital markets but also in the national macroeconomy level. Indeed, this prediction seems dominantly to be accurate in developed countries only.

1 Introduction

If aggregate earnings as the bottom line can predict GDP growth [1], aggregate earnings components should also. Earnings disaggregation provides additional explanatory power [2]. It implies that each earnings component provides different information relevance. This study investigates the predictive value of aggregate earnings components to GDP growth.

Earnings informativeness studies at macroeconomy level take the position of "micro to macro" by investigating the relation between accounting and macroeconomic data [3]. The underlying argument is that corporations represent a substantial part of macroeconomy so that their aggregate activities affect it [4]. Aggregate corporate earnings are closely correlated with macroeconomy and affect each other [5]. Studies that associate aggregate corporate earnings and macroeconomic activity are still very limited in quantity [5]. Yet, such studies are important because empirical findings at the firm level may not necessarily be used as a basis to draw inferences about earnings informativeness at macro level [6].

* Corresponding author: sari.atmini@mail.ugm.ac.id

This study differs from previous studies in two important ways. First, it investigates the predictive ability of aggregate earnings components. Previous studies [1, 3] have major shortcomings for not distinguishing earnings before extraordinary items from net income. It can potentially weaken inferences drawn because extraordinary items are non-recurring items that can be ignored in making future earnings forecasts [7]. This study investigates recurring and non-recurring earnings components separately and hypothesizes aggregate earnings components can predict GDP growth. Research findings confirm the hypotheses.

Second, this study is conducted in an international setting. This study suspects that earnings ability to predict GDP growth is merely accurate in developed countries, not in developing countries. This suspicion arises due to the limited number of companies listed in developing countries' stock exchanges. This results in doubts about aggregate corporate earnings' ability in reflecting macroeconomic activity in those countries. This study investigates the predictive power of aggregate earnings and their components at macro-economic level in developed and developing countries, and expects aggregate earnings predictive ability in developed countries is better than those in developing countries. The findings confirm this expectation and emphasize the importance of contextual factors.

The main contribution of this study is, first, to verify the robustness of the test results showing that aggregate earnings are able to predict GDP growth [1, 3]. The findings of this study show incremental relationship between accounting earnings and their disaggregation with GDP growth. Second, contribution from the importance of international setting factors point of view, namely developed and developing countries contexts. Context consideration is extremely important because it results in different research findings.

2 Literature review and hypotheses development

2.1 Accounting earnings and GDP

GDP is an indicator of a country's economic progress [8], which measures current economic activity [9] and reflects the performance of the economy [10]. GDP value can be estimated either using value-added, income, or expenditure approaches [11]. Accounting earnings relate both indirectly and directly with GDP. Indirect linkage occurs because accounting earnings is a primary indicator of economic trends. It is supported by two macro theories: creative destruction [12] and sectoral shifts [13] theories. Both theories basically suggest that productivity at the firm level influences the dynamics of aggregate productivity growth at macro level. This implies that firm level accounting earnings, representing firm's productivity, are related to macro level income reflected in GDP. Meanwhile, direct linkage could be explained by the income approach [17]. According to this approach, GDP is a reflection of the whole income generated by firms and households in a particular country [11]. Accounting earnings are numbers that reflect firms' net income. Firms listed on a country's stock exchange represent a substantial portion of the country's macroeconomy. Previous research provides empirical support for the relation between accounting earnings and GDP [1, 3] which implies that aggregate accounting earnings growth is the primary indicator of incremental growth in nominal and real GDP. This study hypothesizes aggregate net income ($\mathbf{H_{1a}}$) and its change ($\mathbf{H_{1b}}$) affect GDP growth positively.

2.2 Disaggregating Accounting Earnings Components

Earnings components are informative and have predictive power [2]. This study disaggregates earnings into net operating income, non-operating net income, other comprehensive income, total expenses, and depreciation expenses. Net operating income and non-operating net income represents recurring and non-recurring items. Different transactions and events underlying these two items result in different predictive ability [7]. Permanence level of net operating income is higher than non-operating net income [18]. Based on stochastic processes, differences in permanence level of earnings components lead to differences in their predictive ability [19]. The more permanent is the earnings components, the stronger the predictive ability. This study believes in logical reasoning that the predictive ability of net operating income is different from non-operating net income. Meanwhile, accounting earnings aggregation of all firms listed on a country's stock exchange are expected to reflect the country's GDP. Consequently, this study investigates differences in the ability of both items to predict GDP growth, and hypothesizes aggregate net operating income ($\mathbf{H_{2a}}$) and its change ($\mathbf{H_{2b}}$) as well as aggregate non-operating net income ($\mathbf{H_{3a}}$) and its change ($\mathbf{H_{3b}}$) affect GDP growth positively.

Net income has dirty surplus components or other comprehensive income. It is transitory and should not have direct implications for firm's future operating performance. Because it represents changes in fair value of the statement of financial positions accounts, other comprehensive income reflects changes in firm value. Firm value is related to firm's ability to generate future income. If certain accounts related to other comprehensive income are realized, such as sales of assets, settlement of liabilities, or funding of pension plans, there will be an association between other comprehensive income and future firm's performance and cash flows as well. Accordingly, this study hypothesizes aggregate other comprehensive income (H_{4a}) and its changes ($\mathbf{H_{4b}}$) affect GDP growth positively.

Income approach estimates GDP through total domestic income of various production factors including compensation and payments [11]. Compensation and payments are total expenses recognized by a firm. Total expenses are economic benefits sacrificed by a firm to generate revenues. So, it can be inferred that firm's total expense may reflect firm's ability to generate revenues. On the contrary, from the counter party point of view, total expenses of a particular firm are a reflection of revenues generated by counter party. In macroeconomy level, aggregation of total expenses of all firms listed on stock exchange reflects income of the country. Based on this logical reasoning, this study hypothesizes aggregate total expenses ($\mathbf{H_{5a}}$) and its changes ($\mathbf{H_{5b}}$) affect GDP growth positively.

Depreciation expenses are allocation of fixed assets costs to the periods of the assets' economic life. Recognition of depreciation expenses are related to the use of fixed assets in a firm's production process and operational activities to generate revenues. It means that depreciation expenses can be used to predict firm's revenues. This logical reasoning is supported by previous research which states that depreciation expenses is one of the components to estimate GDP based on income approach [11]. This study hypothesizes aggregate depreciation expenses ($\mathbf{H_{6a}}$) and its changes ($\mathbf{H_{6b}}$) affect GDP growth positively.

3 Research Method

3.1 Population, sample, and unit of analysis

The population is 215 countries (http://data.worldbank.org/indicator/NY.GDP. MKTP.CD) while the sample is 73 countries (35 developed and 38 developing countries). Developed and developing countries are classified based on economic conditions of a country according to Development Policy and Analysis Division (DPAD) of Department of Economic and Social Affairs of the United Nations Secretariat (UN/DESA) (www.un.org/en/development/.../2014wesp_country_classification.pd) and the level of the country's GDP according to The Worldbank (http://data.worldbank.org/indicator/NY.GDP. MKTP.CD). In this study, developed countries are developed economies according to DPAD UN/DESA and high income nations according to The Worldbank. Developing countries are developing economies according to DPAD UN/DESA and upper middle income, lower middle income, or low income nations according to The Worldbank. The unit of analysis is the nation-year.

3.2 Data and variables

This study uses 2004-2014 data, excluding 2007-2009 data to minimize bias due to 2008 economic crisis. This study obtains accounting data from OSIRIS database available at Gadjah Mada University, Indonesia (https://osiris.bvdinfo.com), and GDP data from The Worldbank website (http://data.worldbank.org/indicator/NY.GDP.MKTP.CD). Dependent variable is subsequent year GDP growth (GDP_{it+1}). Independent variables are aggregate net income (NI_{it}), aggregate net operating income (OI_{it}), aggregate non-operating net income (NO_{it}), aggregate other comprehensive income (OCI_{it}), aggregate total expenses (EXP_{it}), and aggregate depreciation expenses (DEP_{it}). Accounting earnings and its disaggregation are measured in level and change. Control variable is current year GDP growth (GDP_{it}).

3.3 Hypotheses tests

This study analyzes data with multiple linear regression analysis using these equations:

$$GDP_{it+1} = \alpha+\beta_1 NI_{it}+\beta_2 GDP_{it}+\varepsilon_{it} \quad (1)$$

$$GDP_{it+1} = \alpha+\beta_1 OI_{it}+\beta_2 NO_{it}+\beta_3 OCI_{it}+\beta_4 GDP_{it}+\varepsilon_{it} \quad (2)$$

$$GDP_{it+1} = \alpha+\beta_1 EXP_{it}+\beta_2 DEP_{it}+\beta_3 GDP_{it}+\varepsilon_{it} \quad (3)$$

Three equations are used as certain independent variables are conceptually correlated. This study measures accounting numbers in level and change. Those equations are for level model. For change model, independent variables are measured as previous year difference.

3.4 Sensitivity tests

This study considers international setting to explore contextual differences between developed and developing countries by splitting the sample into two subsamples. Moreover, to obtain more comprehensive conclusions, this study undertakes two sensitivity tests. The first test adds one and two previous-year GDP growth (GDP_{it-1}, GDP_{it-2}) to each regression equation. The second test divides research period into three subperiods, i.e. pre-crisis (2004-2006), crisis (2007-2009), and post-crisis (2010-2013) subperiods.

4 Results and Discussions

4.1 Hypotheses test results and discussions (overall sample)

Hypothesis test obtains significant results for aggregate net income (t-stat. = 3.06, p-value < 0.01), aggregate net operating income (t-stat. = 1.96, p-value < 0.05), aggregate other comprehensive income (t-stat. = 3.26, p-value < 0.01) and aggregate net income change (t-stat. = 2.74, p-value < 0.01) and supports H_{1a}, H_{2a}, H_{4a}, and H_{1b}. Net income, net operating income, and other comprehensive income at aggregate level are predictors of subsequent year GDP growth. These findings confirm previous findings [1, 3] and provide empirical evidence of the informativeness of earnings and their components in macroeconomy level.

These findings are also consistent with [2] and [20] that earnings components provide additional information and are important in predicting and expecting future performance.

This study is unable to find any predictive ability of aggregate non-operating net income (t-stat.=1.37, p-value=0.17). The most likely reason is that it is non-recurring with low permanence level. This finding confirms [19] that permanence level of earnings component affects predictive ability and consistent with the expectation that the predictive ability of non-operating net income differs from that of net operating income. This study fails to document predictive abilities of aggregate total expenses (t-stat.=0.36, p-value=0.72) and aggregate depreciation expenses (t-stat.= -0.88, p-value=0.38). Indeed, the coefficient sign of aggregate depreciation expenses is not consistent with prediction, although statistically insignificant. Depreciation expense is a non-cash expense. Its recognition does not lead to revenue recognition by other party. Yet, its recognition will decrease economic value of assets, implying a decrease in production capacity of the assets. Meanwhile, research findings indicate aggregate earnings components changes lack of additional information.

4.2 Subsamples test results and discussions

To explore the influence of developed and developing countries contextual differences, this study examines these subsamples separately. In developed countries, test obtains significant results for aggregate net income (t-stat.=3.77, p-value<0.01) and its change (t-stat.=2.89, p-value < 0.01), aggregate net operating income (t -stat.= 2.37, p-value<0.05), and its change (t-stat.=2.72, p-value<0.01), aggregate non-operating net income (t-stat.=2.33, p-value < 0.05), aggregate other comprehensive income (t-stat.=2.81, p-value<0.01) and its change (t-stat.=2.37, p-value<0.05), and aggregate total expense change (t-stat.=3.3, p value<0:01). This study reconfirms to support H_{1a}, H_{2a}, H_{3a}, H_{4a}, H_{1b}, H_{2b}, H_{4b}, and H_{5b}. In developing countries, test gets significant results for aggregate other comprehensive income (t-stat. = 1.98, p-value<0.05) and aggregate net income changes (t-stat.=1.68, p-value<0.1), supports H_{4a} and H_{1b}. The results show stronger predictive power in developed countries in both models. These findings reemphasize the importance of contextual factors consideration.

Most firms in developed countries are listed on stock exchange so that their activities reflect macroeconomic activity of the country. In developing countries, the number of firms listed on stock exchanges is still very limited. Moreover, compared to developed countries, underground economy level is higher while macroeconomic-related regulations and institutions are still less-established. Currencies of developing countries often depreciate while ability to anticipate the shock tends to be weak. These developing country-specific conditions result in higher economic uncertainty level. Under high economic uncertainty, macroeconomic indicators prediction is almost impossible and assumed to be nil [21].

4.3 Sensitivity test results and discussions

Results of first sensitivity test (not reported) with GDP_{it} and GDP_{it-1} are consistent with subsamples test results for each subsample and both models. Sensitivity test results with GDP_{it}, GDP_{it-1}, and GDP_{it-2} (not reported) also show consistent results, except for level model of developed countries. In developing countries, only aggregate net income change and aggregate other comprehensive income that can predict GDP growth. Current year GDP growth is a factor that has consistent predictive power of subsequent year GDP growth. These results reemphasize that the predictive power of aggregate net income and its components to GDP growth is higher in developed countries. In developing countries, current year GDP growth is the best predictor of subsequent year GDP growth.

Results of second sensitivity test are not reported but discussed. Pre-crisis test results support H_{4a}, H_{4b}, and H_{5a} for developed countries, but none for developing countries. Crisis subperiod test results support H_{1a}, H_{2a}, H_{1b}, H_{2b}, H_{3b}, H_{4b}, and H_{6b} for developed countries and H_{1b} and H_{6b} for developing countries. Post-crisis test results support H_{1a}, H_{2a}, H_{1b} and H_{2b} for developed countries and H_{1b} and H_{2b} for developing countries. The findings show the predictive power of aggregate net income and its components occurs sporadically among subperiods, with a higher level of consistency is found in developed countries.

Current year GDP growth is consistently able to predict subsequent year GDP growth in developing countries. Special caution should be given to economic crisis subperiod as the relationship sign of these variables is negative. In developed countries, current year GDP growth cannot predict subsequent year GDP growth especially in crisis and post-crisis subperiods. In both subperiods, aggregate net income and its components are generally able to predict subsequent year GDP growth. Sensitivity test results confirm previous findings that the power of aggregate net income and its components in predicting GDP growth of developed countries is higher than that of developing countries. Besides, current year GDP growth is the best predictor of subsequent year GDP growth for developing countries.

5 Concluding remarks

In general, this study finds evidence concerning with the informativeness of aggregate accounting earnings and their components at macroeconomy level. The results support the notion that earnings components provide additional information and are important to be considered in predicting future performance. The power of aggregate net income and its components in predicting GDP growth is documented to be higher in the context of developed countries than in developing countries. Although hypotheses test results for the overall sample support the findings of previous studies [1, 3], the sensitivity test results do not fully agree with their conclusions. The conclusions of both studies can be applied in developed countries context, but not in developing countries. This study demonstrates the importance of contextual factors, which are ignored in previous research [1, 3].

This study has two major limitations. First, it only tests GDP growth prediction for one year ahead so that it cannot give any conclusions about forecast horizons of aggregate net income and its components. Second, it does not consider the quality of aggregate earnings.

References

1. Y. Konchitchki, P.N. Patatoukas, The Acc. Rev., **89**, 669-694 (2014)
2. R. Lipe, J. of Acc. Res., **24** (Supplement), 37-64 (1986)
3. Y. Konchitchki, P.N. Patatoukas, J. of Acc. and Ec., **76**, 76-88 (2014)
4. S. P. Kothari, L. Shivakumar, O. Urcan, http://www.fox.temple.edu/cms/wpcontent/uploads/2013/04/Agg_ern_inflation_forecasts_2013_04_24-LS-1.pdf (2013)
5. L. Shivakumar, J. of Acc. and Ec., **44**, 64-73 (2007)
6. C. Anilowski, M. Feng, D.J. Skinner, J. of Acc. and Ec., **44**, 36-63 (2007)
7. P. Farfield, R. Sweeney, T. Yohn, The Acc. Rev., **71**, 337-355 (1996)
8. R. Costanza, M. Hart, S. Posner, J. Talberth, *The Pardee Papers 4* (Boston Uni., 2009)
9. M. Fleurbaey, J. of Ec. Lit. **47**, 1029-1075 (2009)
10. N. G. Mankiw, *Macroeconomics*, (New York: Worth Publishers, 2000)
11. J. S. Landefeld, E.P. Seskin, B.M. Fraumeni, J. of Ec. Persp., **22**, 193-216 (2008)
12. J. A. Schumpeter, Socialism and Democracy, (Taylor & Francis e-Library, 2003)
13. D. M. Lilien, J. of Pol. Ec., **90**, 777-793 (1982)

14. L. Foster, J. Haltiwanger, C. J. Krizan, *New Developments in Productivity Analysis* (NBER, Uni. of Chicago Press, 2000)
15. L. Foster, J. Haltiwanger, C.J. Krizan, Rev. of Ec. Stat., **88**, 748-758 (2006)
16. R. E. Lucas Jr., E.C. Prescott, J. of Ec. Th., **7**, 188-209 (1974)
17. S. Nallareddy, M. Ogneva, *Working Paper,* http://ssrn.com/abstract=244014 or http://dx.doi.org/10.2139/ssrn.2444014 (2014)
18. S. Sugiri, Gadjah Mada Inter. J. of Bus. **5**, 363-379 (2003)
19. P. A. Griffin, J. of Acc. Res., **15**, 71-83 (1977)
20. B.M. Bratten, *Dissertation* (The Uni. of Texas at Austin, 2009)
21. R. R. Nelson, The Quart. J. of Ec., **75**, 41-62 (1961)

The Effect of IFRS Convergence and the Proportion of Woman in Audit Committee on Earning Management

Dwi Surti Siswanti[1], and *Ataina* Hudayati[1]*

[1]Accounting Department, Economics Faculty, Universitas Islam Indonesia, Yogyakarta, Indonesia

Abstract. This study aims to find empirical evidence the effect of IFRS convergence and the proportion of woman in audit committee on earning management. Respondents used in this study were companies listed in Jakarta Stock Exchange for the year 2011 until 2013. Using moderated regression analysis this study supports the hypothesis that IFRS convergence and the proportion of woman in audit committee have negative effect on earning management. This finding implies that IFRS convergence and the proportion of woman in audit committee will increase the quality of financial reporting.

1 Background of the study

In this globalization era, standardization and harmonization of accounting standards is a necessity that can not be postponed again. Therefore, convergence on International Financial Reporting Standards (IFRS) that has been done in Indonesia and abroad is accordance with the needs of business practitioners. Some authors argue that there are many benefits of IFRS one of which is that IFRS is likely to reduce earnings management. Researches linking the convergence of IFRS and earnings management include [1]; [2]; [3] and [4].

In Indonesia, studies linking the convergence of IFRS with earnings management have also been made include [5]; [6];[7];[8]and[9]. The results of research showing the relationship between IFRS convergence conducted in foreign countries and in Indonesia is inconsistent and therefore opening space for doing similar research. Therefore the aim of this paper was to examine the effect of the convergence of IFRS with earnings management. In addition, this study also adds a variable that can reduce the likelihood of earnings management that has not been examined in previous studies i.e. the effect of the proportion of women members in audit committee on earning management.

*Corresponding author:ahudayati@yahoo.com

 DOI: 10.1051/shsconf/20173407007

2 Theoretical bases

Agency theory explains the problems that occur between the agent and the principal, where the agency problem arises because the agent has a tendency to pay attention to their own interests that do not always correspond with the principal interest[10]. In the field of financial accounting, agency theory argues that the function of financial reporting is to reduce agency problem between the company and shareholders. This argument implies that IFRS is expected to reduce the negative behavior of the managers by publishing good quality of financial reporting.

This paper use three variables i.e. IFRS convergence, the proportion of women members in audit committee and earning management. According to "Indonesian Dictionary" (http://kbbi.web.id/) convergence means "the state toward a rendezvous point". As such, the IFRS convergence is often defined as a process that aims to establish the accounting standards that can be applied worldwide. Earnings management is defined as managers use judgment in preparing financial reporting by allowing the financial reporting so that the financial statements be biased (misleading) or manager affects various contracts using accounting numbers [11].

3 Hypothesis developments

3.1 Effect of IFRS convergence on earnings management

Some previous studies support the agency theory which states that IFRS may reduce earning management. Some authors which found that IFRS convergence is an effective tool to reduce earnings management include[1]; [12];[13]and[4]. Study in Indonesia found empirical evidence that there was a negative relationship between IFRS convergence and earnings management was [5]. Thus it can be formulated hypothesis 1 of this study:
H1: IFRS convergence has a negative impact on earnings management.

3.2 Effect of the proportion of women members in audit committee on earning management

Some preliminary research shows that women in board of director has a role to lower earnings management, namely [14]; [15]; and [16]. Research carried out in Indonesia that show the effectiveness of the audit committee to lower earnings management, namely [17]. Various research which demonstrate the contribution of women in the board of directors include [18]; [19]; [20] and[21] seems to be used to predict the influence of women on the performance of the audit committee. Formally, the second hypothesis in the study can be formulated as follows:
H2: Proportion of women members in audit committee negatively affects earnings management.

3.3 Proportion of woman members in audit committee strengthening the relationship between IFRS convergence and earnings management

According to [22], the audit committee is moderating variables on the relationship between tenure and reputable public accounting firms to audit quality. [9], found empirical evidence that the characteristics of the audit committee in the form of financial and accounting expertise moderate the relationship between IFRS convergence and earnings management. Furthermore,[5]also examine the influence of corporate governance (managerial ownership)

DOI: 10.1051/shsconf/20173407007

as the moderating variable on the relationship between convergence of IFRS and earnings management. Nevertheless, the results of the study cannot prove the hypothesis.

Unlike the two studies conducted in Indonesia, a study done in France [4] found that corporate governance (composed of independence of the board of directors, independence of audit committees, as well as the quality of external audits) reinforce the effect of IFRS convergence in reducing earnings management. Therefore, the third hypothesis of this study was formulated as follows:

H3: Proportion of woman members in audit committee strengthening the relationship between IFRS convergence and earnings management.

4 Population and research sampling

The populations in this study are all manufacturing companies whose shares are listed on the Indonesia Stock Exchange from 2011 until 2013. The elections of the year used in this study (2011-2013) because (2008-2012) is the first phase of IFRS convergence, i.e., while the year 2013 is the second phase of IFRS convergence.

5 Variable definition and measurement

The dependent variable used in this study is the earning management. Earnings management is defined as "managers use judgment in preparing financial reporting by allowing the financial reporting so that the financial statements be biased (misleading) or manager affects various contracts relating to itself using accounting numbers [11]. In this study, the earnings management measured by discretionary accruals which is calculated using the modified Jones model.

The independent variable used this study were the convergence of IFRS and the proportion of woman in audit committee. Convergence to IFRS can be interpreted to make the same accounting standards of a country with IFRS. IFRS Convergence measured using dummy variables to measure when companies converge IFRS rated 1 if it does not converge IFRS rated 0 [12].The proportion of women in audit committee is a comparison between the number of women in audit committee divided by the total members of audit committee [23] and [4].

6 Results

Moderating regression analysis in this study can be seen in table 1 below:

Table 1 Regression Analysis

	Coefficients		T	Sig.
	B	Std. Error		
Constanta	,273	,191	1,427	,158
IFRS	-,222	,100	-2,210	,030
Woman Audit Committee	-1,160	,288	-4,022	,000
IFRS*WAC	,632	,288	2,192	,031

Analysis found that coefficient of determination (Adjusted R^2) was 0.213. These results can be concluded that the variations of dependent variable affected by independent variable were 21.3% and the remaining 78.7% was influenced by other factors not included in the regression model. F test also found significant of F value, then it shows that the regression model can be used to predict the research model.

The first hypothesis stated that the convergence of IFRS has a negative effect on earnings management. Regression coefficient of IFRS convergence was -0.222 and p value was 0,030. which means convergence of IFRS has a negative effect on earnings management. It can be concluded that the first hypothesis of this study was supported.

The second hypothesis of this study states that the proportion of woman in audit committee negatively affect earnings management. The magnitude of the regression coefficient of the proportion of women in audit committee was -1.160 and p value was 0.000., which means the proportion of woman in audit committee negatively affect earnings management so it can be concluded that the second hypothesis was supported.

The third hypothesis of this study stated that the proportion of the woman in audit committee reinforce negative relationship between the convergence of IFRS on earnings management. The amount of interaction regression coefficient was 0.632 and p value 0.031. From the analysis it can be concluded that although the interaction between the proportion of women in the audit committee and the convergence of IFRS have significant influence on earnings management, but the direction of the relationship indicates the opposite of that predicted. It can be concluded that the third hypothesis was not supported.

7 Discussions

This study aimed to analyze the relationship between the IFRS convergence and the proportion of women members in audit committee on earnings management. Preliminary research indicates that the interaction between the IFRS convergence and corporate governance can reduce earnings management. These studies expand the preliminary study on the role of corporate governance in reducing earnings management.

The results of this study prove that the convergence of IFRS has a significant negative effect on earnings management. Thus, this study supports the agency theory that IFRS has a monitoring tool that can reduce earnings management. The findings of this study are consistent with previous studies that examined the relationship between the convergence of IFRS and earnings management[1]; [4];[24];[5].

The results of this study demonstrate that the proportion of woman in audit committee has significant negative effect on earnings management. In addition to supporting the agency theory, this findings also supports the prediction of stakeholder theory. Stakeholder theory predicts women's leadership is seen as more concerned with the interests of more stakeholders than man's leadership.

The results of this study prove that the woman in audit committe strengthen the positive relationship between the convergence of IFRS on earnings management. The results of this study indicate that the interaction between convergence of IFRS and the audit committee will boost earnings management. In other words, the presence of both these monitoring mechanisms (internal and external mechanisms) at the same time will negatively impact on the quality of financial reporting by increasing earnings management. So that the findings of this study are not consistent with the hypothesis predicted.

8 Conclusions

By using regression moderating, the results of this study found that IFRS convergence has significant negative effect on earnings management. In addition, corporate governance mechanism, as measured by the proportion of woman in audit committee has significant negative effect on earnings management. The higher the proportion of woman in audit committee would reduce earnings management. However, this study found that interaction between IFRS convergence and proportion of woman in audit commmitte has positive effect on earning management.The findings of the study on the interaction of the two independent variables do not fit with the theory and it shows the need for further research.

References

1. L. Cai, A. Rahman, S. Courtenay, Available at: http://ssrn.com/abstract=1473571 (2008)
2. S. Callao, J. I. Jarne, Account. Eur. **7**, 159–189 (2010)
3. V. Capkun, A. Cazavan-jeny, T. Jeanjean, L. A. Weiss, Available http//ssrn.com/abstract=1125716 1, pp. 1–63 (2011)
4. D. Zéghal, S. Chtourou, Y. M. Sellami, J. Int. Account. , Audit. Tax. **20**, 61–72 (2011)
5. R. N. Qomariah, Thesis Diponegoro University Semarang Indonesia, (2013).
6. Y. O. Lestari, El Muhasaba J. Akunt. **2** (2013)
7. Y. P. Handayani, Thesis Universitas Negeri Padang (2014)
8. P. Iranto, Thesis Diponegoro University Semarang Indonesia (2014)
9. A. D. Nastiti, Thesis Diponegoro University Semarang Indonesia (2015)
10. M. C. Jensen, W. H. Meckling, J. financ. econ. **3**, 305–360 (1976)
11. P. M. Healy, J. M. Wahlen, Account. Horizons **13**, 365–383 (1999)
12. T. Jeanjean, H. Stolowy, J. Account. Public Policy **27**, 480–494 (2008)
13. C. Liu, L. J. Yao, N. Hu, L. Liu, J. Accounting, Audit. Financ. **26**, 659–676 (2011)
14. A. Klein, J. Account. Econ. **33**, 375–400 (2002)
15. J. Bédard, S. M. Chtourou, L. Courteau, Audit. A J. Pract. Theory **23**, 13–35 (2004)
16. A. Marra, P. Mazzola, A. Prencipe, Int. J. Account. **46**, 205–230 (2011).
17. R. Jao, G. Pagalung, J. Akunt. Audit. **8**, 43–54 (2011)
18. D. A. Carter, B. J. Simkins, W. G. Simpson, Financ. Rev. **38**, 33–53 (2003).
19. M. Huse, A. G. Solberg, Woman Manag. Rev. **21**,113–130 (2006)
20. D. Jamali, A. Safieddine, M. Daouk, Corp. Gov. **7**, 574–585 (2007)
21. R. B. Adams, Daniel Ferreira, J. financ. econ. **94**, 291–309 (2009)
22. I. P. Nuratama, Thesis Udayana University Denpasar Indonesia (2011)
23. Y. Chen, Z. Rezaee, Int. J. Account. Inf. Manag. **20**, 171–188 (2012)
24. W. A. W. Ismail, K. A. Kamarudin, T. van Zijl, K. Dunstan, Asian Rev. Account. **21**, 53–73 (2013)

SHS Web of Conferences **34**, 07008 (2017) DOI: 10.1051/shsconf/20173407008
FourA 2016

Post-implementation of MFRS 8 "Operating Segment" among listed firms in Malaysia

Jalila Johari[1,*]

[1]Department of Accounting and Finance, Faculty of Economic and Management, University Putra Malaysia, 43400 UPM Serdang, Selangor, Malaysia

Abstract. This study examined the level of disclosure among Malaysian listed firms with respect to segment reporting under new accounting standard MFRS 8 (IFRS 8). It also aimed to determine how these firms defined the chief operating decision maker (CODM) during the disclosure of segment information. Study results showed that the way firms disclosed segment information tended to vary between companies, since disclosure was dependent on company management purposes and business activities. The results also showed that only a few Malaysian firms supply information about their CODM. In summary, this study provides a fairly up-to-date description of the status of segment disclosure post-implementation of MFRS 8.

1 Introduction

IFRS 8 *Operating Segments* was issued in November 2006 by the US Financial Accounting Standards Board (FASB) and the International Accounting Standards Board (IASB) in a standard convergence project to supersede IAS 14R, which was previously issued by the IASB. The standard was eventually adopted by the Malaysian Accounting Standards Board (MASB) and is known as MFRS 8 in the Malaysian context. Although MFRS 8 became effective on 1 January 2009, its adoption is considered tardy compared with the US, which introduced SFAS 131 (similar to IFRS 8) to replace the old SFAS 14 (similar to FRS 114) in 1997. Although the IASB's main intention in replacing IAS 14R with IFRS 8 was never to compel non-disclosing companies to eventually comply, the new standard could have an effect on Malaysian firms, as it was envisaged that no public listed firms would be able to avoid providing segment disclosures based on the requirements of the new standard. This is because IFRS 8 is precisely worded in its intention and it is unlikely that any company would be able to give creative excuses to justify not disclosing segment information.

The shift from a two-tier approach [2] to a management approach has created a few predicaments. These include an inconsistency and decline in the number of reportable segment items, and the overwhelming number of firms claiming to operate in just one segment. Thus, many firms continue to use broad and vague geographic groupings. During implementation of the previous International Accounting Standard (IAS), firms tended to not fully provide either the primary or secondary disclosures required under the industry

* Corresponding author: jjohari@upm.edu.my

DOI: 10.1051/shsconf/20173407008

approach. This was particularly true for disclosures related to capital expenditures and liabilities, and disclosures of expenses relating to transactions with other components of the same entity, whose operating results were regularly reviewed by the entity's chief operating decision maker (CODM) in order to decide about the allocation of resources to the segment, to assess its performance, and for which discrete financial information was available.

As under the parallel US GAAP accounting standard, SFAS 131 [4], the CODM designation does not necessarily refer to a single individual but to a function within the reporting entity. The new standard requires that the reported amount of each operating item (revenue, assets, etc.) be the same measure that is reported to the CODM for the purpose of allocating resources to the segment, and for assessing its performance. This "management approach" is less flexible, as firms must report segment data externally the same way they manage it internally [3]. In this regard, it can be imagined that any company big enough to be publicly listed would be unlikely to prepare statements and analyse their performance on only an aggregate level. All public listed firms would have some sort of reporting structure that divides the overall company or group into distinct cost centres or segments for meaningful analysis and measurement of performance. Therefore, it could be reasonable to expect that no firms would be able to avoid providing segment reporting based on the requirements of the new IFRS 8.

As the firms are required to disclose financial information by both line of business and geographical areas after having considered the internal information system of the CODM. In other words, the preparers of the financial statements, need to consider the internal organization of the firm when recognizing operating segments. If the operating segments identified do not correspond to the firm's business or geographic segments, then the company needs to reorganize its complex activities to identify products/services, groups of related products/services, or geographic areas that are subject to the same risk and returns and are different from those of other business segments or geographic areas. Using this method, the IASB promoted disaggregated information reporting based on the internal organization of firms. This could be concurrently compared with external information to define a set of criteria for segment items and their disclosure in reports.

The increase in managerial discretion to define operating segments and the extent of segment information disclosure. This is important for the users of financial information because defining segments consistent with corporate practice enables users to evaluate firms using the same information available to a firm's management. However, in increasing managerial discretion in this area, there is a risk that segment information could be subject to management manipulation or inconsistency due to management fads and vagaries [8].

As a result, the present study was concerned with examining whether the changes to the standard have affected segment reporting disclosures by Malaysia's public listed firms under MFRS 8. In addition, this study was also concerned with reported segment disclosures and whether firm management is using certain discretion in managing the extent to which segment information is reported under MFRS 8. Thus, the purpose of this study was to examine segment reporting disclosures under MFRS 8 in 2012, the year of its implementation, and whether the changes in the standard have had an impact on the information published by Malaysian public listed firms.

2 Methodology

An empirical analysis of the application of MFRS 8 was conducted on a sample of Malaysian listed firms with Bursa Malaysia. This study examined only listed firms because these firms are required to prepare their financial statements and disclosure notes in accordance with IFRS [6]. The sample of firms is those that published financial statements in year 2012, after the mandatory implementation of MFRS 8 in Malaysia. This study also looked at the accounting items represented in the segment reports and whether were changes in the number and typology of items voluntarily published. For this study, a customized checklist of potentially disclosed items after considering all the items required under IFRS 8 was created. These items were then grouped into the main balance sheet categories common to all types of companies. A sample of 250 firms listed with Bursa Malaysia around the time of the introduction of MFRS 8 was identified using stratified random sampling. Data were collected based on MFRS 8 requirements.

3 Results

The results obtained from the descriptive analysis of the firms that were examined regarding the implementation of MFRS 8 is being discussed in detail in this sections. The result based on the research objective to examine segment reporting disclosures under MFRS 8 and whether its implementation of new accounting standard had an impact on the information published by Malaysian public listed firms. Thus the result discussed upon the segment reporting disclosure based on operating segment disclosure and entity wide disclosure, the selection of CODM and the analysis by sector group. The result upon the primary/operating segment disclosures and secondary/entity-wide disclosures is in three panel as stated in Table 1.

Table 1. Disclosures made by firms based on the segment items required under MFRS 8.

Panel A: Primary/operating disclosures		**Panel B: Secondary/entity-wide disclosures**		**Panel C: Companies entity-wide disclosures**	
Item	**%**	**Item**	**%**	**Item**	**%**
Segment revenue to external customers	85	Segment revenue from external customers based their location	53	Segment revenue from external customers by products and services	79
Segment revenue – inter-segment transactions	79	Capital expenditure	12	Revenue from external customers by geographic area	56
Segment results – continuing operations	95	Total carrying amount of segment assets based on their location	44	Non-current assets by geographic area	45

Table 1 results indicate a low disclosure of information regarding geographic areas and major customers, as firms alleged that disclosures of this type of information were commercially sensitive and could provide competitors with information that might damage their future business prospects. Companies in the sample were only willing to disclose information about major customers that contributed not more than 10% of revenue, which is the threshold stipulated in MFRS 8 (para 34). Table 2 shows the number of firms reporting by business and geographic segment. Some 90% of companies (224) disclosed financial information according to business segment, while 53% (132) and 45% (112) of companies disclosed information based on the location of their customers and the location of their assets, respectively.

Table 2. Number of companies reporting by business and geographic segment.

Disclosure by:	No. of firms
Business segment	224
Geographic by location of customers	132
Geographic by location of asset	112

One of the key element of segment reporting under MFRS 8 distinct from the old standard is that an operating segment is a component of an operating results are regularly reviewed by an entity's chief operating decision maker (CODM) to make a decision about the segment (IASB, 2006a). Even though the standard highlights the importance of the review of segment activities by a CODM, the standard does not specify the identity of the CODM. The standard simply states that the CODM "is not necessarily a manager with a specific title. Table 3 reports the results for firm disclosures made in segment notes, including disclosures about the CODM.

Table 3. Disclosures in segment notes.

Disclosures in segment notes	%
Reference to the introduction of MFRS 8	14
Reference to the management approach	37
Reference to the CODM	44
Identification of the CODM	33
- Chief Executive Officer (CEO)	23
- Managing Director	19

Only 14% of companies in the sample made reference to the introduction of MFRS 8 in the segment notes, while 37% made reference to the use of the management approach in disclosing financial information. The results also show that 44% of firms made reference to their CODM, while only 33% actually identified their CODM. Given that the MFRS requires segment disclosures to be regularly reviewed by the CODM, these results are somewhat surprising as they tend to suggest that the users of most companies' financial statements do not know who is reviewing segment information. Although the disclosure of information about the CODM is not a mandatory requirement under the standard, its absence appears to be a lost opportunity for firms to provide useful information to their stakeholders. From the 33% of firms (82) that identified a CODM, a majority described their CODM as a managing director, chief executive director, or a member of their company's board of directors.

4 Discussion of the results

Under MFRS 8, a firm is required to disclose segment information that is consistent with how management views the entity based on its internal reports. However, the findings of this study show that, post-MFRS 8 implementation in 2012, most Malaysian public listed firms disclosed segment information based on business products and services rather than by geographic segments. Moreover, MFRS 8 has had a significant impact on how entities disclose segment information. For example, new items such as interest revenue, interest expense, and income tax expense are disclosed if recognized by the CODM. In addition, the geographic locations were finer for individual country disclosures and broader to continents under MFRS 8. The entity-wide geographic disclosures seemed to improve the flexibility to disclose more geographic segment information with a finer and broader disaggregation of geographic locations.

Segment reporting has often been analysed from the perspective of voluntary disclosure. Voluntary disclosure, in general, refers to the issuer's interest in signalling good news to the users of its financial information. However, some firms tend to limit the information they disclose to avoid drawing new competitors into profitable segments [7]. The choice of segments to disclose depends on a company's determination to protect certain highly profitable segments. As a result, some companies prefer to aggregate them with less profitable segments to ward off the arrival of new competitors [5].

One of the principles underpinning MFRS 8 is enabling users to interpret a firm's position and performance based on internal reports that are regularly reviewed by the CODM. According to this study's findings, many Malaysian listed firms do not identify their CODM or the internal function to which the position was related. While there were variations in the identity of the CODM in some firms, the function was predominantly associated with the entity's Chief Executive Officer (CEO). This provides some interesting insight, as interpretation of firm performance "through the eyes of management" must depend on a user's appreciation of who or what the CODM is and, therefore, the management perspective that is being presented.

The segment reporting disclosure practices of Malaysian listed firms need further improvement, as continuous non-disclosure by these firms will have serious negative consequences with respect to the quality of the overall reporting practices of Malaysian listed firms, compared to those of other countries. Malaysian listed firms that do not fully comply with segment reporting requirements will inevitably find themselves the subject of criticism from related stakeholders and the authorities. The eventual outcome of the effectiveness of MFRS 8 in solving the problem mentioned above would be of great interest to the financial analyst community and general users of financial information and could set the direction of future research in this domain. More importantly, the adoption of the suggestions recommended by this study may solve this reporting problem to a certain extent, if not fully.

5 Conclusion

In summary, this study is to examine the post - implementation segment reporting disclosures under MFRS 8 in the year 2012. Despite, the old standard showing firms tend not to disclose fully, the changes in the standard have had an impact on the information published by Malaysian public listed firms. The result of the study shows that the way Malaysian-listed firms report on segments under MFRS 8 varies between firms, since reporting is dependent on company management purposes and business activities. From the results, it appears that Malaysian-listed firms still have not fully put MFRS 8 into practice and that improvements are needed to ensure that firms really comply with the standard. Moreover, the identity of CODMs was not provided by a sizeable number of firms in the study sample. The disclosure of this information is important to stakeholders, as the identity of CODMs may help them to assess how firms are organized and where key decisions regarding operations are being made.

The point of interest to be discussed based on the result is that to examine any possibility of enforcement mechanisms that could increase the level of segment disclosure.

As the result of the study showing that the firms have failed in playing their role and putting their effort to increase the level of transparency. The role of the firms to ensure that they fully comply with the standard are fall short and in consequences the level of disclosure are not in line with what being portrayed in the new standard. As a result the future study upon this matter must be considered by introducing what is mediating and moderating factor that might enhance the level segment disclosure in Malaysia.

References

1. D. L. Street & N. B. Nichols. *LOB and geographic segment disclosures: an analysis of the impact of IAS 14 revised.* Journal of International Accounting, and Taxation, **1(2),** 91-113. (2002).
2. E. Micheal, Y.K. Soo & S. David. Security market Effects Associated with SFAS 131: reported Business Segment. Review of Quantitative Finance and Accounting, 18 (4), 323-344. (2002)
3. Financial Accounting Boards. *Statement of Financial Accounting Standards No. 131. Disclosure about Segments of an Enterprise and related Information*, Norfolk. (1997).
4. L. M. Hayes & R. Lundholm. *Corporate Disclosure Policy and Analyst Behaviour.* The Accounting Review **71,** 467-492. (1996).
5. M. Lucchese & Ferdinando Di Carlo. *An analysis of segment disclosure under IFRS 8 and IAS 14R: Evidence from Italian listed companies.* World Business Institute Australia, Refereed Paper **108** (2012).
6. N.B. Nichols & D.L. Street. *The relationship between competition and business segment reporting decisions under the management approach of IAS 14 Revise.* Journal of International Accounting, Auditing and Taxation; **16,** 51–68. (2007).
7. N. Veron. *EU Adoption of the IFRS 8 Standard on Operating Segments.* Note for the ECON Committee of the European Parliament. Bruegel Institute. (2007)

DOI: 10.1051/shsconf/20173407009

Is candlestick continuation patterns applicable in Malaysian stock market?

Chee-Ling Chin[1,*] , *Mohamad* Jais[1] , *Sophee* Sulong Balia[1] , and *Michael* Tinggi[1]

[1]Department of Accounting and Finance, Universiti Malaysia Sarawak, Jalan Dato Mohd Musa, 94300 Kota Samarahan, Sarawak, Malaysia

Abstract. Technical analysis is deemed to be an anathema to the modern finance theory as it contradicts with the efficient market hypothesis, typically the weak form market efficiency which forbids the utilization of past prices and trading volume data to predict future market movement. However, the technical indicator of candlestick trading strategy is widely applied by traders for short term investment. This study thus investigates on the predictive power of candlestick charting which concentrates on the application of continuation patterns in Malaysian stock market from 2000 to 2014. Skewness adjusted t-test is employed to test the statistical significance of candlesticks' profitability. After taking into account the transaction costs, sub-sample, and out-of-sample test, the findings show that only Falling Window pattern after a prevailing downtrend shows predictive power with bearish signals indicated significantly during the 5-day holding period.

1 Introduction

The debating issue revolving around efficient market hypothesis and technical analysis has yet to be resolved until now. It is proposed that all sorts of readily available information including the public and private information have been fully incorporated into the prices of financial assets [1]. Hence, all three forms of market efficiency signify the impossibility to obtain abnormal returns in an efficient market despite any source of information received. Technical analysis on the other hand believes that market psychology can be traced by analyzing market trend through the aids of chart patterns and numerical calculation from past prices and trading volume data [2]. Technical indicators generate forecasted trading signals to assist traders in market timing. In particular, technical analysis is rejected by the weak form efficient market hypothesis which asserts that the use of historical information to forecast future market movement is useless.

The oldest known form of technical analysis is believed to be the Dow Theory which can be traced back to 1884. Notwithstanding the continuous criticisms from the proponents of efficient market hypothesis, technical analysis is indeed widely applied by traders in the real time securities market. Positive evidence which supports the application of technical analysis are even been accumulated [3, 4, 5, 6, 7]. Recent findings from behavioral finance

[*] Corresponding author: cheelingchin26@gmail.com

which show that investors do not always make decision rationally has added value to the practice of technical analysis method [8].

Among all the technical indicators, the Japanese candlestick charting is one of the most popular trading rules which is time-tested since 1700s in the Japanese futures market for commodity. Its popularity is even highlighted in the present time with the inclusion of candlestick charting technique in every real time technical service providers. In fact, candlestick charting technique acts as a leading indicator with its capability to provide trading signals earlier than other technical indicators. Besides, the flexibility and versatility of candlestick charting is underscored with its ability to function as a stand-alone technique or to combine with other Western technical indicators [9].

Mixed results are obtained for the studies on candlestick charting whereby negative evidence towards the predictability of candlesticks is gained on DJIA stocks [8], in the Japanese equity market [10] and for Brazilian stock market [11]. As the effectiveness of candlestick charting has been proven empirically in the Taiwan stock market [12, 13, 14, 15] while positive evidence is provided for the European stock market [16, 17] and even in China [18], there is no known research in the past that has tested specifically on the predictability of continuation patterns. The significance of continuation patterns shall not be ignored even though it might not be as profitable as the reversal patterns. While reversal patterns typically allow traders to gain windfall from trend reversal signal with the concept of 'buy low and sell high', bullish continuation patterns on the other hand provide continuous buying signals which shall allow investors to hold their stocks longer before selling them at the optimal price to increase the probability of gaining higher profit while bearish continuation patterns function as persistent short-selling signals.

The controversy of candlestick trading strategy with the weak form market efficiency leads to a research issue which is worth study. In the context of Malaysian stock market, it was mentioned that the application of candlestick charting technique is deemed to be very suitable as it is observed to exhibit short term rallies followed by short term corrections with trading cycle ranges from three to fifteen days [19]. Therefore, this paper focuses solely on continuation patterns which have been overlooked by past researchers by examining the predictability of candlestick continuation patterns in Malaysian stock market from one day until fifteen days holding period.

2 Candlestick charting

Candlestick charting is formed from the opening, high, low, and closing price as shown in Figure 1. A white candlestick represents a bullish signal, whereas a black candlestick indicates a bearish signal. Candlestick patterns require a definite trend prior to the pattern to enable future market direction to be forecasted more accurately [9]. Continuation patterns are formed from a sequence of single line candlesticks. According to the Sakata Method, continuation patterns represent a time of rest in the market [20], which thus suggest that the prevailing trend shall persist. The common continuation patterns are identified from practitioner books [9, 19, 20] and documented accordingly in Table 1.

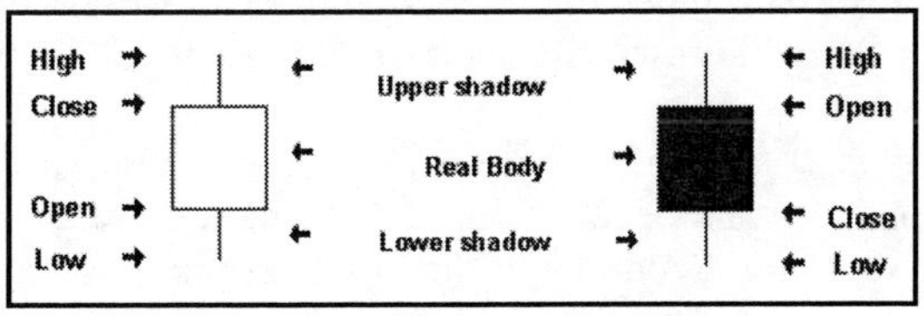

Figure 1. Single line candlesticks.

Table 1. Continuation patterns.

Continuation Pattern	Prior trend	Characteristics	Signal
Falling Window 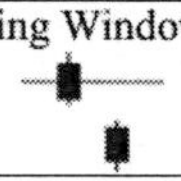	Downtrend	The first and second day can be made up of any basic candles with short or long line except for Four-Price Doji. A gap appears between the first candle's low with the second candle's high.	Bearish
Rising Window 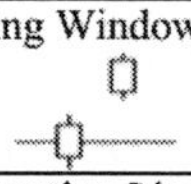	Uptrend	The first and second day can be made up of basic candles with short or line except for Four-Price Doji. A gap appears between the first candle's high with the second candle's low.	Bullish
Separating Lines- 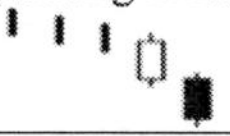	Downtrend	The first day is a white candlestick. The second day is a black candlestick with the same opening price as previous candlestick and then closes lower.	Bearish
Separating Lines+ 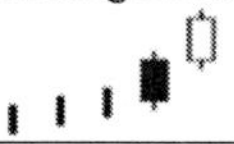	Uptrend	The first day is a black candlestick. The second day is a white candlestick with the same opening price as previous candlestick and then closes higher.	Bullish
On Neck 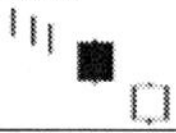	Downtrend	The first day is a black candlestick. The second day is a white candlestick which opens below the low of the previous day and closes at or near the low of the first day's black candlestick.	Bearish
In Neck 	Downtrend	The first day is a black candlestick. The second day is a white candlestick which opens below the low of the previous day and closes at or above the low of the first day's black candlestick.	Bearish
Thrusting 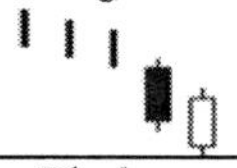	Downtrend	The first day is a black candlestick. The second day is a white candlestick which opens at a lower price than the low of the first day and closes within the body of the first day's body, but still below the midpoint of the first day's black candlestick.	Bearish
Two Black Gapping Candles 	Downtrend	The first day is a black candle with opening price gaps lower from the previous candle. The second day is also a black candle with opening price below prior candle's opening but higher than its low and with closing price below the prior candle's closing.	Bearish
Side-by-Side White Lines- 	Downtrend	The first day is a long black candle. It is then follow by two consecutive small white candles of similar size that gap below the first day's black candle.	Bearish
Side-by-Side White Lines+ 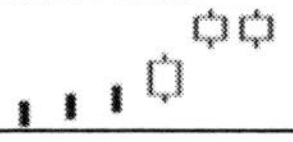	Uptrend	The first day is a white candle. It is then follow by two consecutive small white candles of similar size that gap above the first day's white candle.	Bullish

3 Data and Methodology

The main sample encompasses a total of 420 firms from 3 January 2000 to 30 December 2011 for in-sample analysis while the out-of-sample test utilizes a different set of data which are not reused in the in-sample data from 3 January 2012 to 31 December 2014 and comprises a total of 422 companies. The sub-sample test on the other hand only includes a total of 59 companies with average trading volume that exceeds one million unit shares from 2000 to 2007 after withdrawing the crisis period of 2008 to 2009 and post-crisis period from 2010 onwards.

The whole study is conducted based on three steps research design [15]. Firstly, all possible candlestick patterns that occur within the sample period is scanned and recognized with Candle Scanner software. Secondly, a three-day simple moving average (SMA_3) over six inequalities is employed to identify trends prior to the occurrence of patterns:

$$MA_3(t) = [\, P(t-2) + P(t-1) + P(t) \,] / 3 \qquad (1)$$

A set of inequalities is defined to help identifying market trends with acceptance of at most one violation of the inequalities. A downtrend is detected on day t when:

$$MA_3\,(t-6) > MA_3\,(t-5) > \ldots > MA_3\,(t-1) > MA_3\,(t) \quad (2)$$

Whereas an uptrend on day t is defined as:

$$MA_3\,(t-6) < MA_3\,(t-5) < \ldots < MA_3\,(t-1) < MA_3\,(t) \quad (3)$$

Thirdly, the stock returns is measured as the natural logarithm of the closing price on day t + n divided by the opening price on the day following a continuation pattern computed as:

$$\text{Returns} = \ln\,(C_{t+n} / O_{t+1}) \quad (4)$$

,where n=number of holding period

As candlestick trading strategy is more suitable for short term investment, thus this study examines the raw returns of candlesticks rather than the abnormal ones. The stock returns is computed based on a buy-and-hold investment strategy and it is assumed that a trade is simulated at the opening price on the day following a continuation pattern and then sold at the end of the closing price after holding for a specific holding period. The profitability of continuation patterns is measured from 1-day until 15-day holding period to observe the predictability besides determining the effective holding period of continuation patterns in the context of Malaysian stock market. Binomial test is used to test the null hypothesis H_0: p=0.5 for the true returns rate of candlestick while skewness adjusted t-test is employed to test the null hypothesis H_0: μ=0 for the average returns with the formula stated below:

$$t_{sa} = \sqrt{n}\,[s + (1/3)(\gamma)(s^2) + (1/6n)(\gamma)] \quad (5)$$

,where s = standard mean of returns and γ = estimated skewness of returns.

A transaction cost of 1% is assumed in this study as stated under Bursa Malaysia which includes the brokerage fees, stamp duty, and clearing fees. The continuation patterns have correctly predict the direction of price movement if positive returns rate is obtained by bullish continuation patterns while bearish continuation patterns show negative returns rate. Meanwhile, a prevailing uptrend and downtrend is required for bullish and bearish continuation patterns respectively in order to confirm the predictive power of continuation patterns.

4 Empirical Results

4.1 Candlestick statistics

The statistics from Table 2 show that Falling Window is the only continuation pattern which has obtained significant bearish rate of more than 50% both before and after trend. It shows that Falling Window is more effective in reflecting bearish signals during the 10-day holding period as the highest bearish rate are recorded simultaneously at 51.50% before trend and at 52.30% after downtrend. It is also noticed that continuation patterns with trend identification are more effective in revealing correct signals as the true returns rate have increased following by a prevailing trend.

4.2 Comparing bearish continuation patterns

Table 3 shows that prior to trend, negative returns are attained more often during the 9-day, 10-day, and 11-day holding period while bearish signals tend to be reflected by bearish continuation patterns more frequent on 5-day holding period after a prevailing downtrend. Higher returns rate tend to indicate greater strength of signals. In overall, the bearish

 DOI: 10.1051/shsconf/20173407009

continuation patterns reflect stronger bearish signals after trend as compared to before trend as shown from higher magnitude of negative returns rate obtained after a prevailing downtrend. However, only the predictive power of Falling Window as bearish continuation pattern is found to exist after a prevailing downtrend. Bearish signals are shown significantly from negative returns obtained by Falling Window in both the in-sample and sub-sample test with both concentrating on 5-day holding period.

Table 2. True returns rate of continuation patterns before and after trend from 2000 to 2011.

Continuation Patterns	No.	Before trend 1 day	Before trend 5 day	Before trend 10 day	Before trend 15 day	No.	After trend 1 day	After trend 5 day	After trend 10 day	After trend 15 day
Bearish Rate (%):										
Falling Window	16803	39.43	50.54	51.50	50.49	5088	41.39	51.73	52.30	49.98
		[0.00]***	[0.16]	[0.00]***	[0.20]		[0.00]***	[0.02]**	[0.00]***	[0.99]
In Neck	22	45.45	59.09	59.09	59.09	16	50.00	62.50	62.50	62.50
		[0.83]	[0.52]	[0.52]	[0.52]		[1.00]	[0.45]	[0.45]	[0.45]
On Neck	276	36.96	50.72	51.45	52.90	148	37.84	43.24	45.27	49.32
		[0.00]***	[0.86]	[0.67]	[0.37]		[0.00]***	[0.12]	[0.28]	[0.94]
Separating Lines-	279	30.11	41.94	45.52	48.03	101	32.67	44.55	41.58	40.59
		[0.00]***	[0.01]***	[0.15]	[0.55]		[0.00]***	[0.32]	[0.11]	[0.07]*
Thrusting	340	39.12	45.59	50.59	47.35	196	41.33	41.84	48.47	45.92
		[0.00]***	[0.12]	[0.87]	[0.36]		[0.02]**	[0.03]**	[0.72]	[0.28]
Two Black Gapping Candles	1642	45.25	51.40	51.46	50.00	846	43.85	48.94	50.83	49.53
		[0.00]***	[0.27]	[0.25]	[1.00]		[0.00]***	[0.56]	[0.65]	[0.81]
Bullish Rate (%)										
Rising Window	13231	35.63	42.80	41.93	40.77	3123	37.24	43.42	39.35	39.58
		[0.00]***	[0.00]***	[0.00]***	[0.00]***		[0.00]***	[0.00]***	[0.00]***	[0.00]***
Separating Lines+	197	37.06	51.78	45.18	46.70	93	41.94	53.76	43.01	48.39
		[0.00]***	[0.67]	[0.20]	[0.40]		[0.15]	[0.54]	[0.22]	[0.83]

Bearish rate is obtained by dividing the number of bearish signals (Returns < 0) by the total observed signals (No.) while bullish rate is derived by dividing the number of bullish signals (Returns > 0) by the total observed signals (No.) and presented in percentage form. The p-values for Binomial test are provided in brackets. * Indicates statistical significance at 10% level, ** at 5% level, and *** at 1% level.

Table 3. Significant returns rate of bearish continuation patterns in various samples.

Bearish Continuation Patterns	Before trend: In-sample (2000-2011)	Before trend: Sub-sample (2000-2007)	Before trend: Out-of-sample (2012-2014)	After downtrend: In-sample (2000-2011)	After downtrend: Sub-sample (2000-2007)	After downtrend: Out-of-sample (2012-2014)
In Neck	4-day -3.16 (0.07)*			5-day -4.10 (0.09)*		
	5-day -3.34 (0.06)*			7-day -5.52 (0.08)*		
	6-day -4.05 (0.05)*			9-day -6.83 (0.03)**		
	7-day -4.92 (0.03)**			10-day -6.15 (0.06)*		
	8-day -4.29 (0.08)*					
	9-day -5.48 (0.03)**					
	10-day -5.50 (0.03)**					
On Neck	4-day -1.25 (0.00)***					12-day -3.45 (0.08)*
	5-day -1.24 (0.00)***					13-day -3.28 (0.10)*
	6-day -1.46 (0.00)***					
	9-day -1.20 (0.02)**					
	10-day -1.33 (0.01)**					
	11-day -1.38 (0.01)**					
	12-day -1.17 (0.05)**					
	13-day -1.20 (0.06)*					
	14-day -1.16 (0.07)*					
	15-day -1.30 (0.04)**					
Two Black Gapping Candles	4-day -1.05 (0.00)***					
	5-day -1.20 (0.00)***					
	7-day -1.18 (0.00)***					
	8-day -1.24 (0.00)***					
	9-day -1.32 (0.00)***					
	10-day -1.40 (0.00)***					
	11-day -1.28 (0.00)***					
	12-day -1.15 (0.00)***					
	14-day -1.00 (0.00)***					
	15-day -1.00 (0.00)***					
Falling Window	9-day -1.00 (0.00)***			5-day -1.18 (0.00)***	4-day -1.09 (0.00)***	
	10-day -1.04 (0.00)***				5-day -1.78 (0.00)***	
	11-day -1.04 (0.00)***				7-day -1.12 (0.00)***	
Thrusting		11-day -2.23 (0.07)*			11-day -3.10 (0.08)*	
		12-day -2.52 (0.06)*			12-day -3.31 (0.07)*	
		15-day -3.14 (0.04)**			15-day -3.58 (0.09)*	
Separating Lines-		7-day -2.61 (0.08)*				
		8-day -3.36 (0.06)*				
Side-by-Side White Lines-	3-day -5.50 (0.08)*					

The p-values of skewness adjusted t-test are provided in parentheses. * Indicates statistical significance at 10% level and ** at 5% level.

4.3 Comparing bullish continuation patterns

The predictive power of bullish continuation patterns are ambiguous as the results from robustness tests appear to be inconsistent with the in-sample result as shown from Table 4. Prior to trend identification, only the Bullish Side-by-Side White Lines reveals bullish signals with significant positive returns focusing on short holding period. After a prevailing uptrend, Bullish Separating Lines also portray bullish signals with positive returns rate during short holding period of one and two days but only for the sub-sample and out-of-sample test.

Table 4. Significant returns rate of bullish continuation patterns in various samples.

Bullish Continuation Patterns	Returns rate for different holding period (%)					
	Before trend			After uptrend		
	In-sample (2000-2011)	Sub-sample (2000-2007)	Out-of-sample (2012-2014)	In-Sample (2000-2011)	Sub-sample (2000-2007)	Out-of-sample (2012-2014)
Side-by-Side White Lines+	2-day 1.51 (0.04)** 6-day 2.07 (0.09)*					
Separating Lines+					2-day 1.51 (0.09)*	1-day 1.29 (0.07)*

The p-values of skewness adjusted t-test are provided in parentheses. * Indicates statistical significance at 10% level and ** at 5% level.

5 Conclusion

Continuation patterns suggest a persistence of current trend whereby bearish continuation patterns indicate continuous downtrend or bearish signal whereas bullish continuation patterns shall resume the prevailing uptrend or bullish signal. As limited number of continuation patterns are identified in Malaysian stock market, the results from this study show that Falling Window is the only continuation pattern which is found to be applicable after transaction costs. Following a prevailing downtrend, the predictive power of Falling Window as bearish continuation pattern is proven consistently on 5-day holding period with robustness test. Weak evidence is gained for effectiveness of candlestick charting technical indicator with only one continuation pattern proven to be significant, which thus suggest that Malaysia stock market is in weak form efficient market. As the current study is limited to Malaysian stock market for the application of continuation patterns, it is suggested that future research may focus on different securities market from other countries for comparison and further divide the sample according to characteristics of companies such as size and earnings ratio to examine the effectiveness of candlestick charting trading strategy.

Financial support from Universiti Malaysia Sarawak and Fundamental Research Grant Scheme [FRGS/SS05(04)/1149/2014(16)] are gratefully acknowledged.

References

1. E. Fama, Efficient capital markets: A review of theory and empirical work. Journal of Finance, **25**, 383-417 (1970)
2. J. J. Murphy, *Technical analysis of the financial markets: A comprehensive guide to trading methods and applications*. New York, Penguin Putnam Inc (1999)
3. J. L. Treynor, R. Ferguson, In defense of technical analysis. Journal of Finance, **40**, 757-773 (1985)
4. W. Brock, J. Lakonishok, B. LeBaron, Simple technical trading rules and the stochastic properties of stock returns. Journal of Finance, **47**, 1731-1764 (1992)
5. A. W. Lo, H. Mamaysky, J. Wang, Foundations of technical analysis: Computational algorithms, statistical inference, and empirical implementation. Journal of Finance, **55**, 1705-1765 (2000)

6. P. Shen, Market timing strategies that worked. Journal of Portfolio Management, **29**, 57-68 (2003)
7. C. J. Neely, D. E. Rapach, J. Tu, G. Zhou, Forecasting the equity premium: The role of technical analysis. Journal of Management Science, **60**,7, 1772-1791 (2011)
8. B. R. Marshall, M. R. Young, L. C. Rose, Candlestick technical trading strategies: Can they create value for investors? Journal of Banking & Finance, **30**, 2303-2323 (2006)
9. S. Nison, *Japanese candlestick charting techniques*. New York, New York Institute of Finance (1991)
10. B. R. Marshall, M. R. Young, R. Cahan, Are candlestick technical trading strategies profitable in the Japanese equity market? Review of Quantitative Finance and Accounting, **31**, 191-207 (2008)
11. H. A. Prado, E. Ferneda, L. C. R. Morais, A. J. B. Luiz, E. Matsura, On the effectiveness of candlestick chart analysis for the Brazilian stock market. Procedia Computer Science, **22**, 1136-1145 (2013)
12. Y. Shiu, T. Lu, Pinpoint and synergistic trading strategies of candlesticks. International Journal of Economics and Finance, **3**,1, 234-244 (2011)
13. T. Lu, Y. Shiu, Tests for two-day candlestick patterns in the emerging equity market of Taiwan. Emerging Markets Finance and Trade, **48**, 41-57 (2012)
14. T. Lu, Y. Shiu, T. Liu, Profitable candlestick trading strategies: The evidence from a new perspective. Review of Financial Economics, **21**,2, 63-68 (2012)
15. T. Lu, The profitability of candlestick charting in the Taiwan stock market. Pacific-Basin Finance Journal, **26**, 65-78 (2014)
16. G. Caginalp, H. Laurent, The predictive power of price patterns. Applied Mathematical Finance, **5**, 181-205 (1998)
17. T. Lu, J. Chen, Candlestick charting in European stock markets. The Finsia Journal of Applied Finance, **2**, 20-25 (2013)
18. M. Zhu, S. Atri, E. Yegen, Are candlestick trading strategies effective in certain stocks with distinct features? Pacific-Basin Finance Journal, **37**, 116-127 (2016)
19. F. K. H. Tam, *Investing in KLSE stocks and futures with Japanese candlestick charting techniques*. Selangor Darul Ehsan, Malaysia, Pelanduk Publications (2001)
20. G. Morris, *Candlestick charting explained: Timeless techniques for trading and futures* (3rd ed.). New York, McGraw-Hill (2006)

The study of sustainability report disclosure aspects and their impact on the companies' performance

Aisyah Farisa Caesaria[1,*], and *B.* Basuki[1]

[1]Faculty of Economics and Business, Universitas Airlangga, Surabaya, Indonesia

Abstract. This research is aimed to invetigate the effect of Sustainability Report Disclosure to the Firm's market performance. Three material aspects disclosed in the Sustainability report such as economics (EC), environmental (EN), and social aspect (SC) are used as the independent variables in this research and, furthermore, the dependent variable is the market performance which is proxied by using Tobin's Q. This researchs was conducted using mainstream positivistic quantitative methods to test the three formulated hypotheses. The samples taken were 44 observations from all listed companies in the Indonesia Stock Exchange (IDX) that reveal sustainability reports using GRI-G4 guidelines. This guideline is the latest version issued by the Global Reporting Innitiative (GRI), which can be implemented starting from 2013. The results showed that economicss, environmental, and social aspects have positively significant influence to the companies market performance. The practical implication of this research is the value given by society in term of the company image to those companies which disclosure their activities related to economics, social, and environment activities affects their company performance.

Keywords: Sustainability report, economics disclosure, environmental disclosure, social disclosure, market performance, GRI-G4

1 Background

Sustainability has been the center of discussion in international publication for almost two decades [1]. The concept of sustainability gives attention not only on profit, but also on long-term business sustainability. Sustainability focuses not only on the interests of investors and shareholders, but also on the responsibility of stakeholders who either directly or indirectly affected or related to the business. Businesses awareness towards the trend on sustainable development is driven by the stakeholders needs. According to [2], there is increasing pressure from many groups of stakeholder into the company to reduce negative impacts on the environment and society, as well as demands for more transparent and accountable.

Three aspects of sustainability efforts were disclosed and reported: people-social, planet-environmental, economics-profit. Sustainability reporting of these aspects help organizations to set goals, measure performances and manage changes in order to make their operations more sustainable [3]. In addition, increasing the company that makes reporting separately through sustainability reporting can be seen as a strategy to communicate the company's performance in the company's efforts to be accountable to stakeholders.

The sustainability report has been practiced since 2000, and the GRI guidelines have been used as a reference to the report of the company. Up to 2014, there are 86 companies in Indonesia that have reported sustainability report based on the GRI Guidelines. These guidelines are used by [4], sustainability report is positioned as an additional report on the financial statements. In Indonesia, the implementation of social and environmental responsibility contained in the [5]. The Law stated that any entity run its business in the field and / or related to the natural resources required to implement the Social Responsibility and Environmental budgeted and accounted for as company's costs. Further provisions stipulated in [18]. [18] states that the company is required to disclose the implementation of social and environmental responsibility in their annual report and be accountable to the Shareholders General Meeting.

The increasing need for sustainability report is also followed by increasing companies reports have created the amount of research on the disclosure on sustainability report grow and become a hot topic to be discussed. However, several studies related to the sustainability report showed inconsistent results. Research by [6], as well as [7] states that the three aspects of sustainability report has a significant positive effect on the enterprise market performance, either partially or simultaneously. On the other hand, research by [8] expressed a negative influence on such things. Moreover, [9] found a neutral influence, and few other studies including [10], [2], as well as [11] found the different effect. All these differences are the research gaps which make researchers interested in conducting this research to examine how the influence of each aspect of the sustainability report disclosure. This study used sample of all companies listed in the

*Corresponding author: aisyahfarisa@gmail.com

Indonesia Stock Exchange in 2013-2014. Research years of 2013-2014 were selected, because guidelines for disclosure of sustainability report were most widely used in these periods; GRI G-4 starts to be implemented in 2013.

The research problem: Is the expression of economics, environmental, and social aspects in the Sustainability Report influence the companies' market performances? The purpose of the study, furthermore, is to obtain empirical evidence regarding the impact of those aspects in the companies' performance.

2 Theoretical and development hypothesis

Based on the Stakeholders Theory, the main purpose of the company disclosure on the performance of sustainability is to increase the availability of information, as a form of responsibility to its stakeholders. These responsibilities, according to the Theory of Legitimacy, arise because of the "social contract" requires that all activities of the company can be accepted by outsiders as a legitimate. Therefore, in accordance to the Signaling Theory, a company whose aim is to give a signal that they are good corporate, will disclose more than before because competitors may also have the same pressures to make disclosure. Disclosure on the sustainability report was initially described by the concept of the Triple Bottom Line (TBL) proposed by [12]. Based on this concept, there are three aspects should be disclosed by companies in the sustainability report: economics (profit), environment (planet), and social aspect (people).

2.1 Economics aspects disclosure relationship with the company's performance

The information contained in the economics dimension (EC) convince stakeholders about potential competitive capital resources and low level of risk [11]. The disclosure of these aspects can prove their company's contribution to the economics development of local communities. Sustainability disclosure report is seen as a positive action by the company to maintain trust and good relations with investors and creditors that will invest in the company. The confidence feeling of investors and creditors in the corporate accountability will enhance the company's reputation or image. Consequently, it will increase in market performance of a company for the following years.

H1: Economics Aspects Disclosure (EC) significantly affects the company's market performance.

2.2 Environmental aspects disclosure relationship with the company's performance

Based on the Legitimacy theory, the company responded to public pressure by using environmental disclosure [13]. Environment category in GRI G4-2013 covers the impacts associated with the inputs (such as energy and water) and outputs (such as emissions, effluent and waste). The use of input and output generated by the company triggers a variety of environmental problems. Then the company is required to prevent, minimize, and repair the damage as a form of concern and responsibility to the environment. Transparency of the company's contribution to protect the environment will make stakeholders know and understand the things of what has been done as a form of concern and responsibility to the environment. Then the company will receive recognition and support from its stakeholders, so that the company's business operations will run more smoothly. The second hypothesis can be formulated as follows:

H2: Environmental Aspects Disclosure (EN) significantly affects the company's market performance.

2.3 Disclosure Social Aspects Relations with the Company's Performance

[14] stated that the company's concern increased in the focus of social issues, while maximizing the economy performance to satisfy shareholders and undertake social responsibility for the benefit of society. By revealing the social aspects of the sustainability report means the company will also support many issues of the international organization concern. Social responsibility is not only for external, but also internal stakeholders. The responsibility to the internal side means the company is required to pay attention to employee health and safety, equality of opportunity in competition between male and female employees, and human rights aspects. Meanwhile, to external parties, the company is required to promote anti-corruption policies, anti-competitive and monopolistic practices that can harm the stakeholders and labeling products for the health and safety of customers. According to [15], implement and report on social responsibility towards stakeholders can not only increase the price of company's stock, but also improve the welfare and employee loyalty, lower turnover rate of employees that can lead to increased productivity. When the productivity increases, the company can further enhance the image or company's value in the eyes of all stakeholders. The third formulated hypothesis is:

H3: Social Aspects Disclosure (SO) significantly affects the company's market performance.

3 Research methods

This study uses a quantitative approach to test those hypotheses. The analytical method used by using multiple linear regressions.

3.1 Company market performance

Market performance is a description of certain conditions achieved by the company as a form of reciprocal market perception. Market performance of companies can also be interpreted as the value of the company using Tobin's Q.

3.2 Sustainability report disclosure

Disclosure in the Sustainability Report is an expression that contains information of financial performance, and non-financial information. In this study, researchers used the latest version of the guidelines for the sustainability report disclosure, namely the 4th version of the Guidelines (GRI-G4) which was released in 2013. Of the three categories, there are a total of 91 items that must be disclosed, which consists of 9 items of economics aspects, 34 items for environmental aspects, and 48 items of social aspects. These three variables disclosure in the sustainability report was measured through the Sustainability Report Disclosure Index (SRDI). The calculation is performed by using a dummy which was given a score of 1 if an item is disclosed and 0 if not disclosed. After scoring on all items in the sustainability report, scores are then summed to obtain the overall score of the company. SRDI calculation formula for each aspect of sustainability report:

$\mathbf{SRDI_{EC} = V_{EC}/M_{EC};\ SRDI_{EN} = V_{EN}/M_{EN};\ SRDI_{SC} = V_{SC}/M_{SC}}$

V is number of item disclosed of every aspect, and M is number of item stated in GRI G4

Data collection procedures used are documentation and literature study. The literature sources used is from journals, through the electronic media, such as softfile financial statements of the BEI's website www.idx.co.id, softfile sustainability report on the website http://database.globalreporting.org, and internet sites.

The research population is all companies listed on the Indonesian Stock Exchange (BEI) in the period of 2013-2014. Purposive sampling method used to select samples based on certain considerations in accordance with the research purpose. The criteria established are: (1). All companies listed in Indonesia Stock Exchange (BEI); (2). Publish a sustainability report; (3). Sustainability reports based on the GRI Guidelines G4 in consecutive period 2013-2014; (4). The Company presents the financial statements other than in IDR. Based on these criteria, 22 companies met to be chosen as samples during two-year research periods since GRI-G4 began to be used in 2013. Then the number of samples became 44 companies. However, there are 2 outliers of data so the number of samples became 42 companies.

4 Results and discussion

The results of statistical test are as follows:

Model		Unstandardized Coefficients		Standardized Coefficients	t	Sig.
		B	Std. Error	Beta		
1	(Constant)	-1,531	,466		-3,286	,002
	EC	1,707	,719	,316	2,375	,023
	EN	1,145	,531	,280	2,155	,038
	SC	1,950	,890	,311	2,192	,035
F		15,570				
Sig.		0,000				
R		0,743				
R^2		0,551				
Durbin-Watson		1,719				

The above table shows that, firstly, the disclosure of Economics aspect (EC) has a positive and significant impact on the market performance of the company. These results are consistent and support the research by [6], and [7]. This proves that the disclosure of the economics aspects in sustainability report improve market performance since the disclosure of these aspects can prove their company's contribution to the development of the global economy and local communities. Secondly, the Disclosure of Environmental aspects (EN) also has a positive and significant impact on the market performance of the company. This is consistent with research [16], [2] and [7]. [16] states that the disclosure of environmental performance will positively enhance the company reputation, in which increase the company's market value. This proves that the disclosure of environmental aspects in sustainability report can improve market performance since the disclosure of these aspects can prove the company's concern to environmental conditions associated with its operations. Lastly, the Disclosure social aspect (SC) has a positive and significant impact on the market performance of companies (Tobin's Q). These results are consistent with studies [17], [2], [7] and [10] which states that disclosure of the social aspect of increasing Tobin's Q. This proves that the disclosure of the social aspects of the sustainability report can improve the performance of the enterprise market. By providing a social contribution, the company will have the image of a good performance in the eyes of the market. The results show that the disclosure of each aspect of economics, environmental and social of the company will increase the company's market performance in the future.

 DOI: 10.1051/shsconf/20173408001

5 Conclusions, implications and suggestions

5.1 The conclusions are:

1. Disclosure of the company on aspects of economics, environmental, and social in the Sustainability report turned out to have a significant influence and the market performance of the company. These three aspects will be able to demonstrate their company's contribution to the economics development of both global and local, to prove the existence of the company's concern for the environment, as well as their social contribution to the community and it will enhance the company image in the eyes of the public, thus increasing the company's market performance.
2. The results support some previous research which has found that the higher the disclosure aspects of the sustainability report, namely the economics, environmental and social will lead to improved performance of the enterprise market. Thus, if the company intends to attract the attention of the market trends in performance, the company can optimize these three aspects of sustainability disclosure report.

5.2 Implication of the study

The practical implication of this study is that is that the company must pay attention to the importance of social, economics, and environmental disclosure in the sustainability report. Business people turned out to pay attention to such disclosure, and they judge to give "value good image". A good image is what helps improve the performance of the enterprise market.

5.3 Suggestion

1. Future studies are expected to use more samples and representative, also the length of study period need to be expanded.
2. In calculating the company's market performance should use more relevant market value data, ie the value after the date of publication of sustainability report.
3. Further research is expected to examine the influence of aspects of sustainability report on the performance of companies using other measurement tools or add new variables and involve not only secondary data, but also involves other data, such as the behavior of the manager / owner or shareholder and investor response.

References

1. Stenzel, P. L. Sustainability, The Triple Bottom Line, and The Global Reporting Initiative. *Global Edge Business Review,* Vol. **4** (6) (2010)
2. Hussain, N. Impact of Sustainability Performance on Financial Performance: An Empirical Study of Global Fortune (N100) Firms. *Ca' Foscari University of Venice Working Paper*, No. 2015/01 (2015)
3. Global Reporting Initiative. G4 *Sustainability Report Guidelines: Reporting Principles and Standard Disclosure.* Global Reporting Initiative. (2013)
4. Ikatan Akuntan Indonesia. *Penyataan Standar Akuntansi Keuangan No. 1 (Revisi 2015): Penyajian Laporan Keuangan.* Jakarta (2015)
5. Undang-undang No.40 Tahun 2007 Tentang Perseroan Terbatas (Law No. 40 Year 2007 Article 74 of the Limited Company) (2007)
6. Waworuntu, S.R., *et al.* CSR and Financial Performance Analysis: Evidence from Top ASEAN Listed Companies. *Procedia Social and Behavioral Science,* **164** : 493–500 (2014)
7. Ioannou, I., and G. Sarafeim. The Consequences of Mandatory Corporate Sustainability Reporting: Evidence From Four Countries. *Harvard Business School Research Working Paper,* No. 11–100 (2014)
8. Wibowo, I. and A.F. Sekar. Dampak Pengungkapan SR terhadap Kinerja Keuangan dan Pasar Perusahaan. *Jurnal Simposium Nasional Akuntansi* XVII. Lombok. 24-27 September (2014)
9. McWilliams, A., and S. Donald. Corporate Social Responsibility: A Theory of The Firm Perspective. *The Academy of Management Review,* Vol. **26**(1): 117–127 (2001)
10. Burhan, Annisa H. N. & W. Rahmanti. The Impact of Sustainability Reporting on Company Performance. *Journal of Economicss, Business and Accountancy Ventura,* Vol. **15** (2): 257–272 (2012)
11. Tarigan, J. and S. Hatane. Pengungkapan *Sustainability Report* Terhadap Kinerja Keuangan. *Jurnal Akuntansi & Keuangan,* Vol. **16** (2): 88-110 (2014)
12. Elkington, J. *Cannibals with Forks: The Triple Bottom Line of 21st Century Business,* Oxford: Capstone (1997)
13. Schadewitz, H., and N. Mikael. Communication Via Responsibility Reporting and its Effect on Firm Value in Finland. *John Wiley & Sons, Ltd and ERP Environment Publication* (2010)
14. Rodriguez, M., & Fernandez. Social Responsibility and Financial Performance: The Role of Good Corporate Governance. *BRQ Business Research Quarterly* (2015)
15. Ernst & Young Global Limited. Sustainability Reporting–The Time is Now. *EY Publication* (2014)

16. Dowell, G., *et al.* Do Corporate Global Environmental Standards Create or Destroy Market Value?. *Management Science,* **46** (8): 1059–1074 (2000)
17. Cormier, H., *et al.* Attributes of Social and Human Capital Disclosure and Information Asymmetry between Managers and Investors. *Canadian Journal of Administrative Sciences,* **26**: 71–88 (2009)
18. Peraturan Pemerintah No. 47 Tahun 2012 Tentang Pelaksanaan Tanggung Jawab Sosial Lingkungan Perseroan (the Government Regulation (PP) No. 47 of 2012 on the Implementation of the Environment Social Responsibility Company) (2012)

The institutional logics of local government budget in Indonesia

Ade Palupi[1*], and *Refinaldo* Febiandra[1]

[1]Department of Accounting, Airlangga University, Indonesia

Abstract. This study asks how and why a local government settings budget. It aims to deliver an understanding on how a government entity perceives that budget is important element to manage public expenditures. It uses the S local government in Indonesia as a case study. This study applies Institutional Logics Theory as a theoretical framework in order to analyze data. Data is collected by interviewing seven key persons who directly involves in budgeting process. This research shows that the budgeting process includes determining the purposes of annual budget, setting up planned programs and actvities, determining the amount of revenue that will be collected, allocating resources to expenditures and setting up standar pricing. Some important points in the budgeting process are identified. Those are considering political aspect to determine local tax rate, focusing on competition with other local governments, fulfilling the political promise of local government leader to the constituents, involving community to determine planned program and activities, and allocating resources based on community basic needs. This study concludes that those points develop the institutional logics of politic and managerial in budget setting process.

1 Introduction

This study is about on how and why a local governemnt should develop a budget. Some previous research show that governments in some countries in southern hemisphere apply accounting tools in government entities because of external pressures rather than internal organization needs [1, 5, 7]. Therefore, this study attempts to find out the logical thinking on why budget is important to government entities.

This research aims to deliver an understanding on how a government entity perceives that budget is important element to manage public expenditures. It is important as some countries still apply incremental and line item budgeting system although government reforms budgeting process into performance based management.

The S local government in Indonesia is applied as a case study. The Indonesian government rules the implementation of performance based budgeting to be applied by government agencies. However, it has not succeeded yet because of a lack of human resources and inadequate IT system. Based on this condition, this study take an Indonesian local government to examine on how performance based budgeting is implemented to develop a local governement budget.

*Corresponding author: adepalupy@yahoo.com

DOI: 10.1051/shsconf/20173408002

2 Literature Review

Diamond [6] suggests on how government should implement performance based budgeting. Measuring the efficiency of budget performance requires a change in mindsets at the highest political level and the measuring process should be accepted by middle and lower management levels. In addition, top level bureaucrats must set up firstly the necessary legislative, financial, organizational, personnel and information frameworks for budget implementation. The involvement of citizen is important to ensure that data used to set up budget is credible and fit with public needs. In addition, a new sistem of accountability and incentives need to be introduced to accelerating the sucess of performance based budgeting.

However, performance based budgeting is problematic. The problems are (1) it sometimes does not provide a clear link between performance and fund allocation; (2) performance measurement does not constitute an effective basis for decision making; (3) and budget fails to identify 'who' is responsible for performance and resource spending [4].

The complexity of organisation characteristic influences on how and why budget is developed. A study of French theathers shows that the teathers that are subsidied by the French government experienced a tension when preparing the budget [3]. A tension is occured because the organisation has two important opposed objectives which are escalating original artistic value in producing cultural arts, on the other hand, it has to promote eficiency to manage its fund. As the theathers are funded by the government, the study finds that the budget also as a media to fulfill the government interest [3].

Ezzamel et al [8] also investigates why budget is important in education sector in England and Wales. Similar to Amans et al [3]'s study, Ezzamel et al [8] finds there is a tension of logical thinking during budget setting process. That the budget relates to the number of students, budget motivates the education institutions to go for competition to get a larger number of students. However, in the name of effieciency, the education institutions have to cut their expenditures including the amount of teachers' salaries, as a consequence, they should reduce the number of teachers. The study also shows that the interest of government in requiring the standard salary rate for teachers and increasing the number of teachers leads a conflict during budgeting process.

Those previous research indicate implementing performance based budget compels the change of organisational context. Indeed, applying performance based budgeting requires profesionalism in planning, financing and accounting. However, some countries that having a lack of resources, apply performance based budgeting as it is motivated by external pressures. For example, the Indonesian government encounters some problems to accelerate the budgeting reform as the reform is pushed by external pressures which are government regulations and donor agencies [7]. The perception that managing external institutions with significant actors is more important than to fulfilling the need of organisation itself is the main reason on why the Bangladeshi state-owned enterprises apply performance based budgeting, as a result the budget is relatively uncoupled from detailed aspects of the organizational tasks [2].

3 Research framework

This study applies the institutional logics theory [9] as a research framework. Thornton and Ocasio [9] defined institutional logics as socially constructed, historical patterns of material practices, assumptions, values, beliefs, and rules by which individuals produce and reproduce their material subsistence, organize time and space, and provide meaning to their social reality. The basic idea of institutional logics is to link between individual agency, cognition and socially constructed institutional practices and rule structures. There is a connection between individual choices to do something and their broader institutional orders. The institutional logics theory is capable to explain on how and why people conduct in particular actions in organisations.

As society institutional order has both material and cultural characteristics, institutional logic recognizes that institutions are developing and changing as a result of interplay between material and cultural substances those embedded in individual who locate in organisation. When the structure of society is changed, people will respond differently in accordance to material and cultural capital they have. Practices are a result of the interplay between individual agency and their institutional context.

Likewise, when organisational culture changes, institutional logics on how organizations live could be changed, as a consequence, there would be competition between prevailing and new institusional order. For example, Ezzamel et al [8] studies the change of budgeting sytem in the English and Wales' schools as the UK government issued a new education reform act. As its institutional order was changed by embracing new public management concept, the prevailing institutional logic of education field that was teaching professionalism logic was changed to business logic. The tension of these both logics is evidenced in school budgetary practices. Business logic that dominantly underlies the budget setting process drives the change individual tasks within schools whilts the logic of teaching professionalism does not become a dominant value anymore.

This study applies institutional logics theory to examine how budgetary practices shifts from tradisional to performance based bugeting in a complex organisation that is a local government in an emerging country. By using the theory, it will identify what kind of logics that brought by new budgetary practices and how the new logics competed with prevailing logics and how actors within organisation encounter the conflict of logics in new budgetary system.

4 Research Method

Data was collected by doing in-depth-interview and documentation. All of interviewees are seven persons who are the member of budget comittee. They are the head of revenue agency who is responsible to determine the amount of revenue in the budget; the head of internal auditor; the manager of tax and retribution who assist the head of revenue to calculate revenue budget; the manager of assets; two managers of public works agency as a key person to determine expenditure budget.

All interviews was recorded and transcribed. Then, all interview transcripts and documentation were read in order to find stories on how local government budget is developed. Data is categorized into themes which are the change of budget system; the budgetary practice; determining revenue and expenditure budget. Each theme, then, is analyzed.

5 Findings

5.1. The change of budget system

The Indonesian government reforms its governement budgeting system based on the law No. 17/2003. The reform includes the change of budgeting system in local government that it is ruled by Domestic Affairs Ministry Regulation No. 21/2011. The regulations state that government budget should apply the method of performance based budget to replace the practice of incremental based budget. It also rules that the process of budgeting includes the

 DOI: 10.1051/shsconf/20173408002

development planning discussion to accommodate community voice in a region. The aim of community discussion is identifying community needs in term to determine target ouputs which are the significant point in performance based budgeting.

The reform brings some changes in local government. It changes the structure of budget, the legal system of budget and the change of information technology. Those changes affect on how government entities to set their budgets. The followings explain how the S local government in East Java Province, Indonesia, set its budget.

5.2. The budgetary practice

Bottom up, top bottom and participative approaches are applied by the S local government to set its budget. Bottom-up approach accumulates community needs through the development planning discussion. Participative approach is conducted by identifying what programs and activities that the government agencies within S local government will do for the following year. Top bottom approach means budget should also accomodate the vision of district head. According to the respondents:

"Our budget preparation links to the district head's vision and missions that have already stated in his political champaign" (The Head of Asset Management Agency)

"We conducted the development planning discussion to identify the advantages and disadvantages of development programs in villages. We should focus on how deploy public fund to accomodate community needs that should be the development priority. However, there are too many needs but our public fund is not enough. Therefore, there is a limitation to fullfill people's needs" (Internal Auditor)

"If we find limitation on the amount of public fund in setting budget, we will cut the programs and activities proposed by the government agencies. We will not delete the program and activities that required by community through the development planning discussion. (Head of Parks and Sanitation Agency)

From the interviews, this study sees that the budget preparation is about to fulfilling the political intention of district leader. The budget reflects the vision of dictrict leader during his/her political campaign. Also, the budget put people satisfiction as a priority instead of what the local government management should do to achieve its goals. It could link that satisfying people is considered to fulfilling district leaders' constituent so that their political position is safe. However, the Indonesia regulation states that local government's vision and objectives are as same as the district leaders' vision that announced during political champaign. This regulation indeed distract on how the budget is made. The budget should bring the value of efficiency on how to consume public fund instead of focusing on satisfying constituents and fulfilling the political promise of district leader.

The interview also shows that the S local government finds the limitation of public fund to finance its programs and activities. It means that the budget should prioritize on how the total amount of revenue is allocated to cover spending.

5.3. Setting revenue budget

The Indonesian regulations state that local government revenue budget should consist of three accounts which are district revenue, transfer and other revenues. District revenue accounts for local taxes and retributions collection. Transfer account records transfer funds from the central and province governments. Other revenues account may record dividend revenue as it is a return from local government investment in government business enterprises. Therefore, district revenue is the most important aspect of local government revenue budget as it is a representation the sustainability of financial capability.

The S local government plans its revenue budget by determining the local tax rate as the Head of Finance Agency says:

> *"we consider other local governments' economic condition such as Banyuwangi, Jember and Bondowoso and the social and economic condition of local people to define local tax rates"*

> *"we asked help to the supervision for development financing body to calculate the potential amount of district revenue budget and the district leader determines the amount stated in revenue budget"*

> *"we also justify the amount of budget revenue in front of the local parliament"*

This study implies that the political aspect of budget is a priority. Local government challenges other local governments as competitors to set local tax rates. The main reason is perhaps to attract more investors or business in S. The potential local taxes and retributions should be the basic reference to determine the amount of revenue budget instead of the district leader's decision. This condition might degrade the quality of budget as the district leader might hesitate to take risk. The number of tax revenue is usually set lower than the potential economic capacity. It means the performance of district leader in attaining revenue target is easier and the performance report would look good. Thus, determining district revenue is a kind of political logic instead of managerial logic.

5.4. Setting expenditures

Another important aspect on how to prepare budget in the Indonesian government is setting the total amount of expenditures. Mostly local governments in Indonesia find challenges covering all expenditures with local taxes. Therefore, they rely on the transfer fund from the central and province governments. According to the head of park and sanitation agency:

> *"the programs of government agencies is defined by the agencies' tasks and responsibilities. For example, waste disposal program. We predict how many garbage trucks are needed so we know how much liters of gasoline are required and how many persons needed to operate waste disposal program. Thus, we know how much fund should be available to conduct waste disposal program"*

> *" for basic needs such as waste disposal program, street light and sanitation, we try all those programs can be covered by available fund so that we do not cut the program activities. But sometimes, it happens"*

This study mentions the managerial aspect of budget is also considered in setting expenditure budget in the S local government. The total amount of expenditure is based on the output target of program that link to basic community needs. Therefore, the value of program effectiveness is a logic in budgeting process.

6 Concluding remarks

Thornton and Ocasio's [9] theory on institutional logics states that practices are motivated by value, beliefs or rules. By applying this theory to find out on how the S local government budget is produced, this study shows that budgeting process is a kind of politic and managerial logic.

DOI: 10.1051/shsconf/20173408002

The logic of politic in budgeting process can be seen on the process of revenue budget setting. The budget is a decision of political leader and as a tool to fulfill political commitment. The logic of managerial is also occurred expenditure budgeting process. Management accounting concept is applied to determine the cost of program by considering the total number of target output as a priority. This study evidences that the S local government budgeting combines the logic of politic and managerial.

Conflict between those logics is happened. First, the conflict is occurred between using the potential calculation of local economic versus the district leader decision to set the total amount of revenue budget. Second, the conflict also occurs in determining the total amount of spending budget. The local government agencies main tasks and public needs should be priorities but it cannot because of funding limitation. This study finds logic of politics is dominant than the logics of managerial in budgeting process.

References

1. Adhikari, P. Kuruppu, C. and Matilal, S. Dissemination and institutionalization of public sector accounting reforms in less developed countries: A comparative study of the Nepalese and Sri Lankan central governments, *Accounting Forum 37* , pp. (2013) 213– 230. (2013).
2. Alam, M. Budgetary process in uncertain contexts: a study of state-owned enterprises in Bangladesh, Management Accounting Research, Vol. 8(2), pp. 147-167. (1997).
3. Amans, P., Chapelon, A. M., & Dubus, F. V. Budgeting in institutional complexity: The case of performing arts organizations. *Management Accounting Research, 27*( ), 47-66. (2015).
4. Andrews, M. Performance Based Budgeting Reforms: Progress, Problems and Pointers, in Anwar Shah (ed) Handbook on Public Sector Performance Reviews, Vol. 1: Ensuring Accountability When There is No Bottom Line. (2003).
5. Azmi, A.H., and Mohamed, N. Readiness of Malaysian public sector employees in moving towards accrual accounting for improve accountability: The case of Ministry of Education (MOE), *International Conference on Accounting Studies 2014*, ICAS 2014, 18-19 August 2014, Kuala Lumpur, Malaysia. (2014).
6. Diamond, J. From Program to Performing Budgeting the Challenge for Emerging Market Economies, IMF Working Paper. (2003).
7. Djamhuri, A. A Case Study Of Governmental Accounting And Budgeting Reform At Local Authority In Indonesia: An Institutionalist Perspective, *Thesis submitted in fulfillment of the requirements For the degree of Doctor Philosophy*, School of Management Universiti Sains Malaysia. (2009).
8. Ezzamel, M., Robson, K., & Stapleton, P. The logics of budgeting: Theorization and practice variation in the educational field. *Accounting, Organizations and Society, Vol. 37*, 281-303. (2012).
9. Thornton, P.H. and Ocasio, W. Institutional Logics, in Royston Greenwood, Christine Oliver, Kerstin Sahlin and Roy Suddaby (eds.) *Handbook of Organizational Institutionalism*, CA: Sage. (2008).

SHS Web of Conferences **34**, 08003 (2017) DOI: 10.1051/shsconf/20173408003
FourA 2016

Does Congruence Between Incentive System and Locus of Control Affect Team Performance?

Siti Mutmainah[1,*] and *Slamet* Sugiri[2]

[1]Diponegoro University, Semarang, Indonesia
[2]Gadjah Mada University, Yogyakarta, Indonesia

Abstract. Teamwork is an essential element in most organizations; however, little is known about the best fit among incentive system, team composition, and team performance. This study examines whether the congruence between incentive system and locus of control (LoC) affects team performance. To reconcile opposite lines of arguments regarding the best incentive system for a team, this paper uses the social identity perspective and person-environment fit theory to understand behavior in a group process. One hundred and five postgraduate students were assigned to three-person work groups, where they completed an independent task under one of two types of incentive—individual and group incentive—after their LoC was measured. Results suggest that group incentive results in an enhanced team performance and team performance is better when there is congruence between incentive system and LoC. Group incentive system combined with external LoC results in the best performance, while individual incentive results in a better team performance when combined with internal LoC.

1 Introduction

This research aims to examine whether the congruence between incentive system and locus of control (LoC) affects team performance. Group incentive system is the best motivator in an organizational setting that relies on teamwork [18, 23, 32, 39] because it encourages the whole team to work together and share information [13, 18]. This can be explained by organizational behavior literatures, for example social categorization theory [19] and goal interdependence theory [10]. Unlike the aforementioned literatures, the agency theory sees a potential for causing conflict among team members in the group-based reward [26]. The agency theory describes human conception (which is opportunistic, self-interested, and pursuing individualistic utility motivation [11]) as harnessing the lack of close relation between individual performance and group reward by bringing out social loafing [1, 29].

Empirical evidence of the relation of incentive system and team performance shows inconsistent results. There are evidences that team performance is better when group incentive is applied [18, 23]. However, other evidence shows that team performance is not

[*] Corresponding author: titikhasyim@yahoo.com

significantly affected by the group incentive structure [26]. Meanwhile, that it is often the case that a team member refuses to collaborate and only give little effort for the team although he or she receives the same reward as others [9]. The inefficiency of group incentive is also revealed by other researchers [24].

In order to resolve the debate between theories and findings related to incentive systems and team performance, this research uses social identity perspective [12] and person-environment (P-E) fit theory [21] and investigate the role of personal traits at the group level [5, 22]. One of the important personality traits, which is stable, well documented, and indicating fundamental individual differences, is locus of control [3, 17]. This view on control affects an individual's accomplishments [1] and managerial effectiveness [3, 5].

Teamwork is an essential element in most organizations [23], so it is increasingly important to identify an effective strategy to build a team [20]. Since an organization consists of cooperating individuals, it is important to acquire the knowledge about interactions among team members and their effectiveness [15]. Meanwhile, empirical findings about the ways incentive affects the exchange of information and employees' performance are still limited and varied in number [35]. Ambrose and Kulik [2] who reviewed more than 200 empirical researches about work motivation conclude that little is known of the way teamwork is motivated. When organizations apply more team-based systems, research about team members' composition (e.g. LoC) and motivation (e.g. incentive) become more important as well.

We conduct an experiment involving 105 participants divided into 37 teams. The statistical testing results support our proposed hypothesis. Group incentive system encourages better team performance compared to individual incentive system. Empirically, it is also supported that the group incentive system could improve the performance of a team consisting of individuals with external LoC. Meanwhile, the team consisting of internal LoC individuals shows a better performance if the individual incentive system is applied.

Exploration for the matching of incentive schemes and locus of control in a team to produce an optimized teamwork is rarely examined in accounting management, especially using the experimental method. Experimental method researches with a team as analysis unit is also relatively scarce. Additionally, this research fills the gap between the disparity in the explanation of the appropriate team incentive systems.

2 Hypothesis Development

While agency theory and social cognitive theory support the application of the individual incentive system to improve team performance [26, 29], social categorization theory [19] and goals interdependence theory [10] strongly support group incentive system. To connect both poles, this research relates the incentive system with team performance based on the social identity perspective. This perspective is focused on conditions that determine self-definitions differently, which show the process of individual and group interaction in determining work motivation [12]. This perspective does not discern which one is more important or valid as source of motivation, but rather discern a more proper condition as source of motivation.

In the social identity framework theory, teamwork is a condition faced by team members to show 'the self' in a collective meaning. The group incentive scheme is a system that encourages employees to show their identities as part of a collective. Group-based performance report will urge the employees to adopt a group frame [34]. The group frame encourages individuals to be more attentive of other individuals' outcomes and to build trust between them, while the individual frame encourages individuals to be self-centered. The individual-based incentive could lessen the information exchange between employees although economic incentive is given for the action [18].

Empirical evidence shows that group-based incentive is effective for organizations as it is mutually accountable, in which the members work for the same purpose as co-workers; in addition to the ease in measurement and evaluation [1, 23]. In contrast, the individual incentive, which serves as individual performance compensation, can bring out 'selfish' behavior, such as the tendency to conceal information [38] and deter team activity such as information sharing and discussion [18]. Therefore, the group incentive system could motivate team members to coordinate and cooperate with each other to produce a much better team performance. Based on these arguments we propose the hypothesis:

H_1: Group incentive system encourages better team performance compared to individual incentive system

The influence of group incentive system on team performance is not entirely and consistently supported by numerous of previous empirical studies [for example 18, 23, 32 versus 24, 26]. The meta-analysis shows the inconclusive results regarding the influence of incentive plans on performance [16]. Failure to reach a definitive conclusion from the aforementioned researches indicates that the main effect comparison is not adequate and that the possibility of mediation variable or moderator variable needs to be considered [37].

Person-environment (P-E) fit is generally defined as the match between individual and the working environment, which occur when both have similar characteristics [21]. For each individual there is an environment (interpersonal and non-interpersonal) which more or less matches his or her personality characteristics. The congruence or "best-fit" of an individual and the working environment can be seen from the good working performance, satisfaction, and the low level of stress in a system [4, 7].

Congruence between incentive system and LoC presumably affects the team performance. LoC personality dimension divides individuals based on the level of a person's acceptance of personal responsibility over what has happened to him or her [33]. Internal LoC individuals perceive a strong relation between their actions and the goals and the expected outcomes [33]. Therefore, the internal LoC characteristics will fit in an environment applying the individual incentive system rather than the group incentive. In the individual incentive system, there is a tight relation between individual performance and the received reward [29]. Additionally, this system also possesses a strong control of the individual over their performance [24]. When internal LoC individuals gathered in a team with the group incentive system, an effective team attribute of interdependence would be relatively difficult to obtain. Interdependence could be formed when team members are not entirely self-directed or not entirely independent [36]. The internal LoC team does not easily fulfill this attribute.

On the other hand, the group incentive system will match external LoC individuals. The group incentive scheme is designed to stimulate the collaboration between team members and the synergy within the team. Cooperation is assumed more attainable by external LoC individuals because they believe the chance and forces outside them affect success [33], in this case including the team, which becomes their closest working environment. Internal LoC individuals tend to be more independent and less appreciative of teamwork, while external LoC individuals tend to behave in contrast to internal LoC [30].

The congruence between incentive system and LoC is presumed to be positively responded by individuals. When congruence or P-E fit condition is attained, individuals will feel more comfortable and competent [8]. Additionally, P-E fit fulfills the fundamental needs of belonging and self-actualization, which in turn produces positive attitude and behavior [6]. These arguments give rise to the following hypotheses:

DOI: 10.1051/shsconf/20173408003

H_2: Incentive system interacts with locus of control (LoC), such that internal LoC teams perform at higher levels at individual incentive than at group incentive, and external LoC teams perform at higher levels at group incentive than at individual incentive.

3 Method

To analyze our hypotheses, we conducted a laboratory experiment with a 2 x 2 design at a large university (Diponegoro University) in Indonesia. The participants were 105 accounting postgraduate students divided into 37 teams. All participants are students of the first semester (75.2%) who have interacted with each other for at least 4 months and of the third semester (24.8%) who have interacted for at least 1.5 years, so it is very likely for participants to cooperate as a team. Participants consist of 50.5% women and 49.5% men, with the average age of 30.97 and 82.9% of participants have worked.

Team performance (dependent variable) is measured by the number of tasks precisely completed by all of the member of the team. The incentive system (independent variable) is categorized into two, the individual incentive system and group incentive system. The individual incentive uses the equity norm basis, while group incentive uses equality norm.

The experiment is conducted in the following steps: (1) measuring participants' LoC (moderator variable) by Rotter's (1966) instrument [33]. (2) The participants are gathered in the same hour in different classes with different experiment conditions to prevent the internal validity threat of history, diffusion, and regression. The participants are divided to incentive system groups randomly, to overcome the selection threat, with three members in every group. To prevent reactivity and experimenter effects, the instruction is written in word-by-word details, and is to be read by the research assistant. To confirm that the participants understand their roles and tasks in the experiment, they are asked to answer four questions. (3) Participants work on the experimental task Raven's (2014) Standard Progressive Matrices (SPM) [31] as instructed in 10 minutes. This short duration of time is an effort to prevent the internal validity threat of maturation. The task chosen has a moderate difficulty level to prevent the internal validity threat of mortality. Each member in a group had to complete the different task. (4) Questionnaire sheets after experiment as manipulation check by giving Likert scale questions of five points (5) Incentive distribution. The hypotheses are tested by analysis of variance (ANOVA).

4 Results

The Cronbach's alpha scale of the manipulation check is 0.75, meaning that this scale is reliable [28]. Additionally, the result of the Pearson correlation test shows that the questionnaire is a valid instrument for manipulation check. The mean of individual incentive condition (n=18) is 3.38 (SD 0.86) and the mean of group incentive condition (n=19) is 4.19 (SD 0.63) which signifies that manipulation design have effects as expected.

According to Table 1, the primary effect of group incentive system on performance (mean = 8.8772) is bigger than the individual incentive system (mean = 8.1852). According to Table 1, the internal LoC team in the individual incentive system (mean = 8.5926) is not too different from internal LoC team within the group incentive system (mean = 8.5455). Meanwhile the external LoC team performs much better in the group incentive system (mean = 9.3333) compared to the individual incentive system (mean = 7.7778).

Results in Table 2 support our first hypothesis that the group incentive system encourages better team performance compared to the individual incentive system ($F = 4.242$, $p < 0.05$). Results also indicate support for the second hypothesis that incentive system interacts with locus of control (LoC) such that internal LoC teams perform at higher levels at individual incentive than at group incentive, and external LoC teams perform at higher levels at group incentive than at individual incentive (F = 4.789, p< 0.05).[†]

Table 1. Descriptive Statistics

Locus of Control (LoC)	Mean and Standard Deviation of Team Performance		
	Individual Incentive System	Group Incentive System	Overall
Internal LoC	8.5926	8.5455	8.5667
(Standard Deviation)	(1.07726)	(0.83030)	(0.92306)
	n = 9	n = 11	n = 20
External LoC	7.7778	9.3333	8.5098
(Standard Deviation)	(1.61589)	(0.69007)	(1.46779)
	n =9	n = 8	n = 17
Overall	8.1852	8.8772	8.5405
(Standard Deviation)	(1.39665)	(0.85318)	(1.18661)
	n = 18	n = 19	n = 37

Table 2. ANOVA Result (Dependent Variable: Team Performance)

	Mean Square	df	F	Sig.
Incentive System	5.193	1	4.242	0.047
Locus of Control (LoC)	0.002	1	0.001	0.971
Insentive*LoC	5.863	1	4.789	0.036
Error	1.224	33		
R Squared = 0,203 (Adj. R Squared = 0,131)				

5 Discussion

This research aims to examine whether the congruence between incentive system and locus of control (LoC) affects team performance. The results support the proposed hypotheses. Findings related to the first hypothesis, that the group incentive system encourages a better team performance compared to the individual incentive system, is consistent with the previous research [for example 18, 23, 32].

The empirical findings supporting the second hypothesis is consistent with the P-E fit theory, in which the best performance is obtained by the best-fit. The small difference of internal LoC teams' performances in both incentive systems could be explained by the LoC characteristic, which perceive themselves as active agents. Internal LoC individuals perceive themselves as competent individuals in maintaining the control over their lives, and they generally have a great need for self-determination and competence [27].

Meanwhile, the improvement of performance in external LoC individuals is bigger than internal LoC in the incentive system shifting. When external LoC individuals cooperate, they could synergize and perform better than they do individually, because the inter-person interaction could decrease negative effects (especially fear and stress) [29]. Such a condition is consistent with the definition of teamwork as a cooperative process, which enables people to obtain extraordinary results [36].

These findings regarding external LoC individuals supplement the previous empirical findings that shows the superiority of internal LoC individuals [see 5, 14]. The findings of

[†] To assure that team performance is not affected by accompanying factors such as subject demographic characteristics, an analysis of covariate (ANCOVA) test was also conducted. The result shows that team performance is not affected by either GPA (F = 1.618, p value = 0.213>0.05) nor the age of participants (F = 0.767, p value = 0.388>0.05).

DOI: 10.1051/shsconf/20173408003

this research also contribute in resolving the debate between theories related to incentive system. From the managerial point of view, this research affirms the importance of team composition design [36] and its compatibility with the incentive system. This research supports the supplementary fit in order to obtain organizational goals [25].

This research has its limitations regarding external validation. This research also did not take into consideration the characteristics of the tasks. When a task requires a high level of information processing, the processes within the group, including open communication and performance strategy discussion, become more complex [15].

References

1. A. Ahlgren, I. Anderson, H. Skold, Essay in Management Control (School of Business Economics and Law, Goteborg University 2007)
2. M. L. Ambrose, C. T., Kulik, J. Mgt 25, 231 (1999)
3. C. Boone, W.V. Olffen, A.V. Witteloostuijn, AMJ 48, 889 (2005)
4. W.C. Borman, Curr. Dir. in Psyc. Sci. 13, 238 (2004)
5. P. Brownell, The Acc. Rev. 56, 844 (1981)
6. D.M. Cable, J.R. Edwards, J. App. Psych. 89, 822 (2004)
7. D.M. Cable, C.K. Parsons, Pers. Psych. 54, 1 (2001)
8. J.A. Chatman, AMR 14, 333 (1989)
9. R.M. Dawes, An. Rev. Psych. 31, 169 (1980)
10. M. Deutsch, The Int'l J. Conflict Mgt 1, 237 (1990)
11. K.M. Eisenhardt, AMR 14, 57 (1989)
12. N. Ellemers, D.D. Gilder, S.A. Haslam, AMR 29, 459 (2004)
13. G. Fisher, B. Sprinkle, L. L. Walker, J. Corp. Acc. & Fin. 19, 35 (2008)
14. V. Frucot, W. T. Shearon, The Acc. Rev. 66, 80 (1991)
15. M.E. Gist, E.A. Locke, M.S. Taylor, J. Mgt 13, 237 (1987)
16. R. Guzzo, R. Jette, R. Katzell, Pers. Psych. 38, 275 (1985)
17. T.A. Judge, J.E. Bono, J. Appl. Psych. 86, 80 (2001)
18. K. Kelly, BRIA 22, 43 (2010).
19. H. Kim, K.L. Sutton, Y. Gong, Asia Pac. J. Mgt. 30, 31 (2013)
20. A. Kramer, D. P. Bhave, T. D. Johnson, Pers. & Ind. Diff. 58, 132 (2014)
21. A.L. Kristof-Brown, R.D. Zimmerman, E.C. Johnson, Pers. Psych. 58, 281 (2005)
22. J.A. LePine, B.R. Buckman, E.R. Crawford, J.R. Methot, HR Mgt Rev. 21, 311 (2011)
23. T. Libby, L. Thorne, BRIA 21, 57 (2009)
24. H.M.McGee, A.M.Dickinson, B.E.Huitema, K.M.Cullig, Perf.Impr.Quar.19,101(2006)
25. P.M. Muchinsky, C.J. Monahan, J. Voc. Behav. 31, 268 (1987)
26. D. Naranjo-Gil, G. Cuevas-Rodriguez, A. Lopez-Cabrales, BRIA 24, 177 (2012)
27. T. W. H. Ng, K. L. Sorensen, L. T. Eby, J. Org. Behav. 27, 1057 (2006)
28. J.C. Nunnaly, Psychometric Theory (McGraw-Hill, New York, 1978)
29. C.C. Pinder. Work Motivation in Organizational Behavior (Prentice-Hall, NJ 1998)
30. C.M. Ramirez. Teams: A Competency Based Approach (Routledge, 2013)
31. J.C. Raven. Raven's Progressive Matrices (Pearson, 2014)
32. S. P. Ravenscroft, S. Haka, BRIA 8, 114 (1996)
33. J.B. Rotter,. Psych. Mono.: Gen.& Appl. 80, 1 (1966)
34. C. Rowe, The Acc. Rev. 79, 1153 (2004)
35. E. Siemsen, S. Balasubramaniam, A.V. Roth, Mgt. Sci. 53, 1533 (2007)
36. P. Taricone, J. Luca. Successful Teamwork: A Case Study (HERDSA 2002)
37. D. Tjosvold, Human Relations 37, 743 (1984)
38. M. S. Van Alstyne, HBR 83, 24 (2005)
39. S. M. Young, J. Fisher, T. M. Lindquist, The Acc. Rev. 68, 466 (1993)

Measuring customer profitability through time-driven activity-based costing: a case study at hotel x Jogjakarta

Garry Billy Ardiansyah[1], *Bambang* Tjahjadi[2], *Noorlailie* Soewarno[3,*]

[1,2,3] Faculty of Economics and Business, Universitas Airlangga, Surabaya-Indonesia

Abstract. Travelling intensity has been higher in the past few years making hospitality industry must be ready to provide the best service at optimum prices to customers who purchase and enjoy the services. Hotel as one of the accommodation service providers should be able to allocate resources accurately according to the actual needed capacity and thorough tariff policy for not losing the competition and obtain profit according to how much is projected. TDABC system is the most updated costing system which can assist the allocation of these resources. TDABC system is considered twice important to be applied in the service industry as service is intangible so costing system must show the exact and accurate number of resources used to provide the service. Through TDABC system, Customer Profitability Analysis is conducted with the aim of knowing the level of profitability every customer type possesses and classifying them as profitable or non-profitable customers. A case study had been conducted in a five-star hotel located in the Special Region of Jogjakarta resulting a profit recognized under TDABC system that traditional method used as the hotel's costing system failed to do so and more proper data on costs and profitability of customers.

1 Introduction

Travelling has become a contemporary lifestyle adopted by the modern society lately. Indonesia, as the largest archipelago country in the world, saves million choices of tourist destination for local and international tourists. One of the famous tourist destinations in Indonesia is the Special Region of Jogjakarta, better known as Jogjakarta. Jogjakarta wonderfully preserves its local cultural wisdom that is very well guarded under distinctive government system called sultanate or Kasultanan Ngayogyakarta Hadiningrat that is currently under the leadership of Sultan Hamengkubuwono X. The growing potential of tourism in Jogjakarta increases the need of service that can accommodate these people in order to have a pleasant and satisfying experience. The continuing development of the hospitality industry in Jogjakarta absolutely sparks the unavoidable competition between service companies.

[1] One of the challenges in running service businesses is the development of cost information to create strategic decisions and set a proper price of services. The rising competition in hospitality industry has also fuelled the attention jet to introduce a wider range of new product and service aims to solely meet customers' satisfaction. Therefore, knowledge of customer cost information is worth having to understand exactly what customers want and how much the cost necessary to provide such services. Customer cost data assists managers to identify profitability level of customers in order to improve decision making and target profit setting [2]. A company which has valuable information in comprehending and identifying customers who are more profitable or not will help advance the overall organizational profitability [3]. Customer cost information is considered very helpful in maintaining the level of profits and retain customer relationship, especially in hospitality industry. Customer Profitability Analysis is in charge of allocating revenues and costs generated from segmented customers with the help of customer costs information to gain knowledge of customer profitability.

Time-Driven Activity-Based Costing system has been proven to be capable of producing reliable financing system without spending significant time and cost. Hotel provides more than accommodation for customers in the rapidly growing service market so the application of TDABC system can support the analysis of customer profitability. TDABC systems allows hotel to look at what really happens inside by discovering practical capacity of resources supplied using time as the resource driver and points out the unused capacity where management may eliminate directly or seeking out the benefit they can get.

Hotel X is a newly-open 5-star hotel in Special Region of Jogjakarta which considered having the most outstanding performance in 2014. Not only occupying 17.98% of market and simultaneously beating its 8 competitors, but also achieving Highest Revenue and Highest Average Room Rate (ARR) proved that it could do beyond expectation.

* Corresponding author: noorlailie@gmail.com

However, hotel seemed to still use traditional method of costing system and had not yet to take a look closely at its customer behavior and profitability. Traditional costing system provided less accurate financial information and customer segmentation was only conducted to see the statistics of its guests and no further than that. Author feels the need to introduce Hotel X Jogjakarta to CPA through TDABC system in order to provide more accurate cost information and customer profitability for managers for future effective and efficient strategies. There are two research questions based on the previous introduction:

1. How to design Time-Driven Activity-Based Costing System (TDABC) to determine customer profitability in Hotel X Jogjakarta?
2. How is the comparison of Customer Profitability Analysis using hotel's existing system and TDABC system?

Expected benefits of this research for Hotel X Jogjakarta is as consideration to use more accurate and efficient costing system, to understand the level of customer profitability based on its customer segmentation, to create more effective future strategy for achieving more profit according to customer behavior, and to decide about what hotel wants towards the unused capacity found during the application of TDABC system so that unnecessary cost can be eliminated. Author can also have the practical knowledge of the theories lectured in Cost Accounting and Management Accounting. This research can be used for academics as a new insight and research material related with Time-Driven Activity-Based Costing and its application in the hospitality industry in Indonesia as well as a reference for further research.

2 Literature Review

Cost is the value of resources used to create or obtain something which is a focus in cost accounting [4]. [5] explain that expense is the costs incurred during the process of obtaining revenue in an accounting period, while costs are acquired when resources are used. [6] states that cost accounting is the sum of the total calculated costs assigned to build and control activities, develop quality and efficiency, and support decision-making. Cost Accounting can also be viewed as a combination of financial and management accounting that is made to meet the needs of information by presenting all prominent details about the cost of product or service in which internal party use for planning, control, decision making, and performance evaluation of the company as well for external parties regarding the financial reporting and investment decision making [7]. There are some functions of cost accounting for the company's management mentioned by [8], namely:

1. Create and execute a plan that can motivate employees to work in accordance with the targets set in the company's budget, consistency, and desired economic conditions during the planning process.
2. Design an effective and efficient costing, therefore its quality can be thoroughly controlled and developed.
3. Control the quantity of inventory that company has and provide information regarding the costs assigned to products or services made for pricing determination and performance evaluation.
4. Find out the costs and profit earned during the making of annual financial statements or even in a shorter time depending on how the management desires in accordance with both internal and external reporting rules.
5. Consider the use of short term and long term alternatives that can influence costs and revenues.

[9] introduced Time-Driven Activity-Based Costing (TDABC) system, an upgrade version of Activity-Based Costing system, is simpler, less money-involving and time-consuming yet company will obtain a far more accurate information than the previous ABC system by eliminating the need to interview and conduct employee surveys that are long and complicated only to seek for information about the allocation of resource costs to activities. They listed down the advantages of TDABC system:

1. Easier and faster to build an accurate models.
2. Integrates well with the data now available from ERP and customer relationship management systems (this makes the system more dynamic and less people-intensive).
3. Driver costs to transactions and orders using specific characteristics of particular orders, processes, suppliers, and customers.
4. Can be run monthly to capture the economics of the most recent operations.
5. Provides visibility to Process Efficiencies and capacity utilization.
6. Forecasts of resource demands, allowing companies to budget for resource capacity on the basis of predicted order quantities and complexity.
7. Is Easily scalable to enterprisewide models via enterprise-scalable software applications and database technologies.
8. Enables fast and inexpensive maintenance models.
9. Supplies granular information to assist users with identifying the root cause of problems.
10. Can be used in any industry or company with complexity in customers, products, channels, segments, and processes and large amounts of people and capital expenditures.

Customer cost information is a substantial key managerial decision making [10]. Knowing the amount of actual costs in providing services to customers is no doubt important in all companies by means to know the actual profit earned. Customer Profitability Analysis (CPA) works by allocating all revenues and costs to particular customers with the aim to calculate each individual's profitability level. However, [11] pointed out the need of appropriate costing systems and precise cost techniques if the company wants to obtain accurate information regarding its activities and customers in purpose of conducting CPA. Not only TDABC system has proved sufficient and capable of supporting the need of accurate data, it also has extensive influence such as negotiations with business partners to achieve win-win solutions among the supply chain members. An effective use of CPA obviously helps service companies to be better in increasing both customer satisfaction and profitability [12].

[13-15] had also studied the application of TDABC in service industries, although none of them made customers as their unit of study and conducted the study in Special Region of Jogjakarta. The result shows significant difference between the cost under the existing organizations and under TDABC system, so does the result of this research. Conceptual framework that underlines this research begins with gathering a collection of revenue, cost, and customer segmentation report. Those data were processed to allocate the revenue and cost to customer groups under hotel's costing system and TDABC system. The result which came up under both systems was made a comparison of customer profitability using the CPA program.

3 Research Method

In accordance to the research objectives, this study possesses a qualitative approach using case study method. [16] stated that qualitative approach concentrates more on the process rather than the result and not aim for data or evidence as proof of whether research hypothesis is right or wrong.[17] also argues that qualitative descriptive approach turns out to be commonly used in research that features case study which focuses on the characteristics of a particular unit among various phenomenon. [18] argues that findings and conclusions obtained from a case study based on several sources of information would be more convincing and correct.

This research was conducted at Hotel X Jogjakarta in order to determine the level of customer profitability. All data owned by hotel of revenues and costs from its main activities during 2014 is allocated based on its consumer segmentation. The data obtained is processed and analysed in accordance with the qualitative approach as following:

1. Figuring out what data is needed to create the Customer Profitability Analysis using TDABC system of the activities of the hotel.
2. Collecting the data by field observation, documentation, and interview with manager and key person of each related department.
3. Processing the data which has been collected into TDABC system.
4. Analysing customer profitability from the calculation results using CPA.
5. Comparing customer costs and profitability according to the existing system and TDABC system.

4 Results and Discussion

4.1 Customer Segmentation

I. TRANSIENT : individual booking or less than 10-person booking in published rate who check in directly to reception desk with special rate from DOR or DOSM, booking in promotional or seasonal rate, corporate or government representatives, booking from travel agent or online travel agent, individual advance purchase (room voucher gift), owners of the hotel, individual who applies special rate approved by DOSM or GM, and barter.

II. GROUP : Group booking consists of 10 or more persons from corporate or government representatives, from travel agent, from commercial and social associations, and social groups.

III. CONTRACT : individual or group booking for fourteen consecutive nights and Airlines Contract Rate (crew and non-crew of Airlines Company).

IV. WALK-IN : individual or group coming to hotel only to enjoy having meals at restaurant.

4.2 Revenue Allocation to Customer Groups

Main revenue of Hotel X Yogyakarta is derived from sales of rooms and food and beverages. In details of room revenue, revenue has been classified according to the segmented customers every month. Total customer who stayed at the hotel in 2014 is obtained by multiplying the total of rooms sold in 2014 (59.068 rooms) with room : customer ratio of 1.8 suggested by Director of Sales and Marketing, reaching a total of 105.188 customers during 2014.

Food and beverages revenue is still based on cover; manager assumption is used to allocate food and beverages revenue of Hotel X Jogjakarta in 2014 to customer groups. Total customer of breakfast is grouped by the total in-house guests in 2014 and the difference between the total in-house guests with total breakfast customers goes into group IV. Total customer of lunch and dinner is grouped using the percentage of 10% group I, 80% group II, 0,5% group III, and 9,5% group IV. Group I managed to earn Rp45.229.171.806 in total, group II earned Rp24.521.710.163 in total, group III earned Rp183.426.176 in total, and group IV earned Rp970.431.664 in total.

DOI: 10.1051/shsconf/20173408004

4.3 Cost Allocation to Customer Groups under Hotel's Costing System

Because the operating costs has not been allocated yet to customer groups by Hotel X Jogjakarta, author assumed that the cost allocation is based on the percentage of revenue that every customer group possesses assuming that a group who generates higher total revenue also gathers higher total cost. Total cost of group I is Rp43.172.937.034, total cost of group II is Rp24.382.158.361. total cost of group III is Rp172.247.356, and total cost of group IV is Rp1.317.086.628 under hotel's costing system.

4.4 Cost Allocation to Customer Groups under TDABC System

4.4.1 Activities Performed in Hotel

Front Office	: Welcoming customers, check-in process, escorting guests in, billing review, closing bill, and escorting guest out.
Food and Beverages	: Welcoming customers, taking lunch and dinner order, preparing lunch and dinner order, stand-by during having meal, after-service cleaning, and buffet preparation.
Housekeeping	: Controlling reserved room, turndown service, and cleaning after customers leave.
Marketing	: Arranging appointment, visiting customers, and giving information.

4.4.2 Cost of Capacity Supplied to Main Activities

In the consolidated profit and loss report of Hotel X Jogjakarta in 2014, all costs had not been allocated yet to main activities. Therefore, it is a prominent process to attribute these facility-sustaining costs fairly one by one to the activity centres first than to be traced directly to individual customer. Since this type of costs cannot be determined as specific costs for particular customers, they are first apportioned to the activity-cost pools. Total cost of capacity supplied to front office activities is Rp7.086.785.409, to food and beverages activities is Rp14.054.764.886, to housekeeping activities is Rp30.296.098.999, and to marketing activities is Rp8.148.877.980.

4.4.3 Allocation of Costs of Front Office Activities to Customer Groups

Costs of front office include direct and indirect costs such as rooms, utilities, depreciation, etc. The costs pooled in front office activities were allocated to customer groups based on the practical capacity of receptionists. The time equation represents the estimate time required for front office activity:

Front office activity = (8,5*#customers)[if customer group 1] + (8*#customers)[if customer group 2] + (13*#customers)[if customer group 3] + (2*#customers)[if customer group 4]

Hotel employs 10 receptionists to perform front office activities in 2014. All receptionists are fixed employee. The theoretical working time for one fixed employee corresponds to total of 146.400 minutes per year (8 hours x 60 minutes/hour x 305 days/year). However, every day one fixed employee gets one hour to rest and 12 days of paid leave every year so, the practical working time is 123.060 minutes (7 hours x 60minutes/hour x 293 days/year). Total practical working time of 10 receptionists is 1.230.600. To assign cost according to TDABC, time equation is multiplied with capacity cost rate (Rp7.086.785.409/1.230.600) which is Rp5.759. Total cost of front office activities amounted Rp3.286.065.244 is allocated to group I, Rp1.738.296.560 is allocated to group II, Rp24.631.243 is allocated to group III, and Rp106.011.672 is allocated to group IV. Around 73% of allocated cost had actually been used for productive work in 2014 so unused capacity of front office activities is around 27% of allocated cost.

4.4.4 Allocation of Costs of Food and Beverages Activities to Customer Groups

Costs of food and beverages include direct and indirect costs such as food and beverages, depreciation, etc. The costs pooled in food and beverages activities were allocated to customer groups based on the practical capacity of related personnel. The time equation represents the estimate time required for front office activity is developed per meal cover. Buffet was only served once per cover so there was no need to multiply it with the number of customers.

Breakfast = 30 + (32*#customers)[if customer group 1] + (32*#customers)[if customer group 2] + (32*#customers)[if customer group 3] + (32*#customers)[if customer group 4]

Lunch = 60 + (48*#customers)[if customer group 1] + (48*#customers)[if customer group 2] + (48*#customers)[if customer group 3] + (48*#customers)[if customer group 4]

Dinner = 60 + (48*#customers)[if customer group 1] + (48*#customers)[if customer group 2] + (48*#customers)[if customer group 3] + (48*#customers)[if customer group 4]

Buffet prep. = 30 (breakfast buffet) + 60 (lunch buffet) + 60 (dinner buffet)

Hotel employs 17 fixed personnel and 80 waitress trainees to perform food and beverages activities in 2014. Trainees only worked for 6 months with 30 days off in total and 4 hours/day so their practical working time is 2.937.600 minutes (80 trainees x 153 days x 4 hours x 6 minutes/hours). Fixed personnel are fixed employee which practical working time has been calculated before so the total practical working time in food and beverages activities is 5.029.620 minutes [(17 fixed employee x 123.060) + 2.937.600]. To assign cost according to TDABC, time equation is multiplied with capacity cost rate (Rp14.054.764.886/ 5.029.620) which is Rp2.794. Total cost of food and beverages activities amounted Rp8.705.747.074 is allocated to group I, Rp8.057.556.213 is allocated to group II, Rp61.904.582 is allocated to group III, and Rp1.242.844.489 is allocated to group IV. Around 95% of allocated cost had actually been used for productive work in 2014 so unused capacity of food and beverages activities is around 5% of allocated cost.

4.4.5 Allocation of Costs of Housekeeping Activities to Customer Groups

Costs of housekeeping include direct and indirect costs such as rooms, food and beverages, depreciation, etc. The costs pooled in housekeeping activities were allocated to customer groups based on the practical capacity of housekeepers. The time equation represents the estimate time required for housekeeping activity:

Housekeeping activity = (43*#customers)[if customer group 1] + (57*#customers)[if customer group 2] + (43*#customers)[if customer group 3]

Hotel employs 12 housekeepers, 8 six-month contract housekeeper, and 80 housekeeper trainee to perform front office activities in 2014. The theoretical working time for one six-month contract employee corresponds to total of 73.440 minutes per year (8 hours x 60 minutes/hour x 153 days/year). However, every day a six-month contract employee gets one hour to rest and 6 days of paid leave so the practical working time is 61.740 minutes (7 hours x 60minutes/hour x 147 days/year). Total practical working time of housekeeping activities is 4.908.240 minutes [(12 x 123.060) + (8 x 61.740) + (80 x 36.720)]. To assign cost according to TDABC, time equation is multiplied with capacity cost rate (Rp30.296.098.999 / 4.908.240) which is Rp6.172. Total cost of housekeeping activities amounted Rp16.572.807.520 is allocated to group I, Rp12.109.217.120 is allocated to group II, and Rp81.223.520 is allocated to group III. Around 95% of allocated cost had actually been used for productive work in 2014 so unused capacity of housekeeping activities is around 5% of allocated cost.

4.4.6 Allocation of Costs of Marketing Activities to Customer Groups

Costs of marketing activities include direct and indirect costs such as rooms, depreciation, property insurance, etc. The costs pooled in marketing activities were allocated to customer groups based on the practical capacity of associates. The time equation represents the estimate time required for marketing activity:

Marketing activity = (45*#visits)[if customer group 1] + (105*#visits)[if customer group 2] + (45*#visits)[if customer group 3]

Hotel employs 5 associates to perform marketing activities in 2014. All associates are fixed employee so the total practical working time is 615.300 minutes (5 x 123.060). To assign cost according to TDABC, time equation is multiplied with capacity cost rate (Rp8.148.877.980 / 615.300) which is Rp13.244. Total cost of marketing activities amounted Rp6.277.353.266 is allocated to group I, Rp4.151.615.236 is allocated to group II, Rp106.668.493 is allocated to group III, and Rp383.144.130 is allocated to group IV. Around 76% of allocated cost had actually been used for productive work in 2014 so unused capacity of marketing activities is around 24% of allocated cost.

4.5 Customer Profitability Analysis

To calculate the profitability of every customer group, the assigned revenue of customer groups is reduced with the assigned costs under both hotel's costing system and TDABC system. As can be seen in the following table, only group I managed to generate profit for Hotel X Jogjakarta under TDABC system and all these groups are losable ones under hotel's costing system. In overall however, TDABC system recognized profit of almost three billion rupiahs where hotel's costing system failed to do so.

Table 1. Customer Profitability Analysis in 2014 under Hotel's Costing System and TDABC System

Group	Revenue (a)	Hotel's Costing System (b)	Profit / Loss (c = a-b)	Profitability (c/a)	TDABC System (d)	Profit / Loss (e = a-d)	Profitability (e/a)	Deviation (f = b-d)
I	41.279.484.940	43.172.937.034	(1.893.452.094)	-4,6%	34.841.973.103	6.437.511.837	17,0%	8.330.963.931
II	22.301.790.579	24.382.158.361	(2.080.367.782)	-9,3%	26.056.685.128	(3.754.894.549)	-15,0%	(1.674.526.767)
III	164.240.922	172.247.356	(8.006.434)	-4,9%	274.427.838	(110.186.916)	-65,4%	(102.180.482)
IV	970.431.664	1.317.086.628	(346.654.964)	-35,7%	1.732.000.292	(761.568.628)	-78,5%	(414.913.664)
Unused Capacity					6.139.343.018			
Total	Rp64.715.948.105	Rp69.044.429.379	(Rp4.328.481.274)	-6,7%	Rp69.044.429.379	(Rp4.328.481.274)	-6,7	Rp6.139.511.837

5 Conclusions and Suggestions

It can be concluded that:

(1) Hotel X Jogjakarta still uses traditional system as its costing system which resulted in an inaccurate cost of customers and their profitability level. The conduct of CPA showed loss generated from all customer groups but only group I who were the profitable one and the remaining groups actually consumed resources more than hotel's costing system could realize.

(2) TDABC system application had been proven to be more accurate since its result of a higher profitability and information of unused capacity in every main activity.

(3) Information retrieved from CPA can be used as consideration in future pricing policy and strategy making decision to distinguish profitable customers from non-profitable ones.
(4) Author suggests data of revenues, direct cost, and total quantity of every activity have been allocated in group by the company according to its customer segmentation for further research.

References

1. A.Terungwa. Int. Jr. Fin. Acct. **2**, 6 (2013)
2. D. Kuchta, M. Troska. Jr. Cst. Mgt **23**, 4 (2007)
3. E.M. Van Raaij. Mkt. Int. & Plg. **23**, 4 (2005)
4. W.N.Lanen. *Fundamentals of Cost Accounting* (McGraw-Hill Education, 2010)
5. D.R. Hansen. *Cornerstones of Managerial Accounting* (Cengage Learning, 2013)
6. W. K. Carter, M.E. Usry. *Cost Accounting* (Salemba Empat, 2009)
7. C.A. Raiborn, M.R. Kinney. *Cost Accounting Principles* (South-Western: Cengage Learning, 2013)
8. W.K. Carter. *Cost Accounting*. (Mason: Thomson, 2006)
9. R.S. Kaplan, S.R. Anderson. *Time-Driven Activity-Based Costing: A Simpler and More Powerful Path To Higher Profit* (Harvard Business School Publishing Corporation, 2007)
10. I. Dalci, T. Tanis, L. Kosan. Int. Jour. of Cont. Hosp. Mgt. **22**, 5 (2010)
11. O. Dejnega. Journal of Applied Economic Sciences, **1**,15 (2011)
12. B.Cotton. Ch. Acct. Jour. **84**, 3 (2005)
13. S.R. Dewanta. *The Application of Time-Driven Activity-Based Costing (TDABC) in Determining Cost Per Room and Efficiency Level of Costs and Resources: Case Study in Hotel Y Surabaya* (FEB Universitas Airlangga, 2015)
14. F.H. Fahrizal. *Analisis Penerapan Time-Driven Activity-Based Costing Dalam Perhitungan Harga Pokok Produksi Pada Perusahaan Biro Wisata Umrah PT. Mina Wisata Islami* (FEB Universitas Airlangga, 2015)
15. M.D. Riediansyaf. *Designing Time-Driven Activity-Based Costing (TDABC) In Room Division at Hotel X Malang* (FEB Universitas Airlangga, 2013)
16. S. Wahyuni. *Qualitative Researh Method Theory and Practice* (Jakarta: Salemba Empat, 2012)
17. B. Bungin. *Penelitian Kualitatif: Komunikasi, Ekonomi, Kebijakan Publik, dan Ilmu Sosial Lainnya* (Jakarta: Kencana Prenada Media Group, 2008)
18. R.K. Yin. *Case Study Research: Design and Method* (California: Sage Publication, 2003)

Leadership change: A case study analysis of strategy and control systems development

Ahmadasri Alaudin[1,*], *Zarifah* Abdullah[1], and *Marhaiza* Ibrahim[1]

[1]Tunku Puteri Intan Safinaz School of Accountancy, Universiti Utara Malaysia (TISSA-UUM), College of Business 06010 Sintok, Malaysia

Abstract. This paper reports on how a change of leadership at the CEO-level influences strategic practices and control systems development. The case study describes how the new chief executive developed and communicated his vision and strategy through control systems and structural change. The findings indicate that senior employees' involvement in strategic plan development (through SLT mechanism in this case) brought clarity and assurance to them. Meetings are important control mechanisms to structure the sharing of information and to enhance employee commitment and help decision making. It is argued leadership manifests through the interactions of leader with employees in many control practices. The findings have implications for family-owned businesses that are facing imminent change in leadership. They benefit founders/top managers that are about to change leadership of entrepreneurial firm to the next level by highlighting the importance of preparing leader's capacity for an effective leadership role. The study also highlights some important factors which are seldom discussed in control theories.

1 Introduction

Research on leadership role in accounting has started since 1980s especially in budget process (see e.g. Brownell, 1983; Kyj & Parker, 2008) and the more recent works are by Abernethy, Bouwens, & van Lent (2010). Unfortunately, despite the growing interest in the topic, we do not see much investigation been done in developing economies such as Malaysia. With the exception of Abernethy et al., these studies only examine leadership styles influence on specific accounting tool such as budgeting (e.g. Brownell 1983; Chenhall, 2003). They also viewed leadership from typology perspectives which limit the understanding of the relations between the two concepts.

The present study attempts to fill the gap by examining leadership role and control systems in a Malaysian company using case study to increase our understanding of the subject as a whole [1]. The study stems from an interest in control systems development after the change in top leadership of a family-owned business. The initial question arose was: *what influence does leadership change have on the strategy and control systems development*? We reviewed the literature on leadership and accounting to explore the leadership role on the organizational design. Further questions emerge as to how leadership change manifests in an organization; and how it interacts with other contextual factors.

This paper is organized as follows: the next section presents the concepts on which the study is based. This is followed by a description of the study method and the case company's background. The findings are then presented followed by a section on discussion and limitation. Conclusion is provided in the final section.

2 Leadership and Control Systems

In general there are two opposing views of leadership. The first strand argues that leaders through their visions and strategic actions are able to affect firm's performance. It is well accepted that top management has the power and authority to dictate structure, set direction and targets, and implement formal controls to monitor implementation.

* Corresponding author: ahmadasri@uum.edu.my

DOI: 10.1051/shsconf/20173408005

CEOs and top management decisions have crucial implications on firm outcomes and "ultimately they account for what happens to the organization" [2]. The other view argues that top leaders could not do much due to the constraints as characterized by industrial forces and culture that act as inertia [3]. However, our understanding on leadership change and its influence on the particular changes brought is not well documented. Research on CEO-level management has drawn on transformational and transactional leadership framework [4]. Bass (1990) argues that behavioral tendencies of leaders influence how they use controls. In accounting literature the typology frequently used for examining leadership is of Stogdill & Coons, 1957 (see Abernathy et al., 2010; Otley & Pierce, 1995; Brownell, 1985). Under this typology leadership styles can be conceptualized into "initiating structure" and "consideration".

The theoretical and empirical literature, as discussed above, highlights the importance of CEO leadership styles and control systems as a means of communication and control play important roles in organizations. However, what is lacking is the limitation imposed by examining leadership using typology in relations to strategy and control systems (see Western, 2013 for a critical review of leadership) [5]. Beside behavioral traits and tendencies what other aspects influence how a leader effecting change. What would be the focus of the new CEO when he attempts to introduce change? How the important elements interact to support the change?

3 Case Study Approach

To understand the leadership change and the influence it has on strategy and control systems development requires an in-depth study such as using a case study method. To this end, the study aims to examine the subject as a whole in its context. In this study leadership is viewed as behaviors of the leader in introducing changes and his related actions in the context of events. These involve interactions with and reactions to managers and employees. Hence, leadership is not broken into types based on a set of skills or competencies or to a particular way of being

3.1 Case Study Company

The case study company, MProp Group[2], which involved in property development business, provides a valuable research site for two reasons. First, it is a private company which involves in only one business hence, the complexity of ownership and structure is not too complicated. This provides an opportunity to examine the direct impact of leadership change at the top level introduced by the new leader on the organization. Second, the organization also provides a unique research site to examine strategic change because past research has shown that chief executive succession brings greater organizational change [6].

The research design was planned to capture the change that has taken place during the first year of the new CEO tenure. The data gathered was of three types – interviews, direct observation, and documents. Interviews were conducted with the new CEO himself, three chief operating officers (COOs), one CEO of a joint-venture company, who was the ex-COO of MProp Group under the founder, one general manager (GM) of Operations, and two heads of department. The interviews with subordinates provide information to interpret the nature and magnitude of change. The interviews were recorded and lasted between one to one and half hours conducted at the case study company. Company documents and public data such as business press coverage are also used to corroborate and cross-check sources. During the data collection we were invited to attend one full session of SLT meeting where two of the researchers managed to involve. The direct observation during the meeting provides invaluable insights for the two researchers.

4 Findings

This section presents the findings from the case study that illustrates the antecedents, changes introduced by the new CEO and how he designs and uses control systems to direct and empower his managers.

4.1 The New CEO and Vision

In 2009 MKTS, the founder of the company, mentioned in one of interviews with media about a new CEO's taking over the helm of MProp Group gradually. The successor is his own son, Alif, who has successfully lead and completed two significant projects worth over RM600 million. Alif in the sense is not new in property development industry and has been accumulated experience and continuing.

[2] This is a pseudonym of the real company. Throughout the case the authors also use pseudonym for the individuals mentioned in the case.

As a qualified, young, and the only son[3] of the company's founder, Alif, the new CEO brings knowledge, new aspiration and hope to MProp Group. Having graduated with MBA from one of Australian universities, he is quick to share his goal: "30 years from now MProp Group will achieve global property footprint, the brand with a worldwide reputation for developing projects across the region. It will be customer-centric organization with customers being pivotal in everything we do".

Previous to that he spent several years working[4] in KProp Land Holdings Bhd, a public-listed property company where MKTS is among the major shareholders, focusing in the area of finance. It was in 2012 that he became the CEO of the MProp Group of companies. He still remains active in the running of his private company and also sits on the Boards of several other property companies. He believes in the importance of having core values that guide daily interactions with employees and customers. For example, he mentioned about making employees understand their role and contribution to not only company but society. He described:

> "Once there was this customer – a contractor involved in the MRT[5] project - who rented some office space in one of our buildings. So I told my employees: 'if our client is happy you guys have actually been part of a national project' ... I told them they were contributing to the nation."

He related the company's service to the contribution to the nation. In doing so he extended the definition of company's role (thus, his role and employees' roles) and give meaning to it – matching company's mission to that of the nation. The new CEO's concern is to set a new direction – "to bring [the company] to the next level". Based on our analysis the three main themes that appear from the data are the changes that he makes in the first year after taking over which can be categorized into three – strategy, structure, and control. These changes are driven greatly by his vision that first translated from a training session lead by an external consultant.

4.2 Strategy

The new CEO brings with him a new vision to the MProp Group which guides the formation of the company's strategy. The knowledge and skills gained through education and experience in previous organizations, as discussed in previous section, may to a certain extent shape his perspective. For example, it was revealed in the interview that the company had a formal process of SWOT analysis lead by an expert from the university he graduated. This has resulted in a formal 10-year strategic plan of MProp Group. Apart from training, the output of the session has become the major input for the formulation of its business strategy resulting in two immediate decisions that the company saw as important in order to create value to the company.

First was the branding exercise which goal was to differentiate MProp Group from the public-listed company, KProp Land Bhd. In line with the branding exercise, the new chief executive has initiated a shift of focus to giving more value to the company's products and customers. He wants employees, regardless of their functions and levels, to understand what a product and service (e.g. a unit of apartment) means to customers. The customer-centric initiative is captured in the company's 10-year strategic plan document. For example, an "ambassador program" was introduced to enhance customers' experiential satisfaction. He explained:

> "We want to create joy from the experience of owning great properties. So this tagline or the core purpose is something that is very important for me and very important for everyone who works in MProp Group."

The new CEO also was aware that the employees who have involved in the ambassador program talked among them about the excitements. This he believes creates wave of excitement to other employees who have not involved. MProp Group does reinforce the employees by publishing their stories in company's publication and announcing their names in company's event. In this way, the new CEO sends a strong and consistent message to employees about what is important to the company.

---

[3] The founder has another three daughters who are involved in other associated companies both public and private in different capacity such as board members.

[4] He worked at the senior management position from 2002 to 2005.

[5] Mass Rapid Transit Corporation Sdn Bhd ("MRT Corp.") is a company wholly owned by the Malaysian government established by the Ministry of Finance to facilitate, undertake and expedite public infrastructure projects approved by the government.

In line with the shift to more product-and-customer focus, the marketing activities also have undergone some changes in its business practice. The marketing function comes under the purview of COO (Operations & Sales) and it recognizes the importance of targeting at specific segment. The COO, Azrita explained:

> "Previously product exposure was considered less important and the focus was more on corporate communication. [But] marketing activities are now being outsourced to get optimum impact on sales."

Based on the 10-year strategic plan, MProp Group has started to think its market beyond domestic. It has plans to spread its wing into neighboring countries in 2016. In its early analysis of venturing into international markets, countries such as Indonesia, Myanmar and Sri Lanka were under its radar before finally selected Indonesia to start with. This important step reflects the real intent of MProp Group to compete in international market. This comes from the new CEO's vision, as stated earlier, to build worldwide reputable brand across the region.

4.3 Structure

For analysis purposes, we categorize MProp Group's structure into three eras – the founder's era, transition, and the new CEO's era. During the founder era he had only a COO who was PT and all the GMs of operations. As part of the succession planning, PT was made a business associate and holds a certain stake in a joint-venture company JF Star Sdn Bhd. During this period Azrita was appointed as a COO and to support her in the technical area a position of deputy COO was created.

In the 3rd quarter of 2012, just after a couple of months of his appointment as the new CEO, Alif changed the company's organization structure to streamline its business process. Under the new three-COO structure, the COO (Operations & Marketing) focuses on the overall running of all operations companies whereby the respective GM has to report directly to her. The other two COOs (Finance and Project Management & Business Development) act as support services departments. A new post of COO (Project Management & Business Development) was created to provide technical service and coordinating projects. It seems that the new CEO wanted to introduce more functional responsibility at the top to support him in managing and making decision. This contrasts with what the founder's management style as remarked by PT[6], the ex-COO:

> "[MKTS] called the shots. He decided on things … just direct … [and] we had to conform. The son is more open in the sense that he does source for opinion from the management."

Being the founder, MKTS made almost all the decisions and has made himself as the focal point in the process. This is not uncommon in an entrepreneurial firm in which locus of decision-making and control lies with the entrepreneur [7].

4.4 Control

The new CEO gives more attention on how to get information from his senior managers. In line with the structural change the new CEO has introduced some change to the control systems design features by implementing a number of formal periodic meetings. The goal is to have proper platform for the CEO to get timely report and managers to communicate information effectively. The types of meeting and participants are summarized in Table 1.

Table 1 Types of formal periodic meetings

Meeting	Participants	Frequency	Scope	Main topics
COO	CEO, All COOs	Weekly	Company	
SCORE	CEO, COOs, corporate dept heads, all GMs of Operations companies	Weekly	Company	
SLT	CEO, COOs, corporate dept heads, all GMs of Operations companies, CEOs of associated companies	Monthly	Company	Discuss and debate opportunities and strategic priorities; learning/training using case study (question ideas)

[6] PT was the longest serving employee of MKTS which goes back to the early days when the founder started his business. As part of the succession planning, PT vacated the COO position and was made a business associate of MKTS. PT currently holds some stake in one of JV companies, JF Star Sdn. Bhd.

 DOI: 10.1051/shsconf/20173408005

CCM (client-consultant meeting)	COO (PM & BD)	Weekly	Project	Technical and design issues with consultant/architect
Site Meeting	CEO, COOs, GM of the particular operations company	Weekly	Project	Construction progress and technical issues with contractors
COO-CEO meeting	CEO and COO	As needs arise	Company	Informal way of catching up with each other

It is evidenced from the data collected that the meetings have become important control mechanisms to support the strategy implementation. SCORE meeting is a weekly meeting where all COOs and GMs of operations companies have to attend. It is observed that the SCORE meeting has induced a different effect on the GMs. They are now made to listen to other projects and also have to report theirs in the meeting. This expectation of the new CEO has actually forced the GMs to give attention to what is going on in other people's project. The new CEO mentioned about allocating 15 minutes making the rounds among the meeting members sharing customer-related issues they encountered. He termed this as "stories on customers". He uses this interactively to bring managers' attention to customers' issue.

The most striking is the SLT, a monthly meeting attended by all department heads, which has been perceived by several senior managers as a very good avenue to allow middle-level employees to contribute. The interesting fact about this meeting is that it does not only develop the participants' knowledge and skills but more importantly is their involvement in giving new ideas for the improvement of the company. There are a couple of ideas which came from this meeting that have been implemented successfully. For example, the rebranding exercise was resulted from the SWOT analysis session and the 5-day working days whose idea came from the employees during SLT have been adopted by the company. It has become a new space for the managers to discuss ideas and to get them involved in the company. On the SLT meeting, one of the COOs gave the following remarks:

> "The SLT meeting is good for prompting discussion among the whole team. Meetings also create overlap in knowledge. Everyone should know what is going on in other departments.

Through meetings and all the interactions problems can be heard. On top of COOs meeting with the CEO there is a monthly COO-CEO meeting. This is a one-to-one meeting between Azrita and the CEO. It is not based on specific agenda but a sharing of thoughts and anything that the COO thinks fit and would like to discuss. Azrita reflected:

> "I found [the meeting] as helping me because [the CEO] allows you to clarify on particular issue. It helps me to get the trust. It's like 'letting out' ... like a 'pouring session'. Sometimes it was just more than work issue – asking how you are doing and family?"

5 Discussion and Limitations

The findings from the current study indicate that the new chief executive officer is central in instituting change in the organization's strategy and is consistent with what was argued by Ocasio (1993). The case evidence has shown that senior employees' involvement in strategic plan development (through SLT in this case) brought clarity and assurance to them. Abernethy et al. (2010) found the difference in the intensity of use of planning and control systems which they argued may be explained by the difference in leadership styles. They provided a plausible explanation that top management uses the control systems to structure the planning process and to personally interact with subordinates. The findings of this study support this argument by providing evidence that formal periodic meetings, as control systems, do provide structure for communicating information and in turn the flow.

It is evidenced from the case that the new organizational structure introduced shapes the control systems and in turn the decision making process. For example, the creation of three COO positions entails the need for COOs – CEO meeting and becomes an important platform for coordinating and making decision. The frequency of the meeting which is weekly and conducted right after the SCORE meeting allows room for timely input sharing from all three COOs and in turn help CEO-level decision making.

The findings also highlight the leader's characteristics and values do influence control systems design and use. The new chief executive combines both transformational and transactional leadership styles as evidenced by the findings discussed earlier. It can be argued that the introduction of SLT reflects the transformational dimension of the new chief executive leadership style. He also portrayed transactional style as evidenced by the structure change and more decision making power to the COOs. Next, both leadership styles (initiating structure and consideration) are equally important. Top managers need to strike the balance and know when to use what in order to optimize the effectiveness. The study does not intend to measure and suggest how to strike the balance, but the findings provide rich descriptions to understand the two notions.

The study contributes to our understanding on control choices by highlighting elements which are not often discussed in the accounting literature. For example, structured meetings are considered as initiating structure type of leader and thus, would use less personal relationship with subordinates. The findings, however, point that many meetings (e.g. SLT, COO one-to one meeting) can in fact be used to show consideration side of a leader. The key is how a top manager chooses to use it whether interactively or otherwise.

This study is subject to certain limitations. First, it uses case study approach which may be exposed to the element of recall bias of respondents and researchers' biasness in interpretation. However, we have taken several steps to mitigate it through triangulation of data and analysis. Second, although SLT is considered the best platform/space to discuss and debate new ideas and strategic priorities, the study did not investigate the implementation stage. How MProp Group managers coordinate both across and within departments to implement the strategy (output from SLT) is beyond the scope of this study. Finally, this study did not examine external contingencies that may influence how the new CEO effects change in the organization. This is a contingent view that suits cross sectional type of study.

6 Conclusion

The current study contributes to extant literature by describing what exactly a new chief executive would do to enact changes and how they were adopted. The evidence presented in this study is particularly important in understanding the interactions among leadership, strategy and control systems. The arguments are based on one case study and therefore the authors do not wish to generalize the views to other settings. However, it is worth to note that the findings underline the close relationship between leadership change and the design and use of control systems. Change in organizational structure plays an important role in shaping the control systems design. In particular, meetings are important mechanisms to structure the sharing of information and help decision making.

The paper provides implications for family-owned businesses facing imminent change in leadership. . In particular, it benefits founders/top managers that are about to change leadership of entrepreneurial firm to the next level. It shows the importance of preparing leader's capacity for an effective leadership role. In developing economies, many small-and-medium businesses either get stuck at the same stage or are not well prepared for the leadership change to the next stage. It also highlights some important factors which are seldom discussed in control theories.

We acknowledge the financial support from the IMBRE Universiti Utara Malaysia. We would like to thank Rose Shamsiah, Rosni and the two anonymous reviewers for helpful comments.

References

[1] G. Thomas, How to do your case study, London: Sage Publications, 2016.

[2] D. Hambrick, "Guest editor's introduction: Putting top managers back in the strategy picture," *Strategic Management Journal,* vol. 10, pp. 5-15, 1989.

[3] N. Wasserman, N. Nohria and B. N. Anand, *When does leadership matter? The contingent opportunities view of leadership,* Harvard Business School Working Paper, 2001.

[4] B. M. Bass, "From transactional to transformational leadership: learningn to share the vision," *Organizational Dynamics,* vol. 18, no. 3, pp. 19-31, 1990.

[5] S. Western, Leadership: A critical text, 2nd ed., London: SAGE Publications Ltd, 2013.

[6] W. Boeker, "Strategic Change: The influence of mangerial characteristics and organizational growth," *Academy of Management,* vol. 40, no. 1, pp. 152-170, 1997.

[7] C. M. Daily and D. R. Dalton, "Financial performance of founder-managed versus professionally managed small corporations," *Journal of Small Business Management,* pp. 25-34, 1992.

[8] R. Simons, "How new top managers use control systems as levers of stretegic renewal," *Strategic Management Journal,* vol. 15, pp. 169-189, 1994.

[9] M. A. Abernethy, J. Bouwens and L. van Lent, "Leadership and control system design," *Management*

Accounting Research, pp. 2-16, 2010.

[10] L. Kyj and R. Parker, "Antecedents of budget participation: leadership style, information asymmetry, and evaluative use of budget," *Abacus,* pp. 423-442, 2008.

[11] P. Brownell, "Leadership style, budgetary participation and managerial behavior," *Accounting, Organizations and Society,* pp. 307-321, 1983.

[12] S. Hoozee and W. Bruggeman, "Identifying operational improvements during the design process of a time-driven ABC system: The role of collective worker participation," *Management Accounting Research,* pp. 185-198, 2010.

[13] E. Jansen, "The effect of leadership style on the information receivers' reaction to management accounting change," *Management Accounting Research,* pp. 105-124, 2011.

[14] R. H. Chenhall, "Management control systems design within its organizational context: findings from contingency based research and directions for the future," *Accounting, Organizations and Society,* vol. 28, no. 2/3, pp. 127-168, 2003.

[15] J. Mundy, "Creating dynamic tensions through a balanced use of management control systems," *Accounting, Organizations and Society,* pp. 499-523, 2010.

[16] R. K. Yin, Case study research: Design and methods, Beverly Hills: Sage, 2003.

[17] B. J. Avolio, B. M. Bass and D. I. Jung, "Re-examining the components of transformational and transactional leadership using the Multifactor Laedership," *Journal of Occupational and Organizational Psychology,* pp. 441-462, 1999.

[18] J. J. Gabarro, The dynamics of taking charge, Boston: Harvard Business School Press, 1987.

[19] J. P. Kotter, Leading Change, Boston, MA: Harvard Business School Press, 1996.

[20] J. Roberts, The Modern Firm: Organizational Design for Performance and Growth, Oxford: Oxford University Press, 2004.

[21] M. B. Miles and A. M. Huberman, Qualitative Data Analysis: An expanded sourcebook, Sage Publications, 1994.

[22] R. M. Stogdill and A. E. Coons, "Leader Behavior: Its Description and Measurement," Columbus Bureau of Business Research, Columbus, 1957.

[23] D. Otley and B. J. Pierce, "The control problem in public accounting firms: an empirical study of the impact of leadership style," *Accounting, Organizations and Society,* vol. 20, no. 5, pp. 405-420, 1995.

[24] R. Simons, Levers of Control: How Managers Use Innovative Control Systems to Drive Strategic Renewal, Boston: Harvard Business School Press, 1995.

DOI: 10.1051/shsconf/20173408006

The impact of minimum pay implementation on small businesses operating cost and sustainability: A case of service business

Fariza Hanim Rusly[1,*], *Yurita* Yakimin Abdul Talib[1] , and *Danilah* Salleh[1]

[1]Tunku Puteri Intan Safinaz School of Accountancy, Universiti Utara Malaysia, Malaysia

Abstract. The implementation of minimum wage requirement in Malaysia beginning mid 2016 seems to impact both employees and employers in different ways. While this implementation could increase household income and claimed to boost employees' productivity, employers or business operators, on the other hand are experiencing stringent effect on their business operating cost. The effect is more significant for small business operators, including the childcare centers. Childcare industry operates in the service sector, which represents the main contributor of Malaysian SMEs. Unfortunately for the industry, there is an increasing numbers of childcare centers have to cease their operation due to inability to comply with the minimum pay requirement. In the absence of thorough understanding of the phenomena, the small businesses, particularly among institution-based childcare, is at the risk of losing their businesses. This exploratory study intends to assess how the implementation of minimum pay requirement affects the existing operating cost structure, and consequently business sustainability of Malaysian childcare industry.

1 Introduction

The minimum pay requirement while could benefits the workforces at large could also results in a stringent job opportunity for employees in certain industries that is negatively affected by its implementation. The implementation of the minimum wage requirement, while at early stage, has been a worrying issue for small businesses, including the childcare operators. Since its implementation in July this year, many issues have been raised by the childcare operators, largely concerning changes in their operating cost. Increases in salary particularly, which represents a major components of the childcare operating expenses have causes a spike on their operating cost [1]. As a result, an increasing number of childcare centers have ceased their operations, and some others are struggling to survive by absorbing losses, which affect the sustainability of childcare industry. While this problem has been highlighted as a national issue in media, empirical findings are lacking in the literature. This study will look at how the implementation of the new minimum pay policy exerts changes on the current cost structure and affect sustainability of childcare centers operation in Malaysia. Findings from this exploratory study could offer empirical evidence towards

* Corresponding author: hanim@uum.edu.my

 DOI: 10.1051/shsconf/20173408006

better understanding of the phenomenon and to propose ways for effective cost management strategy and ensure sustainability of the childcare industry.

2 Literature Review

2.1 Minimum Wage Policy

Minimum wage is the lowest remuneration employers pay to employees, according to the governmental legislation. Minimum wage policy was long implemented in various parts of the world as a poverty mitigation tool [2]. Until 2013, many developed countries have implemented minimum wage policy, such as Australia, Luxembourg, France, Ireland, US, Japan, Canada, Korea, Turkey, Greece and many more.

2.1.1 Minimum Wage Policy in Malaysia

The second wave of minimum wage policy in Malaysia was fully gazetted on July 1, 2016, continuing from the first wave that was implemented in January 1, 2014. Under the Minimum Wage Order 2016 (MWO2016) the implementation of the policy increases the monthly minimum wage from RM900 to RM1,000 for Peninsular Malaysia and from RM800 to RM920 for Sabah, Sarawak and Labuan. On average, the affected workers are expected to receive a 13% increase in wages. The minimum wage policy is in line with government's aim to become a high income nation by 2020 [3]. Apart from improving government's labour standards, the increase of the minimum wage will enhance the wellbeing and welfare of low income bracket employees, improve their motivation, and hence, their productivity. While the policy will benefit the employees, many employers express their objection towards the minimum wage. According to Malaysia Employers Federation (MEF) executive director, Datuk Shamsuddin Bardan, the enforcement of MWO 2016 will affect 30,000 lower level employees, where they may be retrenched [4]. In addition, the vice-president of SMI Association of Malaysia claimed that most SMEs in Malaysia would not be able to absorb the cost spike as most of their profit margin is about 3% to 5%, leading to cease of operation [5]. The implementation of the policy is like a double-edged sword.

2.1.2 The Impact of Minimum Wage Policy

According to Efficiency Wage Theory (EWT) it may benefit firms to pay workers a wage higher than their marginal revenue product. The theory purports that when workers are paid higher wage, it will lead to increase in productivity. With higher pay, a worker may feel loyal and attach to the company. It is also possible that with higher pay, a worker may fears of being retrenched, and hence, he will work harder to keep his job. Although the company pays higher, it gets more employees' productivity. Riley and Bondibene [6] found that, upon the introduction of National Minimum Wage (NMW) in Britain, low-paying companies responded to the increase in labour costs from minimum wages by increases in the labour productivity. They suggested that these labour productivity increases may have been associated with decrease in employee turnover and improve in training. Further, Mayneris et al. [7] provided support that recent minimum wage reform in China forces the least productive firms to raise their competitiveness. Stronger and higher minimum wage will raise employees' salary, and hence, contribute to employees' motivation, healthier employees, attract high quality employees, and increase work effort [8] [9]. These will change the present situation of low labour productivity [9]. As the increase in wages could

 DOI: 10.1051/shsconf/20173408006

be mitigated through enhancements in productivity, the overall impact on business cost will likely be insignificant.

Notwithstanding, the minimum wage policy may negatively affects both the employees and employers. A study conducted in China shows that minimum wage reduces the employment of low-skilled employees. This indicates that Chinese employees with low education level (i.e., below vocational school degrees) were directly and adversely affected by the policy. The results support prior study by Fang and Lin [10] that minimum wage increases in Eastern and Central regions of China negatively affects young adults, females and low-skilled employees. With the increase in minimum wage, firms are forced to increase their labour cost. The labour cost increases may force firms' to cut the labour hours, or to change from full time to part time, which negatively affects employees' wages [4] [11]. In developing countries, such as Colombia, increases in labour costs by 13% significantly associated with a reduction of employment rate by 6% [12]. Other firms may adjust to the increase of the labour cost by replacing low-skilled employees with high-skilled employees [13]. These worsen the poverty level of the affected workers. In fact, some firms, especially small and medium enterprises (SMEs) may force to cease their operation due to inability to absorb the cost spike. In Thailand, 110,000 SMEs was closed down due to minimum wage policy [14]. The same situation may be experienced by Malaysian SMEs too as their profit margin is around 3% - 5% only and hard for them to grip the labour cost increases [5].

2.2 Childcare industry in Malaysian SMEs Service Sector

The nation's economy is largely composed of Small and Medium Enterprises (SMEs) operations. SME contributes 65.5% to the overall employment as in 2015. Further, service sector is a major player in the SMEs establishment, which represents 90% of the overall SMEs in Malaysia. Out of 580,985 SMEs in the service sector, 79.6% consists of micro enterprises [15]. The childcare business operations is commonly categorized as a small service business with owner or known as operator employing limited number of employees. Nevertheless, changes in the current economic development offer platform for these small businesses to flourish through knowledge utilization and innovation, hence strengthening its competitiveness and presence in the market [16]. Since small businesses or micro operation play a significant role in the nation's economy, their presence should be recognized by the policy makers. From the entrepreneurial economy perspective, government should play a crucial role as an enabler, rather than a controller, to support small business survival and sustainability [16]. For instance, by reducing administrative burden and facilitating small business ability to cope with changes.

2.2.1 Childcare operation and its future in Malaysia economic development

Childcare operation provides caretaking and custody services to four or more children below the age of four with certain amount of fee to be charged. The operation of the childcare center in Malaysia is under the administration of the Social Welfare Department, and subjected to the Akta Taman Asuhan Kanak-Kanak 1984 [Akta 308]. It provides guidelines for the registration, inspection and enforcement of childcare centers to ensure the prosperity, security and overall development of the children [17].

Early care industry could contribute significantly to the nation economy development on both short term and long term basis [18]. According to The Cornell Linking Economic Development and Child Care project, childcare industry contributions can be classified into three main aspects: its impact on the regional economy or place, impacts on parents by providing social support to the labour force, and impacts on children strategic human capital development. While the impact on children is vital, the benefits of childcare service on parents deserves similar merits.

Despite the importance of the childcare industry, the recent implementation of minimum pay seems to affect the institution-based childcare operation, in particular. For instance, media has reported nearly 70% of registered childcare centers in Pahang, one of the largest states in the country, are unable to comply with the minimum wage requirement (Sinar Harian, 16 September 2016). Out of the total 203 registered centers, 36 centers were closed due to unbearable operating cost as a result of the minimum wage pay. On the other hand, only 12 percent (25 centers) comply with the requirement, although have to absorb loss from the operation.

The child care industry is known as a low skill–low pay environment [19]. As such, constant changes in the technical requirements imposed on licensed operators, and recently the minimum pay implementation, imposed pressures to continue their operations. If the cost for childcare keep increasing, it might affect women involvement in the labour force and lead to withdrawal of professionals among female employees since accessibility to quality childcare would become an economic burden [19]. Consequently, the nation economic development could be negatively affected.

To address the issue, this study proposes the importance of cost management such as revising costing and pricing strategy, which allow childcare businesses to cope with changes and ensure sustainability of its operations, as discussed in the following section.

2.3 The Importance of Managing Effective Cost Management and Accurate Price Setting For Business Sustainability

Sound cost management is the index of the success of an enterprise, its existence and growth. Cost management makes possible the use of available resources in the form of workers, materials and machines more effectively. Thus, it helps in preparation of plans for development and expansion and their successful executions [20]. For instance, cost management permits evaluation of the business profitability in a given circumstances so that a proper decision may be taken to minimize the risk of high operating costs.

Operating costs are the cost of resources used by a business just to maintain its existence. It falls into three broad categories, which are fixed costs, variable costs and semi-variable costs. Further, a business's cost structure refers to the relative proportion of fixed and variable expenses. It is necessary to classify business expenses as either fixed or variable if the management or owner want to understand the profitability of its business. In order for a business to gain a reasonable profit, determining accurate sales per unit or price setting of goods or services is essential [21]. Many businesses are not making a reasonable profit because they do not know the basic concepts of costing and set prices. The situation is most serious in the service business because each service performed has a different cost. Frequently, the service business must bid for jobs by making a price quotation in competition with similar businesses. It is difficult to calculate the costs for services and quote a price that is competitive and returns a profit. Without realizing what they are doing, some business owners set their selling price below their total cost. This may result a loss incurred. Occasionally, a business owner who lacks a knowledge of costing will try to compensate by setting prices very high. The end result is that the business is not price competitive and does not attract sufficient customers to survive.

2.3.1 The effect of salaries and benefits of small businesses

Salaries and benefits are a large portion of any business' operational cost. This is particularly expensive for a small business, as the cost can easily take up a large percentage of revenue. In the current situation of coping with the minimum pay implementation, business could either increase the selling price of its services or make a necessary reduction to its fixed and other variable costs in order to absorb the increment of salaries and benefits expenses. In conjunction to that, a solid understanding of accurate costing and price setting strategy is vital for business to continue generate a reasonable profit for its sustainability. Therefore, this study will assess the existing cost structure adopted in the childcare businesses to understand the impact of minimum wage implementation on their operating cost, and to develop the operating cost model for small business services.

Figure 1 below depicts the importance of effective cost management to ensure business sustainability as a result of minimum pay policy implementation among Malaysian childcare operations.

Fig. 1. Conceptual Model of the impact of minimum pay implementation on operating cost and business sustainability.

3 Conclusion

A qualitative case study is proposed to explore the phenomenon. Data gathering will involve semi structured interviews among childcare operators in Malaysia. Findings from the study could benefit both service business operators and policy makers to better understand the effects of minimum wage implementation on the small businesses' cost structure. It is hoped that findings could enlightened business operators in making improvement in their business and costing strategy; and for the policy makers to introduce a better mechanisms that could facilitate small businesses adapting with changes in relation to minimum pay requirement, hence to ensure the industry sustainability.

References

1. Sinar Harian (2016). http://www.sinarharian.com.my/edisi/pahang/lebih-65-peratus-taska-tak-mampu-bayar-gaji-minimum-1.563894. 19 September 2016.
2. Zhang, Y. (2016). Urbanization, Inequality, and Poverty in the People's Republic of China.
3. The Star Online (2016). http://www.thestar.com.my/metro/views/2016/05/06/minimum-wage-maximum-delay-the-long-overdue-legislation-is-better-late-than-never/
4. Malaysia Digest (2016). http://www.malaysiandigest.com/features/610360-new-minimum-wage-of-rm1-000-to-be-enforced-this-july-what-is-the-real-effect.html
5. The Star Online (2012). Retrieved from www.thestar.com.my.
6. Riley, R. & Bondibene, R. The impact of the national minimum wage on UK business. National Institute of Economic and Social Research and Centre for Macroeconomics (2015).
7. Mayneris F, Poncet S, Zhang T., The cleansing effect the of minimum wage: Minimum wage rules, firm dynamics and aggregate productivity in China, *CEPII Working Paper 2014-16*, Paris, (2014).

8. Bewley, H., Portes, J., Riley, R., Rosazza, B. & Wilkinson, D. Press release: The impact of the national minimum wage on productivity, business performance and employment. National Institute of Economic and Social Research. (2015).
9. Lin, Y., Shang, K., & Li, X. Study on the effect of minimum wage increases in China, Management Science and Engineering, 7(2), 45-49. (2013).
10. Fang, T., & Lin, C. Minimum wages and employment in China. *IZA Journal of Labor Policy*, *4*(1), 1. (2015).
11. Archive (2011), 50 Years of Research on the Minimum Wage', Retrieved date Oct 8, 2015from:http://web.archive.org/web/20110629183749/http:/www.house.gov/jec/costgov/regs/minimum/50years.htm
12. Kugler, A., and M. Kugler. Labor market effects of payroll taxes in developing countries: Evidence from Colombia. Economic Development and Cultural Change, **57(2),** 335–358. (2009).
13. Wilson, M. *The Negative Effects of Minimum Wage Laws*. Cato Institute.(2012).
14. Sin Chew Jit Poh. Minimum Wage in Thailand caused 110,000 SMEs to close down, 3 May, 20, (2013).
15. https://www.statistics.gov.my
16. Burns, P. *Entrepreneurship and Small Business: Start-up. Growth and Maturity.* Palgrave Macmillan. (2010).
17. https://www.jkm.gov
18. Liu, Z., Ribeiro, R., & Warner, M. *Comparing child care multipliers in the regional economy: Analysis from 50 states*. Linking Economic Development and Child Care Research Project, Cornell Cooperative Extension, Department of City and Regional Planning. (2004).
19. Bretherton, T., Davis, S., & National Centre for Vocational Education Research (NCVER). Interview with Tanya Bretherton on 'Developing the child care workforce: understanding 'fight'or 'flight'amongst workers'. (2010).
20. Garrison, R.H., Noreen, E.W. and Brewer, P.C., Cheng, N.S.,and Yuen, K.C.K. Managerial Accounting: An Asian Perspective, McGraw Hill Education (Asia). (2012).
21. Mowen, M.M., Hensen, D.R. and Heitger, D.L. *Cornerstone of Managerial Accounting* (5th Ed.). Cengage Learning. (2014).

DOI: 10.1051/shsconf/20173409001

The Role of Board of Directors in the Establishment of Risk Management Committee

Suhaimi Ishak[1,*], *Mohamad Naimi* Mohamad Nor[1]

[1]Tunku Puteri Intan Safinaz School of Accountancy, Universiti Utara Malaysia, 06010, Sintok, kedah, Malaysia

Abstract. The objective of the study is to discuss the roles of board of directors in the establishment of risk management committee for Malaysian's public listed companies. In Malaysia, based on the Malaysian Code on Corporate Governance (MCCG), (2007; 2012) clearly stated the role and responsibility of the board of directors toward the risk management activities. At the Corporate Governance Week 2010 and 2011 as well, the chairman of the Security Commission Malaysia highlighted the responsibility of the board of directors in risk management processes and she expressed concern over the failure of the board of directors to establish appropriate measures for the risk management process in the company. The statement of the chairman of Security Commission Malaysia is in line with MCCG's best practices (2000; 2007; 2012), i.e., the board of directors should identify principle risks and ensure the implementation of an appropriate system to manage these risks.

1 Introduction

This paper aims to discuss the concept and roles of board of directors (BOD) in the establishment of risk management committee (RMC) for Malaysian's public listed companies (PLCs). This study excluded the banking and finance companies due to this kind of industry have special and extra requirements by authorities especially on risk management practices. BOD is an integral part of every business organizations. They have a role for the future direction of the organization and company. In Malaysia, the regulators frequently ask for the BODs to have the strong position and role in corporate activities. They have another job portfolio that relating to the risk management. The establishment of RMC is one of the company initiatives in risk management practises [11] the formation of RMC is still voluntary in most countries in the world including Malaysia [9]. In Malaysia, based on the Malaysian Code on Corporate Governance [6, 7] clearly stated the role and responsibility of the BODs toward the risk management activities. The code highlighted for the BODs to have a system which effectively monitor and manage risks. This is an indicator of the importance of risk management and the oversight function of the BODs, even though there is no mandatory requirement for the establishment of the RMC. In June 2010, the chairman of the Security Commission Malaysia (SC) at its Corporate Governance

* Corresponding author: suhaimiishak@uum.edu.my

Week 2010 together with Bursa Malaysia (BM), mentioned about the importance of proper risk management in a company. Tan Sri Zarinah Anwar reminded that poor risk management is a symptom of poor corporate governance practices. The chairman cautioned the BODs to fully and deeply understand the risks associated with the business operations, products and the market.

At international level, [2] introduced the Thought Leadership in ERM known as guideline for Understanding and Communicating Risk Appetite. This new guideline stresses the role of management and board oversight function in risk appetite activities for organisations; and the effectiveness of board oversight function is crucial. The Federation of European Accountants, Institute of Chartered Accountants Australia and the Centre for Audit Quality [4] jointly sponsored the roundtable discussion in New York City, Brussels and Hong Kong, where it was agreed that specific aspects of risk oversight responsibility should be allocated to a specific board committee, such as a RMC. The participants in that discussion also suggested for the establishment of a separate RMC that could focus on the consideration and identification of 'unknown risks, since the existing audit committee may be only familiar with the 'known risks'.

2 Review of past literatures

There are some previous studies that examined the BODs characteristics and the establishment of RMC but the samples limited to the financial year ended before the amendment of MCCG 2007 and 2012. Therefore, this study is expected to give knowledge and literature on how far the amendment of MCCG gives effect to the role and responsibility of BODs relating to risk management issues. Previous study such as [11] reveals that BODs with more independent willing to form or establish RMC. The researcher also argues that more independent BODs demonstrate good corporate governance. Meanwhile the result of study by [5] reveals the different result. The BODs independence is not significant to the establishment of RMC. This result supported the findings of [1] which there is no relationship between the BODs independence and the establishment of other BODs committee such as AC, remuneration and nominating committee. However, after the amendment of MCCG in [6] and [7], especially on risk management issues, the independent BODs is expected to be more aware on risk profile of the company.

Normally, the big size of BODs is easily to establish the other board sub-committee such as RMC since there are member with various skill and talent. This argument is supported by the study by [9] which larger BODs offer the resources to operate BODs committee. Further research also supported the findings whereby [5]) also find positive relationship between larger BODs and the establishment of RMC. Hence, larger BODs offer more experiences, skills and diversifications. The understanding on accounting and financial aspects is more crucial. BOD members with accounting and financial literate are more aware on activities related to financial and accounting performance [5] This argument is supported by [11] that BOD members with accounting and financial background have positive association to the establishment of RMC. They play active role in risk management activities.

Usually, BODs will discuss the matter pertaining to the company in BODs meeting. Frequent meeting encourage open discussion and improve communication among BODs members. Hence, the number of BODs meeting during the financial year is one of the initiatives by BODs itself. In term of risk management activities, [5] find no relationship between BODs meeting and the establishment of RMC. However, the earlier study by [11] finds positive relationship between BODs meeting and the establishment of RMC. Theoretically, if the BODs aware and diligent in discussing the risk issues, they intend to establish another board sub-committee to discuss more on the related issues such as the establishment of RMC. Multiple BODs is a common phenomenon in corporate world. More BOD members also become the BOD members in another companies at the same time even at different industry. This situation creates a broader range of knowledge, experiences and skills. The situation also brings them more sensitive and responsive to the company's issues including risk management. [3] argue multiple directorships tend to have motivation to involve in risk management activities. This argument also supported by [5] in the study that BODs outside directorships has positive and significant relationship between the voluntary establishments of RMC.

3 Discussion and conclusion

The increasing concern by the companies on risk management practices has reformed significant emphasis on the role of risk management. BODs as a key governance structure in an organization have a significant role for the implementation of risk management in a company. The establishment of RMC as a board committee is an example of the initiatives in risk management practices. This study intends to investigate which natures or features of BODs structures that established the RMC. Hence, the structure of BODs refers to the characteristics of BODs itself such as BODs independence, BODs size, BODs expertise, BODs diligence and BODs outside directorships. These characteristics are the indicators for effectiveness of BODs as a whole. This situation creates a motivation for the researcher to study on the relationship between the BODs characteristics and the establishment of RMC.

At the Corporate Governance Week 2010 and 2011 as well, the chairman of the SC highlighted the responsibility of the BODs in risk management processes. Tan Sri Zarinah Anwar expressed concern over the failure of the BODs to establish appropriate measures for the risk management process in the company. The statement of the chairman of SC is in line with MCCG's best practices [6, 7] i.e., the BODs should identify principle risks and ensure the implementation of an appropriate system to manage these risks. While the scholars of field such as [9, 10] suggested for future research to study the role and function of RMC, as well as its interaction in corporate governance mechanism. [11] convinced that the formation of RMC is a commitment of the company to improve the internal control environment and reduce the operational, financial and reputational risks. This area is still unexplored and the finding might contribute to new knowledge.

References

1. L. T. De Luca, Propulsion physics (EDP Sciences, Les Ulis, 2009) Carson, E. Factors associated with the development of board sub-committees. Corporate governance, **10**(1), 4-18 (2002)
2. Committee of Sponsoring Organizations of the Treadway Commission. Enterprise risk management-understanding and communicating risk appetite. New York: COSO. (2012)
3. Fama, E. F., & Jensen, M. C. Separation of ownership and control. Journal of Law and Economics, **26**(2), 301-325 (1983)
4. Federation of European Accountants, Institute of Chartered Accountant Australia and Center for Audit Quality. Global observations on the role of audit committee, a roundtable discussion. New York, Brussels and Hong Kong: FEE. (2013)
5. Liew, C. L., Mat Zain, M., & Jaffar, N. Board of directors and voluntary formation of risk management committee: Malaysia evidence. International Journal on Social Science Economics & Art, **2**(2), 67-73 (2012)

DOI: 10.1051/shsconf/20173409001

6. Security Commission of Malaysia. Malaysian code on corporate governance (Revised 2007). Kuala Lumpur: Malayan Law Journal Sdn Bhd. (2007)
7. Security Commission of Malaysia. Malaysian code on corporate governance (Revised 2012). Kuala Lumpur: Malayan Law Journal Sdn Bhd. (2012)
8. Sekome, N. B., & Lemma, T. T. Determinats of volunatry formation of risk management committees: evidence from an emerging economy. Managerial Auditing Journal, **29**(7), 649-671. (2014)
9. Subramaniam, N., McManus, L., & Zhang, J. Corporate governance, firm characteristics and risk management committee formation in Australian companies. Managerial Auditing Journal, **24**(4), 316-339 (2009)
10. Yatim, P. Audit committee characteristics and risk management of Malaysian listed firms. Malaysian Accounting Review,**8**(1), 19-36 (2009)
11. Yatim, P. Board structure and the establishment of a risk management committee by Malaysian listed firms. Journal of Management & Governance, **14**(1), 17-36 (2010)

SHS Web of Conferences **34**, 09002 (2017) DOI: 10.1051/shsconf/20173409002
FourA 2016

CSR initiatives by companies for employees through social protection measures

Noor Akma Mohd Salleh[1,*], *Rusnah* Muhamad[2] , and *Siti Hajar* Mohd Roffeei[3]

[1]Department of Operations and Management of Information Systems, Faculty of Business & Accounting, 50603 University of Malaya, Malaysia
[2]Department of Accounting, Faculty of Business & Accounting, 50603 University of Malaya, Malaysia
[3]Institute of Graduate Studies, 50603 University of Malaya, Malaysia

Abstract. Poverty has always been the main concern globally as it alienates and distresses particular groups of citizen that affects their emotion and psychological behaviour, and may have impact on the socio-economic aspect of the country. Hence, it is critical for the government to address the poverty problem, as this would lead to inequality. One way to minimise income inequality is through the provision of social protection to maintain the basic consumption and living standard of the needed members of the society. The government highlights the importance of increasing involvement of the private sector to improve social protection through their CSR programmes. Thus, the main purpose of this paper is to unearth the concept of social protection and its role in minimising inequality in the society as well as to develop an instrument to measure the social protection dimension of companies' CSR activities.

1 Introduction

The direct and indirect effects of poverty lead to inequality, as there is disparity of wealth distribution between the rich and poor. It is therefore vital that the necessary actions are effectively implemented to eradicate poverty and subsequently minimise inequality, which in turn will address the issue of equity. Equity means addressing the concern of the society for a fair and equal treatment of opportunities to everybody, regardless of their background. Once we are able to address equity, then equitable society can be achieved.

Utting (2007) considered equity and equality as prerequisites to achieving equitable society. One way of increasing chances towards achieving equitable society is through the Corporate Social Responsibility (CSR) initiatives by companies [1]. The focus of CSR initiatives should then be on improving working conditions, community support, labour and human rights, community empowerment, and stakeholders' participation. It is proposed that the four dimensions of equity and equality are social protection, rights, empowerment, and redistribution [1]. Both the public and private sectors can promote these four dimensions through socially responsible activities.

* Corresponding author: akmasalleh@um.edu.my

This paper focuses on the social protection dimension of equity and equality. If CSR agenda by companies is to complement the effort of the government in promoting equity and equality via the social protection dimension, there is a need to understand the gap between CSR activities and the welfare of employees and communities. Hence, the aim of this paper is to unearth the concepts of social protection and how companies could contribute in promoting the social protection for the needy as well as to propose the appropriate measure of social protection in relation to CSR initiative by companies. Within the scope of CSR activities by companies, this study notes that the social protection measures discussed in this paper is limited to those related to employees and their family members which may not be inclusive of the other general members of the society.

The remainder of the paper discusses the concept of CSR and social protection followed by a brief description on the research methodology adopted in Section 3. Section 4 examines and determines the definition and measurement for the social protection dimension through CSR activities. The final Section 5 concludes with recommendations for further study.

2 CSR and social protection

CSR is a set of policies, practices, and programmes, integrated into business operations, supply chains, and decision-making processes of a company [2]. It also encompasses the economic, legal, ethical, and discretionary (philanthropic) expectations that a society has from business organisations [3]. CSR can be considered as a vehicle for companies to interact with their stakeholders, designed to deliver sustainable value to the society by which activities are based on ethical and philanthropic obligations, and respect for the community, employees, the environment, shareholders, and other related stakeholders [2-6].

Review of the extent literature revealed that research in CSR were mainly concerned with CSR disclosures [3, 7, 8], corporate governance and strategic management [9, 10], and the emerging political CSR [2, 11]. Debates on CSR revolved around what theories are most suitable for companies to justify the basis of their CSR activities. It is argued that CSR should firstly be viewed as “corporations’ obligations” to work for social betterment [12, 13]. As such, CSR should move from a philosophical approach to a managerial approach by enabling CSR through “corporate social responsiveness” [12]. Embracing this form of CSR will allow the corporation to react to social forces that introduces the notion of “corporate social rectitude” [13]. This form of CSR considers that in business activities, social and ethical decisions are expected. Therefore, in choosing a social cause for CSR activities, the social cause should be about opportunities for shared value between companies, employees, and the society.

The objective of the social protection on the other hand, is to reduce economic and social vulnerability by providing income or consumption transfers to the poor, protecting the vulnerable against livelihood risks, and enhancing the social status and rights of the marginalised groups [14-15]. At a company level, social protection for employees and their family members often circulates around topics of wages and other fringe benefits. Findings of a study revealed that the most common provisions are different types of benefits including health care benefits and old age pensions [16]. In contrast, it was argued that any expenditure for employees’ benefit regarded as compensation for work is not considered a social benefits [17]. These expenditures include cost of living allowances, transport allowances, saving schemes payments for employees, free housing or housing allowances, crèches for the children of employees, holiday package for official holidays, and sports, recreation and holiday facilities for employees and their families [17].

Even so, the objective of the International Labour Organization (ILO) is the promotion of decent work, described as opportunities for people to obtain decent and productive work in conditions of freedom, equity, security, and human dignity [18]. The ILO has included the element of social protection in the measurement of decent work, which often provided against life contingencies such as ill health, old age, unemployment, and disability [18]. Another term, social security, also was adopted and nine classes of benefits were established: medical care, sickness benefit, unemployment benefit, old-age benefit, employment injury, family benefit, maternity benefit, invalidity benefit, and survivors' benefit. Mostly employment based, social security systems providers are institutional structures that may be private, public and mixed, compulsory and voluntary, universal and targeted. Despite the institutional diversity, the goal of social security is to provide the basic protection for employees and their families [18]. The objective of social protection in this case is similar with the argument of the social cause which is providing opportunities for shared value between companies and employees [12, 13].

Based on this argument, the "Shared Value" concept was adopted as the underlying theory to understand the CSR initiatives and social protection concept. According to the Shared Value concept, the CSR agenda should be about policies and operating practices that enhance the competitiveness of companies, whilst simultaneously advance the social conditions in the communities which they are operating [2, 19]. Shared value focuses on integrating social and business advantage that results in new business models which are social and inclusive [20]. With this underlying theory, it is proposed that social protection can be promoted through CSR activities because both social protection and CSR activities focus on providing opportunities for both companies and employees that enhances social status and rights of the marginalised groups in the society.

Primarily, social protection can be promoted through CSR initiatives by focusing on the following components: 1) social insurance, 2) social assistance, and 3) social services. Social insurance provides protection against risks of income loss and shocks [21] whereby it is commonly related to employment and often contributory. The compensatory support of social insurance is provided in the occurrence of illness, injury, disability, death of parent or breadwinner, unemployment (or retrenchment) and old age. Even with its objectives being for protection and promotion, the social insurance system in developing countries tends to cover certain populations that benefit only those employed in the formal sector and their families. As a result, the vast majority of the poor is excluded from the systems [15]. The usually non-contributory social assistance programmes aim at raising the living standards of employees through transfers in cash and in kind [21-23]. Social assistance is a critical mechanism for poverty reduction amongst the developing countries [15]. Social services through CSR initiatives for the vulnerable groups (such as the disabled, women and girls, children and elderly) are designed to provide these groups of people with special care and needs, and access to basic services [21]. Such services may include healthcare, education, and child protection [23].

Companies through its CSR initiatives encourage employees to be involved and engaged in meaningful activities for their companies and the society. The companies in turn, reward employees by paying fair salary and providing other employment benefits. Therefore, a reference on the set of measure related to the above three components of social protection through CSR initiatives need to be explored and examined.

3 Research methodologies

A search for literature regarding the three components of social protection through CSR initiatives was initiated by a keyword search around words and phrases such as "social protection", "CSR", "social insurance", "social assistance", and "social services". The retrieved articles included academic and non-academic research papers published between the year 2000 and present. The purpose of the literature review was to develop an instrument on the social protection construct based upon the three components (as discussed in Section 2) through CSR initiatives. The published papers were extracted from various databases including relevant studies in psychology, sociology, corporate governance, CSR, and equitable society within the context of the Shared Value concept. Thirty-five (35) papers pertaining to the definitions and concept of social protection were mostly concentrated in the non-academic publications. Content analysis method was used in developing the instrument which consisted of three steps. Step one required identification, review, and screening of the definition and concept of social protection components. When such refined definitions were finalised, step two involved assessment of the pertinent constructs of social protection. Finally, step three established the measurement items for social protection construct (through companies' CSR initiatives focusing on employees).

4 Social protections: Definition, concept and measurement

4.1 Definition of social protection

There were many descriptions on concept and definition of social protection in the academic and non-academic literature. One of the papers reviewed defines social protection based on the World Bank definition; *Social protection is a collection of measures or interventions to improve or protect human capital (i.e., individuals, households, communities), ranging from labour market interventions and publicly mandated unemployment or old-age insurance, to targeted income support* [14]. In another paper it was stated that social protection enables vulnerable groups (i.e., the poor as well as non-poor) to prevent, reduce, and/or cope with risks [24]. The United Nations defined social protection as a set of public and private policies and programmes in response to various contingencies due to absence or substantial reduction of income from work, to assist families with children, and provide basic healthcare and housing [25]. In a similar way, the ILO described the various contingencies as sickness, maternity, employment injury, unemployment, invalidity, old age, and death of breadwinner [25]. A working definition for social protection is a set of actions undertaken through the provision of social insurance, social assistance, and social inclusion efforts [15, 26]. Other recognised forms or mechanisms of social protection include social transfers, social services, and social transformation [27].

Surprisingly, there was limited academic literature that discussed social protection in terms of definition and concept. Social protection was broadly defined as the welfare and wellbeing of workers and other groups of people or stakeholders [1]. Social protection is also described as public measures that provide income security for individuals [28]. Social safety nets is defined as the informal or less formal systems of social protection and the formal social security or social insurance system [29].

The key trend observed from all these definitions focused on the objectives and the instruments used to understand the concept and measurement of social protection. The objective is generally to tackle poverty, risks, and vulnerability. Noting how social protection definition was based on its objectives, this paper thus defines social protection as the set of measures being provided for the welfare and wellbeing of employees and other groups of people or stakeholders which aims at protecting these groups against economic and social distress of various contingencies that would be caused by the absence or a substantial reduction of income from work as described by ILO [25]. Social protection also includes the provision of healthcare and benefits for families with children. Hence, the ultimate goal of social protection is in achieving equity and equality by minimising deprivation and enhancing income and labour opportunities to all.

4.2 The concept of social protection

Social protection measures can have different and often overlapping objectives and impacts. For example, social protection objectives can be about promoting incomes and at the same time preventing deprivation. The review of the extant literature revealed that there are three main measures of social protection: protective measures, preventive measures, and promotional measures [14, 15].

Protective measures are meant to relief deprivation by which the initiatives are usually designed for the poor and can be in the form of cash and in-kind. Such measures may include disability benefits, non-contributory pensions, fee waivers, work schemes, supplementary diets, and food aid (forms of safety net), especially needed in smoothing consumption during crisis [18].

Preventive measures avert deprivation through social insurance (such as pensions and maternity benefits), and social services (such as family counselling, juvenile justice services, and protection and prevention against family violence). The social insurance initiatives aimed at protecting people against risks of shocks whilst social services targeted the marginalised groups [14, 15].

The third measure of social protection is known as the promotive measures. It is designed to enhance real incomes and capabilities of employees, social assistance as springboard aimed at providing employment opportunities to reduce poverty. The provision of social assistance may be in the form of cash and in-kind transfers [14, 15].

4.3 Measurement of social protection

Companies of all sizes share an important role in the society. Despite the focus on economic and investment returns, these companies have roles to play in relation to social, human, technological, and environmental development [30]. Therefore, these four aspects of development were included in the establishment of the measurement of social protection via CSR initiatives. An ILO document was the main resource for developing the measurement [31]. The chosen items were adopted with respect to the concept of social protection as discussed in the Section 4.2. Nine items were identified as the appropriate measures of social protection via CSR initiative, categorised into three main components (i.e. social insurance, social assistance, and social services) as described in Section 2. Table 1 shows the components and items of measurement.

Table 1. Components and items for measuring social protection by CSR initiatives

Component	Item
SI	Companies should establish a pension plan to sustain employees' lives after retirement
SI	Companies should establish a retrenchment plan (e.g. involuntary separation, voluntary separation-VSS, mutual separation-MSS) to appropriately compensate retrenched employees
SI	Companies should contribute to employment injury insurance to provide for employees' injuries and disabilities caused by work-related accidents
SI	Companies should provide paid leaves for employees (e.g. sick leave, maternity leave, compassionate leave, etc.) to ensure a work-life balance and improve productivity levels at work

SI	Companies should contribute to employment insurance scheme which provides financial supports (e.g. funeral expenses, scholarship, pension) for employees' next of kin in the event of death and permanent disabilities of employees
SA	Companies should provide facilities (e.g. accessible toilets, ramp, parking bays, emergency lights, signage, resting place, transport to work) for disabled employees/customers at companies' premises
SA	Companies should contribute to invalidity pension scheme to provide for employees' disabilities arising from chronic diseases (e.g. heart attack, kidney failure, cancer, mental illness, chronic asthma, etc.)
SS	Companies should provide childcare services (e.g. community childcare, kindergarten)
SS	Companies should provide healthcare scheme for employees (e.g. generalist outpatient, dental/optical subsidy, maternity benefits, hospitalisation benefits)

Source: ILO (2007)
SI=Social Insurance, SA=Social Assistance, SS=Social Services

5 Summary and conclusions

The concept and measurement of social protection through CSR initiatives by companies discussed in this paper can be considered as a contribution to the body of knowledge. The paper defines social protection focusing on the welfare and wellbeing of employees, their family members, and other related stakeholders. Social protection can be categorised into three main components, namely, social insurance, social assistance, and social services. These components are measured based on the protective, preventive, and promotive objectives of social protection. Based on the review of the extant literature, we have proposed nine items to appropriately measure the contribution of companies in improving the social protection dimension among employees through their CSR initiatives. It is recommended that in future, an empirical research be conducted adopting the propose instrument to assess the roles and impact of social protection on equity and equality.

Acknowledgments

We would like acknowledge the financial support provided by University of Malaya under the Equitable Society Research Cluster ((ESRC) research grant RP030B-15SBS.

References

1. P. Utting, Third World Q. **28**, 697-712 (2007)
2. S. R. Abdulrazak, S. F. Ahmad, Int. J. Global Bus. Manag. Res. **6**, 210-218 (2014)
3. A. B. Carroll, K. M. Shabana, Int. J. Manag. Rev. 85-105 (2010)
4. F. Z. Abdul Hamid, R. Atan, Int. J. Bus. Soc. Sc. **2**, 198-208 (2011)
5. A. Bidin, Mal. J. Law Soc. **12**, 175-189 (2008)
6. M. Ismail, J. Int. Soc. Res. **2/9**, 119-209 (2009)
7. M. A. Omran, D. Ramdhony, Int. J. Acc. Finan. Report **5**, 38-55 (2015)
8. S. van der Laan, Australas. Acc. Bus. Finan. J. **3**, 2 (2009)
9. S. Brammer, G. Jackson, D. Matten, Soc. Econ. Rev. **10**, 3-28 (2012)
10. D. Crowther, G. Aras, *Corporate Social Responsibility* (2008)
11. J. G. Frynas, S. Stephens, Int. J. Manag. Rev. **17**, 483-509 (2015)
12. W. C. Frederick, Bus. & Soc. 33, 150 (1994)
13. W. C. Frederick, California Manag. Rev. 28, 126-141 (1986)
14. S. Deveraux, R. Sabates-Wheeler, *Transformative Social Protection* (2004)

15. P. White, A. Hodges, M. Greenslade, *Measuring and Maximising Value for Money in Social Protection Systems* (2015)
16. International Labour Organization, Legal basis and reporting tools (2006) Retrieved from http://www.social-protection.org/gimi/gess/ShowTheme.action?id=3446
17. European Union, *ESSPROS Manual and User Guidelines* (2012)
18. International Labour Office (ILO), *Measuring Decent Work with Statistical Indicators* (2002)
19. K. Dembek, P. Singh, V. Bhakoo, J. Bus. Ethics (2016)
20. L. Michelini, D. Fiorentino, Soc. Responsibility J. (2012)
21. S. Abdul Samad, N. Mansor, Malaysian J. Econ. St. **50**, 139-156 (2013)
22. S. Cecchini, J. Southeast Asian Econ. **31**, 18-39 (2014)
23. S. Cook, J. Pincus, J. Southeast Asian Econ. **31**, 1-17 (2014)
24. S. W. Handayani, *ADB Sustainable Development Working Paper Series* **32** (2014)
25. R. Yemtsov, *The World Bank and Social Protection Overview* (2013)
26. P. Brunori, M. O'Reilly, *Social Protection for Development: A Review of Definitions* (2010)
27. Institute of Development Studies (IDS), *Adaptive Social Protection* (2012)
28. R. Holzmann, S. Jorgensen, Int. Tax Public Finan. **8**, 529-556 (2001)
29. R. Mat Zin, ASEAN Econ. Bul. **29**, 197-217 (2012)
30. J. Aston, C. Anca, *Socially Responsible Small and Medium Enterprises (SMEs): Guide on Integrating Social Responsibility into Core Business* (2011)
31. International Labour Office (ILO), *Toolkit for Mainstreaming Employment and Decent Work* (2007)

DOI: 10.1051/shsconf/20173409003

Empowering the society through companies CSR agenda

Noor Adwa Sulaiman[1,*], *Rusnah* Muhamad[1]
[1]Department of Accounting, Faculty of Business and Accounting, UM, Malaysia.

Abstract
Topic of socioeconomic issue is one of the most widely discussed globally as it gives effects to occupation, education, income, wealth, and place of residence of individuals. These social challenges should be addressed and resolved because to enhance individuals' contribution to economic and social life of their society and reduce social tensions and conflicts that negatively affects country's economic development. For this reason, in the Eleventh Malaysia Plan (2016-2020), the Malaysian Government stresses on the importance of participation of companies in empowering society to improve socioeconomic that could support equitable society. The empowerment programs aim to improve the education, quality of life and wellbeing of individuals and groups in society through reducing wealth gap, racial imbalance and promoting employment equity. One way to initiate greater involvement of the companies in socioeconomic development of the society is through CSR agenda. Specifically, the CSR agenda through empowerment activities (such as trainings programs, educational sponsorship mentorship program and learning and development programs) is believed to have a positive implication on society by way of improving wealth, education and skills of the individuals. Hence, this paper aims to develop measurement of empowerment in companies CSR agenda.

1 Introduction

One of the most significant discussions globally is on issue of socioeconomic inequality because it has effect on an individuals' economic activity including lack of education, poverty, wealth distribution, race discrimination and unemployment. Inequalities in socio-economic can take place when individuals and groups in society may not have equal access

* Corresponding author: adwa@um.edu.my

to rights, opportunities (e.g.: employment), or resources (e.g. healthcare, education). This exclusion may have prevented individuals and groups from participating in country's economic, social and political matters. Consequently, the government of Malaysia has identified inclusivity as one of key principle in achieving its national socio-economic development to achieve equitable society [1]. The principle focuses on the need for participation and commitment of all citizens to benefit economic growth and development of the country. Accordingly, in the 10th Malaysia Plan, the government has emphasized about the importance of balance growth and equitable access to economic opportunities to all citizens regardless of gender, ethnicity, socio-economic status and geographic location [2]. Consequently, it should have resulted in a more inclusive society through poverty reduction, increased household income, reduction in income disparity, and improve quality of life. Nonetheless, income gaps for main ethnic groups remain large - per capita income for the ethnic Chinese and Indians were 64% and 27% higher than for the ethnic Malays [3]. Ownership of share capital among ethnic groups' (Malay, Chinese and Indian) remains disproportionate [4]. In addition, the 11th Malaysia Plan (2015) reports distortion in wage structure – in 2014, 77% of wage recipients received salaries of less than RM3,000 per month, inadequate skilled jobs in market, youth unemployment and low women participation in labour market.

For this reason, Malaysia has encouraged more active participation from companies to carry out empowerment programmes that could enhance greater access to education and skills training, infrastructure and employment opportunities for all segments of society [1]. Companies CSR agenda has a role to play in the empowerment of the society through activities that not only desirable in terms of the business objectives but also pursue development of society and satisfy social needs [5, 6]. The companies' CSR agenda could empower individuals or groups of people by providing them with skills that they need and improve their sense of self confidence which effectively could reduce unequal education, income disparity and increase training opportunities, and more skilled categories of occupation. Despite numerous calls by government to encourage greater participation of private sector companies to conduct empowerment programmes in their CSR activities for development of society, evidence still shows that the situation has yet to improve [7]. CSR and empowerment has become an interdisciplinary area involving psychology, sociology, economic and business concepts with varying views of empowerment within the research areas, there is little consensus to the definition and the concept of empowerment. The lack of agreement among scholars concerning definition and concept of empowerment cause inconsistent and incommensurable empirical measures. For this reason, the main objective of this paper is to develop measurement constructs for empowerment in companies CSR agenda. It is hoped that this will encourage a better understanding of the significance of empowerment in CSR while providing recommendations for future research.

The rest of this paper continues as follows. Section 2 reviews relevant prior research that underlies the study. Section 3 outlines the research methodology. Sections 4 ascertain definition and measurement for empowerment through CSR agenda. Final section provides conclusion and recommendations of the study.

2 Prior Research

2.1 Concepts of CSR

Review of prior literature shows that research in CSR have covered various aspects such as CSR disclosures [8, 9] and corporate governance [10, 11] and CSR and strategic management [12] as such, CSR has been studied in interdisciplinary area involving psychology, sociology, economic and business ideas. In addition, CSR is multifaceted concept and researchers have made several attempts to measure specific dimensions for it. In addition, for example, [13] proposes that CRS principles are makeup of two units: business and society that is interlinked with each other that can be measured in three dimensions. First dimension recognizes responsibilities of companies at three different levels (institutional, organizational and individual) and associated principles (principles of legitimacy, principles of public responsibility and principles of managerial discretion). Second dimension concerns with how the principles are related to action of companies that purposely examine issue of 'process of social responsiveness' that cover aspects of stakeholder management, environmental scanning and issues management. Third dimension looks at the outcomes of the companies' actions that can be analyzed through social policies, social programs and social impacts. Comparatively, [14] has divided CSR model into four key dimensions, namely economic, legal, ethical and philanthropic based on his earlier CSR model in year 1979 [14]. Later in year 2004, he proposes refined CSR model that focuses on responsibilities of companies, social issues that companies ought to address and social responsiveness or action of companies that are embedded in the areas of economic, social and legal to measure companies CSR [15]. In general, the model proposes these three domains as key facets of CSR and a number of measurements have been developed to measure CSR which is derived from Carroll's models. Generally, CSR is referring to business activities or practices of a company that is integrated with social, economic and environmental aspects that pursue social values and satisfy social needs of its various stakeholders [16, 17]. In other words, CSR is concerned with outward expression of the company's value to its stakeholders: activities that a company demonstrates its focus other than company's bottom line- financial performance. That is to say those CSR activities are related to how business takes account of its economic, social and environmental impacts in the way it operates [18]. Overall, much of the discourse in concept of CSR in the literature is associated with analyzing companies' responsibilities and actions and effects of CSR on financial performance that fail to address social and developmental implications of CSR.

At the present time, discussion on CSR has also embraced issue of CSR role in society development [19-22]. Prior research suggests that contribution of CSR towards equality and equity of society can be achieved through four key components: redistribution, social protection, empowerment, and rights [19]. His proposed CSR framework considers how empowerment through companies CSR agenda is related to rights of the society with respect to issues such as minimizing disparities or leveling equity, enhancing equality of opportunity, correcting gross imbalances in the distribution of income, wealth and power and eventually social justice. Certainly the role of empowerment in CSR agenda by companies is important towards development of the society. Companies' participation in the economic and social life of the community has been somehow successful in reducing or eliminating inequities and inequalities [23]. There is an increase focus on the importance of empowerment in helping making people aware of and enabling to claim their fundamental rights [21]. It is crucial in achieving better community in terms of efficiency, growth, poverty reduction as well as attaining justice.

Various theoretical standpoints have been used in CSR research. In an analysis of the theoretical standpoints, [20] identifies research in CSR has employed utilitarian theory, managerial theory and relational theory to explain issues in CSR. This study utilizes relational theory to understand the CSR and empowerment. The relational theory is emphasized on various approaches in understanding CSR issues such as business and society stakeholders approach, corporate global citizenship and social contract theory. The social contract theory posits that, there is a social contract between the society and companies, which is an implicit social agreement that lays out the duties and rights of individuals or groups in the society on companies which they consider as a moral agent [22]. At the heart of the theory, society is recognized as legitimate stakeholder that companies are morally obligated to serve upon to legitimize its actions or behavior. The concept is closely related to stakeholder theory that explains about legitimacy of a claim on a firm. It describes that claim on a firm can be based upon a contract, exchange, legal title, legal right, moral right, at risk status, or moral interest in the harm and benefit generated by a firm's actions [24]. In this case, it is argued that companies are socially responsible towards socio-economic development of the society by considering them in companies' broader strategy In effect, companies are responsible to have proper motivation, method or manner of engaging with the society. Here, the concept of empowerment comes into play. Companies should be able to recognize fundamental human rights of the society to basic needs such as food, clean water, education, work, and health services [25]. Conceptually, empowerment is explained as a distinct approach for developing interventions and creating social change. It directs attention toward health, adaptation, competence, and natural helping systems [26]. From this aspect, empowerment is a process in which efforts are made by the individual or the society to exert control of their rights. The society, with the help of certain socially responsible companies, should be informed, trained and able to claim its rights under certain circumstances and events. The companies' CSR can leverage empowerment to enhance the society rights and stability, gain legitimacy and increase their accountability toward stakeholders [27]. Empowerment gives rights to different groups in the society to be included in the socio-economic development and thus increase equity and equality of policies [19].

2.2 Concepts of Empowerment

Empowerment is a multidimensional concept. Its definition commonly associated with economic, human and social empowerment, political and cultural empowerment. Economic empowerment is defined as the right of the society to have appropriate skills, capabilities and resources and access to secure and sustainable incomes and livelihoods [28]. As such, access to assets and resources would be the focus. The human and social empowerment has been defined as multidimensional of social process that helps people gain control over their own lives. This process would foster power in society, for use in their own lives, their communities and their society, by being able to act on issues that they define as important [29]. Social empowerment is about the process of developing a sense of autonomy and self-confidence of individuals [30]. In comparison, [31] characterize political empowerment as the capacity to analyse, organize and mobilize. It refers to collective action that is needed for collective change for citizens to claims their rights and entitlements. Comparatively, cultural empowerment is about redefining rules and norms and the recreating of cultural and symbolic practices in the society [32]. This study would be focusing on concept of economic and social empowerment as a basis in developing conceptual model and measurement for empowerment in companies CSR agenda.

Research in empowerment has covered broad spectrum of groups that include employee, customer, supplier youth and women. As a result, concept of empowerment is diverse that bring different meaning for different authors. Accordingly, this study will only review concepts of empowerment from the economic and social aspects because it is aligned with the focus of the study. In all, prior studies have documented that employee empowerment has been one of the topic that has been extensively investigated. Two key dimensions have been identified in this aspect of empowerment. First, empowerment is connected to psychological construct that is manifested in four cognitions: meaning (fit between the job task and one's own beliefs, values, and behaviors); competence (belief in one's ability to perform a job well); self-determination (feeling like one has control over one's work); and impact (feeling one can affect one's work outcomes) [33, 34]. Scholars such as [35, 36] regard psychological empowerment as motivational construct that involve process of enhancing feelings of self-efficacy among individuals by way of enhancing skills and abilities to do his/her job effectively. The psychological empowerment is considered to be a cognitive state characterized by a sense of perceived control, perceptions of competence, and internalization of the goals and objectives of the organization. Second, construct of the empowerment is related to relational aspect that is perceived control (includes beliefs about authority, decision-making, latitude and availability of resources, autonomy in scheduling, etc), perceived competence (reflects role mastery that in addition to successful completion of assigned tasks also requires coping up with the non-routine tasks) and goal internalization (captures the energizing property of a worthy cause or exciting vision provided by the organization leadership) [34]. Similarly, the literature on economic empowerment is vast, and a large part of this focuses on the economic empowerment that centers around four broad areas: a) the promotion of the assets of people; b) transformative forms of social protection; c) microfinance; and d) skills training. Economic empowerment is concerned with providing equal rights, opportunities, asset ownership and responsibilities to individual that eventually improving equality [37].

Table 1. Prior Research-Measurement of Empowerment

Author(s) and year	**Construct(s) and items (s)**	**Study design and research context**
Kosar & Naqvi (2016)	Meaning: 3 items Competence: 3 items Self-Determination: 3 items Impact: 3 items	Survey Employee empowerment
Kumar (2015)	Meaning: 3 items Competence: 3 items Self-Determination: 3 items Impact: 3 items	Survey Employee empowerment
Ugwu et al (2013)	Meaning: 3 items Competence: 3 items Self-Determination: 3 items Impact: 3 items	Survey Employee empowerment
Dimitriades et al (2007)	Perceived competence: 5 items Goal internalization: 5 items Perceived control: 5 items	Survey Employee empowerment
Cyboran (2005)	Meaning: 1 items Competence: 1 items Self-Determination: 1 items Impact: 1 items	Survey Employee empowerment
Hui et al (2004)	Self-determination: 3 items	Survey Employee and Customer empowerment
Hancer (2001)	Meaning: 3 items Competence (self-efficacy): 3 items Influence: 6 items	Survey Employee empowerment

Spreitzer (1995)	Meaning: 3 items Competence: 3 items Self-Determination: 3 items Impact: 3 items	Survey Employee empowerment

Study by [33] has elaborated and validated a multidimensional measure of psychological empowerment at the workplace. He identified four constructs with 12 items in measuring empowerment, which are meaning (fit between the job task and one's own beliefs, values, and behaviors); competence (belief in one's ability to perform a job well); self-determination (feeling like one has control over one's work); and impact (feeling one can affect one's work outcomes). His model of psychological empowerment was developed based on earlier studies by [35] [40]. Prior research has extensively used empowerment scale by (33) to measure empowerment. For example, [39] examines the impact of empowerment on service employees and found that psychological empowerment has positive influence on employee job satisfaction and loyalty to organization. Comparatively, [38] modified the scales done by [33] by adopting only three of the constructs which are meaning, competence (self-efficacy) and influence. This study investigates the reciprocal relationship between psychological empowerment and job satisfaction of restaurant workers. This study shows that psychological empowerment was positively and significantly related to job satisfaction. In addition, [41] in their study shows that there is higher employee satisfaction when they are empowered. In this study, empowerment was measured by [33] self-determination sub-scale using 3 sample of questions. This three-item subscale was designed to measure perceived autonomy at work.

Attempts to measure empowerment can be also seen through study done by [42] which examined the influences of reflection on the self-perception of empowerment in the workplace using four components of psychological empowerment. The empowerment approach used is again based on the Spreitzer's model which is based on the approach that empowerment is defined as increased intrinsic motivation manifested in four cognitions [40] namely: meaning, competence, self-determination and impact. This study provides sample questions for each components of empowerment. This measure describes the extent to which employees believe they are empowered in their jobs. On the other hand, [43] contributes to the stream of research on empowerment by investigating the construct of psychological empowerment among public employees. Psychological empowerment was measured with a 15-item scale developed by [34], for the exact wording of the items and tapping the three dimensions of perceived competence, goal internalization and perceived control. This author applied the concept of empowerment as the employee-experienced power. According to him, review of the major approaches to the experience of power in conjunction with the various streams of empowerment research described in prior studies reveals that, at on individual level, the three main dimensions of the experience of power underlying empowerment process are: a) power as perceived control, b) power as perceived competence, and c) power as being energized toward achieving valued goals. The study done by [43] followed fairly closely the empowerment measurement that in the [34] study.

Prior research has employed Spreitzer's measurement scale to examine the effect of psychological empowerment. For example, [44] examine the association between psychological empowerment and behavioral outcomes of employee. In their study, they found out that psychological empowerment has positive significant relationship with the employee engagement. The same scale was also used by [45] in his study to measure relationship between empowerment and leadership experience and education. Other example of study was done by [46] show the moderating role of psychological empowerment on trust and employee engagement. In another study, [47] investigating the effects of psychological empowerment dimensions of job satisfaction. Other studies have

DOI: 10.1051/shsconf/20173409003

also adopted empowerment measurement scale by Spreitzer to measure psychological empowerment [48-53].

Based on preceding research stated above, it can be concluded that Spreitzer's measurement scale has been widely adapted in measuring psychological empowerment [33]. Although a variety of approaches have been taken by different authors, and many of the studies are evolvements of previous studies, it is evidenced that Spreitzer's measurement scale appear to be at the forefront of most research in this literature [33]. Table I provides an overview of selected prior research to highlight the numerous measures and method applied in various context.

Overall, the concept of empowerment is diverse. In general, it focused on management practices or activities that are designed to "empower" various stakeholders through access to information and resources and enhance autonomy, responsibility and participation of the various groups in the society thus greater equality and equity of in the society can be achieved. Thus, empowerment in this study is defined as practices or activities that carry out by companies that provide opportunities, resources, knowledge and skills, and motivation to allow individuals to gain authority and ability to improve their competence and enhance their satisfaction and self-efficacy. Society can be empowered through knowledge and skills by giving adequate access to resources, knowledge and skills. This study proposes assessment scale to measure impact of empowerment to equitable society.

3 Research Methodology

A quantitative approach is adopted in this paper whereby relevant literature and documents have extensively been reviewed. The purpose of the review is to ascertain definitions and concepts of empowerment as well as to establish appropriate items to measure empowerment within the context of CSR. There are several types of literatures and documents that have been reviewed which includes academic journals, books, reports, etc. This study uses medium for the literature search such as ABI/INFORMS Proquest, Emerald and Google Scholar so that a pool of studies for the review purposes can be obtained. The literature search has focused on the concept of empowerment with related to CSR activities. Previous studies on empowerment, CSR, and equitable society are extensively been reviewed in this paper in order to identify the themes and frequent issues in this area. The important key terms regarding empowerment such as meaning, concept, dimensions, construct, measurement, as well as questionnaires have been widely used throughout the search. This paper has reviewed related literatures to propose definition of empowerment within the context CSR. Research framework was established to coincide the concept of empowerment and CSR activities so that it will be much easier to generate the items. Related sample of questionnaires from past researchers were used as a reference to measure the dimension of empowerment and the relevancy to CSR agenda.

4 Measuring Concepts of Empowerment through CSR Agenda

Literature recognized the important role of companies in the society. Essentially, on the important contribution of companies CSR agenda toward socioeconomic development of the society [19]. The empowerment dimension of CSR of companies is about giving rights and opportunities to the individuals or groups in the society on occupation, education, income and wealth that promote social inclusion and equitable society. The measurement items for empowerment through CSR agenda by companies were adapted from [33, 43].

The proposed measurement items also took into consideration study and report by [28] and [30] on conceptualization of empowerment. Based on the literature review the empowerment measurement items consist of 45 items that were further scrutinize based on argument from the literature. As a result, only 15 of the total 45 items were chosen for use in this study. The 15 items measure the three dimensions are Perceived competence, Goal internalization, and Perceived control in the context of CSR initiatives by companies in their effort to achieve equitable society through socioeconomic equality. The list of items is shown in Table II.

Table II: Proposed Definition and Attributes of the Study

Definition of Empowerment	
Empowerment refers to practices or activities that carry out by companies that provide opportunities, resources, knowledge and skills, and motivation to allow individuals to gain authority and ability to improve their competence, confidence and enhance their self-efficacy.	
Attributes of the Study	
Perceived Control	
1	Companies should provide individual with relevant trainings programs and workshops to improve individual's knowledge and skills.
2	Companies should have awareness programs to improve the individual knowledge on products and the organization.
3	Companies should provide mode of communication (e.g. direct engagement, proxy voting and social reporting) that allow the individual to participate in organizational decision making process.
4	Companies should provide educational sponsorship to elevate the standard of education of the individual.
5	Companies should provide financial aid to underprivileged individual to elevate their standard of living.
6	Companies should provide their employee with volunteerism program with the community to boost employees' motivation and social skills.
7	Companies should provide mentorship program to nurture individual talents' professional and personal development.
8	Companies should provide learning and development programs via internships to ensuring that individual are able to fulfill their potential.
Perceived Competence	
9.	Companies should assist individual in achieving necessary capabilities that are required to do job well.
10.	Companies should assist individual in achieving the skills and abilities to do job well.
11.	Companies should assist individual to be competence to work effectively.
12.	Companies should assist individual to work efficiently.
13.	Companies should assist individual to handle the challenges at work.
Goal Internalization	
14	Companies can inspire individual to achieve objective in life.
15.	Companies can inspire individual to be enthusiastic about his or her contribution to the organization.

5 Conclusion and Recommendation

DOI: 10.1051/shsconf/20173409003

This paper has attempted to extend the knowledge on empowerment within CSR domain by critically examining the definitions, concepts and measures of empowerment as relevant for the CSR agenda. The review of the literature reveals that the scope of the conceptualization of empowerment within companies CSR agenda could be discussed within psychological perspective of the individuals in the society. In the future, empirical data can be collected through questionnaires in order to test the validity of the instruments.

Acknowledgements

We would like to acknowledge the financial support provided by University of Malaya under the Equitable Society Research Cluster (ESRC) research grant RP018C-13SBS.

References

1. E. P. U, *11th Malaysia Plan (2016-2020)* (Economic Planning Unit, Prime Minister's Department, 2015)
2. E. P. U, *10th Malaysia Plan 2011-2015* (Economic Planning Unit, Prime Minister's Department, 2010)
3. M. Y. Saari, E. Dietzenbacher, & B. Los., *Sources of Income Growth and Inequality across Ethnic Groups in Malaysia, 1970-2000,* (World Development, 2015)
4. Department of Statistics Malaysia & Companies Comission of Malaysia 2016
5. H.R. Bowen, *Social Responsibilities of the Businessman* (Harper and Row, NY, 1953)
6. R. Jatana, & D. Crowther, Soc. Resp. J., **3**, 40-48 (2007)
7. N. N. Nik Ahmad & M. Sulaiman, Int. J. of Com. & Mngmt., **14**, 44-58 (2004)
8. M. C. Chan, J. Watson, D. Woodliff, J. of Bus. Ethics, **125**, 59-73 (2014)
9. T. Shavit & A. M. Adam, *A Preliminary Exploration of the Effects of Rational Factors and Behavioral Biases on The Managerial Choice to Invest in Corporate Responsibility* (Managerial and Decision Economics, 2011)
10. G. Giannarakis, Int. J. of Law & Mngmt., **56**, 393 (2014)
11. M. A. Harjoto, H. Jo, J. of Bus. Ethics, **100**, 45-67 (2011)
12. O. Flak & A. Pyszka, *The Impact of Corporate Social Responsibility (CSR) On Companies' Competitiveness - Proposed Research Directions* (Equilibrium, 2011)
13. D. Wood, *Corporate Social Responsiveness Revisited* (Academy of Management Review, 1991)
14. A. B. Carroll, *A Three-Dimensional Conceptual Model of Corporate Performance* (Academy of

    Management Review, 1979)
15. M. Schwartz, & A. Carroll, *Corporate Social Responsibility: A Three-Domain Approach* (Business Ethics Quarterly, 2003).
16. D. R. Lichtenstein, M. E. Drumwright, B. M. Braig, J. of Marketing, **68**, 16–32 (2004)
17. C. B. Suresh & P. Sudeepta, J. of Transnational Mngmt., **16**, 20-35 (2011).
18. C. Nwaneri, Int. J. of Bus and Mngmt, **10**, 60. (2015)
19. P. Utting *CSR And Equality* (Third World Quarterly, 2007)

20. M. Ismail, The J. of Int. Res., **2,** 199-209 (2009)
21. S. Mena, M. de Leede, D. Baumann, N. Black, S. Lindeman, L. Mcshane, J. of Bus. Ethics,**93**, 161-188 (2009)
22. A. Amran, M. M. Zain, M. Sulaiman, T. Sarker, S K. Ooi, *Empowering Society for Better Corporate Social Responsibility (CSR): The Case of Malaysia* (Kajian Malaysia, 2013)
23. J. Hayes, E. Hannold, J. of Health and Human Serv. Administration, **30**, 352-77 (2007)
24. R. Mitchell, B. Agle, D. Wood, *Towards a Theory of Stakeholder Identification and Salience: Defining the Principle of Who and What Really Counts* (The Academy of Management Review, 1997)
25. International Covenant on Economic Social and Cultural Rights (2002)
26. M. A. Zimmerman, American J. of Community Psy., **23,** 81-99 (1995)
27. N. D. Çakar & A. Ertürk, J. of Small Bus. Mngmt., **48**, 325-359 (2010)
28. C. Luttrell, S. Quiroz, C. Scrutton, K. Bird, *Understanding and Operationalising Empowerment* (London: Overseas Development Institute, 2009)
29. N. Page, & C. E. Czuba, J. of Ext., **37**, 24-32 (1999)
30. http://www.gsdrc.org/topic-guides/voice-empowerment-and-accountability/supplements/social-and-economic-empowerment
31. L. H. Piron, & F. Watkins, *DFID Human Rights Review: A Review of How DFID Has Integrated Human Rights Into Its Work*. (DFID, 2004)
32. N. P. Stromquist, *The Theoretical and Practical Bases for Empowerment in: Women's Education and Empowerment* (Report of International Seminar, 1993)
33. G. M. Spreitzer, Academy of Mngmt. J., **38**, 1442-6 (1995),
34. S. Menon, *Employee Empowerment: An Integrative Psychological Approach* (Applied Psychology: An International Review, 2001)
35. J. A. Conger, & R. N. Kanungo, *The Empowerment Process: Integrating Theory and Practice* (Academy of Management Review, 1988)
36. K. Y. H. Fock, *The empowerment of service workers: Conceptualization and impacts across cultures* (The Chinese University of Hong Kong, H. K., 2004)
37. T. Aind & S. Oraon, Economic empowerment of tribal women in Jharkhand, an analysis. *Anusandhanika, 5*(1), 85-92. (2013)
38. M. Hancer, *An analysis of psychological empowerment and job satisfaction for restaurant employees* (Order No. 3011067). Available from ABI/INFORM Collection. (2001)
39. M. D. Fulford & C. A. Enz, J. of Managerial Issues. **7**, 161-175 (1995)
40. K. W. Thomas & B. A. Velthouse, *Cognitive Elements of Empowerment: An Interpretive Model Of Intrinsic Task Motivation* (Academy of Management Review, 1990)
41. M. K. Hui, K. Au, H. Fock, J. of Int. Bus. Studies, **35**, 46-60 (2004)
42. V. L. Cyboran, V. L. *The influence of reflection on employee psychological empowerment: Report of an exploratory workplace field study* (Performance Improvement Quarterly, 2005)
43. Z. S. Dimitriades, & T. Maroudas, *Internal service climate and psychological empowerment among public employees* (Transforming Government: People, Process and Policy, 2007)

44. R. Kosar, & S. M. M. R. Naqvi, S. M. M. R. *Psychological Empowerment and Employee Behaviors Employee Engagement as Mediator and Leader-Member Exchange as Moderator.* (Economic and Social Development: Book of Proceedings, 2016).
45. S. Solansky, Ldship & Org. Dvlpmnt J., **7**, 637. (2014)
46. F. O. Ugwu, I. E. Onyishi, & A.,M. Rodríguez-Sánchez, *Personnel Review,* **3**, 377-400. (2014)
47. K. M. Kumar, & R. Moorthy, *J. of Contemporary Res in Mngmnt*, **1**, 1. (2015)
48. H. Luoh, S. Tsaur, & Y. Tang, *Int.J. of Cntmpry Hsptlty Mngmnt,* **7**, 1100-1117. (2014)
49. S. Jha, *J. of Mngmnt Res,* **2**, 105-117. (2013)
50. L. Lin, & C. Tseng, *Int. J. of Orgztnal. Innov.* **4**, 21-29. (2013)
51. C.C Schermuly, R. A. Schermuly, & B. *The Int. J. of Eductnal. Mngmnt,* **3**, 252-264. (2011)
52. P. Gkorezis, L., Hatzithomas, & E. Petridou, *J. of Mng. Issues,* **1**, 83-95, 7-8. (2011)
53. F. Wei, X. Yuan, & Y. Di, *Effects of transactional leadership, psychological empowerment and empowerment climate on creative performance of subordinates: A cross-level study* (Frontiers of Business Research in China, 2010)

DOI: 10.1051/shsconf/20173409004

Accounting for inflation: arguments of recognition in Islamic accounting standards

Ahmad Bello

Tunku Puteri Intan Safinaz Scool of Accountancy, Universiti of Utara Malaysia, Malaysia

Abstract. The current Islamic Accounting Standards gives little prominence to changes in prices, instead of developing a full blunt standard to address the menace of changes in prices fair value accounting was used. Changes in prices otherwise referred to Inflation may affect and substantially injures sustainability of Islamic financial institutions. An economy that is characterized also as Inflationary may have its financial statements produced under AAOIFI derailing in quality. From transactionary angle, creditors are going to be shortchanged and Debtors benefits; debtor- creditor Hypotheses. In this paper effort was made to explore the fundamental reasons for non- recognition of inflation in Islamic Accounting Standards and fundamental Islamic framework that may permit that.

1. Introduction

Expansion and continuing growth and development of Islamic Financial Institutions and Sharia products witnessed from the last quarters of twentieth century has created an impetus into considerable research interest in Islamic accounting. Early researches were tilted toward principles and development of conceptual framework. However with the coming of Accounting Standards for Islamic Financial Institutions in 1990 refers to Accounting and Auditing Standards of Islamic Financial Intuitions (hence forth, AAOIFI), research interest is shifted towards practical application and value relevant tests of Islamic Standards. While still, the AAOIFI places greater reliance on IFRS that do not conflict with Shari'a principles [1]. Such reliance should not be seen and ought not to hamper research effort in Islamic accounting principles and conceptual framework, otherwise, Islamic Standards will be a mere hatched from Islamisation of conventional standards. The implication of that as pointed by Gambling [10], is that accounting and business ideas and methods developed in a Western environment influenced by Judeo-Christian ethical notions would not necessarily operate effectively in a Muslim environment. Early writers to the Islamic accounting literature, Gambling and Karim[10] and Tomkins and Karim [11] stress on the need to develop accounting standards based on the Shari'a principles against mere Islamisation of conventional accounting. However, the formalization of rules did not mean elimination of all contradictions and conflict that plague accounting exposition in the past. Rather, the set of rules or conventions have removed the traditional practice that was described as "learning by fingers exercise or by doing" [12:471]. What the rules serve to do is to limit both the nature of acceptable solutions and the steps by which they are obtained.

This paper revisits one area that received greater attention in conventional accounting but has being negated under Islamic accounting framework and standardization, "Accounting for changes in price level". The sole objective is to dig out Shari'a principles and laws guiding business transactions and the extant to which the beneath principles seek to protect interest of both parties and lastly relates them with modern means and factors of production and how these can provide shield to all parties without shortchanging or making either to suffer.

2. Materials and methods

The paper is purely conceptual and analytical. Analysis of normative documents, systematization and generalization of facts and concepts, and more importantly synthesizing fatwas (jurists views) and relating them with modern ways of doing businesses formed the basis of conclusion.

3. Foundation of Islamic Accounting

Tracing the essence of man creation that's is as vicegerent on earth, one can appreciate from the Qur'an that mankind would be asked to account for their deeds at the Day of Judgment.

"We shall set up justice scales for the Day of Judgment, not a soul will be dealt unjustly in the least. And if there be (no more than) the weight of mustard seed, we will bring it (to account); and enough are We to take account"(Al-Qur'an Chapter 21, verse 47)[12].

The Holy Qur'an emphasize on writing down any covenant or contract that involves obligation and assets among parties as:

"O you who believe! When you deal with each other, in transactions involving future obligations in a fixed period of time, reduce them to writing" and "Let a scribe write down faithfully as between the parties"(Al- Qur'an, Chapter 2, verse 282)[12].

The principles and rules, which govern financial contracts in Islam, as summarized by Ibn al A'rabi, are:

a) Prohibition of interest and legitimacy of trade. Although trade is allowed, there are restrictions on the conditions and practices of trade.
b) Prohibition of 'unjustified' enrichment (Akl Amwdl al-Nds Bi al-Bdtil); "do not eat up your property among yourselves for vanities" (Qur'an, 4:29)[12].
c) Prohibition of 'dubious circumstances' and uncertainty in trade contracts (Bai' al-Gharar), that is, a sale, which involves fraud or unfair exchange.
d) Giving consideration to intentions and aims (al-Maqd~id) and to welfare (al- Masdlih). Consideration of aims indicates that the intentions of dealers must conform to Islamic teachings. Therefore, it is not permitted to commit an illegitimate act through a trick.

The central venture of accounting is to communicate economic information to different users on the underlying economic state and changes of economic state of an undertaking or venture within a definite period to inform decision of users. In conventional sense financial accounting is seen as process of identification, recording, summarizing, analyzing reporting to inform users decision. Islamic accounting also composes of all those process plus compliance with Shari'a. The major distinction between Islamic and conventional is that of divinity and humanism.

4. Theoretical Literature and Islamic Principles on Accounting

4.1 Changes In Prices and Quality of Accounting Information

The central theme of accounting is to measure and report periodically the economic wealth and changes in this wealth of an economic entity for effective decisions. Whereas economic wealth and unit of measurement varies with time and conditions, accountants remain conservative in assuming the stability of its measurement unit. By implication this assumption ignored completely the possible effect of changes in price on the periodic entity's report. The informative quality of economic wealth reported could be damaged or even destroyed when an economy is not stable. As a result the value of accounting as a "practical language of economics", is lost when changing user needs [value] are not catered for. However, with the introduction of fair value measurement, the negative effect of price changes in accounting quality is minimizing from reporting point of view. Contrary to that still fair value is just at reporting level, but from transactionary level, parties to transactions are bound to suffer from prices changes. IFI's are required to apply IFRS that does not contravene Shari'a principles in absence of AAOIFI. In this case fair value principles are used in valuation of non-monetary assets. This does not extend to monetary assets, which are more vulnerable to changes in prices.

4.2 Islamic Accounting Principles on Changes in Prices

Conventional accounting was based on cost principles as a golden yardstick of valuation and measurement. This principle assume stable unit of measurement. Although despite strong debates and counter arguments in different world forum, Historical Cost accounting principles (HCA) continue to serve as fundamental accounting practices. The HCA was based on the assumption on stable unit of measurement, however in the period of price rise such assumption ceased.

Contrary to HCA, AAOFI endorses the use of fair value accounting to IFI's. One of the questionable principles from the Islamic point of views the cost principle. It conflicts with justice and fairness if valuation was based on it in an inflationary environment. Furthermore, financial information based on it will passively affect depositors who withdraw from investment in Islamic banks. In addition, unrealistic information as a result of inflation tends to reduce the Zakah base.

4.3 Corollary and Discussion

There are three major Islamic Principles that guides business transactions as spelt out in the Qur'an:1).Prohibition of interest.2). Misar(Gambling).3).Garar (deception).Moreover, the guiding principle that governs financial dealings stipulated in Qur'an and Sunnah are:

i. *Realisation of fairness and Justice*: Several verses of the Qur'an stressed that measures and weights should be given with justice and without withholding from people what is theirs. (Qur'an 6:152, 7:29, 11:85, 17:35, 55:9, 57:25).[12]. More emphatically is Qura'an 4:29, "do not eat up your wealth among yourselves for vanity". This verse prohibits shortchanging one party in a transaction advertently.

ii. *Preservation the rights and dues of all parties*: Transaction between two or parties should be duly made written and well documented precisely, fairly and accurately (Qur'an 2:282)[12]. Precision requires exactness and fairness may mean valuation of debt among parties while accuracy requires faithful representation. Conventional reporting emphasise also on fair & faithful presentation on financial statements but to the favor of financiers (Owners and Creditors). In Islam rights due to all parties should be treated with equity and justice. Therefore, irrespective of weather the parties are owners, creditors or debtors the accounting system, standards and techniques must be designed to ensure no party is cheated or suffers to the detriment to other.

iii. *Paying Zakah*: As one of the five pillars of Islam, Zakah plays a vital role in fiscal policy and income redistribution among muslim and even non-muslims. Having accurate, just, fair, and faithful financial statements becomes imperatives. Both the amount payable and threshold in Zakah are subject to conversion using "real value" of wealth. In nutshell, the wealth of individuals in paper money and other monetary possessions are duly converted to gold and silver equivalent to arrive at threshold (NISAAB) and payable amount. Deducing form this Zakah valuation, one can argue that corporate institutions that operate under the ambits of Islam need to convert their monetary assets using Zakah valuation method. (Qur'an:12:41, 78; 24: 56; 33: 33; 58: 13)[12]. From and Hadith and Sunnah: It stipulated that "there should be no cheating or suffering among the parties" (La Dharirah wala Dhirarah)

iv. *Islamic Jurists and Fatwa*: Two prominent fatwas on the issue of inflation and changes in purchasing power of currency regarding debt settlements useful in this work. The ruling by the Economic Fiqh council in its eighteenth conference session held in Riyadh, KSA, 23-28 September 2000) were:
a)."...repayment of undisputed debt should be made in the same currency of the debt, not the equivalent value. That's it is not permissible to connect debts owed, regardless of the nature of the debt, to changes in the level of prices"[13].
b)...."in the event that inflation is expected at the time of signing of the contract, the contract may be drawn up based on a currency or form of wealth other than that which is expected to fall, such as basing the contract on the following: Gold & Silver, Similar product, a basket of similar product, more stable currency, basket of currencies" "[13].

Based on the two rulings, a). disallowed changes in prices in debt settlement and b). allows it debt provided it's pre-mentioned in the contract on fear of loss in purchasing power.

Subsuming the Islamic provisions, rulings and positions from i-v, one can deduce that there is window to address price changes in financial statement using a stable means of exchange, prominently Gold and Silver standard. This will provide a hedge to both parties and also ensure sustainability of IFI's.

5. Conclusion

Recognizing the defects of financial reports in changing prices situations a lot of debates, observations, reservations and opinions were forwarded in different world forum, however AAOIFI is still yet to develop a standard on changes price effect. However, the effect of changes in process is more on monetary rather than non-monetary assets. This paper argues that if the valuation and measurement of Zakah in terms of Value and Nisaab (threshold) is always done on "real wealth" Gold and Silver not fiat or paper money. Then why shouldn't transaction in debt be converted to such real wealth? Secondly FAS16 issued by AAOIFI requires conversion of foreign currency and recognition of translation difference in capital reserves. Inflation accounting adjustments have similar effects on assuming the same purchasing power of different currencies in a within a giving accounting period.

6. Recommendations and policy implication

Islamic accounting is divine and guided by non- normative assumptions. Fundamental Islamic principles provided that Justice and fairness should be exercised among parties in business transaction and events. In addition, Qur'an strongly prohibits shortchanged/outsmarting/unfair treatment of a party to businesses. It's unfair and a violation of principles of justice to makes creditors suffers during changes in prices while it can be redressed by changes price adjustments. Unlike in conventional accounting, Partial and Full adjustment through use of Indexes and Methods, General Purchasing Power (GPP) and Current Cost Accounting (CCA), and recently use of Fair Value to address the menace of changes in prices. Islamic accounting should use "Real Value" concept to adjust for inflationary effects. Real Value is that value that is equivalent to GOLD & SILVER value at each period.

References

[1] AAOIFI, Statement of Financial Accounting (SFA) 1 (Accounting Objectives), Manama: Bahrain.
[2] AAOIFI, Statement of Financial Accounting FAS 16 Accounting for Foreign Currency Transaction. Manama: Bahrain.
[3] Abdul Rahman, A.R. (2006) "An Islamic Perspective of Accounting Objectives and Concepts", New Horizon, Institute of Islamic Banking and Insurance, London.
[4] Akram Khan, M (undated). "Accounting Issues and Concepts
[5] Bello, A. (2009). "The Marginal Value Relevance of Inflation Accounting Adjustments". *Nigerian Journal of Accounting Research.* Vol.5 (1) p 1-11. ISSN: 978-125-193-x.
[6] Bello, A. & Zaid, A. (2013). "Value Relevance a Tradeoff between Inflation and Historical Cost Accounts". *Accounting Frontier, Vol. 15(1). ISSN 0189-1743.*
[7] Baydoun, N. and R. Willett. (1994). Islamic Accounting Theory. *Proceeding of AAANZ Annual Conference*, Sydney, Australia.
[8] Christoper, N. & Roszani, H. (2012). Islamic Accounting, Edward Pub. U.K.
[9] Gambling, T. and R.A.A. Karim. (1991). *Business and Accounting Ethics in Islam.* Mansell, London.
[10] Gambling, T. (1974), *Societal Accounting*, London: George Allen & Unwin.
[11] *Kuhn, T.S. (1970). Structure of Scientific Revolution, 2nd ed University Chicago. USA.*
[12] *Qura'an & Hadiths*
[13] www.Islamqa.info

Auditor's Risk Assessment of Independent Directors in Nigeria

Abdulmalik Salau*, and *Ayoib* Che Ahmad

Tunku Puteri Intan Safinaz School of Accountancy, Universiti Utara Malaysia

Abstract. The objective of this study is to investigate the external auditor's risk assessment of independent directors in Nigeria. The study utilized data from 94 non-financial listed companies on the Nigerian Stock Exchange for the periods 2008-2013. The study used cross-sectional time-series feasible generalized least square regression, which account for heteroscedasticity and autocorrelation to test the influence independent non-executive director on auditor pricing decision in Nigeria. Our result indicates that the proportion of independent non-executive director has a positive relationship with audit fees, suggesting that this class of directors is priced high by the Nigerian auditors. These findings have both policy and practical implication on corporate governance. For instance, future regulatory reforms could consider collaborative board model instead of the insistence on more independent director presence in the boardroom.

1 Introduction

An important corporate governance mechanism which corporate governance codes have given more emphasis since the Enron scandal in the early 2000 has been the independence of the board of directors. Board independence through the separation of the position of the CEO and Chairman and the composition of the board with majority of non-executive independent directors is highlighted in codes of corporate governance and some empirical studies to improve board oversight function. [1, 2, 3; 4 & 5] all provide empirical assertions on the monitoring prowess of non-executive independent directors from different perspective of organisation outcome consistent with the agency theory. Nevertheless, in the recent time, there are countervailing arguments with respect to monitoring prowess of independent directors. Notably, Fogel et al. asserted that, "there is no predicate either in logic or in experience to suggest that a majority of independent director will guarantee good corporate governance or better financial return of shareholders" [6] (pg. 35). Based on their findings, [6], found an inconclusive evidence that the dominance of independent directors improves firms' financial performance. Similarly, [7] finding revealed that the presence of independent directors does not improve board monitoring in family owned companies While, [4] documented that independent directors reduce earnings ability to predict future cash flows in the Malaysian context. According to [8], the independent directors do not have an adequate understanding of the daily operation of the company. This is because in

* Corresponding author: abdmaliksalau@gmail.com

most cases their presence is only at the instance of the board meeting. As a result, their monitoring prowess is weakened. Based on the foregoing countervailing argument on the monitoring prowess of non-executive independent director, this study investigates the impact of non-executive independent directors on the assessed inherent risk perceived by the Nigerian auditors as revealed in the audit fees charged from the supply perspective like prior studies [9; 10]. For over three decades the factors that determined audit fees has stimulates researchers interest due to the possibility of auditors' independence impairment [11]. However, despite the abundant studies on audit fees, governance and regulation, the overall results have produced conflicting results due to the complex nature of the variables [12]. Logically, from the audit supply point of view, sound corporate governance in companies indicates that the overall control system is strong, effective and efficient towards ensuring sound reporting quality. Accordingly, there will be a negative relationship between governance mechanisms and audit fees which implies that strong control environment reduces audit fees because the auditor perceived client inherent risk low [9].

Contrary to the findings of [9, 10] our finding suggests that auditor assessment of board composition risk affect audit pricing and that the proportion of independent directors tends to increase audit fees. Our findings and argument as well contradict the findings of [5] that argue from the point of view of audit demand and documented that the presence of increase audit fees due to their high demand. Although independent directors supposedly enhance board monitoring, in the recent time critics argues that their presence does not necessarily enhance board independence, as they might not be truly independent. As documented in the findings of [12]" the traditional role and the overbearing influence of family owners in the context of Nigeria might limit the oversight function and independence of the board. Consistent with this postulation, our findings suggest that the presence of independent directors might not be a panacea for weak governance.

The contribution of this study is that it provides further evidence from a less studied market, Nigeria and thus provides an avenue for future studies most especially from the perspective of developing countries whose corporate governance structure are entirely different from those of the developed nations. Although, there are studies in this area from some emerging economies however, the corporate governance landscape in countries varies hence, limiting the generalizability of the frequently researched Anglo-American system in the Nigeria setting. In the sense that the indigenization policy embarked upon by the Nigerian government as some point in time lead to majority share ownership mostly held by founding families who have significant influence on board and on management [13]. Accordingly, the assumptions of market efficiency, competitive market and market mechanisms which agency theory presuppose are not validly applicable in the Nigerian where ownership concentrated [13]. Moreover, the code of corporate governance that was revised in 2011 just codified the requirement for the inclusion of non-executive independent directors and some companies are yet to comply with this requirement. Notwithstanding the poor compliance rate, the independence of the purported independent director is susceptible to impairment due to the control retain by founding family. Apparently, the distinctiveness of the Nigeria corporate governance environment from the corporate governance environment of previous study as highlighted above provides an additional opportunity to extend the body of knowledge.

The remaining part of the study is as follows: section 2 outline prior literature and hypotheses development. We discuss the sample selection procedure and research design in section 3 and section 4 contains the presentation of result. The final section concludes the study.

DOI: 10.1051/shsconf/20173410001

2 Literature review and hypothesis development

Based on [14] traditional audit fees model several audit fees studies established the scope of audit, the complexity of audit work and the risk associated with an audit as the primary determinants of audit fees. Consistent with these three factors, audit fees-governance related studies suggest that the strength of client governance environment significantly influence the scope and risk associated with an audit engagement [9]. For instance, a strong governance environment will reduce audit risk that will subsequently translate into less audit scope hence a reduction in audit fees and vice versa. Several characteristics of a strong governance environment have been identified in literature and Codes of Corporate Governance among which is the extent of board independence (i.e. non-executive independent directors). The advocates of independent non-executive directors claimed that their presence on board enhance the board of director's incentive to monitor and discharge their oversight function because of the absence of any contractual relationship between them and the management. Some of the empirical evidence that support this view are [16] that reported that independent directors improve corporate reporting quality and [9] whom noted that the presence of non-executive independent director in boardroom enhance auditors' reliance on client accounting system and thus reduce the amount paid as audit fees. While there are several empirical studies that established the positive outcome of the presence of non-executive independent directors on board other study have found that their presence does not necessarily improve board outcome but rather contribute negatively. For instance, [17] found a positive relationship between corporate failure propensity and the proportion of independent non-executive director on corporate board. Since empirical findings regarding the effectiveness of independent non-executive director are inconsistent, this present study posits that:

H1: There is a relationship between the proportion of independent non-executive director in boardroom and audit fees.

3 Methodologies

3.1 Sample selection and data

This study used the sample of Nigerian public listed companies with the exclusion of all financial companies. In addition to financial companies, this study as well excludes companies with missing annual report and missing information. Accordingly, the final sample consist 94 companies listed on the Nigerian Stock Exchange between the years of 2008 and 2013 leading to 426 unbalanced observations.

3.2 Regression model and specification

This study employs panel estimation approach to examine the hypotheses regarding the relationship between corporate governance characteristic and audit fees. This approach is considered more suitable for a group of firms listed on the Nigerian Stock Exchange due to the problem of bias caused by unobserved heterogeneity. Due to the presences of contemporaneous correlation and heteroscedasticity as characterised in panel data, this study employs Feasible Generalized Least Square (FGLS) to correct for this problem. However, before then, the Breusch-Pagan Lagrange Multiplier test rejects the null hypothesis of no effect in the cross sectional unit over the period of time as the P-value is 0.000 suggesting that pool OLS is not appropriate. The Hausman test result shows that FE model is preferable to the random effect model. Accordingly, the model as adopted from Sminunic [14] is developed as follows:

$$\begin{aligned} \mathrm{LOGAF}_{it} = {} & \alpha_{it} + \beta_1 \mathrm{IND\_}_{it} + \beta_2 \mathrm{LOGTA}_{it} + \beta_3 \mathrm{BIG4}_{it} \\ & + \beta_4 \mathrm{CA}_{it} + \beta_5 \mathrm{QUICKRATIO}_{it} + \beta_6 \mathrm{GEARING}_{it} \\ & + \beta_7 \mathrm{BUSSEG}_{it} + \beta_8 \mathrm{INVT\_}_{it} + \beta_9 \mathrm{ROA}_{it} + \beta_{10} \mathrm{LOSS}_{it} \\ & + \beta_{11} BUSY_{it} + \beta_{12} \mathrm{DELAY}_{it} + \beta_{13} \mathrm{BLOCKSHR}_{it} \\ & + \beta_{14} \mathrm{INDUSTRY\ EFFECT}_{it} + \beta_{15} \mathrm{YEAR\ EFFECT}_{it} + \varepsilon_{it} + \mu_{it} \end{aligned}$$

Where, log audit fees (LOGAFit) = the amount paid as audit fees by client i in year t; independent non-executive director (IND_it) = percentage of directors in the boardroom that are independent. Consistent with earlier studies we control for client size, complexity, risk and auditor related attribute. Natural log of total asset in thousand (LOGTAit) = client size. Auditor related attribute proxy by audit quality (BIG4it) = dummy variable representing big four audit firm, coded 1 if the firm is audited by one of the big audit firm or 0 otherwise; busy period (BUSYit) = dummy variable representing client that has March and December as their financial year end, coded 1 if the client financial year end falls in these month or 0 otherwise; audit lag (DELAYit) = the number of days between financial year end and the day the audit report was signed. Client risk is proxy by (CAit) = current asset to current liability; quick ratio (QUICKRATIOit) = current asset minus inventory scaled by current liability and loss (LOSSit) = dummy variable representing client who recoded negative earning after interest and tax. Client complexity is proxy by business segment (BUSEGit) = number of client business segment; Inventory (INVT_it) = balance sheet value of closing inventory scaled by total asset. Measure of client performance (ROAit) = net income divided by total asset. Form of ownership (BLOCKSHRit) = company outstanding share more than 5% held by individual block shareholders. Finally, we control for variation between year and industry.

4 Empirical analyses

Table 1 present the result of the cross-sectional time-series FGLS used to examine auditors' risk assessment of independent directors in Nigeria and the result indicates that that audit fees is positively related to the proportion of independent director (IND_) on board (P<0.01) and this is consistent the current study postulation. With respect to the control, variables with the exception of few variables are consistent with theory [11]. Client size measured by log total asset (LOGTA) has a positive relationship with audit fees. Similarly, client complexity measured by number of business segment (BUSSEG) and inventory (INVT_) as well has a positive relationship with audit fees, suggesting that complex client pay more consistent with agency theory. Client risk proxy by current asset ratio (CA) and GEARING reveals a negative coefficient while quick ratio (QUICKRATIO) exhibit positive relationship with audit fees. Profitability measured by LOSS has positive relationship with audit fees. BLOCKSHR exhibit a negative relationship with audit. Auditor related attributes BIG4 and BUSY as expected have a positive relationship with audit fees.

DOI: 10.1051/shsconf/20173410001

Table 1. Regression Result for audit fees

Variable	Model
CON	1.16
	(13.20 ***)
IND_	0.20
	(5.12***)
LOGTA	0.37
	(33.37***)
BIG4	0.32
	(18.29***)
CA	-0.06
	(-2.66***)
QUICKRATIO	0.06
	(2.41***)
BUSSEG	0.02
	(3.63***)
INVT_	0.05
	-0.86
ROA	0.00
	-0.16
LOSS	0.06
	(3.12***)
GEARING	-0.01
	(-2.04***)
BUSY	0.06
	(3.94***)
DELAY	0.00
	(-1.77**)
BLOCKSHR	-0.01
	(-12.69***)
YEAR EFFECT	YES
INDUSTRY EFFECT	YES
Obs	426.00
ADJUSTED R-SQUARE	0.89
F-STATISTIC	0.00

Notes: *** Significant at 1 per cent level; **, significant at 5 per cent level; *, significant at 10 per cent (two-tailed). The dependent variable is log of audit fees (LOGAF). A continuous variable that represent the amount collected as auditor's remuneration in Naira.

5 Discussion of findings and conclusion

In this study, we investigated the Nigerian auditor risk assessment of non-executive independent directors of public listed companies on the Nigerian Stock Exchange. Towards the set objective, we employ cross-sectional time-series FGLS regression which controls the issues of heteroscedasticity and autocorrelation on a sample of 94 public listed Nigerian companies. The findings indicate that an increase in the proportion of independent director increase the amount paid as audit fees. Overall, the result reveals some weakness in independent directors' performance of their advisory and monitoring role has consequence for auditor risk assessment hence the audit fees. The findings of this study have implication for audit fees literature. Against prior studies that examined the relationship between independent board and auditor pricing decision from the supply perspective, the present study finding indicates that the presence of non-executive independent director does not necessary guarantee good corporate governance. Therefore, future research most especially those undertaken in the less developed market with prevalence of family ownership could consider the collaborative board model. Because, recent literatures [15] have begun to fault the one-dimensional view of board role as dominant in literature and recommends that studies should consider other board responsibility in relation to auditing. In addition, future study could embark on a more in-depth study by segregating the so called independent director by the extent of their true independence to determine how it affect the risk perception of external auditors. This is imperative due to recent literatures that questions the genuineness of the so called independent directors on corporate board.

References

1. S. T., Petra, The international journal of business in society, **6**, 2, 107-115 (2006).
2. W., Beekes, P., Pope, S., Young, (2004). Corporate Governance: An International Review, **12**, 1, 47-59 (2004).
3. A. S., Ahmed, S. Duellman, Journal of accounting and economics, **43**, 2, 411-437 (2007).
4. R.A., AL-Dhamari, K. N. I. K. Ismail, Jurnal Pengurusan, **41**, 43-55 (2014).
5. C.S., Armstrong, J. E., Core, W. R. Guay, Journal of Financial Economics, **113**, 3, 383-403, (2014).
6. E. M. Fogel, A. M. Geier, Del. J. CorP. l., 32, 33 (2007).
7. N., Abdullah, S., Z.M., Yusof, & M. N. M Nor, Managerial Auditing Journal, **25**, 6, 526-552 (2010).
8. R.J., Gilson, R. Kraakman, Stanford Law Review, 863-906, (1991).
9. J.S. Tsui, B. Jaggi, F.A Gul, Journal of Accounting, Auditing & Finance, **16**, 3, 189-208 (2001).
10. B, Muniandy, (2007). Managerial Auditing Journal, **22**, 7, 716-728 (2007).
11. O.S. Abdulmalik, A. Ahmad, (2016). DLSU Business & Economics Review, **26**, 1, (2016).
12. D. C. Hay, W.R Knechel, N. Wong, Contemporary accounting research, **23**,1, 141-191, (2006).
13. E., Adegbite, International Business Review, **24**, 2, 319-330, (2015).
14. D.A., Simunic, Journal of accounting research, 161-190, (1980)
15. J. R., Cohen, G., Krishnamoorthy, A.M., Wright, A., Auditing: A Journal of Practice & Theory, **27**,2, 181-198, (2008).
16. L., Setia-Atmaja , J., Haman, G., Tanewski, The British Accounting Review, **43**, 3, 230-246 (2011).
17. H. H., Hsu, C. Y. H, Wu, The British Accounting Review, **46**, 3, 215-227 (2014).

The Classification of Corruption in Indonesia: A Behavioral Perspective

Rizki Hamdani[1]*, Kumalahadi[1], and *Dekar* Urumsah[1]

[1]Accounting Department, Faculty of Economics, Universitas Islam Indonesia, Indonesia

Abstract. This research is aimed to investigate and identify the pattern and classification of corruptors in Indonesia, especially the state officials being imprisoned. This research used the qualitative method. The data were collected through documentations and interviews. The source of the data was chosen by purposive sampling technique. The researcher interviewed deeply 9 suspects of corruption cases being imprisoned. The results of the research show that the classification of corruptors in Indonesia includes all types of corruptions constructed by the Association of Certified Fraud Examiners (ACFE), namely: conflict of interest, bribery, illegal gratuities, and economic extortion. Based on the interview, it is found that the interviewees perform different types of corruptions as follows: there are some suspects perform more than one type of corruptions; there are some suspects perform single corruptions with the same type, and there are some suspects perform single corruptions with the different type. In Indonesia, it is not only the people of executive, legislative, and judiciary who can perform corruptions, but also the people of private sections.

1 Introduction

Corruption is an act of utilizing a public function to self-benefit that extremely leads to bad impacts to the development of social and economy condition of a country [1, 2, 3]. Some literature show that the practice of corruption affects negatively to the economical development [4], productivity [5], investment [2], entrepreneurship [6], international and domestic trades [7, 8], informal sections [9] based on the transmission of public services [10], and frame a country up from a long poverty [11].

Ironically, the Index of Corruption Perception in Indonesia in the last ten years is not more than 4. This condition is categorized as bad. If it is compared with some countries such as Malaysia and Singapore, Indonesia is left behind in 2014. Malaysia gets 52 and is ranked 50, while Singapore gets 84 and is ranked 7 [12].

The Audit Board of the Republic of Indonesia in Semester Examination Results Summary I *(IHPS I)* of 2014 reveals 4.900 cases valued Rp 25.74 million billion which harm this country [13]. The Indonesian Corruption Watch (ICW) releases the results of an investigation on semester I, a period of 1st January – 30th June 2014, that there are 308 corruption cases revealed by the Police, Prosecutor, and The Corruption Eradication

* Corresponding author: rizki.hamdani@uii.ac.id

DOI: 10.1051/shsconf/20173410002

Commission of Indonesia *(KPK)*. They state that the trend of corruptions tends to increase. Meanwhile, on the same semester in 2013, there are 257 cases revealed [14]. The results of the investigation on semester II, a period of 1st July – 27th December 2014, show that there are 629 cases. Meanwhile, in the same semester in 2013 there are 560 cases. It means that there are 69 more cases [15]. If it is summed up, the total numbers of corruption cases during 2014 are 629 cases with 1.328 suspects and country's loss of Rp 5.29 million billion. During 2013, there are 1.271 corruption cases revealed [16]. It can be concluded that compared with 2013, the corruption cases during 2014 increases. The data from the surveys of Indonesian International Transparency show that the members of the House of Representatives from Partai Golkar are 19 people, Partai Demokrasi Indonesia Perjuangan is 17 people, Partai Demokrat is 4 people, Partai Amanat National are 3 people, Partai Persatuan Pembangunan is 3 people, Partai Kebangkitan Bangsa is 1 person, Partai Keadilan Sejahtera is 1 person, and Partai Bintang Reformasi is 1 person [17].

This research is aimed to deeply investigate the classification of corruptors in Indonesia. By viewing the corruptors' sides, it can be revealed a deeper understanding of corruption potentials which can be used as a reference to coping with corruptions [18]. Moreover, [18] shows that if potential suspects are forced, motivated, getting a chance, and having rationales to do a corruption, they are (consciously or not) considering the benefits and costs gotten before deciding to do or not to do the corruption. Viewing and understanding the mechanism in which the potential suspects decide to be involved or not will provide a strong basic for the government and decision makers to construct effective strategies to eradicate corruptions.

2 Literature Review

This part provides the literature review on the corruption issues. The review discusses the types of corruptions, the rationales of corruption, the effects of corruption, the role of against-corruption organizations in the world and Indonesia, and the pattern of corruptions in Indonesia.

2.1 The Types of Corruptions

The professional organization auditing fraud, Association of Certified Fraud Examiners (ACFE), in the United States describes occupational fraud in the form of fraud tree [19, 20]. The diagram is shaped like a tree with branches and twigs so that it is named fraud tree. The three main branches of the fraud tree are corruption, asset misappropriation, and fraudulent statements. Moreover, there is also a formula for the types of corruptions which consists of conflict or interest, bribery, illegal gratuities, and economic extortion [19, 20].

2.2 The Pattern of Corruptions in Indonesia

As written in the Acts Number 31, 1999, the pattern of corruptions happened here is in the context of bureaucracy. In each case of corruptions in Indonesia, the mass media always reveal the name of the individuals doing corruptions and some of the corruptors are proven in a red handed operation by *KPK*. In fact, there is a pattern of corruptions done by the corruptors in all of their deeds. Therefore, the pattern of corruptions in Indonesia is not randomly done and is not incidentally conducted. The pattern is systematically done and is able to be constructed [21]. The corruptors do the action in a system which consists of Principal-Agent-Client-Middlemen (PACM). Figure 1 shows the pattern of corruptions in a bureaucracy taking the state budget.

DOI: 10.1051/shsconf/20173410002

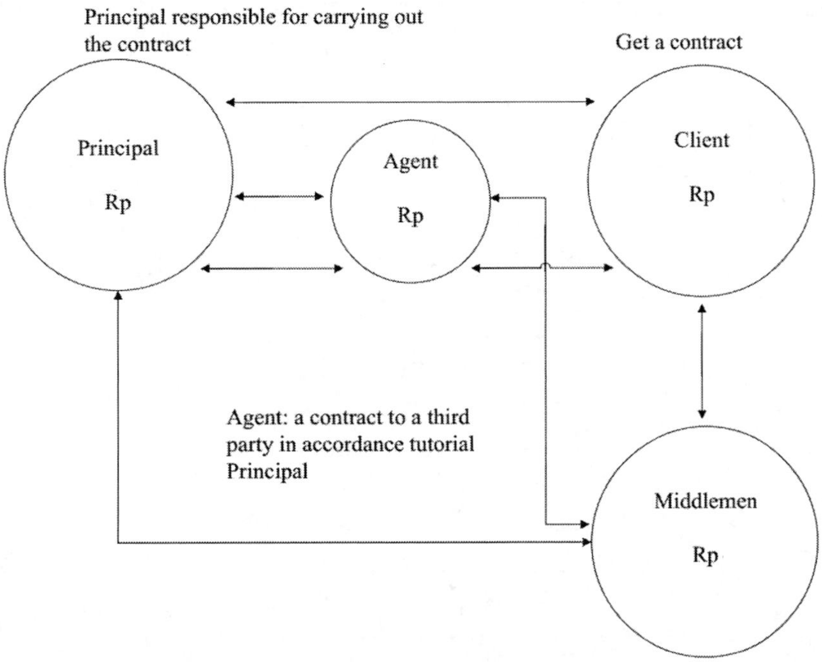

Fig. 1. The Pattern of Corruptions in a Bureaucracy [21]

In Figure 1 above, it is shown that the network of corruption happens systematically and is organized neatly. The pattern shows that corruptions cannot be done by only oneself. In the Principal circle or the government as the ruler or budget, users are responsible for conducting the contract and giving charge to the Agent (division head/division of government procurement) to do the contract to third party – the Client. Before doing so, the Principal has ordered the Agent that the tender must be given to the Client and or the Principal-Agent uses Middlemen's help to link to the Client. In short, the Middlemen act as the connector of the corporate network (Client) and the network inside the government (Principal-Agent).

3 Research Method

The source of the data was selected based on the purposive sampling technique. The data were collected through a semi-structured interview with 9 interviewees of suspects in the cases of corruption being prisoned. The research used the qualitative method because the objectives were to investigate the pattern and the classification of corruptors in Indonesia viewed from the corruptors in the prison. [22] explain that qualitative method is used to give a deeper understanding of the issues that involve the perspective of the population and the context in which they live. Specifically, qualitative research is also beneficial to explore new topics or to understand complex problems, to explain human's attitudes and beliefs, and to identify social norms or cultures [22]. Therefore, the qualitative method fits this research because the research dealt with corruptions in the context of accounting which is believed as a new topic and is worth considering. Figure 2 illustrates the procedure of this research.

 DOI: 10.1051/shsconf/20173410002

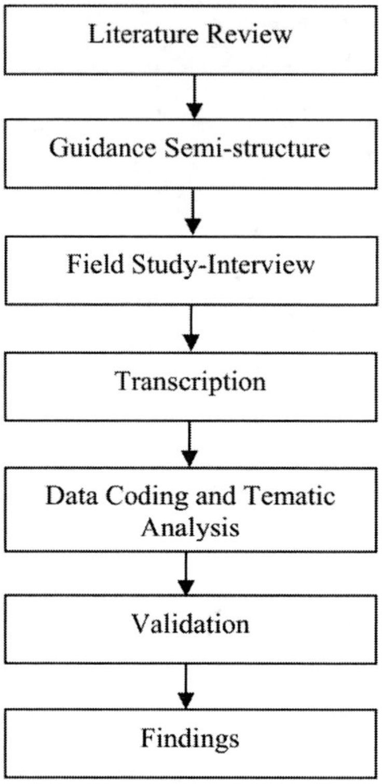

Fig. 2. The Process of Qualitative Research Method

4 Results and Discussion

Table 1 shows the background of the respondents based on their affiliation and their position.

Table 1. The Profile of the Interviewees

No.	**Respondent**	**Position**	**Institution**
1	Interviewee A	Legislative	Member of *DPR*
2	Interviewee B	Legislative	Head of *DPRD*
3	Interviewee C	CEO	Private
4	Interviewee D	Executive	Regional Government
5	Interviewee E	Executive	Regional Government
6	Interviewee F	Judicative	Judiciary
7	Interviewee G	Executive	Regional Government
8	Interviewee H	Legislative	*DPRD*
9	Interviewee I	CEO	Private

Table 1 shows that three people come from the executive position, three people come from

DOI: 10.1051/shsconf/20173410002

the legislative position, one person comes from the judiciary, and two people come from the private company.

The results of the interview are classified based the ACFE as shown in Table 2 below.

Table 2. The Classifications of Corruptors in Indonesia

Code	Conflict of Interest	Bribery	Illegal Gratuities	Economic Extortion
A	–	–	√	–
B	√	–	–	–
C	√	√	–	–
D	√	√	–	–
E	–	√	–	–
F	–	–	√	√
G	√	–	–	–
H	√	–	–	–
I	√	–	–	–

Based on Table 2 above, corruptors classified in the conflict of interest are six people, bribery is three people, illegal gratuities are two people, and economic extortion is one person. From the nine suspects, it is identified that C, D, and F perform two types of corruption: C and D perform conflict of interest and bribery, while F performs illegal gratuities and economic extortion.

The interviewee C is identified performing conflict of interest and bribery based on C's statement as follows:

"I confess that I gave the regent and some persons in the ministry money. They asked the money and promised me the tender. Corruptions are very organized in Indonesia. If we want to get the tender, we need to use money in the governmental bureaucracy. In the ministry, in the province, and in the region, there are many corruption practices by asking for money to the company given the tender by the government. If we do not do that, we will not get the tender from the government."

From the interview with interviewee D, the executives give a guarantee for their workers who have been accused to move them from prison into the city custody by employing a fee lawyer to accompany the accused ones so that the workers can be free. By doing so, the executives will not be involved in the court. Interviewee D tends to confess that he has helped his workers in the judiciary process. He explains his involvement in the corruption practice that he has done conflict interest and bribery:

"In the report of the Audit Board of Indonesia about the financial report of the Regional Government in which I governed in 2010, the country's loss is Rp 40 million billion in the use of social finance during 2009-2010 which should be accountable. However, I had ordered the inspectorate at that time to send the budget user – the region secretary and the head of financial and asset management to report it. At that case, I was involved in the social support. My staff once told me that he was stressful because the judges always asked for money. I helped him give a guarantee for the accused to move him from prison into city custody. I employ fee lawyer to assist the accused ones and was helped by the city government as my moral responsibility as a chief."

Different from interviewee C and D, interviewee F performs two types of corruptions. Interviewee F performs the illegal gratuities and economic extortion even though he does not admit it. However, the judge judges him provenly guilty and assures him receiving money from the entrepreneur and doing extortion to the former chief in the central government. The following is the explanation from interviewee F.

"Both *KPK* and the Judiciary have already investigated the case I am handling, so there is no any factor violating the law as they accuse. It is clearly recommended that the team does not find any enough proof to move this case into the investigation level. Moreover, there is a clear mistake in presenting the proof in the court. It relates to the witness' explanation which comes from only one person without other witnesses or proofs. Ironically, it is used for the proof to accuse me."

The differences of corruptions performed by C, D, and F are: (1) C is a CEO who has the conflict of interest to a person in the government to win the tender, so D must bribe the person in the government who asks for money; (2) D is an executive in the regional government who has the conflict of interest to his workers and the judges so that he is not involved in the case, so D bribes to the judges, and; (3) F is a person in judiciary who does the economic extortion to the chief of central government so that the case being investigated will not be processed. The money gotten is used for illegal gratuities in the form of business deal.

Three of nine suspects have a similar pattern of corruptions which is a conflict of interest. The suspects are interviewees G, H, and I.

In the court, the judges judge interviewee G as personally and systematically enrich himself, others, or a corporation, so it causes a billion losses for the country [23]. Interviewee G is proven to violate the law with the fund of billion rupiahs from the regional government to his personal bank account. The fact is proven by the invoice which cannot be denied by him in the court. He is proven using the money for his personal interest such as buying a house. However, in the interview he says that:

"As a chief of the regional government, I only want to fasten the regional development. I work for my people in my region. I never order my staff to perform corruptions. The judiciary is not doing a fair investigation in this case, so there are many bizarre things and unfairness. The regional secretary and commitment makers should have been the one who is responsible for this, but they have never been investigated. I am a passive person in this project. It can endanger the judiciary and the corruptions will never end."

The corruption performed by interviewee H is about fiction insurance of the members of Regional People's Representative Assembly *(DPRD)*. The case makes the regional government lost billions of rupiahs. The case is started from the insurance program between *DPRD* and insurance company. The program offers one year premise with the total amount of it of Rp 2 billion. However, the implementation of the cooperation is never conducted. Interviewee H is judged guilty. He explains that:

"In the previous meetings discussing budgets, actually, there is no budget for the life insurance. The budget for the members appears when *DPRD* and other members meet the city major. At that time, the city major promised the members grate after retired, but the form was not clearly thought. After some meetings, it was agreed that it was in the form of life insurance. The members of the representatives never knew and were never involved in the cancellation because there was no formal cancellation from the *DPRD* secretary. The cancellation was done by the order of the chief of *DPRD*."

Interviewee I is a CEO in one of the private companies in Indonesia. He is judged as making the country lose because of violating frequency streamer. As the result, the company gets significant benefits. According to the report of the Financial and Development Supervisory Agency *(BPKP),* the company makes the country lose tip to Rp 1 million billion. However, in the interview, he explains that he does not perform the corruption as accused by the Judiciary. The following is interviewee I' explanation:

".... from where the corruption is. The cooperation between the companies is based on the law because it suits the Acts and there are operational regulations. The cooperation agreement between the central company and the company I lead is a corporation and is not a performance of the director."

The differences between G, H, and I are: (1) G is a regional chief who performs conflict of interest with his staff members by violating regional budgets in the infrastructure project, (2) H is a chief of *DPRD* who performs conflict of interest with the members of *DPRD* by creating fiction project of insurance premise, and (3) I is a CEO of a private company who cooperate with the government in making frequency streamer, but unwillingly it causes the country loss. The similarity between G, H, and I is that they perform conflict of interest systematically and involve many people.

There are three suspects who perform a single corruption, but slightly different. A performs illegal gratuities, B performs conflict of interest, and E performs bribery.

Interviewee A is a member of House of Representatives *(DPR)* who is proven guilty of performing corruption and illegal gratuities many times. The Judges judge him guilty because he performs corruption as mentioned in Article 3 of the Acts Number 8, 2010 about the prevention and the eradication of money laundry. However, interviewee A convinces that he is not guilty and does not perform corruption as judged by the Judges. The following is interviewee A' statement:

"…. I am accused of receiving a gratification by the judiciary in the court because of one suspect's statement. They should have proven that I really receive it, but they cannot. I am asked to prove myself that I am not guilty. In my case, the prosecutor cannot prove in the court, but judges me guilty. There is an influence from the media in the decision of corruption cases in Indonesia nowadays. Thus, the judges' judgment is really influenced by the news in the public and do not take a look at the the fact and proof in the court. The judgment is unfair because it is not based on the facts in the court."

Interviewee B is a suspect of the corruption case of the official tour in *DPRD*, instrument of representatives, and the discussion of regional regulation for hundreds of billion rupiahs when he governs as the chief of *DPRD*. He states that there is an agreement in performing the corruption in *DPRD*. For example, he allows the members not to join the official tour, but still get the fund. In another case, there are some members who receive money from the official tour warrant, but they do not do the managed tour. In every official tour, they always use the travel agent recommended by other members of *DPRD*. The following is interviewee B' explanation:

"I admit that there are some members who do not join the official tour but still receive the money. Because this is a systemic mistake, the chief and the members should be suspected. The case of the official tour is known by all chiefs and members, so it is impossible if others do not know. The chiefs are collegial and collective, so it cannot be blamed on one person only. It becomes our responsibility, all chiefs."

Interviewee E is a suspect in the corruption case of social fund in the city government and bribery to persons of judges handling the case. He explains the bribery he performs as follows:

"My case is the social fund. It is purposed for the society in their programs. It is for a good purpose, basically. It helps them. Every region provides the fund. The chief orders us to give money to the chief of the Judges. The city major orders me to handle the coordination of the fund collection with the regional chiefs. We only conduct the order. There is a record of my talk with the city major and I open all of the cases in the court. I help *KPK* in this case because I become the justice collaborator."

The differences between A, B, and E are: (1) A performs illegal gratuities in the project of infrastructure development, (2) B performs conflict of interest in the form of violating the fund of official tour and instruments of representative members, and (3) E performs bribery for himself and not for his staff in the court. The similarity between A, B, and E is that they only perform one type of corruptions.

The results of the research show that the classification of corruptors in Indonesia includes all types of corruptions constructed by ACFE [19, 20], namely conflict of interest, bribery, illegal gratuities, and economic extortion. Based on the interview, it is found that 9 suspects perform different types of corruptions. There are some suspects perform more than one types of corruptions, single corruption in the same type, and single corruption in the different type.

The researcher also finds that in Indonesia, it is not only the people of executive, legislative, and judiciary who can perform corruptions, but also the people of private sections. All of them cause the country loss as the pattern investigated by Indriati [21].

5 Conclusions and Suggestions

All the types of corruptions constructed by the Association of Certified Fraud Examiners (ACFE) namely conflict of interest, bribery, illegal gratuities, and economic extortion are found in the corruption cases in Indonesia. The nine interviewees can be categorized into there types of classification as follows: (1) 3 people perform more than one types of corruption; (2) three people perform single corruption with the same type, and (3) 3 people perform single corruption with a different type.

Based on the pattern investigated by Indriati [20], in Indonesia, it is not only the people of executive, legislative, and judiciary who can perform corruptions, but also the people of private sections who can make the country loss.

The researcher suggest that the central and regional government in Indonesia design or strengthen the fraud control plan on the sections being corrupted, especially on the sections used as the field to corrupt as found in this research, namely license, budget violation, and goods or services procurement; fix and strengthen the system of corruption prevention in every government institutions such as executive, legislative, and judiciary; support all regional government in Indonesia to create e-budgeting and e-procurement in constructing regional budget and procurement of goods and services; and strengthen and employ Government Internal Supervisory Apparatus *(APIP)* such as regional inspectorate in provinces and cities and *BPKP*.

This research has some limitations as follows: (1) this research is only focused on the interviewees of suspected in the corruption prisons and (2) the data related to the potential of the country' loss, judges' judgment, and the period of being suspected are gained from the internet.

References

1. P. Bardhan. Corruption and Development: A Review of Issues. *J. Econ. Lit.* **35**, 1320–1346 (1997).
2. K. M. Murphy, A. Shleifer, R. W. Vishny. Why is rent-seeking so costly to growth? *Am. Econ. Rev. Pap. Proc.* **83**, 409–414 (1993).
3. A. Shleifer, R. W. Vishnyi. Corruption. *Q. J. Econ.* **108**, 599–617 (1993).
4. P. Mauro. Corruption and Growth. *Q. J. Econ.* **110**, 681–712 (1995).
5. A. Dreher, T. Herzfeld. The Economic costs of corruption: A Survey and new evidence. no. June 2005 (2005).
6. A. Dreher, M. Gassebner. Greasing the wheels? The impact of regulations and corruption on firm entry. *Public Choice* **155**, 413–432 (2011).
7. C. Bjørnskov. Can bribes buy protection against international competition? *Rev.*

World Econ. **148**, 751–775 (2012).
8. A. O. Krueger. The Political economy of the rent-seeking society. *Am. Econ. Rev.* **64**, 291–303 (1974).
9. A. Dreher, F. Schneider. Corruption and the shadow economy: An empirical analysis. *Public Choice* **144**, 215–238 (2010).
10. Transparency International. *Corruption Perception Index for 2007*. Transparency International, Berlin (2007).
11. T. S. Aidt. Corruption, institutions, and economic development. *Oxford Rev. Econ. Policy* **25**, 271–291 (2009).
12. Transparency International. Corruption Perceptions Index. *Transparency International* (2015). Available at: www.transparency.org/cpi. (Accessed: 16th April 2015)
13. The Audit Board of the Republic of Indonesia. Examination results in summary 1st half of 2014. BPK RI, Jakarta (2014).
14. Kompas. ICW: Corruption cases increase in 2014 (2014). Available at: http://nasional.kompas.com/read/2014/08/17/16011471/ICW.Kasus.Korupsi.di.2014. Meningkat. (Accessed: 21st April 2015)
15. Media Indonesia. The dominance of local government corruption in 2014 (2015). Available at: http://www.mediaindonesia.com/mipagi/read/8452/Pemda-Dominasi-Korupsi-di-2014/2015/02/18. (Accessed: 21st April 2015)
16. detikNews. Meet the chairman of the council, ICW report surging corruption (2015). Available at: http://news.detik.com/read/2015/03/10/141209/2854525/10/temui-ketua-dpd-icw-laporkan-melonjaknya-kasus-korupsi. (Accessed: 21st April 2015)
17. Merdeka.com. This is a list of members of the House of Representatives last 10 years dragged corruption. (2014). Available at: http://www.merdeka.com/politik/ini-daftar-anggota-dpr-terseret-korupsi-10-tahun-terakhir.html. (Accessed: 23rd April 2015)
18. H. Y. Prabowo. To be corrupt or not to be corrupt understanding the behavioral side of corruption in Indonesia. *J. Money Laund. Control* **17**, 306–326 (2014).
19. ACFE. 2008 Report to the nation on occupational fraud & abuse. The Association of Certified Fraud Examiners, Austin, USA (2008).
20. ACFE. Report to the nations on occupational fraud & abuse: 2016 Global fraud study. The Association of Certified Fraud Examiners, Austin, USA (2016).
21. E. Indriati. *Patterns and roots of corruption: Destroying public vicious cycle of Sin*. Jakarta: PT Gramedia Pustaka Utama (2014).
22. M. Hennink, I. Hutter, A. Bailey. *Qualitative research methods*. London: SAGE Publications (2011).
23. Tempo. Regent tegal sentenced to 5.5 years in prison (2011). Available: http://nasional.tempo.co/read/news/2011/11/24/063368297/bupati-tegal-divonis-5-5-tahun-penjara. (Accessed: 15th May 2015).

Long Tenure and Punishment Effect on Corrupt Behaviour

Rijadh Djatu Winardi[1*], *Wisnu Setiadi* Nugroho[1], and *Amanda* Wijayanti[1]

[1]Faculty of Economics and Business, Universitas Gadjah Mada, Jl. Humaniora No. 1, Bulaksumur, Yogyakarta, 55281, Indonesia

Abstract. This study implements behavioural economics and fraud auditing approach. This study looks at several factors that have been cited in previous studies as determinants of bribery. The result shows that long tenure increases the probability of corruption between civilian and public servant and penalty can be a good disincentive for corrupt behaviour but it is more effective towards the recipient instead of briber.

1 Introduction

Corruption is a serious problem that has been huge concern for the government since the impact hurts many parties. One type of corruption which regulated in Law No. 20 of 2001 is bribery. Bribery is defined as giving something to the officials in order to influence the decision making of the officials. Bribery is closely related to the bureaucracy. The longer the bureaucracy officer in charge of a position, the more likely bribery occurs. According to research by Transparency International Indonesia in 2013, four out of ten people of Indonesia pay bribes and approximately 71% of it paid as bakhshes to speed up the service. According to the same research, people pay bribes to obtain public services in the police department, courts, civil registration, and licensing. Hence bribery is a manifestation of corruption that should be eradicated.

2 Literature Review

2.1 Corruption and Bribery

The term corruption is a general term that can lead to various definitions. Corruption can be defined as financial crime; there is also a call fraud. Definition of corruption itself can vary. Law No. 20 of 2001 defines corruption as 'act of enriching oneself or another person or a corporation that can be detrimental to state finance or economy of the country'.

According to [5], they include the corruption as one of the three financial crime and fraud in addition to misappropriation of assets in the financial statements. Then, Law No. 20 of 2001 lays out seven corruption-related corruptions among other state assets, bribery, embezzlement in office, extortion, cheating, conflict of interest in procurement, and

* Corresponding author: rijadh.winardi@ugm.ac.id

DOI: 10.1051/shsconf/20173410003

gratuities. Of the seven types of corruption, bribery the most widely dealt corruption form by KPK during the year 2004-2015 [4]. There are a total of 192 cases of bribery handled by KPK. Beside that bribery is also the corruption form that is easily found in Indonesia and tends to be easily implemented. Due to this reasons, bribery could quickly become a culture and increasingly difficult to combat and make it more dangerous among other types of corruption. To understand the bribe then we observed a few things. First, a bribe will happen if there are two parties involved i.e. those who give bribes and those who receive bribes. Second, bribery can occur in two directions. Bribes can be started from the giver who gives bribes own initiative or at the request initiated from the recipient of bribes. Third, bribes are selling influence like buying and selling. Fourth, there are expectations for the return of the bribe giver and the bribe recipients benefit from the optimization.

2.2 Power

Power according to this study is defined as the ability to determine the size of bribes. Power is not restricted as structural position or positions in the organization. A person considered to be in power if he had bargaining power in determining how much should be paid for by the bribe giver. Previous research [14] also argues that a person may have power without status or status without power, but only when there is a clear distinction between formal and informal authority. According to [19] [20] the power plays a role in social relationships and focus on the person's intentions or actions to carry out for example the dominance and influence, [14]. Research [14] defined power as an individual person's capacity to modify others to give or withhold resources or punishment. Example of resources and punishment in the context of bribery is economic opportunity or material. Power is considered as one of the factors driving the emergence of an act unethically. The more powerful a person the stronger the urge to ask for bribes. Although not directly related to bribery however, power has an influence on personal decision-making. Examples [11] found under generally when the individual power increases, the assessment of performance against others decline and self-assessment tends to rise. There are allegations under the authority also will affect the bribe.

2.3 Punishment

One deterrent factor of corruption is the fear of punishment. Punishment becomes a disincentive for corrupt behaviour [13]. Another theory supports that the penalty would reduce bribery behavioural intention is Theory of Planned Behaviour (TPB). TPB proposed by [2] predicts that the positive consequences of an action, the person will tend to want to do. Conversely, if one considers the consequence of an action is negative (i.e. penalties) then the person will decrease his intentions. Research by [24] stated that the main driver of an action is the reward not a punishment. Kind of punishment also has a different effect on the behaviour of bribes. If placed in a continuum that there is a relationship with the deterrent effect of the low to the high. In this study we believe that penalty will be variable moderation of three independent factors namely power, motivation, and culture.

2.4 Bribery Game

Bribery game is of plenty manifestation of laboratory experimental economics. Laboratory experiments enable us to test ideas and theories that may be impossible to test naturally in real world. It also allows us to simulate what can possibly happen in real world into a human-made laboratory environment, it nearly impossible to be done in real world due to ethical value. Numbers of researches have done research on corruption, however using

DOI: 10.1051/shsconf/20173410003

laboratory experiment to determine the behavior of briber and bribe receiver is not yet very common. Research [8] categorize experiments on corruption corruptibility as those involving bilateral settings and unilateral settings. In one of the first investigations, [10] focused on individual tendencies to engage in corruption in procurement and whether economists are more likely to accept bribes. Later on, [10] extended their analysis to state how intrinsic motivations are affected by threats of penalties. [17] conducted a laboratory experiment on free riding: individuals were asked to allocate a given financial endowment to private use and contributions to a public good. Research [6] used an ultimatum bargaining game, where the 'proposer' offers a division of a given amount of $10 between him and the 'responder'. [10] conducted a prisoner's dilemma experiment and found that economics majors defect significantly more often (60 percent) than non-majors (39 percent). This paper mostly inspired by previous researches that were conducted specifically to analyze corrupt behavior like [1] [10], 16[] and [12].

3 Methodology and Experimental Design

According to [1] there are three main characteristics of bribery the first is reciprocal relationship between the briber and the bribed, secondly the existence of negative effects for the social welfare of society, and the third is punishment when action is known by law enforcement agencies. Referring to research carried out by [1], we replicate the conditions of treatments in lab experiments to see whether the three main characteristics have impact on bribery. The experimental design by [1], contained two actors, people who are bribed (in this case are government employees) can accept or reject the bribe. Bribers have the option to transfer some money to influence the decisions of government employees. The game is done in sequence, where the first player is the briber and the second player is the recipient of the bribe (denotes as government officials). Following the design of experiments by [1], this study will conduct three types of treatment situations. The number of subjects in our experiment is 32 people who play in pairs, which means we have 16 pairing in the experiment. The whole experimental situation is more easily understood through this Game Tree.

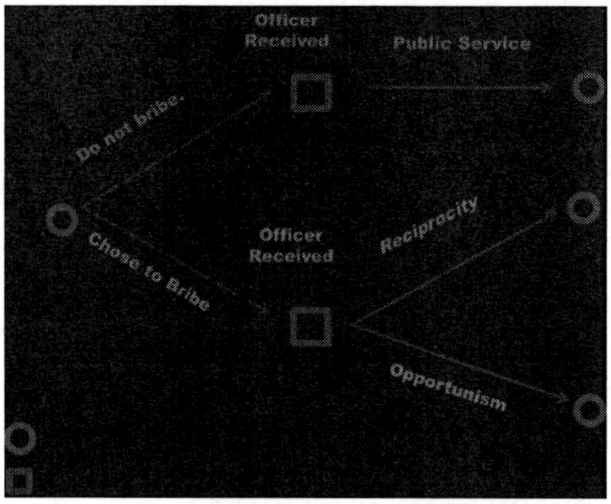

Figure 1. Game Tree

The game tree shows all possible outcomes for this experiment. The general rule of the experiment is the first player must pay mandatory public service fee and optional bribe then player two, whether accept or reject the bribe, must provide player one with the mandatory public service. The mandatory public service fee is 2$ and public service value is 3$. Besides that, everything the second player receives will be tripled.

In stage 1, the first player (briber) must decide whether to transfer a sum of money amounting to 2$ as administration fee and value of bribes (between 0 and 8$) to the second player. After that, the second player (government officer) must decide whether to accept or reject the transfer. If refused, the transfer did not take place and the multiplier simply tripled the value of 2 to 6. Value 6 will be divided equally, 3 points back to the first player, and 3 points remain stored second player. Stage 2 is similar to Stage 1, the only different is we induce random pairing for the subjects. Meaning the subjects will exchange partner in every round of this stage. The different pairing system denotes power. Fixed pairing is metaphor for long tenure where public officer is in charge in his position for so long hence obtains certain power in bribery matter. The random pairing denotes the absent of power where public officer is constantly rotated therefore has no ability to obtain power. In real world we often witness that bribery is most likely to happen when public officer has been around for so long and his presence is well known by potential bribers. The goal of this stage is to see how potential bribers react if the public officer is changed a lot. The random pairing is called "Random" in the analysis. Stage 3 is similar to Stage 1, the only different is we introduce instrument of penalty. All subjects will be audited randomly, if they get caught engaged in corruption they will lose all of the payoff in the current round as punishment.

We built these following hypotheses for our research:

H1: The power will positively influence the behaviour of bribery

H2: The penalties will reduce the impact of variable power, motivation, and behavioural culture of bribery against bribery

4 Analysis

We conducted t-test on the different value of average bribe and giveback from all subjects to see if there is difference among the treatments. The amount of bribe indicates the willingness to bribe from first player. Meanwhile the amount of giveback indicates the willingness to cooperate with briber from second player. Result of the t-test is showed in the table 1. below.

Table 1. T-test Result for Average Bribe and Giveback

Average Difference Test for Bribe					Average Difference Test for Giveback				
Treatment	Mean	Std. Dev	Diff	Significant	Treatment	Mean	Std. Dev	Diff	Significant
Paired Reg	2.625	2.94	0.975	YES	Paired Reg	4.150	4.789	1.350	NO
Random Reg	1.65	2.359			Random Reg	2.800	3.156		
Paired Reg	2.625	2.94	1.138	YES	Paired Reg	4.150	4.789	2.350	YES
Paired SD	1.488	0.263			Paired SD	1.800	2.839		
Paired Reg	2.625	2.94	0.775	NO	Paired Reg	4.150	4.789	1.700	YES
Random SD	1.85	2.537			Random SD	2.450	1.28		
Paired SD	1.488	2.349	-0.16	NO	Paired SD	1.800	2.839	-1.000	YES
Random Reg	1.65	2.359			Random Reg	1.650	2.359		
Random Reg	1.65	2.359	-0.2	NO	Random Reg	1.650	2.359	0.350	YES
Random SD	1.85	2.537			Random SD	2.450	1.28		
Paired SD	1.488	2.349	-0.36	NO	Paired SD	1.800	2.839	-0.650	NO
Random SD	1.85	2.537			Random SD	2.450	1.28		

According to the Table 1. above we can see that not all of the t-test showed significant result. In terms of the bribe value, only treatment "Paired Reg and Random Reg" and "Paired Reg and Paired SD" have significant difference. Significant difference between "Paired Reg and Random Reg" shows that instrument randomized pairing does affect behaviour of briber. This fact might indicate the importance of rotation among public officers. Significant difference between "Paired Reg and Paired SD" shows that instrument penalty does affect behaviour of briber. This fact might indicate the importance of penalty and regulation in bribery. Meanwhile in Table 1 we can also see that not all of t-test showed significant result. In terms of the giveback value, only treatment "Paired Reg and Paired SD", "Paired Reg and Random SD", "Paired SD and Random Reg", and "Random Reg and Random SD" have significant difference, when it comes to giveback, any treatment including SD or penalty shoes significant different. This might indicate that second players are more reluctant to bribery when introduced to penalty.

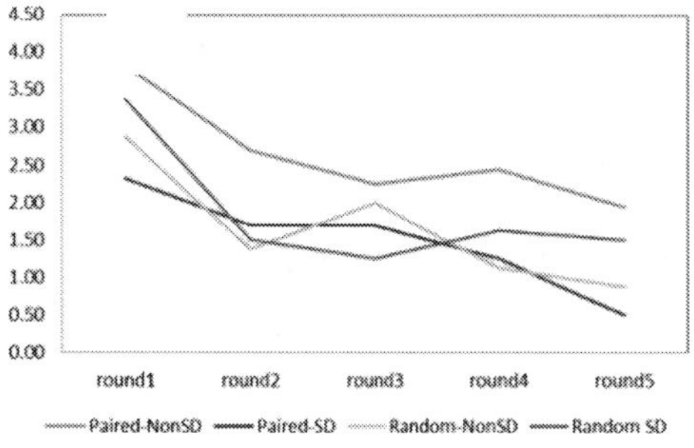

Fig. 2. The Amount of Average Bribe

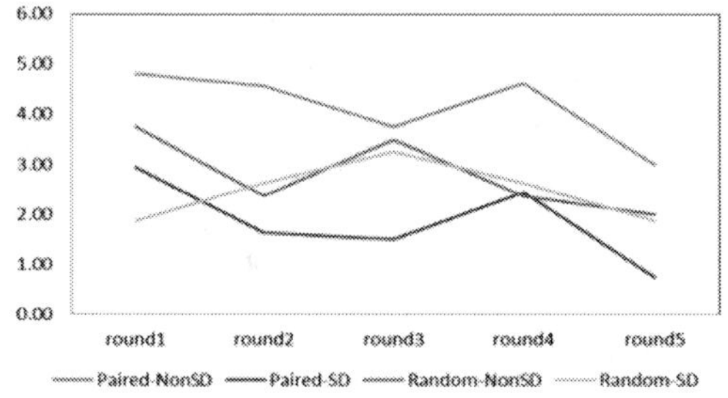

Fig. 3. The Amount of Average Giveback

In Figure 2. and Figure 3., we can see the average value of bribes and giveback that are chosen by the participants. Based on these data, we can see for the case of bribery, on average, the highest is Paired NonSD treatment. It is not surprising, because in the treatment Paired NonSD there is no risk of punishment. Beside that in the treatment Paired NonSD subjects did not change partners. By having fix partner, there might be trust between them. The same thing applies in the case giveback. On the other hand, the average bribe and giveback is lowest for the treatment of Paired SD. This suggests that the risk of punishment makes someone reluctant to commit bribery.

5 Conclusion

According to the analysis above we can draw several conclusions. Bribery is more likely to occur in treatment fixed pairing (Pairing), this indicates that long tenure in public service might attract bribery. Hence periodic rotation among public officials is important. The implementation of punishment or penalty that impoverish can reduce the amount of bribe that occurs from the side of the bribe giver and the bribe receiver (which do giveback). This is shown in the average value of treatment with sudden death (SD) which has an average Bribe and giveback lower than the treatment without SD. But based on the results of this research, procurement punishment is more effective if applied at the receiving end of bribery. This may be due, the public officials "have more to lose" than those who give bribes.

References

1. K. Abbink, B. Irlenbusch, and E. Renner. An Experimental Bribery Game. *Journal of Law, Economics, and Organization*, **18**(2), pp.428-454. (2002)
2. I. Ajzen. From Intentions to Actions: A Theory of Planned Behavior, in J. Kuhland and J. Beckman (eds.), *Action-control: From Cognitions to Behavior (Springer, Heidelberg)*, pp. 11–39. (1985)
3. L.R. Akers. Social Structure and Social Learning. *Los Angeles: Roxbury*. (1988)

4. Anti Corruption Clearing House (ACCH). Penanganan TPK berdasarkan jenis perkara. [online]. accessed at 19 April 2015 di: http://acch.kpk.go.id/en/statistik-penanganan-tindak-pidana-korupsi-berdasarkan-jenis-perkara. (2015)
5. Association of Certified Fraud Emaniners (ACFE). Report to The Nation on Occupational Fraud and Abuse. Diunduh 19 April 2015, http://www.acfe.com/rttn/docs/2014-report-to-nations.pdf. (2014)
6. J. R. Carter and M. D. Irons. Are economists different and if so, why? *Journal of Economic Perspectives*, Vol. 5, pp. 171-77. (1991)
7. J. R. Cohen, L. Holder-Webb, D.J. Sharp, and L.W. Pant. The Effects of Perceived Fairness on Opportunistic Behavior. *Contemporary Accounting Research* **24**(4). 1119–1138. (2007)
8. L. Dušek., A. Ortmann and L. Lízal. Understanding Corruption and Corruptibility through Experiments: A Primer. *Prague Economic Papers*, No. 2. (2004)
9. J. H. Evans, R. J. Hannan, R. Krishnan and D. V. Moser. Honesty in Managerial Reporting. *The Accounting Review* **76(**4), 537–559. (2001)
10. B. Frank and G. G. Schulze. Does economics make citizens corrupt? *Journal of Economic Behavior & Organization*. 43, 2000, Vol. 4, pp. 101-113. (2000)
11. J. C. Georgese & M. J. Harris. Why's My Boss Always Holding Me Down? A Meta-Analysis of Power Effects on Performing Evaluaton. *Personality and Social Psychological Review*, **2**, 184-195. (1998)
12. M. Giamattei. To Bribe or Not To Bribe… An Extension to Lambsdorff & Frank (2007) – Corruption Game. *Seminar Experimental Economics*. (2010)
13. W. H. Hegarty and H. P. Sims. Some determinants of unethical decision behavior: An experiment. *Journal of Applied Psychology*, Vol **63**(4), Aug 1978, 451-457.http://dx.doi.org/10.1037/0021-9010.63.4.451. (1978)
14. D. Keltner, D. Gruenfeld, and C. Anderson. Power, Approach, and Inhibition. *Psychological Review*, **110**, pp. 265–284. (2003)
15. D. Kusumasari. Perbedaan Antara Suap dengan Gratifikasi. [online]. http://www.hukumonline.com/klinik/detail/cl3369/perbedaan-antara-suap-dengan-accesed at 19 April 2015. (2015)
16. J. G. Lambsdorff and B. Frank. Corrupt Reciprocity – an experiment. *Passauer Diskussionpapiere*. 2007, V-51-07. (2007)
17. G. Marwell and R. Ames. Economists free ride, does anyone else? *Journal of Public Economics*, 15(3), pp.295-310. (1981)
18. W. L. Miller, B. G. Åse, and Y. K. Tatyana. A Culture of Corruption: Coping with Government in Post-Communist Europe. *Budapest: Central European University Press*. (2001)
19. D. S. Moskowitz. Cross-situational Generality and the Interpersonal Circumflex. *Journal of Personality and Social Psychology*, **66**, 921-933. (1994)
20. D. S. Moskowitz. Does Elevated Power Lead to Approach and Reduced Power to Inhibition? *Psychological Review*, **111**, pp. 808-811. (2004)
21. Public Company Accounting Oversight Board (PCAOB). Consideration of Fraud in a Financial Statement Audit. AU Section 316, www.pcaobus.org. (2005)
22. E. H. Sutherland, D. R. Cressey and D. F. Luckenbill. Principles of Criminology. **11th** Edition. *General Hall Inc, New York, NY*. (1993)
23. G. Schulze and B. Frank. Deterrence versus intrinsic motivation: Experimental evidence on the determinants of corruptibility. *Economics of Governance,* Vol. 4 (2): 143-60. (2003)
24. L. K. Trevino and S.A. Youngblood. Bad Apples in Bad Barrels: A Causal Analysis of Ethical Decision Making Behavior. Journal of Applied Psychology, 74:378-385. (1990)

The Implementation of E-procurement System: Indonesia Evidence

Aditya Pandu Wicaksono[1,*], *Dekar* Urumsah[1], and *Faishal* Asmui[1]

[1]Accounting Program, Faculty of Economics, Universitas Islam Indonesia, Yogyakarta

Abstract. Indonesia has implemented e-procurement system to struggle fraud in the procurement process. The purpose of this research is to examine e-procurement in a relationship for reducing fraud. This research uses quantitative approach by distributing questionnaires to 159 employees, 100 usable. The main finding suggests e-procurement system has a positive relationship and significant to fraud prevention that emerging expectation to eradicate fraud in the procurement process. The research gives a knowledge to the government to use e-procurement in all of procurement processes in order to prevent fraud.
Keywords: fraud, e-procurement, Indonesia.

1 Introduction

Information technology has changed rapidly institution's behavior in buying and selling the goods and services. In the digital era, the Indonesian government has adopted technology for many purposes including procurement process. E-procurement is an online system that can streamline the procurement process [1]. The potential benefits of e-procurement are greater transparency and improve accountability [2] and reducing corruption risks [3]. The Indonesian government in 2008 has created INAPROC (Indonesia Procurement), a national e-procurement system to procure and deliver goods and services electronically [4]. Then, Presidential Regulation No 54/2010 has been issued to regulate procurement processes are executed electronically. The purposes are to improve effectiveness, transparency, accountability, and the procurement process can be accessed by stakeholders.

However, the fact in Indonesian public sector, many cases of corruption have taken place in the procurement process. According to Indonesia Corruption Watch (ICW), in 2015, Indonesia suffered Rp. 3.107 trillion (US$ 230 million) loss caused corruption in procurement process [5]. The Corruption Eradicate Commission (KPK) reports corruption in procurement is the second largest type of corruption in Indonesia [6]. Indonesia government focuses on combating corruption in the procurement process. Presidential Instruction No 7 Year 2015 has issued to prevent corruption in all government levels.

* Corresponding author: aditya.pandu@uii.ac.id

Indonesia government wants to advance e-procurement system for better monitoring procurement process. He reminds that building a good, effective, and comprehensive system is important and help to be corruption prevention [7] .

Research has been conducted on Indonesian e-procurement. According to [8], the implementation of e-procurement was visible but not accountable. It gives an opportunity to conduct fraud such as corruption and bribery, because e-procurement system is not implemented effectively. Therefore, internal audit and internal control are necessary to complete the fraud prevention [9]. Internal audit is established to monitor the activities of organization and to control irregularity actions. Internal control is ensuring the procurement process is performed in compliance with statutes and regulations [10]. This paper is organized with the following structure. The following part describes the literature review and development of hypotheses. The next part contains research methodology, then explains the results and research discussions. The last part is the conclusion, implication, and future research suggestion.

2 Literature review and hypotheses development

2.1 Fraud and Prevention in Procurement

Fraud in public procurement is increasing especially in developing country including Indonesia. As stated by Transparency International Report, Huge sum of money from taxpayer's are spent by governments to procuring goods and services with so much money changing hands. Hence, government activities offer temptations and opportunity to fraud [11]. Indonesia is ranked as one of the most corrupt countries in South East Asia with corruption in public procurement as a serious problem. Procurement fraud is established by perverting awarding public contract and tenders. At bidding process, the bidders have used coercive power to win the process. The bidders can not submit bidding documents and be ignored by public official because of bidder's influence. According to [3], public official and bidders involve directly or indirectly to create a monopoly power in procurement process. According to KPK, ineffective budget and collusion emerge fraud in government procurement process [12].

In order to tackle fraud in procurement, Indonesia government concern to governing public procurement process which emphasize on transparency and accountability. Indonesia has created procurement online system namely e-procurement to prevent collusion and corruption in procurement process. E-procurement covers electronically bidding process which provides real-time information for the bidders to check their status, approval, or rejection reason. E-procurement system offers the clean procurement process which close opportunity to collusion and corruption [13]. To complete procurement fraud prevention, Indonesia government needs to increase surveillance at procurement process. Nowadays, Lack of surveillance in procurement process has been easily used by corrupt public official and outside parties.

2.2 E-procurement

Procurement is the acquisition of goods or services by an organization [1]. Prior to e-procurement is implemented, the procurement process is executed traditionally. The government and the vendors or suppliers can meet physically to deal for buying goods and services. This condition has a weakness that emerge fraud occurs in the government. The weakness can be used by vendors/suppliers to bribe the public official to select or appoint their company as a vendor that provide goods or services. It makes the procurement system is not transparent and accountable [3].

The reformation on procurement system has been conducted after many cases of procurement create fraud opportunities. The procurement process should be done transparently and can be monitored. E-procurement is a new way to procure the goods and services electronically. E-procurement is also described as a procurement automation process which improves competition among the suppliers or providers. The implementation of e-procurement can improve the effectiveness and efficiency over traditional process [4]. E-procurement is also implemented to reduce the opportunity of occurrence of fraud in government area [3].

As regulated by Presidential Regulation No 54-2010, there are three types of electronic procurement in Indonesia, they are *e-tendering*, *e-catalog*, and *e-purchasing*. *E-tendering* is process to select suppliers, conducted transparently and all of the suppliers have a chance to offer their goods and services via online. *E-catalog* is electronic information system which contains information about the product in detail. *E-purchasing* is process to purchase goods and services through *e-catalog* system. The procurement process is conducted through the National Procurement Agency (LKPP) and supervised by National Development Planning Board (Bappenas).

Research in procurement system finds that e-procurement system is the good way to struggle fraud. As explained by [3], government must create and implement e-procurement system to combat several fraud issues such as monopoly power of government and information asymmetry. Transparency, accountability, and trust are the main issues. People need to understand the cycle process how to procure goods and services in government area. Lack of transparency and accountability create more opportunities for public official and vendors to break the rule for corruption purposes. Through e-procurement system, people understand the process and oversee procurement process to be free from fraud. Specific for the bidders, they want a transparency in bidding process, fairness of competition, fixed-price contract. E-procurement system provides electronically bidding process known all the bidders which mean selecting the bidding winner is known all the bidders. Therefore, the public official and vendors can not perform collusion in procurement.

H_1:E-procurement system is positively associated with fraud prevention of procurement.

2.3 Internal control and internal audit

Internal control can be described as policies and procedures established to provide reasonable assurance that the specific entity objectives will be achieved [14]. Internal control is expected to ensure integrity, accountability, and transparency in the procurement process. Internal control is commonly regarded as a key deterrent of fraud [15]. The effectiveness of internal control is always evaluated to ensure there is no hole or weakness that can emerge fraud behavior [16]. Internal control should be suitable and compliance with regulations and policies to combat fraud. In contrast, internal control weakness provides an opportunity to commit fraud [14].

The Association of Certified Fraud Examiners (ACFE) in their report found that a lack of internal control leads to fraud activities [17]. In this context, the government is mandated to create an internal control that reduces the opportunity in the procurement process. Today, the procurement process in Indonesia should be executed electronically under LKPP's supervision. For the government, implementing e-procurement system help them to monitor and control the quality of the procurement process. For the suppliers, they can monitor bidding process online at anytime [4].

After the internal control works, Internal auditing should be held to ensure there is not an unusual behavior. The role of internal auditing is to review the effectiveness of internal control and to understand what the system works properly. Internal auditing has been used to help all government areas execute their responsibility effectively with gives the objective analysis, judgment, recommendation, and opinion associated with the works have been reviewed. Internal auditing helps the government to prevent, detect, and investigate fraud in the procurement process [18].

H_2: Internal control is positively associated with fraud prevention of procurement.
H_3: Internal auditing is positively associated with fraud prevention of procurement.

3 Research method

A quantitative apporach is conducted in this research. This research explores the potential of e-procurement system, internal control, and internal auditing to reduce corruption in the procurement process. A survey is used for collecting data using a questionnaire that distributed to participants. To select the participants, a convenience sample design is adopted. The participants are the employees of Electronic Procurement Agency (LPSE) and Procurement Services Unit (ULP) at six districts in West Java, Indonesia. A total of 159 paper-based questionnaires were distributed, 100 returned and usable. The collected data were analyzed by the smartPLS.

3.1 Measurement

This measurement scale in this research uses six-point Likert scale ranging from 1= strongly disagree to 6= strongly agree to measure all the items. E-procurement variable is measured by 3 item questions [19], while each of internal control [20], [21] and internal auditing [22], [23] are measured by 4 item questions. Fraud prevention is measured by 5 items [24].

4 Results and discussion

4.1 The measurement model

The research model is tested through two steps: first, the validity and reliability of the measurement model; second, the structural model for testing the hypotheses. Validity requires average variance extracted (AVE) to be greater than 0.5 [25]. Reliability requires internal composite reliability (ICR) to be greater than 0.7 [3]. The result of measurement model is shown in Table 1 indicates that validity and reliability is met, which means all the variables are different from each other and reliable to use in this research.

Table 1. AVE and ICR

	AVE	ICR
EP	0.782	0.915
IC	0.644	0.879
IA	0.696	0.919
FP	0.609	0.885

DOI: 10.1051/shsconf/20173410004

4.2 The structural model and test of hypotheses

The bootstrap method is used to test the hypotheses. Table 2 shows the results of path coefficient and t-statistic. The critical t-value for two-tailed test is 1.96 (significance level = 10 percent) [3]. The results show that H_1 and H_3 are supported, while H_2 is not supported. The model research explains 31% of the fraud prevention, while 69% is explained by other variables.

Table 2. Path coefficient and t-value

Hypotheses	Path	Path coefficient	t-value
H_1	EP ➔ FP	0.242	2.118
H_2	IC ➔ FP	0.053	0.627
H_3	IA ➔ FP	0.374	4.026

4.3 Discussion

This research finds the implementation of e-procurement system will prevent and reduce fraud or corruption in the procurement process. This result supports previous research by [3],[4]. However, internal control do not influence to fraud prevention of procurement. This finding is consistent with prior research by [14]. The research result also find that internal audit has a positive impact to fraud prevention. This result is in line with [18] study.

Looking at website infrastructure, the electronic procurement systems such as e-tendering, e-catalogue, and e-purchasing is ready at LKPP website (lkpp.go.id). These systems support all procurement process starts from tendering process until purchasing process conducted electronically. If these systems are consistently used, fraud in procurement can be decreased. The procurement process can be transparently established in bidding process such as selecting the winning bidder. However, classic problem in developing countries, infrastructure in information technology in government area has a gap between metropolitan and non-metropolitan city. Combating fraud through e-procurement system need a strong infrastructure. According to [29], procurement reform in Indonesia strong association with the geographical location and the wealth of a district. Therefore, many districts especially on non-metropolitan take partial implementation of e-procurement.

In addition, according to [4], decentralization policy has made procurement process less much transparent. Abusing power will slowly eliminate the transparency and accountability of e-procurement system. Essentially, e-procurement system should prevent and eradicate vertical and horizontal collusion. Vertical collusion is an official action that selects vendor because he or she has close relationship with the vendor. Public official abuses their power to appoint a vendor because their relationship with vendor. Other reason, public official may get personal benefit from vendor if vendor win the bidding process. Another collusion type namely horizontal collusion is committed among vendors by ignoring objective requirements and will be selected responsively to supply goods and services to government. The deal has been reached among vendors to decide who to be supplier in one government project.

According to [26], fraud still happens because of internal control weaknesses. Although the bidding processes are conducted electronically, the winning bidder has been selected in some cases for creating a benefit not only for individual but also organizational purposes.

The decentralization policy may not unsuitable for developing country such as Indonesia that has wide range of human resource capability. According to [4], In most of non-metropolitan area there is a problem at government's human resource that impede the implementation of e-procurement system in Indonesia. The civil servants have not ability to conduct procurement process through the technology. Indonesia government should aware that civil servant in government are getting "shock" because the changes of procurement process. To heal it, government creates a training program to improve civil servant's skill when faced with technology like e-procurement. In technology acceptance theory [28], training is an important influence to ensure an information technology can be operated by people. In this case, Bappenas with non-government institutions has held many training programs to introduce a new system and train civil servant how to use e-procurement system.

Although the government provide training program, the main challenge is the public official does not have intention or willingness to be a procurement official. It is reasonable if a civil servant rejects this position. First, to be procurement official, public official should pass a very complicated exam to achieve certification of procurement official. Hence, much public servant fail to pass the test and never be procurement official. It is will be very hard if e-procurement system applied without a competent user to operate it. Now, appointing procurement official does not consider their qualification and tend to be formality reason. Without a competent procurement official, procurement process risks to be abused and fraud committed. Second, procurement official position contains very high risk and pressure with low incentives. This is the main reason why civil servant does not want to be. Civil servant does not want to involve in fraud behavior. The temptation comes when vendors have no doubt to give money or privilege in order to influence procurement official to select their product. Then, if government official abusing their power to appoint a vendor as bidding winner, procurement official does not have choice and reason to reject it. It is why procurement official is not comfort position for civil servant. Actually, Indonesia government require a lot of procurement official to carry out e-procurement in the right track and fraud in procurement going down. Although e-procurement applied, with high risk and pressure for procurement official may difficult to prevent fraud in procurement process.

This research finds positively significant between internal auditing and fraud prevention. This finding is supported by the Regulation of Minister of Home Affairs No. 71 Year 2015 that should be applied to enhance the surveillance all of the province and local governments. The inspectorate has been mandated to supervise internal activities and speed up the good government governance. The Inspectorate's role should be maximized to monitor and control procurement activities in order to prevent the occurrence of fraud. The Inspectorate will complement the role of Bappenas for monitoring procurement process. In other words, the procurement process need to be audited by the internal body. In fact, internal audit does not prevent fraud procurement yet. It may be caused that internal audit focuses on big amount of procurement transaction.

5 Conclusion

The implementation of e-procurement process empirically will prevent and reduce the procurement fraud. But, the implementation of e-procurement in government areas is under questions. The fact shows the procurement fraud is still high, seems like this system is unemployed. Decentralization policy which to make a transparency procurement process is not good choice. It contains a big risk because of additional power of official province or local government to procure their own needs. Then, the decentralization emerges the

DOI: 10.1051/shsconf/20173410004

flexible interpretations among government level which makes e-procurement implementation is not effective. Central government need to rethinking its policy to decrease fraud with decentralization. Indonesia may need to bring back centralized procurement process with e-procurement system to complete the process. It is better than give the freedom to province and local government to anticipate power abuse.

Human resource in government area do not support to execute e-procurement system. It is not easy to appoint a competent user for this system. Higher risk, complicated certification exam, and uncomfort position make the procurement official position not interesting. It is becoming a challenge to get permanent officials in dealing with procurement [4]. Then, the most important is to increase integrity of procurement official to execute e-procurement process clear from fraud.

Internal control and internal auditing need to conduct in order to complete the prevention of fraud. The Minister of Home Affairs aware the importance of monitoring process in the province and local governments. Although applied in 2016, Regulation of Minister of Home Affairs No. 71 Year 2015 is the one step ahead to enhance monitoring and evaluating procurement process by the government.

5.1 Implication and future research suggestion

The result of this study has practical implications for the governments. E-procurement system should be implemented in order to prevent the occurrence of fraud. But, the government needs to ensure that the problems associated with implementation should be solved first such as human resources and infrastructure. One urgent point of the central government is to create legal regulation and official system to prevent multiple interpretations among the province and or local government. This research also suggests to enhancing monitoring process through internal control and auditing. This research future suggestion recommends the future research to investigate the acceptance of technology at province and or local Indonesian government. This is important to measure readiness of province and or local government to use technology like e-procurement. In Indonesia phenomena, the adoption of technology strong associated with geographical area and district characteristic.

References

[1] S. Rotchanakitumnuai, "The governance evidence of e-government procurement," *Transform. Gov. People, Process Policy*, vol. 7, no. 3, pp. 309–321, 2013.

[2] J. C. Bertot, P. T. Jaeger, and J. M. Grimes, "Using ICTs to create a culture of transparency: E-government and social media as openness and anti-corruption tools for societies," *Gov. Inf. Q.*, vol. 27, no. 3, pp. 264–271, 2010.

[3] A. Neupane, J. Soar, K. Vaidya, and J. Yong, "Willingness to adopt e-procurement to reduce corruption," *Transform. Gov. People, Process Policy*, vol. 8, no. 2, pp. 283–308, 2014.

[4] A. Nurmandi and S. Kim, "Making e-procurement work in a decentralized procurement system: A comparison of three Indonesian cities," *Int. J. Public Sect. Manag.*, vol. 28, no. 3, pp. 198–220, 2015.

[5] Republika, "Kerugian Negara Akibat Korupsi Menurun pada 2015," 2015. [Online]. Available: http://nasional.republika.co.id/berita/nasional/umum/16/02/24/o321yb383-kerugian-negara-akibat-korupsi-menurun-pada-2015.

DOI: 10.1051/shsconf/20173410004

[6] P. K. Komisi, "Ivestigasi korupsi berdasarkan jenis perkara," 2016. [Online]. Available: http://acch.kpk.go.id/berdasarkan-jenis-perkara?inheritRedirect=true.
[7] T. J. Post, "Jokowi Advances E-Governance System to Prevent Corruption," 2015.
[8] Kodar, "Pelelangan Barang dan Jasa Elektronik di Kota Yogyakarta (e-Procurement in Yogyakarta city: accountability in procurement," *J. Stud. Pemerintah.*, vol. 1, no. 2, pp. 34–45, 2010.
[9] J. S. Zanzig and D. L. Flesher, "Internal Auditors Speak Out on Controlling Employee Fraud," *Res. Prof. Responsib. Ethics Account.*, vol. 15, pp. 225–250, 2015.
[10] J. M. Rendon and R. G. Rendon, "Procurement fraud in the US Departmen of Defense: Implications for contracting process and internal controls," *Manag. Audit. J.*, vol. 31, no. 6/7, pp. 748–767, 2016.
[11] Transparency International, "Curbing Corruption in Public Procurement: A Practical Guide," 2014.
[12] Kompas, "Negara Rugi Hampir Rp 1 Triliun dari Korupsi Pengadaan Barang dan Jasa," 2016.
[13] N. Oye, "Reducing Corruption in African Developing Countries: The Relevance of e-governance," *Greener J. Soc. Sci.*, vol. 3, no. 1, pp. 6–13, 2013.
[14] K. M. Zakaria and A. S. Azlin, "Internal controls and fraud - Empirical evidence from oil & gas company," *J. Financ. Crime*, vol. 23, no. 4, 2016.
[15] K. Rae and N. Subramaniam, "Quality of internal control procedures Antecedents and moderating effect on organisational justice and employee fraud," *Manag. Audit. J.*, vol. 23, no. 2, pp. 104–124, 2008.
[16] Y. Tong, M. Wang, and F. Xu, "Internal control , related party transactions and corporate value of enterprises directly controlled by Chinese central government," *J. Chinese Manag.*, vol. 1, no. 1, pp. 1–14, 2014.
[17] Association of Certified Fraud Examiners, "Report To the Nations on Occupational Fraud and Abuse," 2016.
[18] F. H. Fadzil, H. Haron, and M. Jantan, "Internal auditing practices and internal control system," *Manag. Audit. J.*, vol. 20, no. 8, pp. 844–866, 2005.
[19] M. Jasin, A. R. Zulaiha, E. J. Rachman, and N. Ariati, *Mencegah Korupsi Melalui e-Procurement: Meninjau Keberhasilan Pelaksanaan e-Procurement di Pemerintahan Kota Surabaya*. Jakarta: Komisi Pemberantasan Korupsi, 2007.
[20] O. N. Joseph, O. Albert, and J. Byaruhanga, "Effect of Internal Control on Fraud Detection and Prevention in District Treasuries of Kakamega County," *Int. J. Bus. Manag. Invent.*, vol. 4, no. 1, pp. 47–57, 2015.
[21] C. of S. O. of T. C. (COSO), *Internal Control Integrated Framework*. New York: AICPA Publication, 1992.
[22] A. H. Mohamud, "Internal Auditing Practices and Internal Control System in Somali Remittance Firms," *Int. J. Bus. Soc. Sci.*, vol. 4, no. 4, pp. 165–172, 2013.
[23] O. Monisola, "Effect of Internal Audit on Prevention of Frauds, Errors And Irregularities In Corporate Organisation," *Res. J. Financ. Account.*, vol. 4, no. 19, pp. 103–107, 2013.
[24] Amrizal, "Pencegahan dan Pendeteksian Kecurangan oleh internal Auditor," Jakarta, 2004.
[25] C. Fornell and D. Larcker, "Evaluating Structural Equation Models with Unobservable Variables and Measurement Error," *Amerrican Mark. Assoc.*, vol. 18, no. 1, pp. 39–50, 1981.
[26] Kompas, "E-Procurement Belum Menjamin Pengadaan Barang dan Jasa Bebas Korupsi (E-Procurement Have Not Guarantee Procurement Free From Corruption)," Jakarta, 2016.

[27] T. I.-U. and C. for I. P. Enterprise, "APEC procurement transparancy in Indonesia," Jakarta, 2011.
[28] V. Venkatesh, "Technology Acceptance Model 3 and a Research Agenda on Interventions," *Decis. Sci.*, vol. 39, no. 2, pp. 273–315, 2008.
[29] A. Sacks, E. Rahman, J. Turkewitz, M. Buehler, and I. Saleh, "The Dynamics of Centralized Procurement Reform in a Decentralized State: Evidence and Lessons from Indonesia," 2014.

DOI: 10.1051/shsconf/20173411001

The Relationship between Tax Rate, Penalty Rate, Tax Fairness and Excise Duty Non-compliance.

*Perabavathi Sinnasamy** and *Zainol Bidin*

Tunku Puteri Intan Safinaz School of Accoutancy, Universiti Utara Malaysia, Malaysia

Abstract. The rise of indirect tax non-compliance by taxpayers became the main concern of most of the tax authorities around the globe. In Malaysia, non complaince such as smuggling and illegal trade activities by importers involving cigarettes, liquor and imported vehicles bound under Excise Act 1976 have caused revenue losses in monetary and non-monetary aspects. Therefore, the objective of this study is to examine the relationship of tax rate, penalty rate and tax fairness of excise duty non-compliance. This study uses the Deterrence Theory as a basis theory to investigate the phenomenon of excise duty non complaince. A total of 500 excise duty offenders throughout Malaysia responded to the survey. The model was empirically tested by using Partial Least Squares (PLS) with disproportionate stratified random sampling technique. The results indicated that the perception of tax rate and penalty rate are positively related while tax fairness is negatively related to excise duty non-compliance among importers.

1 Introduction

Royal Malaysian Customs Department (RMCD) as indirect tax authority administering exports and imports of commodities and collect indirect taxes. The taxes collected contribute as a source of revenue to the federal government. However, the tax non-compliance occurs through unscrupulous importers by falsifying customs documents, under-declaration and hiding goods. Malaysia like any other countries faces a serious problem of Customs frauds yearly. This Customs frauds involved smuggling and illegal trade activities carried out by importers to avoid paying excise duty to RMCD.The taxpayers' non-compliance is the main issue encountered by RMCD. This agency looses revenue caused by the taxpayers' failure to pay excise duty. RMCD's statistics indicate that the additional excise duty detected from illegal trade activities and smuggling offences totalled almost more than RM1,800 million from the year 2010 to 2014 from the breakdown of 14,661 cases for cigarettes, 2,592 for imported vehicles and 4,662 for liquor [1]. These statistics shows that the problem of non-compliance need to be highlighted.

The tax non-compliance issue need to be investigated, in Malaysian context as developing country. Therefore, this study contributes several merits. First, it uses actual smuggling and illegal trade offenders (importers) to measure the non-compliance behaviour in Malaysia. Whereas, most of the researches are done studies on tax non-compliance using

* Corresponding author: mbasrpraba@yahoo.com

annual reports which do not reflect the real situation. Secondly, this study contributes to the body of knowledge in indirect tax non-compliance in Malaysia. Finally this study tested the use of deterrence theory in Malaysian indirect tax environment. Furthermore, the understanding on excise duty non-compliance is still incomplete. This study will provide a fundamental awareness to RMCD as well as other researchers around the globe based on smuggling and illegal trade activities prevailing as a universal phenomenon.

2 Literature review and hypotheses development

2.1 Deterrence Theory

The Deterrence Theory by Becker [2] emphasized about economics of crime. Taxpayers are assumed to act under uncertainty position either to maximize their tax savings or face the punishment by comparing the cost incurred. This will be determined the compliance behaviour by weighing the chance of being detected and penalty acquisition. Based on explanation, this theory highlighted that by increasing tax rate and penalty rate will deter the non-compliance behaviour.

2.2 Tax Non-compliance

Tax non-compliance is defined as an illegal conduct of importers as taxpayers either done intentionally or unintentionally to evade taxes from tax authority. In Customs context tax non-compliance or tax frauds are mis-classification of tariff code, producing double invoices, falsifying country of origin [3].

2.3 Tax rate

Tax rate in Customs context is defined as an effect of the dollar paid for the import / export based on the Customs Harmonized System-nomenclature classification. In Customs perspective Fisman and Wei [4] found the higher the tax rates of the products the greater the revenue loss. Whereas In India, the effectiveness of tariff policies of customs duties evasion in 1990's tax reform examined by Mishra, Subramaniam and Topalova [5]. They found a positive effect between import tax evasion and Customs tariffs. Kubo and Lwin [6] investigated on smuggling offences related to import duties in Myanmar and identified increasing import duties caused smuggling. Therefore, the hypothesis developed for this study is:

H_1 : Tax rate is positively related to excise duty non-compliance.

2.4 Penalty Rate

Penalty is defined as a deterrent measurement used by the tax authority on the taxpayers' [7]. The first systematic non-compliance behavior model originated by Becker [2]. Becker [2] assumed evaders will weigh the rewards and cost if they perform criminal behaviour by considering the severity of penalties and the risk of detection. Virmani [8] indicated that tax evasion and penalty rate have positive relationship, the higher the penalty rate, will induced people to behave dishonestly. Hence, the hypothesis developed is :

DOI: 10.1051/shsconf/20173411001

H_2 : Penalty rate is positively related to excise duty non-compliance.

2.5 Tax Fairness

Tax fairness in tax compliance can be defined in two different dimensions. The first dimension is on the benefit received with the dollars paid as tax. While the other one is on equity and the taxpayers' burden when referring to other individual in tax matters [9]. Azmi and Perumal [10] undertaken a study to recognized fairness perception within Malaysians taxpayers. The result showed perceived tax fairness positively influence tax compliance behaviour. The related hypothesis developed is as stated below:

H_3: Tax fairness is negatively related to excise duty non-compliance.

3. Research theoretical framework

This conceptual framework expanded from deterrence theory by considering non-compliance determinants such as tax rate, penalty rate and tax fairness.

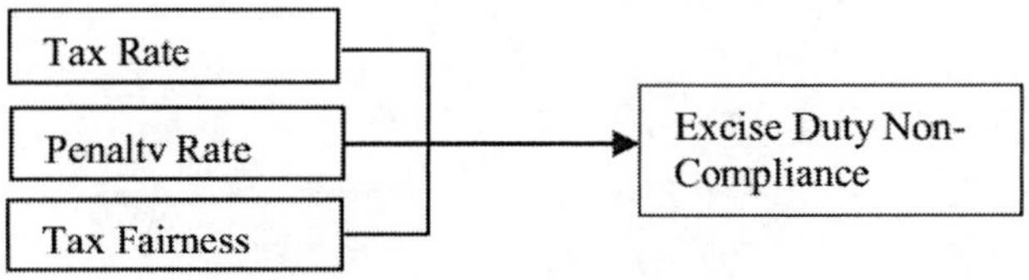

4. Research methodology

This study attempted to test the hypotheses developed in excise duty non-compliance. The population comprises of Managers who were committed in the cigarettes, liquor and imported vehicles offences throughout Malaysia. The survey was administered to the Manager as they are directly involved in importation declaration decision making. The sample obtained from RMCD's database for the year 2014.The questionnaires were distributed personally to 600 companies. After 4 months, 500 useable questionnaires returned. Therefore representing the response rate of 83%. The participating companies' covers cigarette (223), followed by liquor (166) and imported vehicles (111).The items for all the constructs adapted from previous literatures to fit the study's objectives. All the main items were measured based on 7-point Likert Scale ranging from strongly disagree (1) and strongly agree (7). This study's survey items for tax rate, penalty rate and tax fairness adapted from previous literatures. The tax rate items was self employed. While penalty rate and tax fairness adapted from Devos [11] and Gilligan and Richardson [12] respectively.

5. Findings

SmartPLS version 3.2.4 which is variance based structural equation modelling (SEM) was employed to analyze research model based on the two-stage analytical procedures consists of measurement and structural model. Measurement model (outer model) tested for the validity and reliability of the measures. Followed by the structural model (inner model) evaluated by testing the hypothesized relationship. A bootstrapping method was used to test the path coefficient significant and the loadings [13].

DOI: 10.1051/shsconf/20173411001

6. Assessment of measurement model

The measurement model evaluates the relationships among the indicators and the constructs. Two types of validity were examined to assess the measurement model, the first is convergent validity followed by discriminant validity. Convergent validity is established by examining the indicators loadings, average variance extracted (AVE) and composite reliability. As shown in Table 1, indicator loadings for all items surpassed the threshold value of 0.50 as recommended by Hair, Hult, Ringle and Sarstedt [13]. However, the loadings for item tax rate item 1, penalty rate item 4, tax fairness item 3, 6 and 7 and excise duty non-compliance item 2,3, 4 and 6 were deleted due to low loading. The AVE values are higher than cut-off point of 0.50. While, composite reliability greater than value 0.70 as suggested by Hair et.al. [13].

Table 1 Convergent validity

Construct	Items	Loadings	AVE	CR
Excise duty non-compliance (EDN)	EDN1	0.848	0.631	0.894
	EDN5	0.723		
	EDN7	0.877		
	EDN8	0.816		
	EDN9	0.691		
Penalty rate (PR)	PR1	0.680	0.577	0.844
	PR2	0.735		
	PR3	0.738		
	PR5	0.871		
Tax fairness (TF)	TF1	0.853	0.752	0.924
	TF2	0.893		
	TF4	0.869		
	TF5	0.852		
Tax rate (TR)	TR2	0.800	0.675	0.893
	TR3	0.878		
	TR4	0.807		
	TR5	0.799		

Note: CR= Composite reliability; AVE average variance extracted

6.1. Discriminant validity

After confirming the convergent validity, the discriminant validity was assessed by using Fornell and Larcker [14] criterion. Discriminant validity is the degree which each construct differentiate from other construct in the research model. The square root of each construct's AVE should exceeds than all of the correlations between the construct and other constructs as shown Table 2 [15].

Table 2 Discriminant Validity

Construct	EDN	PR	TF	TR
Excise Duty Non-compliance	**0.794**			
Penalty Rate	0.347	**0.759**		
Tax Fairness	0.606	0.286	**0.867**	
Tax Rate	0.256	0.144	0.324	**0.822**

Diagonal values (bolded) are the square root of the average variance extracted while the other entries shows correlations.

7. Assessment of structural model

The structural model assessment involves evaluating R^2, path coefficient (β values) and t-values [13]. In this study, SmartPLS algorithm function is used to obtain the coefficient of determination (R^2) value. The result indicates that R^2 value for excise duty non-compliance is .403 suggesting that 40.3% of the variance in excise duty non-compliance explained by tax rate, penalty rate and tax fairness. A bootstrapping technique with resampling of 1000 for 500 cases was applied to calculate the path coefficient and t- statistics values for hypothesized relationships. Table 3 shows the structural model's analysis results. The results revealed that tax rate (β=0.056, $p<0.05$), penalty rate (β=0.186, $p<0.01$) was positively related while tax fairness (β=0.535,$p<0.01$) negatively related to excise duty non-compliance which supported H_1, H_2 and H_3 hypotheseses.

Table 3 Path Coefficient

Relationship	Path coefficient (β value)	t-value	Decision
Tax rate -> Excise duty non-compliance	0.056	1.664**	Supported
Penalty rate -> Excise duty non-compliance	0.186	4.322*	Supported
Tax fairness -> Excise duty non-compliance	0.535	13.974*	Supported

$p<0.01$* ; $p<0.05$**

8. Discussion

The present study's objective is to examine the relationship between tax rate, penalty rate and tax fairness with excise duty non-compliance. The findings revealed that tax rate is found to have positive relationship with excise duty non-compliance. This result is in line with the findings of previous literatures [4,5].However, most of the theories does not provide clear prediction that the higher tax rate induces tax non-compliance. Therefore, this study sheds the light on excise duty as indirect tax claiming that the higher tax motivates the greater incentives to engage in fraud. Further, the higher tax rate will make the importers (tax payers) weigh between the risk aversion and fines for the successful tax evasion. Therefore, RMCD have to emphasize on the consequences of their tax reform especially on cigarettes, liquor and imported vehicles as the demand increases yearly. The solution for overcoming tax non-compliance is not by increasing excise duty (tax rate) as confirmed in this study.

This result showed that penalty rate is positively related to excise duty non-compliance. The finding is in line with previous studies by Virmani [8] and Alm, Kirchler and Muehlbacher [16]. The penalty imposed by RMCD is quite severe. For instance under Section 74 of Excise Act 1976 the penalty on evasion of excise duty is not less than 10 times the amount of the excise duty of RM50,000 for illegal importation. While, for illegal manufacture of dutiable goods is RM10, 000. However, the importers still commit offences under excise duty. Therefore, in order to deter non-compliance, taxpayers should be able to understand the tax rules and regulations provisions. The excise tariff complexity under self-assessment in excise duty declarations has caused less confidence among importers to deter non-compliance.

The statistical analysis of this study revealed that tax fairness was found to have negative relationship on excise duty non-compliance parallel to the findings of prior literatures [10]. Therefore, it is concluded the contributing factor for this phenomenon is based on importers perception of excise duty paid and the public goods received in return. Importers judged the procedural fairness on excise rules and regulations provisions applied to them and the Customs Officers treatments in the declaration processes. Fair treatment to the taxpayers should not be like 'cops and robbers' that always find faults on taxpayers but by showing trust in them and not only deterrent behavior can solve the tax non-compliance phenomenon.

This study attempted to explore the tax compliance determinants in indirect tax context. Furthermore, based on our understanding this is the first study in Malaysia that highlighted excise duty non-compliance focused on cigarettes, liquor and imported vehicles among importers as taxpayers. Prior studies on indirect tax compliance and non-compliance vastly concentrated on goods and services or value added tax, import tax and sales tax. The application of theory mainly focused on Theory of Planned Behaviour and Theory of Reasoned Action as theoretical framework to predict the taxpayers' behaviour on tax non-compliance. Whereas, this study used Deterrence Theory in determining the importers non-compliance behaviour.

Moreover, this study's findings provides implications to RMCD as tax authority and policy maker. This revealed importers are looking forward for more friendly treatment rather than penalising them. Therefore RMCD should focus more on educating them on excise duty rules and regulations by conducting training to the importers. Most of the training provided emphasized on the tax agents rather than importers.

9. Conclusions, limitations and future research

This study has its strength and limitations as in other studies. There is a shortage of excise duty non-compliance empirical studies by practitioner experiences and observations. Specifically, on using actual excise duty offenders in tax non-compliance study. Therefore, this study has its own merits utilized the actual offenders those breached the tax laws, to investigate tax non-compliance behaviour among importers in Malaysia. However, like any other tax non-compliance studies this study has limitations mainly on using survey questionnaires method. Some of the sensitive and embarrassing issues are unable to be highlighted in structured questionnaire. Therefore as a suggestion, future research should be conducted by applying mix method approach to attain in depth understanding on tax non-compliance phenomenon. Further, the non-compliance behaviour is required to be extended and confirmed by using different methodological approaches. In line with this, several variables could be introduced such as peer influence, tax knowledge and corruption by considering probability of detection as a moderator.

DOI: 10.1051/shsconf/20173411001

References

1. Royal Malaysian Customs Department, *Annual report 2012-2014.* Kuala Lumpur Percetakan Nasional Malaysia Berhad.
2. Becker, G. (1968). Crime and punishment : An economic approach. *The Journal of Political Economy, 76*(2), 169-217.
3. Thanasegaran, H., & Shanmugam, B. (2007). International trade-based money laundering: the Malaysian perspective. *Journal of Money Laundering Control, 10*(4), 429-437.
4. Fisman, R., & Wei, S.-J. (2004). Tax rates and tax evasion: Evidence from" missing imports" in China. *Journal of Political Economy, 112*(2), 471.
5. Mishra, P., Subramanian, A., & Topalova, P. (2008). Tariffs, enforcement, and customs evasion: Evidence from India. *Journal of public Economics, 92*(10), 1907-1925.
6. Kubo, K., Lwin. (2010) Smuggling and import duties in Myanmar.
7. Devos, K. (2013). Do penalties and enforcement measures make taxpayers more compliant? The view of Australian tax evaders. *Far East Journal of Psychology and Business, 12*(1), 1-9.
8. Virmani, A. (1989). Indirect tax evasion and production efficiency. *Journal of public Economics, 39*(2), 223-237.
9. Jackson, B. R., & Milliron, V. C. (2002). Tax Compliance Research. *Taxation: Critical Perspectives on the World Economy, 3*, 56.
10. Azmi, A. A. C., & Perumal, K. A. (2008). Tax fairness dimensions in an Asian context: The Malaysian perspective. *International Review of Business Research Papers, 4*(5), 11-19.
11. Devos, K. (2005). Attitudes of tertiary students on tax evasion and penalties for tax evasion-A pilot study and demographic analysis. *Journal of tax research, 3*(2), 222-273.
12. Gilligan, G., & Richardson, G. (2005). Perceptions of tax fairness and tax compliance in Australia and Hong Kong-a preliminary study. *Journal of financial crime, 12*(4), 331-343.
13. Hair Jr, J. F., Hult, G. T. M., Ringle, C., & Sarstedt, M. (2016). *A primer on partial least squares structural equation modeling (PLS-SEM)*: Sage Publications.
14. Fornell, C., & Larcker, D. F. (1981). Evaluating structural equation models with unobservable variables and measurement error. *Journal of Marketing Research*, 39-50.
15. Chin, W. W. (2010). How to write up and report PLS analyses *Handbook of partial least squares* (pp. 655-690): Springer.
16. Alm, J., Kirchler, E., & Muehlbacher, S. (2012). Combining psychology and economicin the analysis of compliance: From enforcement to cooperation. *Economic Analysis and Policy, 42*(2), 133-151.

Post Implementation of Goods and Services Tax (GST) in Malaysia: Tax Agents' Perceptions on Clients' Compliance Behaviour and Tax Agents' Roles in Promoting Compliance

Izlawanie Muhammad[1,*]

[1]Faculty of Economics and Muamalat, Universiti Sains Islam Malaysia, 71800 Nilai, Malaysia

Abstract. The Malaysian government introduced the Goods and Services Tax (GST) starting from 1 April 2015 to enhance the revenue collections and mitigate the transfer pricing manipulation. Tax agents play a significant role to help businesses to comply with GST law and regulations. After one year of GST implementation, it is vital to understand tax agents' perceptions on clients' compliance behaviour and tax agents' roles in influencing compliance. A total of 30 registered tax agents completed a survey questionnaire. The analysis shows that tax agents devote their time to provide advice to their clients on meeting their GST requirements, and recording and reporting of GST transactions. Tax agents assert that clients pass on their GST responsibilities to tax agents to some extent. Tax agents also perceive that clients' compliance level is low because clients occasionally submit GST03 after the deadline, compromise the accuracy of GST03 in order to get it done on time and intentionally make errors in their records. In terms of tax agents' role in promoting compliance, the tax agents strongly agree that it is important for them to act as trusted advisors for their clients. After one year of GST implementation, this is the first study that explores tax agents' perceptions on clients' compliance and tax agents' roles in promoting compliance. The findings benefit the Royal Malaysian Customs Department (RMCD) in assisting tax agents and the public for future compliance. Similar study should be adopted by countries that have recently implemented GST (for example, India) and it should be conducted to other GST players (i.e. taxpayers and RMCD officers) on annual basis to analyse the behavioural trends and identify weaknesses in GST administration.

1 Introduction

The Malaysian GST replaced the Sales and Services Tax which was in practiced for more than 40 years. The Prime Minister announced the replacement in 2013 and GST is not just a tax change but it will benefit the economy as a whole and have far-reaching impact on businesses. The GST Act was gazetted in 2014 for the public to understand the basic principles of GST and the businesses to make intense preparation in terms of accounting

* Corresponding author: izlawanie@usim.edu.my

system, records and training for staff. The Government appointed the Royal Malaysian Customs Department (RMCD) as the authoritative body of GST. To date, the Government has issued several GST Advance Rulings and GST Orders to improve and support information in the GST Act. On the same remarks, the RMCD has been issuing numerous GST Handbooks and Specific Guideline to reduce information gap and assist the businesses for compliance.

In general, GST is imposed on most goods and services purchased in Malaysia. Similar with many other countries like Australia and Singapore, goods and services are divided into taxable supplies (i.e. standard rate supply and zero-rated supply) and non-taxable supplies (i.e. exempt supply and out-of-scope supply). Several individuals and organisations receive special relief from the government; they are not liable to pay GST and businesses should not impose GST to them. A GST registered person (i.e. individuals and corporations) are responsible to understand the GST rules and regulations, impose and collect GST, keep proper records and documents, and remit tax return (i.e. GST03) to the RMCD before the deadline. The GST03 contains summary of taxable and non-taxable supplies of companies and net tax amount to be paid or claimed. Complying with the GST rules and regulations is challenging for not only the businesses but also tax agents.

2 Issues and problems of GST in Malaysia

The most significant challenge for the Malaysian government was to establish the GST Act and continuously issue supplement rulings that are suitable with the country's economics and business operations of various industries. To date, there are various amendments made to the Act, Advance Rulings and Orders. The amendments are unavoidable for a country that recently implemented GST. Comments, recommendations and suggestions from the industry and public are valuable to ensure that the GST rules and regulations are relevant and applicable. For the businesses, the main challenges are understanding the rules and compliance costs [1]. Tax law is complex, open to different interpretations and often changing. Each taxpayer has unique business operation that they may have interpreted the tax law differently. Tax officers also may have different views on taxpayers' motives and their enforcement regulatory behaviour is influenced by various factors, for example, managerial control and experience [2, 3]. In terms of compliance costs, Ramli et al. [4] found that the estimated tax compliance costs for a Small and Medium Enterprise is RM34,742 which does not include other internal costs for tax calculation, tax planning and tax appeal. The impact of higher compliance costs is increase in prices of goods and services. Although the government assures that the price of goods and services will decrease with the implementation of GST, however, the impact could not be seen yet.

3 Tax agents' roles in assisting compliance

Tax agent is a professional who assists taxpayer comply with tax law and represent taxpayers in their dealings with the revenue authority [5]. Tax agent is necessary when complexity of compliance exists. Devos [6] found a significant relationship between engaging tax agents and taxpayers' compliance behaviour. In assisting taxpayers for compliance, tax agents served three key roles: providing tax advice, correct tax return preparation and lodgement, and risk management for tax minimization [7]. However, Klepper and Nagin [8] assert that tax agents also can contribute to non-compliance by helping taxpayers take advantage of legal ambiguity. The mixed roles of tax agents on compliance could be influenced by tax agents' ethical behaviour, perceptions on government's support and legitimacy of the tax system. The objectives of this study are twofold. First is to explore tax agents' perceptions on taxpayers' compliance behaviour. Second is to explore tax agents' roles in promoting compliance to taxpayers.

4 Research Methodology

To achieve the research objectives, a survey instrument was developed based on studies by the Australian Tax Office (ATO). The ATO conducts a similar study annually; they started the study in 2011 on three groups of respondent which are tax agents, micro or small and medium enterprise and Business Activity Statements (BAS) agents. The ATO's objectives are to inform, support and monitor the effectiveness of GST [9].

In this study, the survey instrument contains 9 sections – Section A until Section H. Section A, G and H contained information about the respondents. Section B and C contain information about the tax agent's clients and GST services provided by the tax agent respectively. Section D of the questionnaire requires the tax agents to evaluate clients' behaviours in complying with GST. Questions related to tax agents' attitudes and beliefs on GST are asked in Section E. And, Section F contains about the information and support that the tax agents use to help their clients with GST requirements.

4.1 Data Collection

Data on RMCD's website shows that as of 19 December 2014, there were 1,002 GST agents approved by the Ministry of Finance. The tax agents range from individual bookkeepers who are working from home to larger businesses that offer management and accounting services. The author strongly believe that the number of GST agents has increased significantly in 2015 because several accounting bodies in Malaysia like the Malaysian Institute of Accountants and Malaysian Association of Tax Accountants are consistently organise GST courses for tax agents. Participants who passed the exam after the 6-day course are qualified to become a GST agent.

The data collection was conducted in May 2016 which was one year after the implementation of GST. The ATO used different approach for the data collection; telephonists were appointed to call tax agents and guided them to answer the survey questions. Instead, the researcher of this study appointed research assistants to collect data. The research assistants contacted GST agents in two states, which are Wilayah Persekutuan Kuala Lumpur and Selangor because the list of GST agents in the RMCD's website shows that the majority of GST agents are located in the two states. Then, the research assistants set an appointment with the tax agents for a meeting and they guided the tax agents to answer the survey questions. This approach was adopted to avoid low response rate. The data collection process stopped after the research assistants managed to get 30 respondents due to time constraint and cost limitation.

5 Results

5.1 Demographic profiles

All the 30 respondents are registered GST agents or who are working with a registered agent firm or companies. The majority of the respondents (36.7%) have one to 5 years of working experience and only one respondent had more than 30 years of working experience. The working experiences are related to taxation for both direct and indirect taxes. The majority of the firms have been operated between six to ten years. The analysis shows an equal percentage of 36.7% for respondents who have 1-10 and 11-20 GST clients, and only 3.3% of respondents who have between 51-60 GST clients. The majority (70%) of the respondents' clients dealt with cash transactions only. This is followed by clients who do importation and exportation of goods and services (56.7%), buy and sell products and services via online (53.3%), and are in property and development industry (46.7%).

5.2 Services provided to clients

Figure 1 shows type of services tax agents provide to their clients. It is apparent that in the first year of GST implementation, tax agents had to do almost everything for the clients; they assist client with GST registration (53.3%), recording transaction (50%), preparing GST03 from records provided by the clients (46.7%) and lodging GST03 statements (46.7%). They also frequently provide advice about recording and reporting of GST transaction (56.7%) and provide clients on meeting their GST requirements (53.3%). Tax agents occasionally audit client data against the records provided (46.7%) and audit data against other financial records (50%).

Table 1. Tax Agents' GST Services for Clients

Services for Clients	Never (%)	Occasionally (%)	Very frequently (%)
Assist client with registering for the GST	10	36.7	53.3
Record transaction for the client	13.3	36.7	50.0
Provide advice about recording and reporting of GST transaction	0	43.3	56.7
Prepare GST03 from records provided by your clients	3.3	50	46.7
Lodge GST03 statements on behalf of your clients	6.6	46.7	46.7
Audit client data against the records provided	23.3	46.7	30.0
Audit client data against other financial records	23.3	50	26.7
Provide advice to your clients on meeting their GST requirements	3.4	43.3	53.3
Provide training to your clients on using software to record GST transactions	26.6	36.7	36.7

5.3 Tax agents' perceptions on clients' GST compliance

As shown in Figure 1, the majority of the respondents strongly agree that many of their clients feel burdened with GST requirements (80%) and they find it difficult to understand how to comply with GST obligations (66.7%). The majority of the respondents are neutral when asked whether grey areas of the GST law make non-compliance more likely (63.3%) and whether clients have tried to intentionally dodge the GST system (46.7%). In terms of penalty, 46.6% of the tax agents strongly agree that RMC penalties are a concern for many of their clients.

5.4 Clients pass on their GST responsibilities to tax agents

Businesses are responsible to prepare all documents and submit GST03 to the RMC before by the deadline. However, the majority of the tax agents (73.3%) agree that clients pass on their GST responsibilities to the tax agents to some extent. Figure 2 shows ways clients pass on their GST responsibilities. 90% of the tax agents stated that they have to ask clients for information and documents. This is followed by reminding clients to lodge GST03 (63%) and asking clients for tax information needs (63%). Some of the respondents agree that clients do not check their records passing them onto tax agents (30%) and not voluntarily providing all relevant information to tax agents (30%). Another way of how clients pass on their GST responsibilities is letting tax agents to decide on the GST03 tax code.

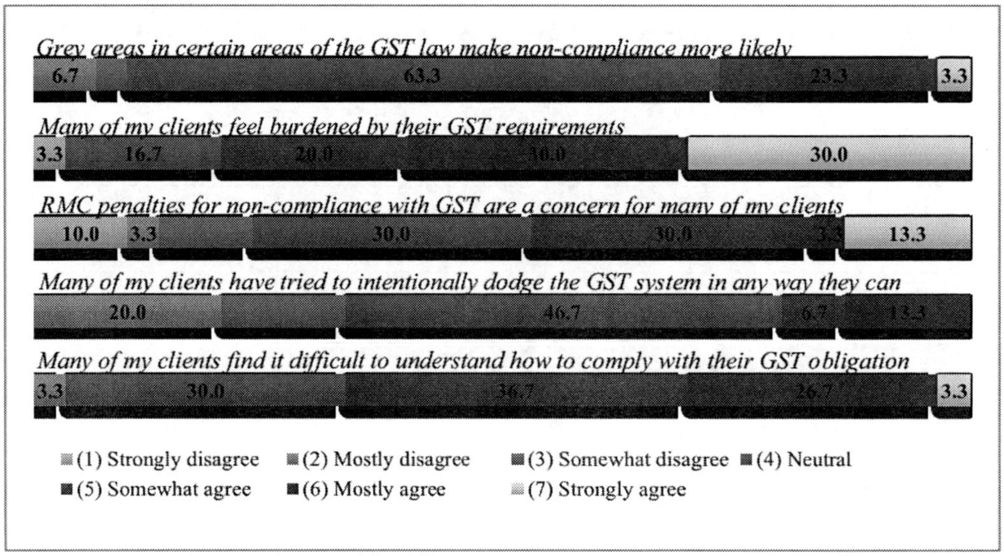

Fig. 1. Tax Agents' Perceptions on GST Clients

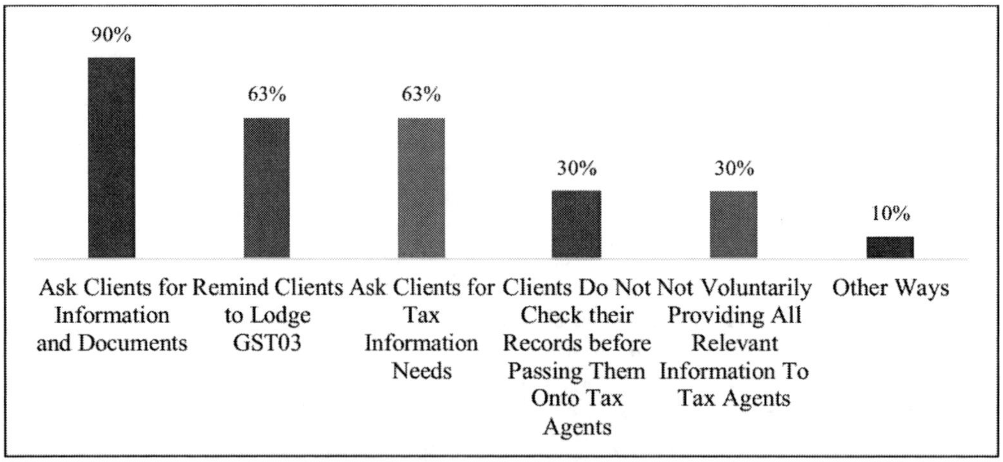

Fig. 2. Ways Clients Pass on Their GST Responsibilities to Tax Agents

 DOI: 10.1051/shsconf/20173411002

5.5 Tax agents' evaluations on clients' documents and GST03

Tax agents' were asked to evaluate clients' documents and GST03. The results are shown in Table 2. The analysis shows that tax agents perceive that their clients occasionally lodged GST03 after the deadline (73.3%), compromise the accuracy of GST03 in order to get it done on time (70%), intentionally make errors in their records (53.3%), under-reporting GST payable (53.3%) and look for grey areas in the GST law to minimise GST debts (53.3%). Tax agents also perceive that clients never de-register to avoid paying a GST (63.3%) and use the GST collection to fund other business expenses and manage cash flow (60%).

Table 2. Tax Agents' Evaluations on Clients' Documents and GST03

Clients' Documents and GST03	Never (%)	Occasionally (%)	Very frequently (%)
Error in your clients' record keeping that you think were intentional	43.0	53.3	3.3
Client GST03 statements being lodged late	23.3	73.3	3.3
Clients compromising the accuracy of their GST03 in order to get it in on time	20.0	70.0	10.0
Under-reporting of GST payable on your clients GST03	36.7	53.3	10.0
Incorrect claims of GST credits on your clients GST03	43.4	50.0	6.7
Clients using GST collected to fund other business expenses and manage cash flow	60.0	33.3	6.7
Clients dealing in cash to avoid paying some GST	50.0	46.7	3.3
Clients looking for grey areas in the GST law to minimise a GST debt	43.3	53.3	3.3
Clients de-registering to avoid paying a GST bill	63.3	33.3	3.3

5.6 Tax agents' perceptions on managing GST

The tax agents were also asked about their perceptions on managing GST. Figure 3 shows that the majority of the tax agents strongly agree that GST is a matter of course in their everyday business (56.6%) and it is easy to keep well-informed and up to date with GST legislation and rules (46.6%). However, 50% of the respondents were neutral when asked whether GST obligation is a simple thing to deal with.

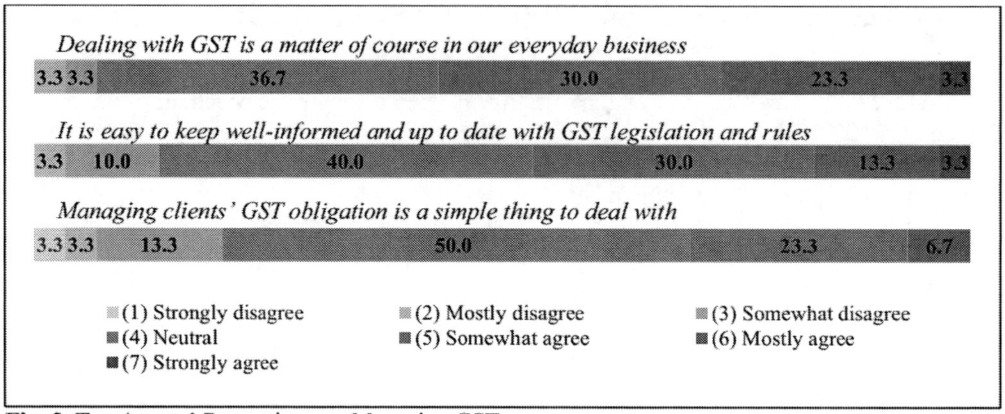

Fig. 3. Tax Agents' Perceptions on Managing GST

5.7 Tax agents' roles in influencing GST compliance

Tax agents play important role in influencing compliance. Figure 4 shows tax agents' perceptions on their roles in influencing compliance. 76.7% of the respondents strongly agree that it is a significant risk in the practice if clients incorrectly repost GST. Correspondingly, tax agents strongly agree that they rely on having accurate records (76.7%) and pay close attention to detail in client records when completing the GST03 (73.3%). The majority of the respondents (86.67%) also agree that it is important for them to act as trusted advisors for their clients. The tax agents conduct their responsibilities ethically as 70% of the respondents would not run a blind eye and let the client take responsibility for the GST lodge.

5.8 Tax agents' perceptions on legitimacy of tax system and the GST

Tax agents' perceptions on legitimacy of tax system is important in promoting compliance. Figure 5 shows that the majority of the tax agents are being neutral when asked about the legitimacy of tax system and the GST. Only 40% of the respondents strongly agree GST supports the wellbeing of the economy and society and 36.7% of the respondents strongly agree GST helps to improve business financial practices. A total of 26.3% of the respondents strongly disagree that GST is a fair tax.

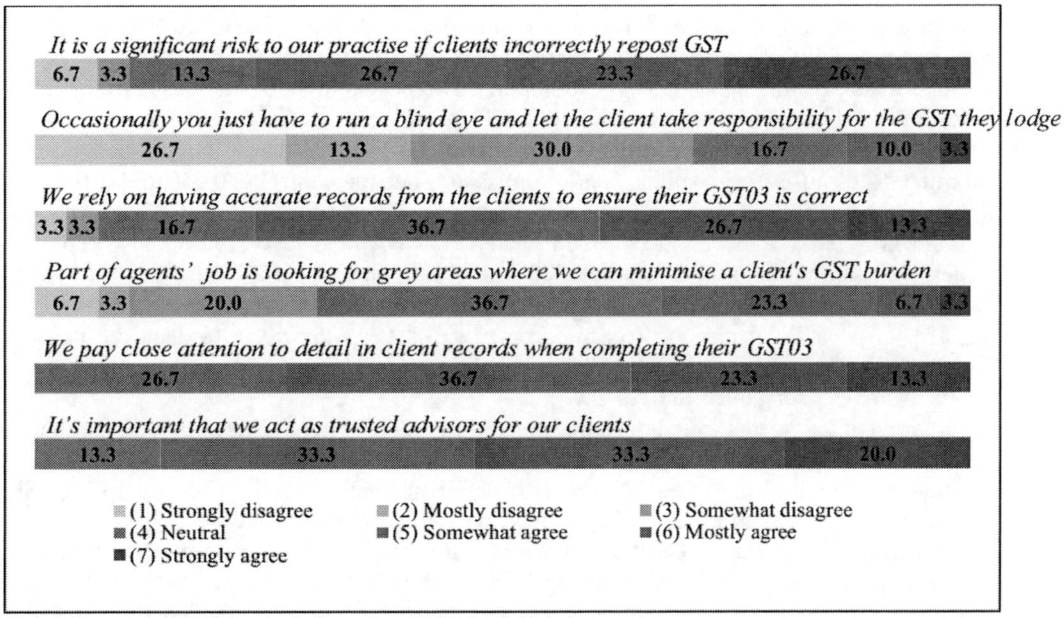

Fig. 4. Tax Agents' Roles in Influencing GST Compliance

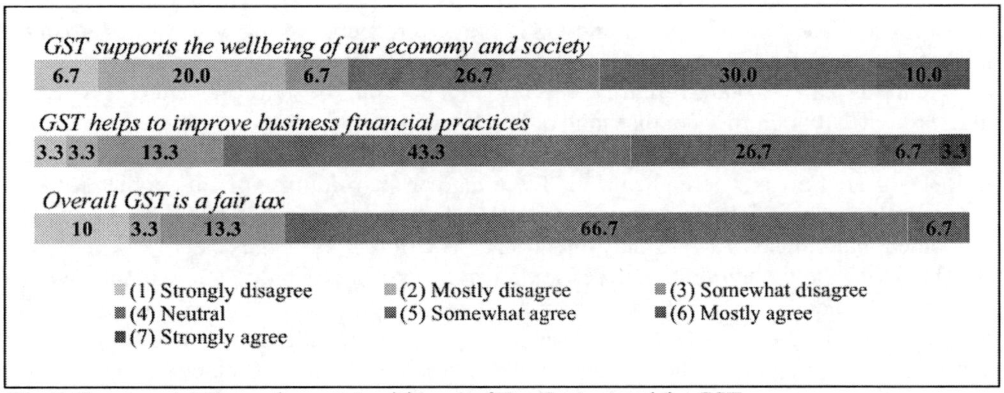

Fig. 5. Tax Agents' Perceptions on Legitimacy of Tax System and the GST

6 Discussions and recommendations

GST is another source of income to the government to build facilities for the public and develop the country. The successful of GST implementation is depending on several factors including tax administration and taxpayers' compliance. Businesses rely on tax agents' expertise to understand and comply with the newly implemented GST rules and regulations. In assisting compliance, tax agents play a significant role to provide tax advice, prepare and lodge correct tax return, and minimise tax risk.

The tax agents strongly agree that many of their clients feel burdened by GST requirements and find it difficult to understand how to comply with GST obligations. Consistent with what has been expected, tax agents confirm that their clients pass on their GST responsibilities to tax agents to some extent. Tax agents not only provide advice to their clients on meeting their GST requirements, but also recording and reporting of GST transactions for the clients. It is understandable that in the first year of GST implementation, businesses may not understand the concept and technical issues of GST. Furthermore, the GST rules and regulations are keep on changing in regards to the tax professionals' comments and suggestions. It is expected that businesses will be more independent after several years or when the GST rules are more stable to comply with.

Tax agents perceive that clients' compliance level is low because clients occasionally submit GST03 after the deadline, compromise the accuracy of GST03 in order to get it done on time, intentionally make errors in their records, under-reporting GST payable and look for grey areas in the GST law to minimise GST debts. However, these non-compliance issues are common in any tax regimes. Studies found that non-compliance is influenced by various factors including economics, psychology, tax administration and behaviour [10, 11, 12, 13]. In tax administration, it is important for the tax authority to reduce the tax gap, have an effective audit and investigation activities, ensure tax evaders are penalised and create awareness for future compliance.

In terms of tax agents' role in promoting compliance, the tax agents strongly agree that it is important for them to act as trusted advisors for their clients. Hence, they rely on having accurate records and pay close attention to detail in client records when completing the GST03. However, when asked about the legitimacy of tax system and GST, the majority of the tax agents are neutral. Only a minority of the respondents strongly agree that GST supports the wellbeing of the economy and society, and helps to improve business financial practices. With more income to the government, the public want to see significant changes on country's development, better economic condition to expand businesses and they also expect more facilities are provided with minimal or no cost to the public. If the government is able to prove the effectiveness of GST to the country, tax agents' may strongly agree with the legitimacy of tax system and GST. However, there are many ways of looking at fairness; perhaps the tax agents were unsure as to which aspect of fairness that was aimed at. Future research could certainly explore the concept of fairness, either procedural fairness, retributive fairness or distributive fairness.

Another interesting finding is that about 30% of the agents agree that part of their job is to look for grey areas where clients' GST burden can be minimised with another 37% did not want to disagree. This may confirm past studies on tax agents' ethical behaviour [14, 15] which found that tax agents play dual roles - as enforcer as well as exploiter.

This is the first study that explores tax agents' perceptions on clients' compliance and tax agents' roles in promoting compliance after one year of GST implementation. The findings benefit the RMCD to understand tax agents' challenges in assisting clients' compliance. For future studies, researchers should examine businesses' compliance behaviour and explore RMCD officers' enforcement regulatory behaviour. These key players in the tax system play significant roles to ensure the effectiveness of GST. They have different issues and challenges which need to be identified and resolved. Similar study should be adopted by countries that have recently implemented GST, for example, India. Similar study should also be conducted annually to analyse the behavioural trends and identify weaknesses in GST administration.

References

1. R. Ramli, M.R. Palil, N.S. Abu Hassan, A.F. Mustapha, Jurnal Pengurusan, **45**, 1-15 (2015)
2. I. Muhammad, International Foundation for Research and Development Conference, 223 (2013)
3. Z.S. Nordin, I. Muhammad, International Conference on Accounting Studies, (2015)
4. R. Ramli, M.R. Palil, N.S. Abu Hassan, A.F. Mustapha, Jurnal Pengurusan, 45, 1-15 (2015)

5. M. Walpole, D. Salter, e-Journal of Tax Research, **12**, 335-358 (2014)
6. K. Devos, Journal of Business and Econonomics, **5**, 265-284 (2014)
7. P. Niemirowski, A.J., Wearing, Journal of Australian Taxation, **6**, 166-200 (2003)
8. S. Klepper, D. Nagin., Political Science, **22**, 167-194 (1989)
9. A.T.O., GST Voluntary Compliance Program (2013)
10. E.C. Loo, C. Evans, & M. McKerchar, Asian Journal of Business and Accounting, **3**, 145-161(2010)
11. J. Pope, H. Abdul-Jabbar, Small Enterprise Research, **16**, 47-60 (2008)
12. M. McKerchar, Australian Tax Forum, **17**, 3-26 (2002)
13. M. Allingham, A. Sandmo, Journal of Public Economics, **1**, 323-338 (1972)
14. S. Klepper, M. Mazur & D. Nagin, The Journal of Law & Economics, **34**, 205-229 (1991)
15. Y. Sakurai. & V. Braithwaite, Journal of Business Ethics, **46**, 375-387 (2003)

The Effect of Volume of Transaction On The Intention Towards Tax E-filing

Saliza Abdul Aziz[2,*], and Kamil Md Idris[1]

[1]Tunku Puteri Intan Safinaz School of Accountacy, Universiti Utara Malaysia, Dean, 06010 UUM Sintok, Kedah, Malaysia
[2] Tunku Puteri Intan Safinaz School of Accountacy, Universiti Utara Malaysia, Senior Lecturer, 06010 UUM Sintok, Kedah, Malaysia

Abstract. Volume of transaction is specifically referring to number of clients or return forms dealing by tax agents/preparers. This variable does play role even though there is less empirical evidence on this matter. This paper is basically to explore the effect of volume of transaction that could possibly influence the acceptance of tax e-filing system as it did in other areas of transaction i.e. court cases. Undoubtedly, it is proved to be significantly influenced the behavioural intention towards tax e-filing acceptability in Malaysia. The tax agents/preparers are randomly selected in this study from the list provided in the Inland Revenue Board of Malaysia (IRBM) website. The responds are analysed via Partial Least Square (PLS) method which produce the significant results among the low and high volume of transaction of tax e-filing which support the fact that volume does matter in the decision of acceptability of tax e-filing system in Malaysia.

Keywords: Volume of transaction, Behavioral intention, Tax e-filing.

1 Introduction

The practice of using technologies in taxation is not new as tax preparation software have been introduced since 1993, but the submission of the return form is done manually. This is because of the clients' requirement who would desire to have physical copy of return form as well as proof of mailing [1]. In 1997, fully utilized the e-filing system among tax practitioners have pushed the percentage of using the tax e-filing to increase. The acceptance is due to the advantages of e-filing obviously more than the manual system, *i.e.* more efficient; save time and money; reduce number of errors; attributable to the one time early of figures; and the checks is performed by tax preparation software. As a matter of fact, the tax practitioners become more efficient, effective, progressive and technologically sophisticated. In addition to that because of a large number of clients meaning to have huge volume of transactions, e-filing is considered worth and smart decision in handling clients. In fact, this volume of transaction is seen as one of the reason that influences decision in accepting technology among tax practitioners in the United States. Indirectly, e-filing help in reducing the labor costs; reduce the volume of paperwork; paperless and saves paper, toner and file storage costs that benefits the company and clients in terms of cost saving [2, 3].

* Corresponding author: saliza@uum.edu.my

Strengthening the fact, besides the big companies, the United States Judicial Courts also has introduced the e-filing system. This is because the e-filing system could help in reducing the ever-growing caseloads. In average, 35 per cent of documents are filed electronically in federal courts, whereas in bankruptcy courts cases are closer to 60 to 70 per cent [4]. The high volume of transaction or cases is one of the factors that led the courts to embrace the technology. Indeed, more than 70 per cent judges, lawyers and clerks supporting the implementation of the e-filing system in handling the increasing of caseloads as well as volume of paperwork that indirectly improve in job performance [5]. As a matter of fact, judges witness that courts benefited a lot from the e-filing system as it could improve access to information; increase efficiency for clerks; and reduce the amount of storage needed for court records. Besides that, due to simple and available software as well as other facilities needed, *i.e.* PDF software, scanner and computer, e-filing is seem as the best alternative to get rid from the increasing problem of managing caseloads. In addition, less effort is required to educate the related parties has attract the users to accept the system introduced.

The significant effect of volume of transaction that could influence the acceptance of e-filing system is well known. However, there is less evidence or study conducted on this matter to conclude the relationship of this element towards a behavioral intention to accept a new technology. Thus, the objective of this paper is to provide an evidence on the relationship between volume of transaction and behavioral intention in accepting tax e-filing among tax agents/preparers in Malaysia.

2 Previous studies

Reports from the United States clearly pictured the scenario of accepting e-filing system that is due to the ability to improve performance expectancy and less effort requirement with facilities available. Unfortunately, support on the level of significant effect of volume of transaction on behavioral intention is empirically unknown. Thus, there is a need to conduct study to test the element of volume of transaction on behavioral intention for future reference.

This variable does play role even though there is less empirical evidence on this matter. It is because big companies as well as courts that overload with work and cases preferred to change to e-filing system in order to overcome their inefficiency problem [2, 4, 5]. Thus, there is a possibility of having a significant effect between volume of transaction and acceptance of tax e-filing. However, there is less evidence or study conducted on this matter to prove the relationship towards a behavioral intention to accept a new technology. Hence, the same effect is predicted in tax e-filing scenario in Malaysia where workload or increasing in volume of transaction could possibly affect the work performance. Then, the intention to use the tax e-filing is directed. Moreover, the effectiveness and efficiency of tax e-filing and be able to improve performance with less effort required is not influenced the intention to adopt tax e-filing system without any pressure from the volume of transaction. This condition is also predicted even when facilities available to ease the tax agents/preparers, tax e-filing is not the alternative until volume of transaction really affected the work performance. Thus, the following hypotheses are derived:

H_1: The influence of performance expectancy on behavioral intention toward tax e-filing among tax agents/preparers will be moderated by volume of transaction.

H_2: The influence of effort expectancy on behavioral intention toward tax e-filing among tax agents/preparers will be moderated by volume of transaction.

H_3: The influence of facilitating conditions on behavioral intention toward tax e-filing among tax agents/preparers will be moderated by volume of transaction.

3 Methodologies

The dependent variable of this research is behavioral intention of accepting e-filing technology. Behavioral intention is the degree to which the tax agents/preparers intend to use the technology of e-filing in preparing and submission of clients' return form [6, 7]. It is crucial to study on the intention as employees and organization acceptance could improve technology efficiency and effectiveness [8]. The independent variables are performance expectancy, effort expectancy, social influence, facilitating conditions and perceived value. In addition to the independent variables, the related moderating variable to the research conceptual framework is volume of transaction. Considering the intervention introduced in TAM3 as moderator is purposely to test the significant of those variables in the Unified Theory of Acceptance and Usage of Technology (UTAUT) model. In fact, Venkatesh [7] give a thought on the design of interventions in UTAUT model as it could give an idea for managers in assessing the intention of adopting new technology introduced and help in increasing user acceptance. Thus, the introduction of this variable as moderator could provide explanation on the effectiveness and efficiency of the variables if introduced as intervention in a future study.

In total, there are 1,871 tax agents/preparers officially register with IRBM scattered in the 15 different states in Malaysia. However, the sample size are limited to 714 which taken into consideration of an additional 70 per cent from the recommended size, *i.e.* 420 tax agents/preparers as according to the table suggested by Krejcie and Morgan [9]. Tax agents/preparers are considered the sample instead of the corporate taxpayers/clients. This is because freedom in choosing the medium of transaction on tax filed to IRBM is fully given by the corporate taxpayers/clients to the tax agents/preparers. Indeed, the influence or factors from point of view corporate taxpayers/clients are not considered.

The questionnaire that is design using 7-point Likert scale is sent via mail to respondents who are selected randomly using simple random sampling technique throughout Malaysia. The selection is made based on the list developed using SPSS software, which excluded the tax agents/preparers who have been participated during the Delphi and pilot test. The statistical tool applied in the result analysis is Structural Equation Modeling (SEM) with additional supports from Partial Least Squares (PLS) and Bootstrapping.

4 Findings and discussions

Volume of transaction is one of the new introduced moderators in this research model. This moderator is analyzed via bootstrapping approach as it is in the form of non-continuous scale. The parameter is estimated based on ML estimation derived on 500 usable bootstrap samples. In the process of analysis, two bootstrap samples are unused because a solution is not found. The output tree that represents the estimation of standard error of each loading parameter for this model is attached in the appendix.

Indeed, compared to the original samples' ML standard error estimation (*i.e.* S.E. - S.E. Column), the approximate standard error of the bootstrapping is within the acceptance level, *i.e.* small from the original samples. On the other hand, the bias results that are the difference of mean between the bootstrap and the original at estimations show that there is no bias due to low differences reported. As for the goodness-of-fit indices, the standardized estimations are almost within the acceptable range. The absolute fitness reported for group one that consist of 60 or less clients is above 0.08 (*i.e.* RMSEA = 0.094). The incremental fit is above *i.e.* CFI is above 0.9 (0.928), TLI is 0.914 and IFI is 0.930. The parsimonious fitness also shows an acceptable level, which the ratio is below five, *i.e.* 1.617 with degree of freedom of 174. As for group two where the number of clients are above 60, the absolute fitness reported is above 0.08 (*i.e.* RMSEA = 0.099). The incremental fit is above, *i.e.* CFI is almost 0.9 (0.893), TLI is 0.871 and IFI is 0.895. The parsimonious fitness also shows an acceptable level, which the ratio is below five, *i.e.* 2.574 with degree of freedom of 174. The indication of significance level for volume of transaction is very much depends on the chi square and degree of freedom values. The basis of comparison is between the constraint and unconstraint models for each group on each construct. Indeed, the degree of freedom with a difference of one and chi square difference of above 3.84 for each group on each constructs ensured the moderation is significant.

Table 1 and 2 contains the standardized parameter estimates (chi square and degree of freedom), standardized indirect effects for all constructs and its effect on the interaction of volume of transaction. Basically, the three hypotheses reported, *i.e.* H_1, H_2 and H_3 indicate having significance effect and successfully moderate the direct relationship between exogenous and endogenous. This relationship simply means that volume of transaction does moderate the influence of performance expectancy, effort expectancy and facilitating condition. Specifically, the volume of transaction is tested via multigroup estimation to determine which group of volume, *i.e.* high or low is more pronounced in giving group effect for the moderator variable. The group that is determined from the standardized beta estimate (slope) indicates that high volume of transaction give more impact on performance expectancy and effort expectancy rather than low volume of transaction group (Table 2). As for the facilitating condition, low group of transaction is more pronounced in affecting the direct relationship of the constructs (Table 1). In addition, the analysis also pointed out the significant effect of volume of transaction in social influence and perceived value constructs (Table 1 and 2). On the whole model, high volume of transaction is affected the relationship toward behavioral intention more pronounced. The ability to explain the relationship of the whole model of behavioral intention towards tax e-filing acceptability is increased slightly to 52.8 per cent. The increment of 0.2 per cent in the power of explained suggest that volume of transaction does improve the best model.

Table 1
Hypothesis testing result of volume of transaction (Group 1: Low volume)

Hypo-thesis	Relationship	Constraint		Unconstraint		Estimate	Result
		x^2	DF	x^2	DF		
H_1:	PE* Volume →BI	285.936	175	281.417	174	.191	S
H_2:	EE* Volume →BI	281.832	175	281.417	174	1.223	NS
	SI* Volume →BI	295.581	175	281.417	174	-.156	S
H_3:	FC* Volume →BI	288.582	175	281.417	174	-.868	S
	PV* Volume →BI	285.915	175	281.417	174	-.057	S

Notes: ***p<0.001, **p<0.05, *p<0.1, S=Significance, NS=Not Significance

Table 2
Hypothesis testing result of volume of transaction (Group 2: High volume)

Hypo-thesis	Relationship	Constraint		Unconstraint		Estimate	Result
		x^2	DF	x^2	DF		
H_1:	PE* Volume →BI	451.997	175	447.872	174	.624	S
H_2:	EE* Volume →BI	462.945	175	447.872	174	.458	S
	SI* Volume →BI	468.300	175	447.872	174	-.174	S
H_3:	FC* Volume →BI	458.209	175	447.872	174	-.201	S
	PV* Volume →BI	463.517	175	447.872	174	-.198	S

Notes: ***p<0.001, **p<0.05, *p<0.1, S=Significance, NS=Not Significance

5 Conclusions

Empirically, the research's finding indicates that tax e-filing intention does influence by the number of clients. Undeniable, the perceived value that tax agents/preparers expect from tax e-filing, *i.e.* save time, effort and cost, has encouraged them to accept the system. This research somehow as to support the element of volume in taxation point of view is successfully achieved its purpose. However, further study should be carried out to strongly support the difference behavioral intention between low and high volume of transaction. This is basically due to differences in significance level achieved between the two groups especially in low group of volume. As expected, the performance expectancy, effort expectancy, social influence, facilitating conditions and perceived value factors are definitely influenced the intention of tax agents/preparers with large number of clients. However, in considering effort expectancy, the tax agents/preparers with low group of clients refused to accept the tax e-filing. Indeed, this volume of transaction factor is need to be emphasized and focused as one of moderating factor and test or include in any study related to new technology introduce in future. This is due to its contribution that possibly plays some role in decision making in choosing any technology application. However, this factor is rarely tested and in reality not much reference on this constructs.

References

1. Warkentin, M., et al., *Encouraging Citizen Adoption of e-Government by Building Trust.* Journal of Electronic Markets, 2002. **12**(3): p. 157-162.
2. Anderson, T., M. Fox, and B.N. Schwartz, *History and trends in e-filing: A survey of CPA Practitioners.* The CPA Journal, 2005: p. 66-69.
3. Saman, W.S.W.M. and A. Haider, *E-Shariah in Malaysia: Technology Adoption Within Justice System.* Transforming government: People, Process & Policy, 2013. **7**(2): p. 256-276.
4. Krause, J., *E-filing made easy.* ABA Journal, 2003: p. 68.
5. Anonymous, *Legal group expects e-filing increase.* The Information Management Journal, 2005: p. 12.
6. Davis, F.D., *Perceived Usefulness, Perceived Ease of Use and User Acceptance of Information Technology.* MIS Quarterly, 1989. **13**(3): p. 319-340.
7. Venkatesh, V., et al., *User Acceptance of Information Technology: Toward a Unified View.* MIS Quarterly, 2003. **27**(3): p. 425-478.
8. Anderson, J.E., P.H. Schwager, and R.L. Kerns, *The Drivers for Acceptance of Tablet PCs by Faculty in a College of Business.* Journal of Information Systems Education, 2006. **17**(4): p. 429-440.
9. Krejcie, R.V. and D.W. Morgan, *Determining Sample Size for Research Activities.* Educational and Psychological Measurement, 1970. **30**: p. 607-610.

DOI: 10.1051/shsconf/20173412001

ISLAMIC SOCIAL REPORTING IN ISLAMIC BANKING: STAKEHOLDERS THEORY PERSPECTIVE

Inten Meutia[1,*], Devi Febrianti [2]
[1] Accounting Department Faculty Economic
[2] Post Graduate Program University Sriwijaya

Abstract. This research aims to compare the level of Corporate Social Responsibility disclosure on Islamic banking in Indonesia and Malaysia. Using data from annual report in seven public Islamic banks in Indonesia and seven public Islamic banks in Malaysia, this research find that the level ISR of Islamic banking in Indonesia are better than the level ISR of Islamic banking in Malaysia. There were significant differences between the two groups were observed with respect to the all theme of reporting.

1 Introduction

The development of the Islamic financial industry in the world, especially Islamic banking is growing rapidly. In 2014, the Islamic banking assets reached $ 1,214,274 of the total Islamic financial assets of $ 1,657,998, or 73.23% [1]. The growth in the financial sector should be followed by the increasing role of Islamic banks in terms of social responsibility. [2] says that social responsibility is very pertinent for discussion on Islamic banking.

[3] explains that the form of these responsibilities is the disclosure of a social responsibility report. In the Islamic reporting viewpoint, it is expected that social concerns would be a principal factor of disclosure in annual reports. Consequently businesses are expected to be more transparent in their disclosure practices ([4], [5]).

Furthermore [6] explains that key stakeholders perceive Islamic banks to have equal social and economic roles. The common of the Islamic finance and economics literature, as well as key people concerned in the development of the industry, announce that Islamic banks and financial institutions adhere to Islamic Sharia principles of social justice and fairness, and deliver an alternative to conventional banking and finance ([7], [8], [9]). These opinions and expectations as well as the claims of the industry itself infer that Islamic banks are supposed to report base on their social activities ([10], [6] Islamic banks are supposed to disclose full information to the ummah (Muslim nation) about how their activities meet

* Corresponding author: inten26@yahoo.com

Sharia' objectives and increase the wellbeing (and cohesion) of society ([11], [12], [13], [6]).

Malaysia is a country that representative of modern Islamic country caused by Internalization Islamic values in the government system. The companies in Malaysian have an experienced pressure to apply corporate government best practices internationally. One of the pressures is the amendment in the Bursa Malaysia's Listing Rules on January 2001 that requires public listed companies to disclose a narrative statement of how they applied principles of corporate governance in their organizations in their annual reports.

Regarding to Islamic Financial Insitutions, Malaysian government established the Shariah Advisory Council in 1996. The task of Shariah Advisory Council is to offer advice Securities Commission on Shariah related matter. Their tasks include providing guidance on Islamic Capital Market (ICM) transactions and activities with the aim to standardize and harmonizing implementations [15].This development has consequent to ISR, where Muslim decision-makers expect companies can meet their spiritual needs by disclosing relevant information [16].

Development of Islamic banking in Indonesia with a marked the establishment of Bank Muamalat Indonesia (BMI) in 1991. One of the functions and its role is to carry out social functions in order to assist people to prosperity. Act No. 21 of 2008 about Islamic banking article 4 points 2 states that one of the functions of the Islamic banks and Sharia Business Units (UUS) are doing a social function. The accounting standards published by AAOIFI (Accounting and Auditing Organization for Islamic Financial Institution) also states that one of the functions and the role of Islamic banks is to implement social activity.

The presence of differences in the rules regarding Islamic financial institutions in the two countries are very likely to result any differences in terms of social responsibility disclosure carried out by Islamic banks. Therefore our focus in this paper is two-fold: (1) to get prove empirically the level of corporate social responsibility disclosure in Islamic bank in Indonesia and Malaysia using Islamic Social Reporting (ISR), and (2) to analyze the difference of corporate social responsibility disclosure in Islamic bank in Indonesia and Malaysia using ISR. The paper proceeds as follows: section two provides the theoretical foundations and literature review of the study. Section three reports on the research methodology, section four discusses detail the findings of this study and their analysis, sections five provide recommendations and conclusion,

2 Literature Review

2.1. Stakeholder Theory

According to Stakeholder theory the sustainability of corporation needs the support from the stakeholders. The more powerful stakeholders, the more the company must adjust [17]. As indicated by [18] stakeholders can influence (directly or indirectly) control of assets required by the corporation. For that reason, stakeholder's domination is determined by the level of control they have over the assets.

Therefore, when stakeholders have a control to significant resources of the organization, the company is such as to react in a manner that satisfies the stakeholders. Moreover, [19] discusses that organizations select the stakeholders that they need to consider. The activities that they will choose in order to realize the desired relationship with those stakeholders. The Government is one important stakeholder for the company; therefore the company will try to make disclosures based on the rules set by the government. Therefore, it is very likely there will be a difference in terms of social responsibility disclosure in Islamic Bank caused by the differences in the rules that exist in each country. Therefore we suppose variances in disclosure level based on rules applied in each Countries. Thus, the hypothesis is: There are differences in levels of ISR between Islamic Banks in Malaysia and Indonesia.

 DOI: 10.1051/shsconf/20173412001

2.2. Islamic Social Reporting

[21] stated that ISR is a social reporting that involves not only a holistic expectation of the society regarding the role of companies in the community but also on spiritual perspective [16]. In ISR, the concern is not just on social justice but goes beyond reporting on the environment, minority interests and employees. This concern is appropriate with the stakeholders' theory whereby management presents information to stakeholders to maintain the success of corporate. The sustainability of company existence and success is dependent on the persistent support from the stakeholders [19]. It is, therefore, very important for the management to disclose information related to social and environmental according to the demand of the stakeholders [23].

A number of studies have surveyed CSR disclosure in Islamic bank (IBs) ([6], [10], and [24]). ([25], [10]) open up additional analysis into this space by provides an ethical identity index. For example, [6] consider the effect of Islam on the social reporting at 29 IBs in 16 countries. They found that level of social disclosure by IBs is far from their expectation (13.3%). [24] using content analysis in analyze annual reports of seven IBs and found that level of CSRD for their sample is thirty-eighth. That is comparatively low comparing with their expectations, objective, and values of IBs.

[24] investigate whether any difference exists between the corporate social activities disclosed by using Islamic business ethics framework in the annual reports of seven Islamic banks from seven countries. The results show that only one Islamic bank out of seven Islamic banks has CSRD index above average and the issues of CSR are not of main concern for most Islamic banks.

[28] compare CSR disclosure in sixteen IBs from three different regions by using a standard supported Islamic value. They find that level of CSR information far below the expected level (49%). [29] investigate the extent of CSRD in twenty-one IBs operational within the Gulf region. They found that the extent of social disclosure (30%) was well below their benchmark index.

[10] investigate the disclosure of information considered crucial to Islamic ethics in business. They notice a major gap (48%) between the disclosure and the ideal disclosure of seven IBs. They conclude that to be competitive IBs have to deliver more information to their stakeholders in order to improve good reputation in society.

[30] examine the level and factors that influence CSRD in the annual reports of forty-seven IBs from fourteen countries. They realize that CSRD is 16.8%. Moreover, [31] explore CSRD in IBs in East Pakistan from 1983 till 2010 to examine the IBs' performance relating to CSR. They found that level of CSRD only 42nd. This result is supported by many studies that found a low-level CSRD for IBs (e.g., [32]).

Recently, the correlation between CSR and financial performance has been investigated in [33]. They found a poor level for CSRD (44%). [30] also found other factors that have influences on the level of CSR disclosure in Islamic banks are the level of political and civil pressure.

Based on the above studies, none of them especially focus on differences in the level of social responsibility disclosure between Islamic Bank in Indonesia and Malaysia. Therefore, this research attempts to give a contribution to Islamic accounting literature.

3. Research Methodology

This study examines social responsibility reporting in 14 Islamic banks in the two countries by using ISR index. The content analysis uses to find if particular topics associated with social justice are disclosed (or not) in annual reports. The annual reports are related to the years 2010-2014 and were downloaded from the banks' web sites.

DOI: 10.1051/shsconf/20173412001

The level of ISR in this research consisted of 43 items grouped into six topics that were adopted from Othman et al (2009). There are 6 topics namely Finance and Investment, Products/Services, Employee, Society, Environment and Corporate Governance. These topics consist of 43 items disclosure. By content analysis, each item had the value of 1 if the items were disclosed by the company. Otherwise, it was 0.

If the entire item ISR Index disclosed in the annual report of Islamic Bank so it means that the Islamic Bank will get full score which are 43 points. After scoring process of IS Index finished to each Islamic Bank, the analysis will be do in order to know each of its social performance. After that, it will be calculated the average for each item. Social performance disclosure index calculation is formulated as follows: ISR index = the number of items disclosed × 100% the total number of items.

Table 1. List of the Islamic Bank Sample

No	Islamic Banks in Indonesia	Islamic Banks in Malaysia
1	Bank Muamalat Indonesia (BMI)	Public Islamic Bank Berhad (PIBB)
2	Bank Syariah Mandiri (BSM)	CIMB Malaysia (CIMB)
3	Bank Mega Syariah (BMS)	Bank Islam Berhad (BIB)
4	Bank BNI Syariah (BNIS)	Hong Leong Islamic Bank (HLIB)
5	Bank BRI Syariah (BRIS)	Affin Islamic Bank Berhad (AIBB)
6	Bank Panin Syariah (BPS)	RHB Islamic Bank (RHB)
7	Bank Syariah Bukopin (BSB)	Hong Leong Islamic Bank (HLIB)

4. Findings

4.1. Disclosure Level of ISR

Picture 1 below provides the average of ISR index scores for 2010 - 2014.

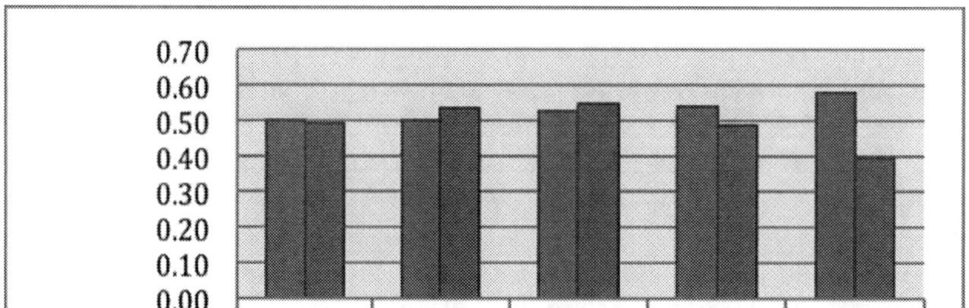

	2010	2011	2012	2013	2014
Indonesia	0.50	0.50	0.53	0.54	0.58
Malaysia	0.49	0.53	0.55	0.49	0.40

Fig 1. The ISR Index of each Country

From the figure 1, it could be seen that level ISR in Indonesia and Malaysia almost same for all year except for year 2014 Islamic bank in Indonesia gets higher level than Malaysia. ISR level in Indonesia increased from 2010 to 2014. While in Malaysia since 2012 ISR level decreases. In general, there are no Islamic banks both in Indonesia and Malaysia who fully (100%) use index ISR to carry out their corporate social responsibility activities. This happens because of the first two possibilities. The first is that because the

DOI: 10.1051/shsconf/20173412001

Islamic bank does not implement CSR activities. The second reason is that the Islamic banks which carry out corporate social responsibility activities, do not disclose their performances in the annual report. These empirical analyses argue that most Islamic banks both in Malaysia and Indonesia do not take serious attention for the CSR issues. Our finding consistent with many prior studies that found the low level of CSR disclosure in Islamic bank ([33], [31], [32], and [34]).

Table 2 summarizes the disclosure level for the each theme. These sections contain Finance and Investment, Product and service, employee, society, environment and corporate governance.

Table 2.

Index ISR of Islamic Banks in Indonesia and Malaysia base on each Themes

No	Theme	Indonesia	Malaysia
1	Finance and Investment	0.53	0.35
2	Product and Service	0.86	0.66
3	Employee	0.32	0.36
4	Society	0.59	0.33
5	Environment	0.17	0.42
6	Corporate Governance	0.69	0.83

Based on each theme Islamic Banks in Indonesia have higher score on the theme Finance and Investment, Product and Service and Society. While Islamic banks in Malaysia have high scores on the theme Employee, Environment and Corporate Governance. Some things that might make Islamic bank in Malaysia put more emphasis on disclosure of those themes are:

On 5 Sept 2006, Bursa Malaysia has launched CSR Framework for public-listed companies. This Framework is a guide to Malaysian companies in order to develop strategies to communicate them effectively to stakeholders through CSR. The CSR Framework focus on 4 main themes namely, environment, workplace, community and the market place.

The emphasis on Occupational Safety and Health (OSHA) for employees mandated by laws in the late nineties; the eradication of poverty; scholarships for human capital development; funds for education; non public universities; training; caring and loving society and environment protection could have motivated the level of ISR.

This is consistent with [23] that found one of the main factors why companies have a tendency to disclose more CSR information is government policy.

After the East Asian financial crisis in 1997, CSR awareness in Malaysia increased [35]. This awareness following by the Malaysian Securities Commission in the middle of 2001. Furthermore, to support this initiative CSR included by the government as one of the public policies in the National Economic Policy, the Malaysian Vision 2020 and National Integrity Plan. The most important all public listed company have to disclose CSR in the annual reports.

In the next table we try to give prove if the differences under each dimension significantly different by using Z test.

Table 3. Differences Analize for Each

	Finance & Investment	Product & Service Theme	Employees Theme	Society Theme	Environment Theme	Corporate Governance
Mann-Whitney U	269,500	340,500	450,500	177,500	327,000	290,000
Wilcoxon W	899,500	970,500	1116,500	807,500	993,000	956,000
Z	-4,243	-3,527	-2,143	-5,256	-3,777	-4,180
Asymp. Sig. (2-tailed)	,000	,000	,032	,000	,000	,000

a. Grouping Variable: Negara

Table 3 indicate that significant two-tailed is lower than 0.05. It means there were significant differences in all themes between the two countries. [36] stated that numerous laws and government regulation affect effectively every aspect of business activities.

There are numerous components which influence CSR disclosure of Islamic banks including socio-political pressures and economic incentives. [30] said that because the activity of Islamic banks is in a broader social and political environment, the interaction of power affecting the rights and responsibilities of all actors within this environment will impact the content of information and dialogue. This will consequently have an impact on the level of CSR disclosure by Islamic banks.

5. Concluding Comments

5.1. Research limitations/implications

The results of this study are subject to several limitations. The major limitation of this paper is the small sample size of only 14 Islamic banking institutions from 2 countries. Future studies may expand the sample size used here. Second, this study focused on annual reports in measuring the extent of the corporate social responsibility disclosure practices by the selected Islamic banks. Future research might be extended by including other forms of csr disclosure such as websites and others media.

This paper is intended to explore the disclosure level of CSR for IBs in two countries using ISR index developed by [37]. Our sample is 14 Islamic banks from Malaysia and Indonesia. Based on our study ISR index scores show that the extent of disclosure is 26% that is out of our expectations. Our results are consistent with [10] and [24] who found a gap between expectations and actual disclosure of CSR and conclude that CSRD for IBs is very low. Our finding also shows that there are differences between two countries and it concludes that the rules and pressure from government have impact on the information disclosed by Islamic banks.

REFERENCES

[1] ICD Thomson Reuters Islamic Finance Development Report, (2014).
[2] A. Wajdi Dusuki, and Humayon Dar. "Stakeholder's Perceptions Of Corporate Social Responsibility Of Islamic Banks: Evidence From Malaysian Economy". *Islamic Business Reseacrh Center.* pp 389-4197. (2005).
[3] I. Meutia., *Menata Pengungkapan CSR di Bank Islam (Suatu Pendekatan Kritis).* Jakarta: Citra Pustaka Indonesia. (2010).
[4] C.H Cho., D.M Patten, The role of environmental disclosures as tools of legitimacy: a research note. Acc., Org. and Soc.E 3 (7/8), pp 639- 647. (2007).
[5] R. Haniffa,. & T.E Cooke, Culture, Corporate Governance and Disclosure in Malaysian Corporations, Ab. **38** 3, pp.317-349. (2002).
[6] B. Maali., P.Casson, & C. Napier, "Social Reporting by Islamic Banks". Ab. **142** 2, pp 266-289. (2006).

[7] M . A . El-Gamal, *Islamic Finance: Law, Economics, and Practice*, Cambridge University Press, New York, NY. (2006).
[9] M . T . Usmani, *An Introduction To Islamic Finance*. Idaratul Ma'arif, Karachi. (1998).
[10] R . Haniffa, and M. Hudaib. "Exploring the Ethical Identity of IB via communication in Annual Reports," *J of Bus Eth*, **76** 1, pp 97-116, (2007).
[11] A. Al-Mograbi. *The Social Responsibility of Islamic Banks*. International Institute of Islamic Thoughts. (1996)[12] N. Baydoun, and R. Willet. Islam and accounting: Ethical issues in the presentation of financial information. Acc Comm and Fin: The Islamic Perspective. **1** 1, pp.1-25, (1997).
[13] M.K. Lewis. 'Islam and Accounting' Acc For , **25** 2, pp 103–27, (2001).
[14] S. Hassan. "Corporate Governance Practices of Shariah Islamiah Approved and Shariah Islamiah Non-Approved Companies:Does it matter?", Paper presented at IIUM International Accounting Conference 3, 26-28 June, Kuala Lumpur, (2006).
[15] Bursa Malaysia. *The Islamic Capital Market*. Kuala Lumpur: Bursa Malaysia. Available at Belal, A., O. Abdelsalam, and S. Nizamee, 2014, "Ethical Reporting in Islami Bank Bangladesh Limited (1983–2010)," J of Bus Eth, DOI 10.1007/s10551-014-2133-8. (2007).
[16] R . Haniffa. "Social Reporting Disclosure-An Islamic Perspective", Ina Mgt & Acc R **1** 2 pp.128-146.(2002).
[17] R. Gray., R. Kouhy. and S. Lavers. "Corporate social and environmental reporting: A review of the literature and a longitudinal study of UK disclosure", Acc. Au and Acc. **8** 2, pp. 47-77. (1995).
[18] C. Deegan., M. Rankin., and P. Voght. "Firms' disclosure reactions to major social incidents:Australian evidence", Acc For **24** 1 pp. 101-130. (2000).
[19]A. Ullmann. "Data in search of a theory: a critical examination of the relationships among social performance, social disclosure, and economic performance of US firms", Ac of Mgt Rev, **10** 3. pp. 540-557. (1985).
[20] R. Gray., D. Owen., and C.A. Adams. *Accounting and Accountability: Changes and Challenges in Corporate Social and Environmental*, Prentice-Hall Europe, London. (1997).
[21] R . Othman., A.M. Thani., E . K . Ghani. Determinants of Islamic Social Reporting Among Top Shariah-Approved Companies in Bursa Malaysia. Res J of Int Stu **12** (October). (2009).
[22] M. Sulaiman. *Islamic corporate reporting: Between the desirable and the desired.* Research Centre. IIUM Publication. (2005).
[23] M. Zain. "Corporate social reporting in Malaysia: The Current State of the art and Future Prospects", Dissertation for the Degree of Doctor of Philosophy, University of Sheffield. (1999).
[24] A . Hassan., S.S. Harahap. "Exploring corporate social responsibility disclosure: The case of Islamic banks". Intl J of Is and Mid East Fin and Mgt. **3** 3 pp 203-227. (2010).
[25] E.R.Gray. and J.M.T. Balmer. "The corporate brand: a strategic asset." SAM: Mgt in Prac. **4**. pp 1-4. (2001)
[26] A. Rahman. Islamic microfinance: An ethical alternative to poverty alleviation. Humanomics, **26** 4, 284-295. (2010).
[27] J. A Mohammed. "Corporate Social Responsibility in Islam". PhD work under Auckland University of Technology, New Zealand. (2007).
[29] Z.A. Aribi. and S. Gao. Corporate Social Responsibility Disclosure: A Comparison Between Islamic And Conventional Financial Institutions. J of Fin Rep and Acc. **8** 2, pp. 72-91. (2010).

[30] S. Farook., M.K. Hassan., R. Lanis. Determinants Of Corporate Social Responsibility Disclosure: The Case Of Islamic Banks.. J of Is Acc and Bus R. **2** 2, pp114-141. (2011).
[31] A . R . Belal.. O Abdelsalam. & S . S . Nizamee. "Ethical reporting in Islamic bank Bangladesh limited (1983–2010)". J of Bus Eth. **129** ,pp 769–784. (2014).
[32] R. Kamla., H.G. Rammal. Social reporting by Islamic banks: Does social justice matter?. Acc Au Acc J. **26** 6. pp 911-945. (2013).
[33] H.K. Farag. and K. Mallin. Corporate Social Responsibility and Financial Performance in Islamic Banks. J of Ec Behav & Org **103**, pp S21–S38. (2014).
[34] Z.A. Aribi. An Empirical Study of Corporate Social Responsibility and Its Disclosure in Islamic Financial Institutions. Doctor of Philosophy, Edinburgh Napier University. (2009).
[35] R. Ismail., R.A. Rahman., N. Ahmad. Risk Management Disclosure In Malaysian Islamic Financial Institutions: Pre- And Post-Financial Crisis, The J of Appl Bus R – March/April **29** 2 pp 419-431. (2013).
[36] A . B . Carroll. A three-dimensional model of corporate performance, Ac of Mgt Rev **4** pp 497-505. (1979).
[37] R . Othman., & A.M. Thani, A. M. Islamic Social Reporting of Listed Companies in Malaysia. Int Buss & Ec R J **3** pp. 135- 144. (2010).

Empirical Research on Rate of Return, Interest Rate and Mudharabah Deposit

Inten Meutia Relasari[1,*], and *Achmad* Soediro[1]

[1] Accounting Department Faculty Economic University Sriwijaya

Abstract: The objective of this study is to analyze the effects of interest rate and the rate of return on mudharabah deposits in Islamic banks. Using data covering the period from January 2012 to September 2015, the study applies the regression analysis to analyze the effect between variable and independence t- test to analyze the mean difference between rate of return and rate of interest. Regression analysis show that rate of return has negative influence significantly on mudharabah deposits, while interest rate has negative influence but not significant. The result of independent t-test shows that the interest rate is not different from the rate of return in Islamic Bank.

1 Introduction

Islamic banking is an activity based on Islamic Syariah principles, which does not allow the paying and receiving of interest and promotes profit sharing in the conduct of banking. The most important difference between Islamic and conventional banking is the prohibition of interest in Islamic banking [1]. Islamic financing works on the concept of buying and selling where the financial institution purchases the property and subsequently sells it to customers above the purchase price rate [2].

In the dual banking system where Islamic banking system operates in conjunction with the conventional banking system, such as in Indonesia, there will be a substantial relationship between the two systems. Changes in interest rates can have a pressure on the rate of return or profit-sharing in Islamic banking. [3] stated that when the interest rate in the conventional system increases, the Islamic banks need to increase their rate of return. The failure to respond this change can have an effect on the liquidity of the bank.

Therefore, the purpose of this paper is to investigate the influence of the rate of return and the interest rate on total time deposits mudharabah in the Islamic bank in Indonesia. In addition, this paper investigates whether the interest rate in conventional banks and the rate of return in Islamic banks in Indonesia are different significantly.

2. Literature Review

2.1. Islamic Bank

The primary and most vital feature of Islamic banks is that the prohibition of interest (riba), no matter its form or source. The prohibition of riba in Islamic social science has received a lot of attention. Several western studies have urged that the prohibition of interest is associated with degreeti-capitalist and an obstacle to the correct functioning of a contemporary economy and a prejudicious issue to economic development and growth [4]. [5] pointed out that for the conventional financial institution, the bottom lending charge (BLR) and prices of return on deposits would adjust according to alterations inside the industry fascination level. A rise on the market curiosity fee would cause the rate of return on deposits to increase. For the reason that Islamic bank is featuring lower deposit rates, it is unable to compete with the conventional bank in attracting new deposits. Consequently, the speed of progress of new deposits will have a tendency to decline. Simultaneously, depositors who would like to get benefit from the higher interest rates will transfer their deposits from the Islamic bank to the conventional financial institution. Customers may want to help keep their deposits for shorter length. Generally, the volume of Islamic deposits could be negatively influenced. The above rationalization theoretically exhibits that any change in the market interest rates would produce a shifting effect from Islamic bank to the conventional financial institution and vice versa.

* Corresponding author: relasari_mk@yahoo.co.iz

2.2. Interest Rate

Rate of interest has always been featured as one of the important considerations in explaining the saving behavior of individual. The higher the rate of interest, the more money will be saved, since at higher interest rates people will be more willing to forgo present consumption. Interest rate risk is a major issue that needs serious attention from the bank. [7] illustrated that changes in interest rates may affect the bank both in terms of income and economic value. As in [3] stated that the constant increase in the interest rate will cause some problems, such as; the rising cost of funds of banks, because the banks have to pay more to attract new customers and retain existing customers. [8] stated that the risk rate of return is the most important risk faced by Islamic banking compared to other risks such as operational risk and liquidity risk.

Islamic banks obtain funding from two sources, namely 1) the public funds in bank deposits and 2) the public funds in non bank deposits. The former is comprised of wadiah demand deposit, Mudharabah saving deposits and Mudharabah time deposit. In the latter, there are received financing, securities issued by banks, interbank liabilities, liabilities to Bank Central and other payables [9].

In relation to the mudharabah time deposits, Islamic banks provide a rate of return to customers. The rate of return varies between banks because it depends on the bank's profit and the profit sharing ratio that determined at the beginning [10]. In theory, the rate of return on deposits is based on the profit sharing ratio agreed between the bank and the customer.

2.3. Profit and Loss Sharing (Rate of Return)

Profit- and loss-sharing (PLS) is a contractual agreement between two or more transacting parties that permit them to pool their resources to invest inside a project to share in financial profit and loss.

Profit sharing is a system for the distribution of net income between the owners (the first party) and fund managers (second party). Both sides agreed that the working capital provided from the first party to the second party will be managed professionally and responsibly [11].

2.4 Mudharabah Deposit

According to [12] mudharabah contract are the cornerstone of Islamic banking. Mudharabah is a profit sharing and loss bearing contract; and it may be used on both sides of the balance sheet assets and liabilities. It is a contract between the capital provider as rab al mal, whereby the capital provider would contribute capital to an enterprise or activity that is to be managed by the entrepreneur as mudarib.

In the case of time deposit, bank act as the mudharib. In fact usually, as a mudharib bank will set proportion for the result based on equivalent rate. So, the rab al mal will get the profit based on equivalent rate promised at the beginning contract. If the rate of return higher then there is a probability that total time deposit will be increased.

In the countries that have dual banking system like in Indonesia, capital provider also has access about interest rate in conventional bank. A rational depositor will take into account the amount of results that will be obtained by comparing the level of interest in conventional banks and the level of return of sharing in Islamic bank.

2.5 Previous Research

The relationship between total Islamic deposits and the conventional rate of return on deposits has been proved by [13]. A negative relationship between the interest rates on fixed deposits of conventional banks and the volume of interest- free investment deposits of Islamic banks was found.

A survey by [14] analysed that the relation between changes in the rate of interest of conventional bank deposits as well as the rate of return on Islamic bank deposits since 1984 to 2003. [15] found that Islamic banks in Malaysia commonly seen in conventional bank interest rate before adjusting the rate of return or profit. Research conducted by [16] also tried to analyse the effect of changes in interest rates on the level of profit-sharing in Islamic banks and the effect of changes in interest rates on savings deposits in conventional banks and Islamic banking.

Another study by [17] found that the interest rate in conventional banking is one of the main factors that determine the savings in Islamic banking in Indonesia. The same thing is expressed by [18] that the Islamic investment rate is closely linked to the interest rate in conventional banking.

[19] found a strong negative relationship between the interest rate and total deposits in Islamic banks. In another study [13] and [17] showed a negative relationship between the interest rate and total deposits in Islamic banks. In addition, [15] and [3] found a significant relationship between the rate of return of Islamic banks and interest rate of conventional bank and proved that changes in interest rates will cause a change in the level of profit- sharing in Islamic banking. Research conducted by [16] found that the total deposits in Islamic banks is significantly influenced by the rate of return.

Another study conducted by [20] also mentions that the Third-Party Funds (TPF) in Islamic banking is influenced by a conventional deposit rates. [9] states that there is an indication the rate of return on deposits mudharabah tend to mimic the interest rate by adjusting the level of financing and deposits. As in [18] find that in Malaysia only a small

portion of Islamic bank financing is based on profit-loss sharing and that Islamic deposits are not interest-free, but closely pegged to conventional deposits, a finding confirmed by [21] for a sample of large Islamic banks across several countries.
The proposed hypothesis are as follows;
H1: Rate of return has influence on mudharabah deposit
H2: Interest rate has influence on mudharabah deposit
H3: There is no mean difference between rate of return and interest rate

3 Methodology

3.1 Empirical Framework

This study describes the relationship among a set of variables using regression analysis. In order to analyse the effect of Islamic banks' rate of return and conventional banks' interest rate to Islamic banks' total deposit in Indonesia, a regression model is utilized.

TOTDEP = a + β1 IRCB + β2 RRIB + ε TOTDEP = Total Mudharabah Deposit IRCB =
Interest Rate Conventional Bank RRIB = Rate of Return Islamic Bank
a = intercept
β 1 and β2 = coefficient
ε = term of error

Next, this study conducts independent t-test to ascertain if there is a difference in mean between the rate of interest and rate of return.

3.2. Data and analysis

This study uses monthly data covering the period from January 2012 to September 2015. The reason for using 1 month deposit data because this deposit is more responsive with the change of interest rate or rate of return because the depositor can swift their fund quickly. Additionally, 1 month deposits is the largest source of deposits for Islamic banks and conventional deposits compared with 3,6, or 12 months.

4. Data and Empirical Result

The descriptive statistics of variables are presented in Table 1 showing the mean, minimum and maximum for variables rate of return, interest rate, and total mudharabah deposit.

Table 1. Descriptive Statistic

	N	Minimum	Maximum	Mean
Rate of return	45	3,34	7,74	6,2082
Interest rate	45	5,43	6,84	6,1011
Mudharabah deposit	45	19,597	69,106	42,3933
Valid N (listwise)	45			

The average rate of return stands at 6,2082 versus 6,1011 for interest rate. Table 1 indicates that minimum rate of return is lower than interest rate. The average Mudharabah deposit is 42,3933 with minimum 19,597 and maximum deposit 69,106. Using regression analysis, the result shows that simultaneously there is a significant influence in rate of return and rate of interest on mudharabah deposit.

Table 2. Anova

Model	Sum of Squares	df	Mean Square	F	Sig.
Regression	3167,939	2	1583,969	10,452	,000[b]
Residual	6365,202	42	151,552		
Total	9533,140	44			
a. Dependent Variable: DPK					
b. Predictors: (Constant), SKBNG, BGHSL					

Table 3 shows that there is significantly influence for rate of return to Mudharabah Deposit (p=0,002). While for rate of interest there is no significant influence (p=0,185) to Mudharabah Deposit. It means that H1 is accepted while H2 is rejected.

Table 3. Regression Analysis

Model	Unstandardized Coefficients		Standardized Coefficients	t	Sig.
	B	Std. Error	Beta		
(Constant)	122,807	21,245		5,780	,000
RRIB	-7,928	2,404	-,465	-3,297	,002
IRCB	-5,113	3,793	-,190	-1,348	,185

a. Dependent Variable: DPK

The significant negative influence between rate of return and total mudharabah deposit is consistent with [16] which found that rate of return influence mudharabah deposit in general. This indicates that the increase in rate of return will be impacted negatively to deposit in Islamic Bank.
The result also shows that there is a negative influence interest rate to mudharabah deposit but it is not significant. Although the finding is not significant but it is consistent with result from [13] and [17] that found a negative relationship between the interest rate and total deposits in Islamic banks. [22] in his research also found same conclusion. Furthermore to determine whether the rate of return and the interest rate are significantly different, we conduct tests of differences and report the results in Table 4 and table 5.

Table 4. Independent t test

BANK	N	Mean	Std. Deviation	Std. Error Mean
ISLAMIC BANK	45	6,2082	,86406	,12881
CONVENTIONAL BANK	45	6,1011	,54764	,08164

Table 4 presents the average rate of return in Islamic Bank and interest rate in conventional bank. It seems that rate of return in Islamic Bank is higher than interest rate in Conventional Bank.

Table 5. Independent Samples Test

		Levene's Test for Equality of Variances		t-test for Equality of Means				
		F	SSig.	t	df	Sig. (2-tailed)	Mean Difference	Std. Error Difference
PROCENTAGE	Equal variances assumed	2,950	,089	,702	88	,484	,10711	,15250
	Equal variances not assumed			,702	74,438	,485	,10711	,15250

The value on Levene's Test shows significant at 0,089 higher than 0,05. It's means that both group (Islamic bank and conventional bank) have same variances, which is requirement for mean differences test. In Table 5, significant value is 0,485 more than 0,05.

It means that there is no mean difference between rate of return in Islamic Bank and interest rate in conventional bank. This is indicate that H3 is accepted. This finding shows that there is no difference between rate of return and interest rate. The result is in accordance with [9] that stated there is an indication the rate of return on deposits mudharabah tend to mimic the interest rate. Our finding also consistent with [15] who found that Islamic banks in Malaysia commonly seen in conventional bank interest rate before adjusting the rate of return or profit. This result emphasized by [16] that found the relationship between Islamic banks' rate of return and conventional banks' interest rate is shown to be very strong and positive.

5. Conclusion

Using regression analysis, the result found that the rate of return in Islamic bank influence mudharabah deposit negative significantly. This study also found that there is negative influence interest rate on Mudharabah Deposit, but it is not significant. This finding indicates profit motive among the Islamic bank depositors as reflected by the significant impact of the rate of return on total mudharabah deposits.

Using independent t-test, the findings conclude that there is no difference in means of returns between Conventional and Islamic banks. This study however, has its own limitation whereby it is conducted within a limited time period and variables. Further extension of this research should consider a wider scope of deposits by conducting the same approach on different type of deposits available in Islamic banks and specific financial institutions categorized under Islamic financial institutions. Further extension of the study could consider the determinants of rate of return of the Islamic banks.

References

1. A. Andriyanti, and Wasilah, "Factors influences Third Party Funds in Muamalat Bank Indonesia (BMI)",SimposiumNasionalAkuntansi XIII Purwokerto. (2009),
2. M.Y. Nasir, and Hafiz, M.N. *"Islamic Banking and Finance"*, 1st Edition. Kuala Lumpur, Malaysia. McGraw-Hill. (2006),
3. I. Bacha, "Dual banking systems and interest rate risk for Islamic banks", Available at www.fivepillarsassc.com/documentation/Dual-Banking-Systems.pdf, 2004).
4. R.N. Kayed, "The entrepreneurial role of profit-and-loss sharing modes of finance: theory and practice", *International Journal of Islamic and Middle Eastern Finance and Management* Vol. 5 no. 3, pp. 203-228. (2012).
5. S.A Rosly, Al-Bay' Bithamin Ajil financing: Impacts On Islamic Banking Performance, *Thunderbird International Business Review*, Vol. 41 no.4, pp 461-481. (1999)
6. M. Hanif, Differences and Similarities in Islamic and Conventional Banking, *International Journal of Business and Social Science,* Vol 2 No. 2, pp 166-175. (2011).
7. Basel Committee on Banking Supervision. *"Principles for the Management and Supervision of Interest Rate Risk"*, Available at www.bis.org/publ/bcbs102. (2004).
8. Khan, T. and Ahmed. H. "Risk Management: An Analysis of Issues in Islamic Financial Industry", IRTI/IDB Occasional Paper, No. 5. (2001).
9. R. Ismal, *"Islamic Banking in Indonesia. New Perspective on Monetary and Financial Issues"*, John Willey & Son. Singapore Ltd. (2013).
10. D. Olson, and Zoubi, T. "Using Accounting Ratios to distinguish between Islamic and Conventional Banks in the GCC Region", *The International Journal of Accounting*, Vol. 43, pp. 45-65. (2008).
11. M.R.N Akhtar,."Musharakah financing for small enterprise in Pakistan", in Harper, M. (Ed.), Partnership Financial for Small Enterprise: Some lessons from Islamic Credit Systems, ITDG Publications, Rugby, pp. 35-41. (1997).
12. A.M El Thiby ,*"Islamic Banking. How to Manage Risk and Improve Profitability"*, John Willey & Son.Inc. New Jersey. (2011).
13. S. Haron, and Ahmad, N."The effects of conventional interest rates and rate of profits on funds deposited with Islamic banking system in Malaysia", *International Journal of Islamic Financial Services,* Vol. 1 No. 4, pp. 1-7.(2000).
14. I.B. Obiyathulla, "The Islamic interbank money market: an analysis of risks", Asian Islamic Banking Summit-Workshop II - Implementing Islamic Money Market: Issues and Challenges, 23 September 2004. Kuala Lumpur. (2004).
15. A. Kaleem , and M. Md Isa."Causal relationship between Islamic and conventional banking instruments ", *International Journal Of Islamic Financial Services*, Vol.4 No. 4, pp. 1-8.(2003).
16. Z. Zainol, and Kassim, S.H. "An Analysis of Islamic Banks' Exposure to Rate of Return Risk", *Journal of Economic Cooperation and Development,* Vol. 31 No.1, pp. 59-84. (2010).
17. R.A Kasri, and Kassim, S.H. "Empirical determinants of saving of the Islamic banks in Indonesia*", Journal of King Abdul Aziz University: Islamic Economics*, Vol. 22 No. 2, pp.3-23.(2009).
18. B.S. Chong, and Liu, M.-H. "Islamic banking: Interest-free or interest-based? ",*Pacific-Basin Finance Journal,*Vol.17 No.1, pp.125-144.(2008).
19. S. Haron, and Shanmugam, B. The Effect of rates of profit on Islamic bank's deposit: a note, *Journal of Islamic Banking and Finance,*Vol.12 No. 2, pp.18-28.(1995).
20. Farikh, Nurdian, *"Analysis factors influences Third Party Fund in Islamic Bank and Conventional Bank in Indonesia",* Unpublished *Thesis* PSKTTI.UI. (2007).
21. O. Khan, "An Examination of the Underlying Rationale of the Profit and Loss Sharing System, With Special Emphasis on the Mudharabah and Musharakah Within the Context of Islamic Law and Banking", *Journal of Finance, Accounting & Management,* Vol. 3 No.1, pp. 23-31. (2012).
22. I.J. Mangkuto, *"The Influence Rate of Return Conventional Deposit and Rate of Return Mudharabah Deposit to Deposit Growth in Muamalat Bank Indonesia",* Unpublished Thesis. PSKTTI UI. (2004).

SHS Web of Conferences **34**, 12003 (2017) DOI: 10.1051/shsconf/20173412003
FourA 2016

The Effect of Financial Performance and Environmental Performance on Firm Value with Islamic Social Reporting (ISR) Disclosure as Intervening Variable in Companies Listed at Jakarta Islamic Index (JII)

Rahma Frida Ratri* and *Murdiyati* Dewi

Economics and Business Faculty, Airlangga University, Indonesia

Abstract. This study examined how far ISR disclosure as the intervening variable could mediate the effect of financial performance and environmental performance on the firm value. The subjects of this study were the companies listed in Jakarta Islamic Index (JII) in 2012-2014. The samples of the study were sixty companies chosen by purposive sampling method, and the analytical technic used path analysis. The results of this study proved that: (1) financial performance had positive effect to the firm value, (2) environmental performance had no effect to the firm value, (3) financial performance had positive effect to the ISR disclosure, (4) environmental performance had positive effect to the ISR disclosure, (5) ISR disclosure had positive effect to the firm value, (6) ISR disclosure could mediate the effect of financial and environmental performances on the firm value.

1. Introduction

Corporate Social Responsibility (CSR) is a commitment way for the companies to report their social and environmental performances. CSR disclosure cannot be culture free and is attributed primarily to government policy [10. This relationship is based on the companies' role to the stakeholders and the legitimacy of the companies to the society. CSR disclosure can be used as an intermediary to report the firm performance to the stakeholders [12]

The concept of CSR is not only for the conventional economics, but also for shariah economics. The companies listed at shariah index should fulfill the shariah aspects in their operating activity and information disclosure. Now there is Islamic Social Reporting (ISR) concept which is based on Islamic Social Reporting Index (ISRI). This index is the development of the CSR disclosure and match with the shariah principles. The concept of ISR is getting better because of the prior studies on it which discussed the importance of ISR for the Muslim decision makers [4]; [9]; [13] In ISR, the stress on social justice goes beyond reporting the environment, minority interests and employees. It concerns with issues related to the well-being of the society pertaining to interest and unfair trading

* Corresponding author: rahmafridaratri@gmail.com

practices such as income distribution or zakat (Othman *et al.* 2009). The objectives of ISR are to demonstrate accountability to God and community, and to increase transparency of business activities by providing relevant information in conformance to the spiritual needs of Muslims decision makers ([9]The basic constructions of ISR are Tawhid (unity of God) and Shari'a. Tawhid means accepting that God is the ultimate owner of everything on earth and universe (Qur'an 57:5), and man has to be accountable to God (Qur'an 2:28). Shari'a is concerned mainly in promoting social justice and realizing benefits for the people in this world and in the hereafter (al-falah) by setting the basic foundations in every aspect of a Muslim's life including the ethical philosophy [9]

2. Literature review

2. 1. Stakeholder theory

Freeman (1983) changed the definition of shareholder became stakeholder by adding more constituents except shareholders itself (including society and government). The main objective of the stakeholder theory is to help the corporate management in creating the value added from operational activity that they do and minimize the possible loss for corporate stakeholders. Company is not an entity whose operational activity only for their interest but also company has to give benefits for the stakeholders. Going concern of the company depends on the support of the stakeholders so the activity of the company is to look for the stakeholders' support. CSR disclosure has been a successful media to make negotiation to that relationship [8]).

2. 2. Legitimacy theory

Legitimacy theory focuses on interaction between company and the society. Information disclosure policy depends on the perception of the society to decide the financial and economic resources, so company tends to use the CSR disclosure to influence the perception of the society about that company. Company obeys the norms and rules from the society to get the acceptance from society so the firm value will be increasing. The base of legitimacy theory is "social contract" between company and the society. If the society is not satisfied with the attitude and legitimacy of the company, society can pull back that social contract and threaten the continuity of the company's operational activity (Deegan 2002).

3. Hypotheses

Previous studies have found that financial performance significantly effects the firm value. The higher financial performance, the higher firm value [5];[7]. Previous studies also have found that environmental performance significantly effects the firm value. The better environmental performance will give better firm value [1]

H1: Financial performance has significant effect on firm value.

H2: Environmental performance has significant effect on firm value.

Previous studies have found that financial performance significantly effects the ISR disclosure. The higher financial performance, the wider ISR disclosure [2[13]; [11]. Previous studies also have found that environmental performance significantly effects the ISR disclosure. The better environmental performance will make the ISR disclosure becomes wider [11].

H3: Financial performance has significant effect on ISR disclosure

H4: Environmental performance has significant effect on ISR disclosure.

Previous studies have found that ISR disclosure significantly effects the firm value. The wider ISR disclosure, the higher firm value (Arshad *et al.* 2012). Previous studies also

DOI: 10.1051/shsconf/20173412003

have found that ISR can be the intervening variable which can mediate the effect of financial and environmental performances on firm [14].

H5: ISR disclosure has significant effect on firm value.

H6: ISR disclosure can mediate the effect of financial performance on firm value.

H7: ISR disclosure can mediate the effect of environmental performance on firm value.

4. Methodology and results

4. 1. Sample selection and data collection

The data used in this research were quantitative data. Data collection procedure was using documentation method. The annual reports of the companies for 3 years period (2012-2014) were collected from the website of Bursa Efek Indonesia on December 15, 2015. The PROPER ranks were collected from the website of Indonesia Ministry of Environment (KLH) on March 6, 2016. The population and sample were the companies listed at Jakarta Islamic Index (JII) in 2012-2014. The companies which are in JII are the selected companies based on these criteria: their operating activities have fulfilled the Islamic principles, have the biggest market capitalization, and have the best liquidity. The sample selection used purposive sampling by setting the criteria. The data gathered were analyzed using SPSS.

Table 1. Population and Sample Selection.

Criteria	Years		
	2012	**2013**	**2014**
Companies listed at JII	33	32	33
Companies were not listed at PROPER ranks	(11)	(13)	(14)
Total Companies per Year	22	19	19
Total Sample	60		

4. 2. Data measurement

4.2.1. Islamic Social Reporting (ISR)

ISR disclosure is measured by Islamic Social Reporting Index (ISRI). The index used based on the prior study by Othman *et al.* (2009). There are 43 items divided into 6 themes: finance and investment, product and services, employees, environment, and corporate governance. Each item will get score, if that item is disclosed then will get score 1, but if that item is not disclosed then will get score 0. This is the formula for ISRI:

$$ISR\ Index = \frac{Total\ Score\ on\ Company}{Maximum\ Total\ Score}$$

4.2.2. Financial performance

The proxy used for financial performance is Return on Assets (ROA). This is the formula for ROA:

$$Return\ on\ Assets = \frac{Net\ Income\ after\ Tax}{Total\ Assets}\ x\ 100\%$$

4.2.3. Environmental performance

The proxy used for environmental performance is PROPER ranks. PROPER is a program from Indonesia Ministry of Environment to supervise and control the environmental performance of the companies in Indonesia through information disclosure instrument. The PROPER ranks consist of 5 colors:

Gold : Very very good
Green : Very good
Blue : Good
Red : Bad
Black : Very bad

4.2.4. *Firm value*

The proxy used for firm value is Tobin's Q. If the point of Tobin's Q is more than 1, it means that company has big competitive advantage so the investment opportunity in that company is also big. This is the formula for Tobin's Q:

$$Tobin's\ Q = \left[\frac{(OS\ x\ P) + (D + I) - CA}{TA}\right]$$

Where: Tobin's Q = Firm Value; OS = Outstanding Shares; P = Stock Price; D = Total Debt of Company; I = Total Inventory; CA = Current Assets; TA = Total Assets

4. 3. Results

Table 2. The Effect of ROA and PROPER on Tobin's Q.

Model		Unstandardized Coefficients		Standardized Coefficients	t	Sig.
		B	Std. Error	Beta		
1	(Constant)	-0.404	1.167		-0.346	0.730
	ROA	0.274	0.031	0.799	8.883	0.000
	PROPER	-0.150	0.354	-0.038	-0.424	0.673

Dependent Variable: TOBINS

Based on the results on Table 2, can be concluded that financial performance (ROA) has significant influence on firm value (TOBINS). This can be seen from Sig. 0,000 < significance rate 0,05 (5%). This shows that financial performance significantly influences firm value so H1 is accepted. Based on the results on Table 2, can be concluded that environmental performance (PROPER) has not significant influence on firm value (TOBINS). This can be seen from Sig. 0,673 > significance rate 0,05 (5%). This shows that environmental performance does not significantly influence firm value so H2 is rejected.

Table 3. The Effect of ROA and PROPER on ISR Disclosure.

Model		Unstandardized Coefficients		Standardized Coefficients	t	Sig.
		B	Std. Error	Beta		
2	(Constant)	0.620	0.033		18.643	0.000
	ROA	0.004	0.001	0.457	4.080	0.000
	PROPER	0.026	0.010	0.292	2.610	0.012

Dependent Variable: ISR

Based on the results on Table 3, can be concluded that financial performance (ROA) has significant influence on ISR disclosure (ISR). This can be seen from Sig. 0,000 < significance rate 0,05 (5%). This shows that financial performance significantly influences

ISR disclosure so H3 is accepted. Based on the results on Table 3, can be concluded that environmental performance (PROPER) has significant influence on ISR disclosure (ISR). This can be seen from Sig. 0,012 < significance rate 0,05 (5%). This shows that environmental performance significantly influences ISR disclosure so H4 is accepted.

Table 4. The Effect of ISR Disclosure on Tobin's Q.

Model		Unstandardized Coefficients		Standardized Coefficients	t	Sig.
		B	Std. Error	Beta		
3	(Constant)	-14.498	3.745		-3.871	0.000
	ISR	22.677	4.906	0.519	4.622	0.000

Dependent Variable: TOBINS

Based on the results on Table 4, can be concluded ISR disclosure (ISR) has significant influence on firm value (TOBINS). This can be seen from Sig. 0,000 < significance rate 0,05 (5%). This shows that ISR disclosure significantly influences firm value so H5 is accepted.

Table 5. The Mediating Effect.

Path	Beta (Indirect Effect)	Beta (Direct Effect)	Total Effect
X1 → Z → Y	0,237	0,799	1,036
X2 → Z → Y	0,152	-0,038	0,114

Based on the results on Table 5, can be concluded that ISR disclosure can mediate the effect of financial performance on firm value. This can be seen from Sig. 0,000 < significance rate 0,05 (5%) and Beta of the indirect effect 0,237. This shows that ISR disclosure can be used as an intervening variable for the effect of financial performances on firm value so H6 is accepted. Based on the results on Table 5, can be concluded that ISR disclosure can mediate the effect of environmental performance on firm value. This can be seen from Sig. 0,012 < significance rate 0,05 (5%) and Beta of the indirect effect 0,152. This shows that ISR disclosure can be used as an intervening variable for the effect of environmental performance on firm value so H7 is accepted.

5. Conclusion

Based on the results, can be concluded that financial performance has significant influence on firm value. This finding supports the prior studies by [5] and [7]. This indicates that the increasing ROA can increase the value added for the firm. It is also can be concluded that environmental performance does not significantly influence firm value. This finding is consistent with prior study by [14] but inconsistent with prior study by [1]). This indicates that the information about corporate environmental performance in the form of PROPER ranks which is issued by Indonesia Ministry of Environment (KLH) does not significantly influence the value added of the firm.

Based on the results, can be concluded that financial performance has significant influence on ISR disclosure. This finding supports the prior studies by [2], [13], and [11] This indicates that the higher profitability, the wider ISR disclosure. It is also can be concluded that environmental performance has significant influence on ISR disclosure. This finding is consistent with prior study by [11]. This indicates that company who has good environmental performance will tend to disclose it in their ISR disclosure.

Based on the results, can be concluded that ISR disclosure has significant influence on firm value. This finding is consistent with prior studies by [2] and [30] This indicates that the wider ISR disclosure, will increase the trust of the Muslim stakeholders. It is also can be concluded that ISR disclosure can be an intervening variable which can mediate the

 DOI: 10.1051/shsconf/20173412003

influence of financial performance and environmental performance on firm value. This indicates that ISR disclosure has the important role to mediate the influence of financial performance and environmental performance on firm value.

References

1. Al-Tuwaijri, Sulaiman A., Theodore E. Christensen & K.E. Hughes II. The Relation Among Environmental Disclosure, Environmental Performance, and Economic Performance: A Simultaneous Equations Approach. *Accounting Organizations and Society*, **2**: 1-31(2004)
2. Anuar, Hairul A., Maliah Sulaiman & Nik Nazli Nik Ahmad. Some Evidence of Environmental Reporting by Shariah Compliant Companies in Malaysia. *IIUM Journal of Economics and Management*, **17**(2): 177-208 (2009)
3. Arshad, R, Suaini Othman & Rohana Othman. Islamic Corporate Social Responsibility, Corporate Reputation and Performance. *International Journal of Social, Behavioral, Educational, Economic, Business and Industrial Engineering,* **6**(4): 643-647 (2012)
4. Baydoun, Nabil & Roger J. Willett. Islamic Corporate Reports. *ABACUS*, **36**(1): 71-90 (2000)
5. Bidhari, Sandhika C., Ubud Salim & Siti Aisjah. Effect of Corporate Social Responsibility Information Disclosure on Financial Performance and Firm Value in Banking Industry Listed at Indonesia Stock Exchange. *European Journal of Business and Management*, **5**(18): 39-46 (2013)
6. Deegan, Craig. The Legitimising Effect of Social and Environmental Disclosures – A Theoretical Foundation. *Accounting, Auditing & Accountability Journal*, **15**(3): 282-311 (2002)
7. Gamayuni, Rindu Rika. The Effect of Intangible Asset, Financial Performance and Financial Policies on the Firm Value. *International Journal of Scientific & Technology Research*, **4**(1): 202-212 (2005)
8. Gray, R., Reza Kouhy & Simon Lavers. Corporate Social and Environmental Reporting: A Review of the Literature and a Longitudinal Study of UK Disclosure. *Accounting, Auditing, & Accountability Journal,* **8**(2): 47-77 (1995)
9. Haniffa, R. Social Reporting Disclosure – An Islamic Perspective. *Indonesian Management & Accounting Research*, **1**: 128-146 (2002)
10. Haniffa, R.M. & T.E. Cooke. The Impact of Culture and Governance on Corporate Social Reporting. *Journal of Accounting and Public Policy*, **24**: 391-430 (2005)
11. Maulida, Aldehita P., Agung Yulianto & Asrori. Analisis Faktor-faktor yang Mempengaruhi Pengungkapan Islamic Social Reporting (ISR). *Simposium Nasional Akuntansi XVII.* Mataram (2014)
12. Nurlela, R & Islahuddin. Pengaruh Corporate Social Responsibility terhadap Nilai Perusahaan dengan Prosentase Kepemilikan Manajemen sebagai Variabel Moderating. *Simposium Nasional Akuntansi XI.* Pontianak (2008)
13. Othman, R., Azlan Md. Thani & Erlane K. Ghani. Determinants of Islamic Social Reporting Among Top Shariah-Approved Companies in Bursa Malaysia. *Research Journal of International Studies,* **12**(10): 4-20 (2009)
14. Rahmawati, A'la & Tarmizi Achmad. Pengaruh Kinerja Lingkungan terhadap Financial Corporate Performance dengan Corporate Social Responsibility Disclosure sebagai Variabel Intervening. *Diponegoro Journal of Accounting*, **1**(2): 1-15 (2012)

The impacts of the antecedents on absorptive capacity and export performance: a meta-analytic review

Wu Wann-Yih[1] and *Permatasari* Yani[2,*]

[1] International Business Department, Chinese Culture University, Taipei, Taiwan
[2] Department of Accountancy, Faculty of Economics and Business, Universitas Airlangga, Surabaya, Indonesia

Abstract. Although exporting performance and absorptive capacity, with its relevant antecedents have been studied extensively, rare of them have integrated these variables into a more comprehensive framework. Not only this study aims to synthesize extant knowledge on the subject based on meta-analysis and empirical verify the influences of absorptive capacity on export performance, but also the antecedents of absorptive capacity are evaluated. The assessment reveals that: all hypotheses are supported and the antecedents of absorptive capacity have small-to-medium effects size with positive relationship.
Keywords: Export performance, Absorptive capacity, Meta-analysis

1. Introduction

As a result from the still fragmented knowledge regarding successful export practice--despite of it has been an important stream of international marketing research for over 25 years--, the importance of theories development in this exporting area is arising. Therefore it is essential to develop theory which focuses on the causal processes that give enlargement to variation in export performance among firms [1,2].

This study proposes three antecedents of absorptive capacity which are regulatory focus, export orientation, and organizational cultural intelligence. This study argues that having promotion focus or prevention focus, being proactive toward customers' needs or being responsive toward competitors' actions in the market, and having ability to adapt will increase the level of absorptive capacity [3-6]. Moreover, this study discusses about under what conditions that these antecedents have stronger or weaker influences toward firms' absorptive capacities in their export activities.

2. Literature Review

2.1 The Effect of Regulatory Focus

Regulatory focus refers to when a person pursues a goal in a way that maintains the person's own personal values and beliefs, with basic principle that people motivated to approach pleasure but avoid pain [3]. Higgins (1997, 1998) [3,4] proposed two distinct self-regulatory systems, one in which people have a promotion focus and the other in which they have a prevention focus. Each regulatory focus influences the nature and magnitude of the people's emotional experience and affects the perception, decision making, and also individuals' behaviors and performances [3,4,7]. As for the context of export, exporters with stronger chronic promotion concern tend to be more "risk-taker", wanting to benefit from high energy and to develop the advantage of creative behaviors [7-10]. On the other hand, exporters with stronger chronic prevention focus are more to be "risk-averse"[8,11]. Furthermore, the purpose of the prevention focus is wanting to avoid the chance of illness or assure their safety and security, to maintain routines, and to preserve the status quo [9,10]. Therefore, these exporters will have intention to form relationships with their partners, produce some innovation, increasing the export performances, and developing its absorptive capacity at the same time. Thus, this study propose:
H1 - There is a positive influence of exporters' regulatory focus on their absorptive capacity
H4 - There is a positive influence of exporters' regulatory focus on (a) organizational innovativeness, (b) export performance, and (c) relationship satisfaction

2.2 The Effect of Export Orientation

In this study, export orientation is developed based on the market orientation, which consists of two behavioral components: customer orientation and competitor orientation [5]. Customer orientation emphasizes the role of sufficient

* Corresponding author: yanipermatasari@feb.unair.ac.id

understanding of target customers.Consequently, the firm will have greater capacity to absorb new information related with its customers, and also have better export outcomes. On the other hand, some firms frequently makes decisions so as to perform well relative to their competitors which Armstrong and Collopy (1996) [13] refers that as competitor-oriented objectives.

Furthermore, the customer and competitor orientations are two distinct approaches that will lead to competitive advantage. Because customer-oriented exporters has detailed analyses of customer needs in order to reach customer's high satisfaction, a customer orientation will result in a differentiation advantage. For a competitor-oriented exporters, they tends to focus on monitoring the costs in order to match the competitors initiative, thus they may have a cost advantage as a result [14]. Therefore, this study propose,

H2 - There is a positive influence of exporters' export orientation on their absorptive capacity

H5 - There is a positive influence of exporters' export orientation on (a) organizational innovativeness, (b) export performance, and (c) relationship satisfaction

2.3 The Effect of Cultural Intelligence

Cultural intelligence (CQ) is defined as a natural ability that an outsider has, to interpret someone's unfamiliar and ambiguous gestures in just the way that person's compatriots and colleagues would, even to mirror them [15]. Ang and Inkpen (2008) [16] stated that there are three components of a firm's CQ that is managerial, competitive, and structural capabilities. They also suggested that as boundary spanners, project managers or exporters should come into daily contact with supplier and/or customer in the international business ventures. Hence, the exporters should have greater level of cultural awareness, knowledge about the other cultures, and the capability to gives appropriate verbal and nonverbal actions with people from different cultures [17].

Moreover, exporters' possession of competitive CQ could be viewed as a meta-capability [18] that transcends technical or operational capabilities. Consequently, exporters who have these competitive CQ can be expected to have better capability to create relationship satisfaction, to have greater absorptive capacity, and good export performance. Thus, this study propose:

H3 - There is a positive influence of exporters' organizational cultural intelligence on their absorptive capacity

H6 - There is a positive influence of exporters' organizational cultural intelligence on (a) organizational innovativeness, (b) export performance, and (c) relationship satisfaction

2.4 The Effect of Absorptive Capacity

Absorptive capacity is the ability of a firm to recognize the value of new, external information, assimilate it, and apply it to commercial ends to produce a dynamic organizational capability. Such ability is critical to its innovative capabilities [19-20].

Lane et al. (2001) [21] developed measures for the three processes within absorptive capacity of recognition, assimilation, and utilization (c/f [22]). In the context of export activity, when exporters have ability to recognize, absorb, and understand new external related to the operations knowledge, then apply that new knowledge toward commercial ends right after combining it with their prior knowledge, the exporters can be expected to have more innovations, greater export-related performance, and higher satisfaction with its partners [19,20,23]. Thus, this study proposes:

H7 - There is a positive influence of exporters' absorptive capacity on (a) organizational innovativeness, (b) export performance, and (c) relationship satisfaction

3. Methodology

3.1 Conceptual Framework

In this study, the following conceptual framework, which is developed based on the research backgrounds, describes the relationship between absorptive capacity and its antecedents, and their impact to the product innovativeness, export performance, and relationship satisfaction.

DOI: 10.1051/shsconf/20173413001

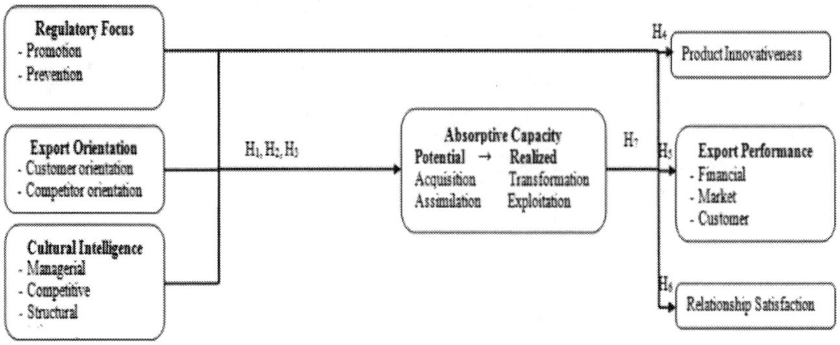

Fig. 1. The Conceptual Framework

3.2 Construct Measurement

After completing the journal-by-journal searching process, a total of 84 studies done in 1984–2016 was obtained in this study. Of these 84 studies, there was a body of 18 that included theoretical papers, and qualitative investigations, which left 66 studies with quantitative validation. These studies produced 15,804 independent samples, 66 effect sizes, and a cumulated n of 28 subjects. A list of the studies used in our empirical meta-analysis is available on request.

All identified studies were examined in terms of the following relevant variables: authors, year, journal, total sample size, the statistic measurement of each variable, and effect sizes. Two alternatives are used for inclusion: (1) correlational studies must present the correlation coefficient (r) or the standardized regression (beta) coefficient or (2) studies must present the related statistic (t-test, p-value) for the relationship between related variables. These statistics could be converted into effect size (r) and Fisher Z effect using Comprehensive Meta-Analysis (CMA) Software. In addition, this paper also incorporates those studies that provide only the standardized regression (beta) coefficients using the formula suggested by Peterson and Brown (2005) to estimate the correlations from the beta coefficients. The formula used is $r = 0.98\beta + 0.05\gamma$, where λ is a variable that equals 1 when β is non-negative and 0 when β is negative.

Moreover, this study also conducts homogeneity analysis which tests whether the assumption that all of the effect sizes are estimating the same population mean is a reasonable assumption. Homogeneity of the effect size distribution was tested by the Q statistic, distributed as a chi-square with degree of freedom calculated by n-1, where n is a number of studies [24].

4. Result

Table 1 presents the result from integration of effect sizes of the variable. Reminds as stated in Lipsey and Wilson (2001) [24] r analyzing the magnitude of effect sizes ($r < 0.10$ as small; $r = 0.25$ as medium, and $r > 0.40$ as large effect size), it is shown in Table 1 that these variable, the antecedents of absorptive capacity have small-to-medium effects size with the relationship varying from 0.018 (Hypothesis 3) to 0.311 (Hypothesis 9). Following is the analysis of the hypotheses result,

Hypothesis 1 shows the coefficient correlation result as 0.279 which shows that there is positive correlation with medium effect size. Different with hypothesis 1, hypothesis 4 has small effect size with coefficient of correlation (r) = 0.025.

Hypothesis 2 has the coefficient correlation result as 0.109. Therefore, this hypothesis is supported with medium correlation. So does Hypothesis 5, with coefficient correlation as 0.224, this hypothesis is supported with category of medium effect size.

Hypothesis 3 and Hypothesis 6 have coefficient correlation 0.018 and 0.038, which means there is a small effect size. Finally, the last hypothesis asserted that exporters' absorptive capacity has positive influences on organizational innovativeness, export performance, and relationship satisfaction. This hypothesis is supported because it has coefficient correlation $r = 0.309$ with category medium effect size.

5. Conclusion

5.1 Conclusion

From the findings above, this study can be concluded as follows: First of all, exporters emphasizing on prevention focus – which tend to be "risk-averse", concerned with duties and obligations, are likely to be sensitive to the presence or absence of punishments, and use avoidance as a goal attainment strategy – and exporters emphasizing on promotion focus, which is wanting to benefit from high energy and to develop the advantage of creative behaviors [9,10], will have intention to form relationships with their partners, produce some innovation, increase the export performances, and develop its absorptive capacity at the same time. Secondly, it is important for exporters to have customer orientation or competitor orientation simultaneously. These are two distinct approaches that will lead to competitive advantage.

Thirdly, firms that have this competitive capability will be able to integrate and combine various knowledge assets within the firm and between the firm and international business partners [21,22]. Consequently, exporters who have these competitive CQ can be expected to have better capability to create relationship satisfaction, to have greater absorptive capacity, and good export performance.

Fourthly, when exporters have ability to recognize, absorb, and understand new external related to the operations knowledge, then apply that new knowledge toward commercial ends right after combining it with their prior knowledge, the exporters can be expected to have more innovations, greater export-related performance, and higher satisfaction with its partners [23,24,27]. Therefore, in order to increase the export performance, exporters could maximizing the using of the antecedents of absorptive capacity.

5.2 Limitation and Future Research Direction

There are several limitations in this study. Though the literatures were comprehensively surveyed, it can not guarantee a full coverage of all available studies. In addition, several studies could not be included due to insufficient information. However, no systematic difference between the studies included and those few not included are anticipated. The challenge for future research is to reach beyond studies which are found here to identify other important contextual variables related to the export performance.

References

1. N.-E. Aaby, S.F. Slater. Intl. Mrktg Rev. **6(4)**, 53–68 (1989).
2. S. K. Chetty, R.T. Hamilton. Intl. Mrktg Rev, **10(3)**, 26–34 (1993).
3. E.T. Higgins. Amrcn Psycl., **52**, 1280–1300 (1997).
4. E.T. Higgins. Adv. in Exp. Soc. Psycl., **30**, 1–46 (1998).
5. J. C. Narver, S. F. Slater. J.of Mrktg, **54(4)**, 20-35 (1990).
6. S. Ang, A. C. Inkpen. Dec. Sci., **39**, 337-358 (2008).
7. J. Brockner, T. E. Higgins. Org. Bhvr. & Human Dec. Proc., **86**, 35-66 (2001).
8. R. Friedman, J. Förster. J. of Prsnlty and Soc. Psycl., **81**, 1001-1013 (2001).
9. J. L. Aaker, A. Y. Lee. J. of Consmr. Rsrch., **28**, 33–49 (2001).
10. R. Kark, D. Van Dijk. Acdmy. of Man. Rev., **32(2)**, 500–528 (2007).
11. E. Crowe, E. T. Higgins. Org. Bhvr. & Human Dec. Proc., 69, 117-132, (1997).
12. A. K. Kohli, B. J. Jaworski. J.of Mrktg, **54(2)**, 1–18(1990).
13. J. S. Armstrong, F. Collopy. J. of Mrktg Rsrch, 33(2), 188–199(1996).
14. K. Z. Zhou, J. R. Brown, C. S. Dev, S. Agarwal. J. of Intl. Busns. Stud., **38**, 30(2007).
15. P. C. Earley, E. Mosakowski. Harvard Business Review, **83**, 139-146(2004).
16. S. Ang, A. C. Inkpen. Dec. Sci., **39**, 337-358(2008).
17. P. C. Earley, S. Ang. Palo Alto, CA: Stanford University Press (2003).
18. D. J. Teece. Strat. Man. J., **28(13)**, 1319–1350(2007).
19. M. W. Cohen, D. A. Levinthal. Adm. Sci. Quartrly., **35**, 128–152(1990).
20. S. A. Zahra, G. George. Acdmy. of Man. Rev., **27**, 185–203(2002).
21. P. J. Lane, J. E. Salk, M. A. Lyles. Strat. Man. J., **22**, 1139-1161 (2001).
22. H. W. Volberda, N. J. Foss, M. A. Lyles. Strat.Man.&Glob.,978-87-91815-51-5 (2009).
23. W. Tsai. Acdmy. of Man. J., **44(5)**, 996-1004(2001).
24. M. W. Lipsey, D. B. Wilson. Thousand Oaks, CA: Sage(2001)
25. Permatasari, Yani. http://ir.lib.pccu.edu.tw/bitstream/987654321/24688/2/fb14120517 4750.pdf. (2011)

Table 1. Relationship between Studies in Meta-Analytic Review

Meta-Analytic Relationship											
Hyp.	Variables		k studies	Total n	Effect Size & 95% Confidence Interval			Heterogeneity			
	Independent	Dependent			r	LCI	UCI	P-value	Chi-Square	Q-value	I-squared
1	Regulatory Focus	Absorptive Capacity	6	324	0.279	0.173	0.379	0.456	11.07	4.679	0.000
2	Export Orientation	Absorptive Capacity	6	730	0.109	0.036	0.181	0.000	20.52	34.853	82.249
3	Cultural Intelligence	Absorptive Capacity	8	10081	0.018	-0.001	0.038	0.000	24.32	149.9	95.33
4	Regulatory Focus	(a) Organizational Innovativeness; (b) Export Performance; (c) Relationship Satisfaction	7	751	0.025	-0.047	0.097	0.000	22.46	46.519	87.102
5	Market Orientation	(a) Organizational Innovativeness; (b) Export Performance; (c) Relationship Satisfaction	7	608	0.224	0.146	0.299	0.000	22.46	17.298	65.315
6	Cultural Intelligence	(a) Organizational Innovativeness; (b) Export Performance; (c) Relationship Satisfaction	8	934	0.038	-0.027	0.103	0.000	24.32	41.367	83.078
7	Absorptive Capacity	(a) Organizational Innovativeness; (b) Export Performance; (c) Relationship Satisfaction	24	2376	0.309	0.229	0.318	0.000	27.88	131.158	93.906

DOI: 10.1051/shsconf/20173413002

Corporate social responsibility and corporate governance in Indonesian public listed companies

Siti Rochmah Ika[1,*],*Titop* Dwiwinarno[2], and *Ari Kuncara Widagdo*[3]

[1]Accounting Department, Faculty of Economics, Janabadra University, Yogyakarta, Indonesia
[2]Management Department, Faculty of Economics, Janabadra University Yogyakarta, Indonesia
[3]Accounting Department, Faculty of Economics, Sebelas Maret University Solo, Indonesia

Abstract. This study aims to investigate whether theree has been a change in the level of corporate social responsibility (CSR) disclosure and to examine whether corporate governance attributes influence CSR disclosure in corporate annual report of Indonesian public listed companies(PLCs). The annual reports of 115 PLC for two years (2011 and 2012) were analysed using content analysis. Multiple regression analysis was utilized to determine factors influencing CSR disclosure in annual reports. Consistent with expectations, the paired sample t-test showed that there was an increase (significant at the 1 percent level) in the extent of CSR disclosure. The multiple regression analysis revealed that audit committee effectiveness and company's size were positively associated with the extent of CSR disclosure (significant at 5 per cent level). The findings appear to suggest that The Indonesian Capital Market and Financial Institutions Supervisory Agency (Bapepam LK) effforts in promoting CSR through the release of Regulation No X.K.6 in 2012 have had some positive impact on CSR disclosure in annual report. The results also suggest that the involvement of audit committee through its effectiveness in overseeing company's financial reporting could lead to better concern in corporate social activities and hence disclosure in annual reports. This study however, has limitation that should be considered in interpreting the results. The regression model documented an R^2 of 21.4 percent, which indicates that almost 80 percent of factors influencing CSR disclosure in Indonesian PLC have not been captured by the model. These other factors may perhaps be indentified in the next research.
Keywords: Corporate social responsibility, corporate social responsibility disclosure, corporate governance, audit committee, audit committee effectiveness.

1 Introduction

The Indonesian government's incentive to further promote CSR among PLCs is very encouraging. The enactment of Law No 40 in 2007 concerning Limited Liabilities Company and Government Regulation No 47/2012 concerning Corporate Social and

*Corresponding author: ika@janabadra.ac.id

DOI: 10.1051/shsconf/20173413002

Environmental Responsibility has made CSR activities mandatory to all companies in Indonesia. In terms of disclosure, Bapepam LK has issued Rule No x.k.6/2012[1]concerning *Submission of Issuer or Public Company Annual Report*. This rule basically provides detailed guideline with respect to the form and content of annual reports of PLC which serve as a source of information for investors and public. In the rule, [1], point h, it is stated that the discussion of CSR includes policies, types of programs, and costs, among others related to aspects: (a) environment, (b) labor practices, (c) social and community development, and (d) product safety.This new rulereplaces the previous Rule No X.K.6/2006 [2] in which one of the main changes is the content of CSR disclosure. In the old rule [2], it was only stated that the CSR disclosure *"is a description of the activities and costs relating to corporate social responsibility towards society and the environment"*. Hence it may be expected that after the issuance of the new rule, the content of CSR disclosure among PLC increases.

The purpose of this study is to investigate whether there has been a change in the level of CSR disclosure after the issuance of the new rule and to determine wheteher corporate governance attributes influence CSR disclosure in corporate annual reports of Indonesian PLCs. Unlike previous studies in Indonesia, [3],[4]which haveonly focused on the amount and theme of CSR in a single year, this study extends prior studies by examining if there had been changes in the extent of CSR disclosure in Indonesian PLCs. Examining the level of CSR discosure would give some indications on whether guidelines as stated in the new rule have had some positive effect on corporate social activities.

The present study also examine the influence of corporate governance attributes in CSR disclosure of which is audit committee effectiveness. A number of studies including [5],[6], [7], have examined the association of audit committee andCSR disclosure. Some of these studies examined single charasteristic ofaudit committee, for example proportion of independent non-executive directors sit in audit committee[5]; the presence of audit committee [6]. Likewise, [7] examined some characteristics of audit committe namely audit committe size, expertise, and meeting in influencing CSR dislosure. The examination of these characteristics however may not be adequate to capture the effectiveness of audit committee in overseeing financial reporting quality [8], specifically in terms of the level of CSR disclosure. The present study fills the gap in the literature by utilizing an index to measure audit committee effectiveness as developed by [8]. The index comprises of audit commitee charteristics related to structure, membership, independence, job duties, and disclosure of auditcommittee in company annual reports.

The remainder of the paper is organized as follows. The next section discuss prior relevant literature and hypotheses development. The research method including sample selection and data collection is then explained. Findings and analysis which are followed by concluding remarks are then discussed in the ensuing section.

2 Literature Review and Hypotheses Development

A number of studies have been done relating corporate governance to CSR disclosure; for example [4-7], [9-11]. Corporate governance factors that have been investigated in these studies include board size, board independence, CEO duality; board meetings, women on board, audit committee characteristics, and ownership structure (i.e. foreign ownership, managerial ownership, government ownership).In Indonesian context,[4]examined the impact of board size, foreign ownership, firms size, profitability, and leverage on CSR reporting. The study found thatboard size board and firm size have a positive impact on CSR disclosure. [4] however, yet to examine the influence of audit committee in their study. Therefore this paper attempts to fill this gap.

2.1 The extent of CSR disclosure

As stated in the introduction section,[1] regulates the disclosure of information in issuers' and public company's annual reports, which serve as a source of information for shareholders and the public. In terms of CSR, this rule provides guideline with respect to content of the disclosure in annual reports of PLCs which the discussion of CSR includes policies, types of programs, and costs, related to aspects: (a) environment, (b) labor practices, (c) social and community development, and (d) product responsibility.The obligation to prepare the annual report in accordance with the new rule applies for the financial years ended 31 December 2012 onwards. Therefore, with the issuance of [1] rule, PLCs are expected to involve more CSR activities and have more disclosure of CSR in the annual report.

[9] scrutinized the level of CSR disclosure after the introduction of CSR guideline, called Silver Book among Malaysian government-link companies (GLCs). The study found that that there was an increase and statistically significant in the extent of CSR disclosure in year 2007 as compared to 2005. Therefore, it may be expected that the level of CRS disclosure among Indonesian PLCs increases after the issuance o the new rule. The hypothesis is expressed as follow.

H1: There is a significant increase in the extent of CSR disclosure in the annual reports of Indonesian PLCs after the issuance of [1] rule.

2.2 Board Size

[5] stated that board of directors is a crucial aspects of corporate governance mechanism relating to its duty in monitoring how the companies business is being properly managed by their agents. Indonesia adopts a two-tier governance model, meaning there are two board structures in the company, board of commissioners (BOC) and board of directors (BOD). BOC is a shareholder representative whose duty is to provide strategic guidance and give advice to management decision, while BOD is in charge to run the business. The two tier board structure ensures that all commissioners are non-executives. The functions of BOC in two-tier model are similar to those of BOD in a one-tier system.

[12] argued that the CEO will be easier to be controlled and monitored by larger board. However, [13] advocated that as boards grow, they are less likely to function effectively because coordination and information- processing costs outweigh the benefit of having more people's expertise. Empirical studies on the relationship between board size and CSR disclosure are mixed. [10] documented that board size has no association with CSR disclosure. Meanwhile [5, 6, 9] suggested that board size is positively associated with the extent of CSR disclosure. In Indonesian context, [4] found that board size has a positive and non-linier relationship with the extent of CSR disclosure. The results imply that large boards are able to oversee business operations better than smaller ones; however too large a board makes the process of monitoring ineffective. Thus it is hypothesized that:

H2: There is a positive association between a board size and the extent of CSR disclosure in Indonesian PLCs.

2.3 Board Independence

In the corporate governance literature, the presence of independent directors is presumed to strengthen the board by monitoring the activity of management. Therefore, it is generally

 DOI: 10.1051/shsconf/20173413002

believed that with a high proportion of independent directors, the interests of the investors are protected, specifically when arriving at board decision [14]. Several studies have examined the association between board independence and CSR disclosure. [6,7] showed a positive and significant relationship between board independence and CSR disclosure. In Indonesian Stock Exchange (IDX), listing requirements state thatthe number of independent commissioners is at least 30% of the total BOC members. This requirement can be interpreted as having more independent boards to be more effective in monitoring role. Some studies in Malaysia [5,9] did not find any association between board independence and CSR disclosure, neither did in the US [10]. With these literatures, it is hypothesized that:

H3: There is a positive association between proportion of independent commissioners and the extent of CSR disclosure in Indonesian PLCs.

2.4 Audit Committee Effectiveness

The role of audit committee in enhancing corporate governance standard has been well documented by prior studies[e.g. 15,16]. As a committee of the board, its existence relates to oversight responsibility in the area of financial reporting, internal control, and external auditing activity. Therefore, in the agency theory context, it may be expected that effective audit committee are likely to improve financial reporting quality, including the extent of disclosure.

A number of studies have examined the association between audit committee and financial reporting quality using some proxies for financial reporting quality. These proxies include timeliness of reporting [8], level of interim financial disclosure [17], and the extent of CSR disclosure [5-7].[8]found that the effectiveness of audit committee may reduce reporting lag It was found that the existence of an audit committee was significantly and positively related to the extent of CSR disclosure [7]. Although, it appears unlikely that audit committee financial expertise may be linked directly to the understanding of audit committee members of CSR-related information, [6] documented that the number of financial expert on audit committee and the number of audit committee meetings per year were positively associated with the extent of CSR disclosure. Hence, it is hypothesized that:

H4: There is a positive association between audit committee effectiveness and the extent of CSR disclosure in Indonesian PLCs.

2.5 Ownership Concentration

According to [11], companies that have dispersed ownership are expected to involvemore in community or social activities since the issue of public accountability is important in these companies. Therefore, it may be expected that there will be more CSR disclosures in annual report of widely held companies compare to closely held companies, as a consequence of the higher level of public accountability. Conversely, in a highly concentrated ownership company, since the public interest is relatively low, the social activities may also be expected less active.

Evidence from previous study [18]supports the existence of a negative relationship between ownership concentration and the extent of voluntary disclosure in annual reports, where CSR disclosure in the study was part of non financial voluntary disclosure.However, [11] found that ownership concentration has no effect on CSR disclosure of Malaysian company annual reports. The hypothesis is as follow

H5: There is a negative association between ownership concentration and the extent of CSR disclosure in Indonesian PLCs.

3 Research Method

3.1 Sample and Data

The sample consists of all 148 manufacturing companies listed on IDX from 2011 to 2012. Thirty three companies excluded from the sample due to incomplete information (e.g. annual reports were not available during window period) and missing data. The final sample comprises 115 with a total of 230 firms-years observations. Data was collected from annual reports of sample companies which are available in IDX website. Social responsibility information was hand-collected from any sources in annual reports wherever the information is available, for example from the CSR disclosures, corporate governance disclosures, directors' report, Chairman's statement, and notes to the financial statement.

3.2 Dependent Variable

The extent of CSR disclosure represents the dependent variable in the study. Following [19], the extent of CSR disclosure was measured by a CSR disclosure index based on the Global Reporting Initiative (GRI) guideline. The GRI provides a framework of the CSR-related issues that should be disclosed in corporate annual reports. We construct a 79 items of checklist which covers 6 themes, namely Economic (9 items), Environment (30 items), Labour (14 items), Human Right (9 items), Society (8 items), and Product Responsibility (9 items). In finalizing the checklist, reference was made to the [1] rule. Some items in the checklist which were stated in the rule, i.e. policies, types of programs, and costs, related to environment, labor practices, social and community development, and product responsibilityindicated as the mandatory CSR disclosure. The numbers of these items were 30. The remaining 49 items were then indicated as voluntary CSR disclosure.

Scoring was based on the existence of the items and equally weighted.A dichotomous procedure was utilized whereby a company is scored a 1 if an item included in the checklist is disclosed and 0 if it is not disclosed. The number of item disclosed was then divided by the maximum possible score (79) to get the CSR disclosure index [7,9,11].

3.3 Independent Variables

Corporate governance mechanisms namely Board Size (BDSIZE), Board Independence (BIND), Audit Committee Effectiveness (ACEF), and Ownership Concentration (OWNC) are independent variables examined in the study. ACEF was measured by an index developed by [8]. The index comprises of four elements, namely composition, authority, resources, and diligence which are divided in to ten requirements. The scoring process of the index refers to [8].Scoring was based on the existence of the requirement (i.e. "1" and "0"), except for audit committee responsibility. If there is a brief statement stating a responsibility of audit committee, it will be awarded a score of 1. If there is also an explanation on act a particular responsibility, it will be awarded a score of 2. If a company does not disclose at all, the score will be 0. The score of audit committee effectiveness is the sum of all requirements of which the maximum is 14. Table 1 presents audit committee effectiveness index while Table 2 shows the operationalization of dependent and independent variables.

DOI: 10.1051/shsconf/20173413002

Multivariate regression analysis was used to examine the relationship between the corporate governance variables and the CSR disclosures. The regression model employs is as follows:

$$CSRD = \beta_0 + \beta_1 BDSIZE_j + \beta_2 BDIND_j + \beta_3 ACEF_j - \beta_4 OWNC_j + \beta_5 ROA_j + \beta_6 SIZE_j + \beta_7 YEAR + e_j \quad (1)$$

The model includes two control variables, i.e. firm's size and profitability. These variables had been widely used by prior studiesin CSR reporting and expected to enhance the association between corporate governance mechanisms and CSR disclosure. Consistent to[4,5,10,19] firms size was measured by total assets,and profitability was measured by Return on Asset [5,6,7,9,11].

Table 1. Audit Committee Effectiveness (AC) Index.

Dimension	Proxies	Scoring
Composition	AC independence ▪ All members shall be external independent parties	1;0
	AC expertise ▪ At least one member of the AC has educational background and experience in accounting or finance	1;0
Authority	AC charter ▪ Proxy statements concerning AC charter	1;0
	AC responsibility/duty ▪ Reviewing company's financial information ▪ Reviewing external auditing activity ▪ Reviewing the effectiveness of company's internal control ▪ Reviewing company's compliance with regulations	1;2;0
Resources	AC size ▪ Comprise at least three members	1;0
Diligence	AC meeting ▪ AC shall has a meeting at least four times in a year	1;0
	AC voluntary disclosure ▪ AC reports voluntary disclosure	1;0

Table 2. The Regression Variables.

Variables	Measurement
CRSD	CSR disclosure index
BDSIZE	Number of board commissioners
BDIND	Proportion of independent commissioners on board
ACEF	AC effectiveness index (refer to Table 1)
OWNC	The proportion of ordinary shares held by substantial shareholders (holding 5% or more)
ROA	Profit before tax to total assets
SIZE	Natural logarithm of total assets
YEAR	1 if data are from 2011, 0 if 2012

DOI: 10.1051/shsconf/20173413002

4 Results and Discussion

4.1 Descriptive Statistic

Table 3 presents descriptive statistic of all sample. As can be seen in the table, the number of board of commissioners was in the range of 2 and 11, and the average board size was 4. The proportion of independent commissioners on board was0.38 (38%), and the minimum proportion was 0.17 (17%) which indicates there were still companies not complying with IDX Rule [21]. It is requiredthat all listed companies to have at least 30% of board members to be independent. Consistent to [20], listing companies in Indonesia are highly concentrated;with mean score of ownership concentration is 75.65%. There are sample companies that disclose maximumin accordance to the CSR index (79 items) and audit committee effectiveness index (14). The average number of CSR itemdisclosedin annual reports is 31.37 (39.70%), whileaudit committee effectiveness score of sample companies is 9.38 (67%) on average.

Table 3. Descriptive Statistics of Continuous Variables.

Variables	Min	Max	Mean	SD
BDSIZE	2	11	4.14	1.726
BIND	0.17	0.75	0.38	0.098
OWNC	29.71	98.96	75.65	16.644
ACEF	2	14	9.38	2.526
SIZE	21.84	32.36	27.88	1.652
ROA	-1.50	63.00	9.73	12.091
CSR	3	79	31.37	15.671

Table 4 shows descriptive statistic of CSR disclosure in 2011 and 2012 as well as results of t statistic test. Table 4 shows that the extent of CSR disclosure generally has increased almost doubled from an average of 17.20 (21.77%) in 2011 to 34.12 (43.18%) in 2012. Table 4 also suggests that there were companies disclosed all the items in the checklist in their company annual reports in 2012. Additionally, results of the paired sample *t*-testindicate that there was a significance difference in the mean score of disclosure in 2011 and 2012 at 1 percent level (sig. = 0.00). The findings show that the release of [1] rule has improve the extent of CSR disclosure in Indonesian PLCs. Hence, H1 is failed to reject.

Table 4. The Extent of CSR Disclosure: Descriptive Statistic and *t*-test.

Year	Min	Max	Mean	SD	t	Sig
2011	3	50	17.20	8.308	-16.357	0.00
2012	5	79	34.12	13.526		

DOI: 10.1051/shsconf/20173413002

4.2 Correlation Analysis

The collinearity between the independent variables is presented in Table 5. The Pearson correlations in the table indicate that the multicollinearity is not serious for the independent variables. As suggested by [22], that collinearity should be harmful if the correlation coefficientscoefficient exceeds 0.7. Multicolinearity was also evaluated using the variance inflation factor (VIF) figured by the regression analysis on SPSS. Table 6 shows that VIFs of all independent variables are less than 2. According to [23], collinearity is considered a problem only when VIF is more than 10. Therefore, these findings suggest that the regression model do not indicate any serious collinearity in the examined model.

Table 5. Pearson Correlation among Continuous Independent Variables

	BDSIZE	BIND	OWNC	ACEF	SIZE	ROA	CSR
BDSIZE	1						
BIND	0.002	1					
OWNC	-0.080	-0.092	1				
ACEF	0.048	0.155	0.135	1			
SIZE	0.506**	0.173	-0.270**	0.179	1		
ROA	0.021	-0.057	0.023	-0.106	0.119	1	
CSR	0.228*	0.138	-0.110	0.246**	0.346**	-0.034	1

Notes: ** Correlation is significant at the 0.01 level (2-tailed), *. Correlation is significant at the 0.05 level (2-tailed).

4.3 Regression Analysis

As shown in Table 6, the model which incorporates six continuous variables and one categorical variable is significant at the 1 per cent level (sig = 0.003) and is able to explain 21.4 per centof the variations in CSR disclosure in Indonesian PLCs (Adjusted *R-squared* = 21.4 per cent) for the two year period 2011 and 2012. Audit committee effectiveness and firm's size is significant at the 5 per cent level.

Table 6. Multiple Regression Analysis Results.

Variables	B	Significance	Tolerance	VIF
Constant	-3.388	0.686		
BDSIZE	0.238	0.379	0.728	1.373
BIND	2.671	0.522	0.937	1.067
OWNC	-0.015	0.545	0.876	1.141
ACEF	0.344	0.041*	0.893	1.119
SIZE	0.670	0.030*	0.621	1.610
ROA	-0.015	0.661	0.955	1.047
YEAR	0.183	0.162	0.848	1.179

R^2		23.2		
Adjusted R^2		21.4		
F statistics		3.635		
Significance		0.003**		

Notes: **Significant at the 0.01 level *Significant at the 0.05 level

The positive association between audit committee effectiveness and the extent of CSR disclosure implies that the effectiveness of audit committee is more likely to lead to better concern in corporate social activities and hence disclosure in annual reports. It may also be interpreted that an effective audit committee encourages the management to involve more in social activities and hence to disclose more in annual reports. An effective audit committee through its monitoring function may push the management to comply with mandatory CSR disclosure as stated in the [1] rule. This relates to one of the audit committee responsibilities as stated in [24, 25] rule that is reviewing company's compliance with regulation. The effectiveness of audit committee therefore, leads to the improvement in the extent of CSR disclosure among PLCs in Indonesia. The results appear to support[8] which found that the effectiveness of audit committee may enhance financial reporting quality in terms of its value relevance (timeliness) of financial reporting. In terms of the relation between audit committee and CSR reporting, the results seem to support [6]which found that audit committee financial expertise and audit committee meetings were positively associated with CSR disclosure.

Firm's size is positively associated and statistically significant with CSR disclosure. In Indonesian context, the significance of firm's size is consistent with [4]. Larger firms basically have more resources in order to be able to bear the cost to do more social activities. The results are consistent to [7, 11] which found that company size has a positive and significant effect on CSR disclosure. According to [11], larger companies may involve in more social activitieshence CSR disclosure in order to build positive image. It may also due to their approach in managing political cost and legitimizing their existence since larger companies face more public pressure and more politically sensitive.

Profitability, board size, board indepence, and ownership concentration are found to have insignificant relationship with the extent of disclosure. The non-significance of the profitability supports [4, 5, 11] which appears to confirm [26]'s view that social disclosure are very much linked to public pressure rather than the market place or economic pressure. Ownership concentration is not statistically significant in explaining the extent of CSR disclosure in annual reports. This implies that the majority shareholders do not influence and intervene in social activities undertaken and hence CSR report prepared by the management.

Board commissioners in terms of their size and independence are not statistically significant in influencing the extent of CSR disclosure. The non-significance of the board variables perhaps due to the role of the board has been represented by audit committee. In this research, audit committee effectiveness variable has a positive association with the extent of disclosure. As a committee of the board, the role of audit committee is to oversee financial reporting quality, including the extent of disclosure. The monitoring function of independent commissioner has been represented in the audit committee since an independent commissioner also acts as a chairman of audit committee [24, 25]. Audit Committee was formed with the primary objective to assist the Board of Commissioners in carrying out supervisory responsibilities over the financial reporting process, internal control system, the audit process, good corporate governance implementation and the process of monitoring compliance with laws and regulations on the company. The results are consistent to [5] which found that board size and board independence has no relationship with the extent of disclosure but the proportion of independent non-executive director sitting on audit committee has a positive relationship with the extent of CSR disclosure in annual reports.

5 Conclusions, Limitations and Suggestion for Further Research

This research has examined the extent of CSR disclosure in Indonesian PLCs with the expectation that after the issuance of the [1] rule which provides guidelines of what should be disclosed related to CSR in annual reports, more CSR activities would be undertaken and hence disclosed in company annual reports. The research also examined whether corporate governance mechanisms have an impact on CSR disclosure in Indonesian PLCs. As expected, the paired sample t-test suggested that there was an increase in the extent of CSR disclosure of Indonesian PLCs in 2011 as compared to 2012 and the increase was statistically significant at the 1 per cent level. This finding implies that the issuance of the [1] rule has had some positive impact on in terms of encouraging the minimum level of social activities should be carried out and hence disclosed in company annual reports. In addition, the multiple regression analysis showed that audit committee effectiveness and size were significant factors in explaining the extent of CSR disclosure in annual reports. It would appear that PLCs with a higher score of audit committee effectiveness and a higher total asset disclosed significantly more CSR information than their counterparts. This findings suggest that the involvement of audit committee through its effectiveness in overseeing company's financial reporting and the availability of more resources (assets) could lead to better concern in corporate social activities and hence disclosure in annual reports.

This study uses two-year period, i.e. 2011 and 2012 which represents the period before and after the issuance of the [1] rule respectively. Year 2012 however, is the year of immediate effect of the rule should be complied. The rule was issued in August 2012 so it took only six months for PLCs to prepare the 2012 annual reports in accordance to the regulation. It may be expected that in 2013 the extent of CSR disclosure increased as the time to engage of social activities is longer than in 2012. Additionally, this study uses manufacturing companies as a sample, hence the findings may not be generalized to all Indonesian PLCs. Future research in CSR disclosure may perform a longitudinal analysis to portray the trend in CSR reporting in Indonesian PLCs. Extending the analysis to included other factors that may relate to CSR disclosure in the model may also be encouraged as the R^2 of the regression analysis is relatively low.

References

1. Bapepam-LK Rule No. No. X.K.6 (attachment to the decision of the Chairman of Bapepam-LK No. Kep-431/BL/2012, dated 1 August 2012) on Submission of Issuer or Public Company Annual Reports (2012)
2. Bapepam-LK Rule No. No. X.K.6 (attachment to the decision of the Chairman of Bapepam-LK No. Kep-134/BL/2006, dated 7 December 2006) on Submission of Issuer or Public Company Annual Reports (2006)
3. E. Mirfazli, Int. J. of Isl. and Midd. East. Fin. and Manag.**1**, 275-284 (2008)
4. S.V. Siregar, Y. Bahtiar, Int. J. Isl. and Midd. East. Fin. and Manag. **3**, 241-252 (2010)
5. R. Said, Y. Hj Zainuddin, H. Haron, Soc. Resp. J. **5**, 212-226 (2009)
6. M. I. Jizi, A. Salama, R. Dixon, R. Stratling, J. Bus. Ethics**125**, 601-615 (2013)
7. A. Khan, M.B. Muttakin, J. Siddiqui, J. Bus. Ethics **114**, 207-223 (2013)

8. S.R. Ika, N.A. Mohd Ghazali, Man. Aud. J. **27**, 403-424 (2012)
9. E. Esa, N.A. Mohd Ghazali, Corp. Gov. **12**, 292-305 (2012)
10. G. Giannarakis, Soc. Resp. J. **10**,569-590 (2014)
11. N.A. Mohd Ghazali, Corp. Gov. **7**, 251-266 (2007)
12. P. Collier, A. Gregory, J. Acc Pub. Pol. **18**, 311-322(1999)
13. M. Jensen, J. Fin. **3**, 305-360 (1993)
14. S.T. Petra, Corp. Gov. **5**, 55-64 (2005)
15. S. Turley, M. Zaman, J. Man. Gov. **8**, 305-332 (2004)
16. S. Turley, M. Zaman, Acct. Aud. Account.J. **20**, 765-788 (2007)
17. M. Mangena, R. Pike, Acct Bus Res. **35**,327-349 (2005)
18. M. Hossain, L.M. Tan, M. Adams, Int J. Acc. **29**, 334-351 (1994)
19. R. Gamerschlag, K.Möller, F.Verbeeten, Rev. Manag. Sc. **5**, 233-262 (2011)
20. W. E. Daniel, Bond Law Rev **5**, 345-347 (2003)
21. IDX. The Decision of Directors of JSX No. Kep-315/BEJ/06-2000 concerning Independent Directors, Audit Committee, and Corporate Secretary, Indonesia Stock Exchange, Jakarta (2000)
22. B.G.Tabahnick, L.S. Fidell, *Using Multivariate Statistics, 4th ed.* (Allyn and Bacon, Boston, 2001)
23. D.N. Gujarati, Basic Econometrics, 3rd ed. (McGraw-Hill, New York, 2005)
24. Bapepam Rule No. 29/PM/2004 concerning Guidelines for Establishment and Working Implementation of Audit Committee (2004).
25. Bapepam-LK Rule No. Kep-634/BL/2012 and the Annex of Regulation IX.1.5 on the Establishment and Implementation Guidance of the Audit Committee (2012)
26. S. Williams, Int. Jour. Acc. **34**, 209-238 (1999)

The Role of Corporate Governance in Firm Performance

Zahroh Naimah[1,*], and *Hamidah*[1]

[1]Accounting Department, Faculty of Economics and Business, Universitas Airlangga, Indonesia

Abstract. The objective of this study is to examine the role of corporate governance to increase firm performance. The measure of corporate governance are corporate governance mechanism and Corporate Governance Perception Index (CGPI). Samples are companies that followed CGPI award at 2005-2014. The examination of the relationship of corporate governance and firm performance is conducted by regression of corporate governance mechanism variables and control variables to profitability. Corporate governance mechanisms are board size, board independence, outside directors, audit committee size, audit committee meeting, audit quality, and CGPI. Control variables are leverage and firm size. The results of this study indicate that board independence negatively influence profitability, audit committee meeting positively influence profitability, audit quality positively influence profitability, CGPI positively influence profitability, leverage negatively influence profitability, and firm size negatively influence profitability.

1 Introduction

Corporate governance is commonly understood as the way and practice to directing, organizing, and control the company. Corporate governance is the system designed to professionally direct the company based on good corporate governance principles. Good corporate governance principles are transparency, accountability, responsibility, independence, and fairness.

The practice of corporate governance is strongly influenced by the parties involved in the management system of a company such as shareholders, investors, creditors, employees, and government. Good corporate governance is expected to increase firm performance. The main objective of the implementation of good corporate governance is to optimize value for shareholders and stakeholders in the long run.

This research tries to prove that corporate governance that is performed well can improve firm performance. Generally, this study is aimed to examine whether corporate governance mechanisms and principles influence firm performance. Specifically, the objectives of this study are examine the influence of board size, board independence, outside directors, audit committee size, audit committee meeting, audit quality, and corporate governance principles to firm performance.

* Corresponding author: zahrohnaimah@yahoo.com

 DOI: 10.1051/shsconf/20173413003

2 Literature Review

2.1 Corporate Governance

Cadbury Report defines corporate governance as "a system in which companies are directed and controlled" [1]. The board of directors is responsible for the implementation of corporate governance. The role of shareholders in governance is to appoint directors and auditors, and ensuring that the directors and the auditor has run the appropriate governance structure. While the auditor's role in providing an assessment of the financial statements presented by management.

Corporate governance mechanisms include institutional ownership in the company, stock ownership by directors and executive officers, board of directors characteristics, age and tenure of the CEO, and the sensitivity of CEO pay-for-performance [2].

Institutional investors in large numbers will have the opportunity, resources, and ability to monitor and influence the manager. Shares and/or options ownership by directors and executive officers will encourage behavior that can increase the value of the firm, but will also encourage managers to use discretionary accruals to increase firm performance in the period around the sale of shares or the use of options, so as to improve the welfare of directors and the executive officer [2]. Directors appointed to manage and control the business of the company. As an agent of the company and its shareholders, directors must protect the interests of shareholders as the owner of the company. Characteristics of the board of directors can be measured by the percentage of independent directors (outside directors), CEO duality, and the size of the board. Age and tenure of CEO influence the effectiveness in managing the company. The older or the longer the tenure of the CEO, the deeper understanding of the CEO of the company and industry, and improve the performance of the company [2].

2.2 Corporate Governance and Firm Performance

P.A. Gompers, L. Ishii, and A. Metrick found a strong correlation between corporate governance and stock returns throughout 1990 and firm value, as measured by Tobin's Q [3]. L.D. Brown and M.L. Caylor found that companies that are managed better would be more profitable, more valuable, and to pay more cash dividends to shareholders [4].

Managers tend to expropriate the company's assets and work on projects that benefit themselves personally [5]. Effective corporate governance that reduces 'right to control' of shareholders and creditors given to managers, increasing the probability that managers invest in projects that generate net positive present value [6]. This shows that companies that are managed better have a better operational performance, the performance measures used by L.D. Brown and M.L. Caylor [4]. Empirically, R. La Porta, F. Lopex-de-Silanes, A. Shleifer, and R. Vishny indicate that investor protection associated with effective corporate governance [7].

2.2.1 Board size and firm performance

There are two opposing ideas on the relationship between board size and firm performance. First, thought that the lesser the board size can greatly contribute to the company's success. D. Yermack found an inverse relationship between board size with company's value and favorable financial ratios such as profitability, asset utilization, and Tobin's Q [8]. Empirical evidence of S. Cheng showed that companies that have more board members, the variability of the firm performance would be lower [9].

DOI: 10.1051/shsconf/20173413003

The second thought argued that a large board size will increase the firm performance. Board size is a determinant of the ability of directors to monitor and control manager. R. Adam and H. Mehran argued that the company should have a large board size to be able to monitor effectively [10]. Large board size will support a more effective management of the company [11]. Large board size will be easier to obtain information [12].

Hypothesis 1: Board size influence firm performance

2.2.2 Board independence and firm performance

Previous studies of the relationship between the independence of directors and firm performance, showed inconsistent results. The higher the proportion of outsider, the more independent board of directors [13]. S. Bhagat and B. Black found no association between proportion of outsider and Tobin's Q, ROA, asset turnover, and stock returns [14].

The composition of board of directors plays an important role in determining the firm financial performance. Board of directors is authorized to monitor managerial activities, evaluate the performance of managers, and give awards to the manager. According to E.F. Fama and M.C. Jensen, board of directors is an internal control mechanism that is essential for monitoring top management [15].

S. Rosenstein and J. Wyatt found a market award to the company that has the outsider [16]. J. Brickley, J. Coles, R. Terry found a positive relationship between the proportion of outsider by the reaction of stock market [17]; and R. Anderson, S. Mansi, and D. Reeb found the opposite relationship between independence of directors and cost of debt [18]. A large number of board and a large proportion of outsider can provide great information. Several studies have found that board size and the proportion of outsider positively related to firm size and complexity [19].

L.D. Brown and M.L. Caylor found no relationship between the independence of directors and Tobin's Q, but found a positive relationship between the independence of directors and ROE, profit margin, dividend yield, and stock repurchases [4]. They found that the company would be worth if the position between the CEO and the board are separated. S. Rosenstein and J. Wyatt found evidence that shareholder value is influenced by the proportion of outside directors as indicated by the positive stock price reaction during the announcement of the appointment of outside directors [16].

Hypothesis 2: Board independent influence firm performance

Hypothesis 3: Outside directors influence firm performance

2.2.3 Audit committee and firm performance

Audit committee as a committee plays an important role in ensuring and monitoring the accounting process so that management can provide information that is relevant and credible to all stakeholders [20]. The existence of audit committee independence is able to provide reliable accounting information, so audit committee independence is expected to improve the company's performance. L.D. Brown and M.L. Caylor found that independent audit committee positively related to dividend yield [4].

The 1999 Blue Ribbon Committee Report recommends that audit committee, as supervisor of financial accounting processes, conduct meetings at least four times a year in order to guarantee the quality of financial reporting [21]. If audit committee size and audit committee meeting frequency can improve financial accounting process, it is expected that audit committee size and meetings frequency can improve firm performance.

Hypothesis 4: Audit committee size influence firm performance

Hypothesis 5: Audit committee meetings influence firm performance

 DOI: 10.1051/shsconf/20173413003

2.2.4 Audit quality and firm performance

Audit quality as a means of monitoring corporate governance from the outside, can improve firm performance. The study results proved that large auditors have a good reputation which reflects the good quality than smaller auditor [22]. Z.V. Palmrose suggest that the big-8 auditors audit charge high fees because of high audit quality [23]. A. Martinez, and A. de Jesus Moraes found that auditors imposes higher audit fees will give a signal to markets that high audit quality may enhance shareholder value [24].

Hypothesis 6: Audit quality influence firm performance

2.2.5 Corporate governance principles and firm performance

The principles of corporate governance consists of transparency, accountability, responsibility, independence and fairness. IICG (The Indonesian Institute for Corporate Governance) create a corporate governance index, known as Corporate Governance Perception Index (CGPI). Implementation of good corporate governance is expected to improve the company's performance [25].

Hypothesis 7: Corporate Governance Principles influence firm performance

3 Methodology

The study period covers 10 years, t = 2005 to 2014. The samples were all companies registered for the Corporate Governance Perception Index (CGPI). Data was obtained from secondary data that are CGPI and financial data operiod of 2005-2014.

The independent variables is corporate governance mechanism and Corporate Governance Perception Index (CGPI). Corporate governance mechanisms include number of board of directors (BoSize), number of independent board (Bolnd), number of outside directors (OutDir), umber of audit committee (AcSize), frequency of audit committee meetings in a year (AcMeet), and audit quality (Audit). The dependent variable is firm performance as measured by profitability. Measures of profitability is return on assets (ROA). This study also incorporate leverage (LEV) and firm size (SIZE) as control variables.

The examination of the influence of corporate governance on firm performance is done by multiple regression test:

$$PROF_{it} = \alpha_0 + \alpha_1 BoSize_{it-1} + \alpha_2 BoInd_{it-1} + \alpha_3 OutDir_{it-1} + \alpha_4 AcSize_{it-1} + \alpha_5 AcMeet_{it-1} + \alpha_6 AUDIT_{it-1} + \alpha_7 CGPI_{it} + \alpha_8 LEV_{it-1} + \alpha_9 SIZE_{it-1} + \varepsilon_{it} \quad (1)$$

where PROF is profitability.

4 Results

The samples were all over the listed company (listed on the Indonesia Stock Exchange) listed as participants of Corporate Governance Perception Index (CGPI) in 2005-2014. Data collection are 133 observations.

Table 1 below is an overview of the results of hypothesis testing. Results shows that board independence negatively influence firm profitability. This shows that the independent board will potentially reduce the profitability of the company. These results are not consistent with previous studies that show that board independence will increase the company's performance. Results of previous studies have not been consistent on the association of board independence and corporate performance.

The number of audit committee meetings is found positively influence firm profitability. This shows that the more frequent audit committee meetings, the higher

DOI: 10.1051/shsconf/20173413003

profitability of the company. The audit committee oversees the ongoing accounting procedures so that management can provide information that is relevant and credible to all stakeholders [20]. More frequent audit committee meetings is expected to improve firm performance. These results are consistent with L.D. Brown and M.L. Caylor were shown a link between audit related governance factors and firm performance [4].

Table 1. The association of corporate governance and profitability

Variable	Prediction	Coefficient	t-statistic	Prob-value
Constant	No	36.13970	2.317005	0.0221
BoSize	+	0.428252	1.091527	0.2771
BoInd	+	-0.116002	-2.209253	0.0289
OutDir	+	0.015278	0.510375	0.6107
ACSize	+	0.329015	0.686524	0.4936
ACMeet	+	0.107269	1.991949	0.0485
AUDIT	+	5.877332	3.193858	0.0018
CGPI	+	0.473773	3.169802	0.0019
LEV	-	-0.139464	-5.132474	0.0000
SIZE	+	-4.982884	-3.180228	0.0018

Audit quality is found have positive effect on firm profitability. Audit quality as a means of monitoring the implementation of corporate governance from the outside, can improve firm performance, consistent with A. Ferguson and D. Stokes [22]. Big Auditor has a good reputation and produce good audit quality, so as to improve firm performance.

Corporate Governance Perception Index (CGPI) is found positively influence firm profitability. This indicates that the implementation of good corporate governance will increase firm profitability, consitent with F. Siagian, S.V. Siregar, and Y. Rahadian that found that GCG practice positivelly affect firm value [26]. Implementation of good corporate governance is expected to improve firm performance the firm is managed transparently and accountable.

5 Conclusion

The results indicates that board size does not significantly influence profitability. Board independence significantly have negative effect on profitability. Outside director found not to significantly affect profitability. Audit committee size did not significantly affect profitability. The number of audit committee meetings significantly have positive effect on profitability. Audit quality significantly have positive effect on profitability. The principles of corporate governance significantly have positive effect on profitability. Leverage and firm size have negative effect on profitability.

References

1. Cadbury Report, (London: London Stock Exchange, 1992).
2. M.M. Cornett, A.J. Marcus, H. Tehranian, *Journal of Financial Economics,* **87**, 357-373 (2008)
3. P.A. Gompers, L. Ishii, A. Metrick, *Quarterly Journal of Economics,* **118(1)**, (2003)
4. L.D. Brown, M.L. Caylor, *Working Paper,* Georgia State University (2004)
5. C.M. Jensen, H.W. Meckling, *Journal of Financial Economics*, **3**, 305-360 (1976)
6. A. Shleifer, R. Vishny, *Journal of Finance,* **2**, 737-783 (1997)
7. R. La Porta, F. Lopex-de-Silanes, A. Shleifer, R. Vishny, *Journal of Financial Economics,* **58**, 3-27 (2000)

8. D. Yermack, *Journal of Financial Economics,* **40**, 185-211 (1996)
9. S. Cheng, *Journal of Financial Economics,* **87**, 157-176 (2008)
10. R. Adam, H. Mehran, *Working Paper*. Federal Reserve Bank of New York (2002)
11. A. Klein, *Journal of Law and Economics*, **41(1)**, 275-303 (1998)
12. D.R. Dalton, C.M. Daily, J. Jonhson, A.E. Ellstrand, *The Academy of Management Journal*, **42(6)**, 674-686 (1999)
13. K. John, L.W. Senbet, *Journal of Banking & Finance,* **22**, 371-403 (1998)
14. S. Bhagat, B. Black, *Journal of Corporation Law,* **27**, 231-274 (2002)
15. E.F. Fama, M.C. Jensen, *Journal of Law and Economics,* **26**, 301-325 (1983)
16. S. Rosenstein, J. Wyatt, *Journal of Financial Economics,* **26**, 175-191 (1990)
17. J. Brickley, J. Coles, R. Terry, *Journal of Financial Economics,* **35**, 371-390 (1994)
18. R. Anderson, S. Mansi, D. Reeb, *Journal of Accounting and Economics,* **37**, 315-342 (2004)
19. K. Lehn, P. Sukesh, M. Zhao, *Working Paper,* Katz Graduate School of Business, (2003)
20. M. Beasley, *Accounting Review,* **71**, 443–465 (1996)
21. Blue Ribbon Committee on Improving the Effectiveness of Corporate Audit Committees (1999)
22. A. Ferguson, D. Stokes, *Contemporary Accounting Research.* **19(1)**, 77-110 (2002)
23. Z.V. Palmrose, *Journal of Accounting Research,* **24(1)**, 97-110 (1986)
24. A. Martinez, A. de Jesus Moraes, *Journal of Modern Accounting and Auditing,* **10(4)**, 442-450 (2014)
25. The Indonesian Institute for Corporate Governance, *Laporan Program Riset dan Pemeringkatan Corporate Governance Perception Index 2014*. (The Indonesian Institute for Corporate Governance, 2015).
26. F. Siagian, S.V. Siregar, Y. Rahadian, *Journal of Accounting in Emerging Economies* **3(1)**, 4-20 (2013).

AUTHOR INDEX

AUTHOR INDEX